Modern Applied Mathematics

Modern Applied Mathematics

Editor
J C Misra

Narosa Publishing House
New Delhi Chennai Mumbai Kolkata

Editor
J C Misra
Department of Mathematics
Indian Institute of Technology
Kharagpur, India

NAROSA PUBLISHING HOUSE PVT. LTD.

22 Daryaganj, Delhi Medical Association Road, New Delhi 110 002
35-36 Greams Road, Thousand Lights, Chennai 600 006
306 Shiv Centre, D.B.C. Sector 17, K.U. Bazar P.O., Navi Mumbai 400 703
2F-2G Shivam Chambers, 53 Syed Amir Ali Avenue, Kolkata 700 019

www.narosa.com

Printed from the camera-ready copy provided by the Editor.

ISBN 81-7319-492-0

Published by N. K. Mehra for Narosa Publishing House Pvt. Ltd., 22 Daryaganj,
Delhi Medical Association Road, New Delhi 110 002 and printed at
Rajkamal Press, New Delhi 110 033, India.

Preface

This really is the golden age of Mathematics, because half the Mathematics ever created has been in the last 100 years and half the mathematicians who have ever lived are alive today. We have seen such achievements as the resolution of the four-colour problem and Fermat's last theorem, with the latter being a special manifestation of a much more general result !

This book is prepared with the aim of providing the readers with recent developments in the realm of modern applied mathematics. All those who deal with engineering education are well aware of the precarious state of mathematical skills and conceptual comprehension which marks a large section of students at present. The mathematical methods encountered in each course do not evolve into unified patterns which the future engineer would be able to recognize and use universally. This smokestack mentality inflicts incalculable damage to engineering practice specially in many areas closely connected to research and development.

It is felt that this situation should be radically improved. We think that it is possible to offer a concise yet comprehensive material which summarizes, unifies and completes a reader's mathematical knowledge by constructing a comprehensive system of operating mathematics and setting up patterns that can be recalled for use in a wide range of engineering disciplines. We strongly believe that the ideas of modern applied mathematics can be understood and successfully used by engineers.

As a senior Professor of Mathematics, I felt encouraged to bring out a series of books covering most major areas of Mathematical Sciences during this golden age of Mathematics. The other volumes of our series bear the titles : Uncertainty and Optimality, Biomathematics : Modelling and Simulation (both brought out by World Scientific, Singapore), Applicable Mathematics in the Golden Age; Computational Mathematics, Modelling and Algorithms; Applicable Mathematics, its Perspectives and Challenges; Recent Trends in Mathematical Sciences; Computing and Information Sciences : Recent Trends; Mathematical Models for Bioengineering and Probabilistic Systems; Industrial Mathematics and Statistics (all by Narosa Publishing House, New Delhi), two volumes entitled Mathematical Modelling of Nonlinear Systems, Vol. 1: Nonlinear Systems and Mathematical Models ; Vol. 2 : Mathematical Techniques and Application Areas (both by IIT Kharagpur); Mathematics and Mathematicians (by Allied Publishers, New Delhi).

The present volume deals with important emerging topics of modern applied mathematics. Throughout the book, topics involving mathematical treatment of various fields of knowledge are presented.

For nearly four decades I have been engaged in teaching and research at several well-known institutions in India, Germany and North America. My primary interests have been in the fields of biomathematics, biomechanics, mathematical modelling and theoretical solid and fluid mechanics. While in the present volume I have contributed two chapters dealing with current topics of the mechanics of solids and fluids, two chapters on mathematical biomechanics, entitled 'Modelling of Head Impact Problems' and 'Mathematical Models in Fracture and Remodelling of Bones' written by me have been included in the volume *Mathematical Models in Bioengineering and Probabilistic Systems* (Narosa, New Delhi). Moreover, I have authored a number of chapters of the book *Biomathematics : Modelling and Simulation* (World Scientific, Singapore).

Publication of the series of books covering various fields of mathematical sciences has been the fruit of a long period of collaboration with eminent fellow scientists together with relentless perseverance. It has been our endeavour to make these books useful to a large section of people interested in mathematical sciences, professional as well as amateur. The volumes have been so designed and planned that illustrative examples and exercises as well as fairly comprehensive bibliography are included in various chapters. This will help strengthen the level of understanding of the learners. Thus the books of this series will be of interest not only to graduate students but also to instructors, research scientists and professionals. The volumes of the series might not exactly serve as textbooks, but will definitely be worthwhile supplements. Our labour will be deemed amply rewarded if at least some of those for whom the volumes are meant derive benefit from them.

I am thankful to Professor S. K. Dube, Director, IIT Kharagpur for his kind encouragement in publishing this volume. I feel deeply grateful to the contributors of all the volumes of the series who have so kindly accepted my invitation to contribute chapters. The enormous pleasure and enthusiasm with which they have responded to my request has touched me deeply, boosting my interest in the publication of the series.

I constantly remember the extent of care my parents have taken to impart proper education to me. I am highly indebted to Srimat Swami Shankaranandaji Maharaj, seventh President of the Ramakrishna Math and the Ramakrishna Mission, Swami Tejasanandaji and Swami Gokulanandaji, the then Principal and Vie-Principal of Ramakrishna Mission Vidyamandira, Belur Math and to the monastic members of the Ramakrishna Mission Calcutta Students' Home, Belgharia for their kind guidance and suggestions and for instilling in me, while I was still a college and university student, a deep sense of total involvement in pursuing academic goals and a strong commitment to human values.

It is a pleasure to acknowledge the moral support, help and encouragement that I have been receiving constantly in all my academic activities from my wife Shorasi and children Subhas, Sumita and Sudip.

Special care taken by Mr. N. K. Mehra of Narosa Publishing House and the members of his team for producing the volumes with utmost efficiency is greatly appreciated.

I.I.T. Kharagpur

J. C. Misra

Contents

Section 3: Mathematical Physics

SECTION 1

MACROSYSTEMS

Modern Applied Mathematics
Editor: J.C. Misra

Mathematical Theory of Thermoelasticity

J.C. Misra

Department of Mathematics, Indian Institute of Technology, Kharagpur-721302, India

(Dedicated to the loving memory of my brother Dines, who left us for Heavenly abode at a very premature age when I was in Germany)

The purpose of this chapter is to present a brief overview of the mathematical foundation of the Theory of Thermoelasticity together with a review of some of the theoretical studies made during the past few decades by different researchers in this important area of Applied Mathematics. The review is not exhaustive. An attempt has been made to present the material in a comprehensive manner in order that the beginners will find an opportunity to have an acquaintance with the present state-of -the- art of the area. Keeping this in view some fundamental concepts and basic equations of the theory of elasticity have also been included along with those of the theory of thermoelasticity.

1. Basic Concepts

The theory of elasticity is concerned with the mechanics of deformable media which after the removal of the applied forces producing deformation, completely recover its original shape and volume. The deformation of the solid media due to the action of external forces or mechanical load is measured by the strain functions defined by the rate of change in dimension. The intensity of the internal forces developed within the body as reaction to the applied load is termed as stress. In the linearized theory of elasticity the relation between the strain and displacement is given by

$$e_{ij} = \frac{1}{2}(u_{i,j} + u_{j,i}), \qquad i, j = 1, 2, 3 \qquad (1.1)$$

where u_i are the components of the displacement vector at a point x_i within the body in the rectangular Cartesian coordinate system. Here e_{ij} is a symmetric tensor of rank 3, so that $e_{ij} = e_{ji}$.

The stress component σ_{ij} is defined as the force acting along the jth direction on unit area in the plane x_i=constant, and hence the system of stress within the body can be represented by nine such stress components from which, using the relation $\sigma_{ij} = \sigma_{ji}$, there remain only six independent stress components.

The first attempt to develop the theory of elasticity on the basis of the concept of a continuous medium, which enables one to ignore its molecular structure and describe macroscopic phenomena by the methods of mathematical analysis, dates back to the first half of the eighteenth century. The fundamental contribution to the classical theory was made by Hooke, Navier, Cauchy, Lame, Clapeyron and a few others.

2. Hookean Elastic Solid

A Hookean elastic solid is one that obeys Hooke's law, which states that at low stress level, the stress tensor is linearly proportional to the strain tensor, i.e.,

$$\sigma_{ij} = C_{ijkl} e_{kl} \tag{2.1}$$

where σ_{ij} is the stress tensor, e_{kl} the strain tensor and C_{ijkl} is a tensor of elastic constants or moduli which are independent of stress and strain.

As a tensor of rank 4, C_{ijkl} has $3^4 = 81$ elements; but in as much as $\sigma_{ij} = \sigma_{ji}$ we must have

$$C_{ijkl} = C_{jikl} \tag{2.2}$$

Further, since $e_{kl} = e_{lk}$, we can write

$$\sigma_{ij} = \frac{1}{2}\left(C_{ijkl} + C_{ijlk}\right) e_{kl} = C'_{ijkl} e_{kl} \tag{2.3}$$

with the property

$$C'_{ijkl} = C'_{ijlk}. \tag{2.4}$$

With such a symmetrization the tensor C_{ijkl}, under the conditions (2.2) - (2.4), has a maximum of thirty six independent constants.

For most elastic solids, however, the number of independent elastic constants are much smaller than thirty six. The reduction is caused by the existence of material symmetry.

3. Classification of Elastic Solids : Elastic Symmetry

As far as the elastic properties are concerned, all bodies may be divided, on one hand, into homogeneous and nonhomogeneous media, and on the other hand, into isotropic and anisotropic media. A homogeneous body is one whose structure and composition are the same at all points of the medium. An isotropic body is one for which the elastic properties are identical in all directions drawn through a given point. More precisely, isotropy for a material is defined by the requirement that the constants C_{ijkl} have exactly the same numerical values, no matter how the coordinate system is oriented. Depending on the structure and composition, a body may be isotropic or anisotropic and at the same time homogeneous or nonhomogeneous. If the structure of a material body has some kind of symmetry, the generalized Hooke's law given by (2.1) is simplified accordingly, since some of the elastic constants are zero, while among others there exist some linear relations. The greatest reduction in the number of elastic constants takes place when the material is isotropic. In this case, exactly two independent elastic constants characterize the material. Hooke's law for an isotropic solid reads

$$\sigma_{ij} = \lambda e_{\alpha\alpha}\delta_{ij} + 2\mu e_{ij} \tag{3.1}$$

The constants λ and μ are called Lame's constants.

The number of independent elastic constants for an anisotropic material are reduced (on the basis of particular symmetry) to 9 for an orthotropic solid, and for a transversely isotropic solid this number is only 5.

4. Viscoelastic Solids

The features of hysteresis, relaxation and creep are common to many materials. Collectively, they are called the features of viscoelasticity. Viscoelastic stress analysis is usually concerned with polymeric materials. Their long-chain molecular structure leads to stress, and hence leads to viscoelastic behaviour. Such a response to stress is highly temperature-sensitive. Viscoeasticity of a material is the principal cause of energy dissipation. In real solids, even instantaneous application of external loads does not lead to immediate deformation. This delayed response of solids to loads can be described as viscoelastic material behaviour. At low stress levels, where heat generation due to viscous dissipation is negligible, the viscoelastic behaviour can be described mathematically as a function of the time operator $\frac{\partial}{\partial t} \equiv D$ and the constitutive relations for a viscoelastic material may be written as

$$P_1(D)\left[\sigma_{ij}(x_i,t)\right] = P_2(D)\left[C_{ijkl}e_{kl}(x_i,t)\right] \tag{4.1}$$

where $P_1(D) = a_0 + a_1 D + a_2 D^2 + ... + a_m D^m$

$$P_2(D) = b_0 + b_1 D + b_2 D^2 + ... + b_n D^n \quad (4.2)$$

where a_i and b_i are constants.

Mechanical models such as Maxwell model, Voigt model and standard linear model are often used to describe the viscoelastic behaviour of materials which can be obtained by a suitable choice of values of a_i and b_i.

5. Effect of Temperature : Thermal Stress

Even in the absence of any external load, most solids undergo appreciable changes in shape and size under the influence of a non-uniform temperature field. This temperature may develop due to external heat sources through conduction of heat energy and/or due to internal sources of heat arising out of straining. With the variation of temperature the elements of a thermoelastic material body expand or contract. Such an expansion or contraction cannot proceed freely in a continuous body and the body becomes stressed. Thermoelasticity describes the behaviour of elastic bodies under the influence of non-uniform temperature fields and the stresses thus developed are known as thermal stresses.

In various fields of engineering sciences especially in the design of nuclear reactors, air-planes, missiles, steam turbines, internal combustion turbines, reciprocating internal combustion engines, high pressure gas compressors and machines which deal with heat conversion, thermal stresses play a very important role. Knowledge of the distribution of thermal stresses is very essential to the designers for studying the strength of the material affected by the temperature distribution. Thermoelasticity embraces a wide field of phenomena. It contains the theory of heat conduction and the theory of strains and stresses due to flow of heat. It helps us to determine the stresses produced by the temperature field and also to calculate the temperature distribution due to the action of internal forces.

6. Stress-Strain-Temperature Relations

For elastic solids, Hooke's law can be modified into Duhamel-Neumann form to account for the temperature change. Let the elastic constants c_{ijkl} be measured at a uniform constant temperature θ_0. Then if θ be the temperature excess over θ_0, we write

$$\sigma_{ij} = c_{ijkl} e_{kl} - \beta_{ij} \theta \quad (6.1)$$

in which β_{ij} is a symmetric tensor measured at zero strains. The coefficients appearing in the above equation do not, however, determine directly the material constants, since their values change with directions of the coordinate axes. If they remain invariant under certain transformations of coordinates, the body exhibits properties of elastic symmetry and the number of constants is reduced. Here we write explicitly the stress-strain-temperature relations in simple matrix notations derived from equivalent forms (6.1) only for the following cases :

a) Orthotropic body :

For an orthotropic body which possesses three mutually orthogonal planes of thermoelastic symmetry (where the number of elastic constants reduces to only 9), we have the stress-strain- temperature relations in the form

$$\begin{bmatrix} \sigma_{11} \\ \sigma_{22} \\ \sigma_{33} \\ \sigma_{23} \\ \sigma_{31} \\ \sigma_{12} \end{bmatrix} = \begin{bmatrix} c_{11} & c_{12} & c_{13} & 0 & 0 & 0 \\ c_{12} & c_{22} & c_{23} & 0 & 0 & 0 \\ c_{13} & c_{22} & c_{23} & 0 & 0 & 0 \\ 0 & 0 & 0 & c_{44} & 0 & 0 \\ 0 & 0 & 0 & 0 & c_{55} & 0 \\ 0 & 0 & 0 & 0 & 0 & c_{66} \end{bmatrix} \begin{bmatrix} e_{11} \\ e_{22} \\ e_{33} \\ e_{23} \\ e_{31} \\ e_{12} \end{bmatrix} - \begin{bmatrix} \beta_{11} \\ \beta_{22} \\ \beta_{33} \\ 0 \\ 0 \\ 0 \end{bmatrix} \theta \tag{6.2}$$

$$\text{where } \begin{aligned} \beta_{11} &= c_{11}\alpha_1 + c_{12}\alpha_2 + c_{13}\alpha_3 \\ \beta_{22} &= c_{12}\alpha_1 + c_{22}\alpha_2 + c_{23}\alpha_3 \\ \beta_{33} &= c_{13}\alpha_1 + c_{23}\alpha_2 + c_{33}\alpha_3 \end{aligned} \tag{6.3}$$

in which α_i ($i = 1,2,3$) denote the coefficients of linear thermal expansion.

b) Transversely isotropic body

Such a body shows thermoelastic symmetry with respect to a plane (the x-y plane, say and is isotropic in that plane; its equations are invariant under a rotation about the z-axis). Consequently further simplifications occur as $c_{11} = c_{22}$, $c_{23} = c_{31}$, $c_{44} = c_{55}$ and the stress-strain-temperature relations can be written in matrix form as

$$\begin{bmatrix} \sigma_{11} \\ \sigma_{22} \\ \sigma_{33} \\ \sigma_{23} \\ \sigma_{31} \\ \sigma_{12} \end{bmatrix} = \begin{bmatrix} c_{11} & c_{12} & c_{13} & 0 & 0 & 0 \\ c_{12} & c_{11} & c_{13} & 0 & 0 & 0 \\ c_{13} & c_{13} & c_{33} & 0 & 0 & 0 \\ 0 & 0 & 0 & c_{44} & 0 & 0 \\ 0 & 0 & 0 & 0 & c_{44} & 0 \\ 0 & 0 & 0 & 0 & 0 & \frac{1}{2}(c_{11} - c_{22}) \end{bmatrix} \begin{bmatrix} e_{11} \\ e_{22} \\ e_{33} \\ e_{23} \\ e_{31} \\ e_{12} \end{bmatrix} - \begin{bmatrix} \beta_{11} \\ \beta_{11} \\ \beta_{33} \\ 0 \\ 0 \\ 0 \end{bmatrix} \theta \tag{6.4}$$

where

$$\beta_{11} = (c_{11} + c_{12})\alpha_1 + c_{13}\alpha_3 \quad \text{and} \quad \beta_{33} = 2c_{13}\alpha_1 + c_{33}\alpha_3. \tag{6.5}$$

(c) Cubic crystalline solid

For a cubic crystalline solid body the elastic constants reduce to only 3 and the stress-strain-temperature relations in matrix form are given as

$$\begin{bmatrix}\sigma_{11}\\ \sigma_{22}\\ \sigma_{33}\\ \sigma_{23}\\ \sigma_{31}\\ \sigma_{12}\end{bmatrix} = \begin{bmatrix} c_{11} & c_{12} & c_{12} & 0 & 0 & 0\\ c_{12} & c_{11} & c_{12} & 0 & 0 & 0\\ c_{12} & c_{12} & c_{11} & 0 & 0 & 0\\ 0 & 0 & 0 & c_{44} & 0 & 0\\ 0 & 0 & 0 & 0 & c_{44} & 0\\ 0 & 0 & 0 & 0 & 0 & c_{44}\end{bmatrix}\begin{bmatrix}e_{11}\\ e_{22}\\ e_{33}\\ e_{23}\\ e_{31}\\ e_{12}\end{bmatrix} - \begin{bmatrix}\beta_{11}\\ \beta_{11}\\ \beta_{11}\\ 0\\ 0\\ 0\end{bmatrix}\theta \tag{6.6}$$

where $\beta_{11} = (c_{11} + 2c_{12})\alpha_1$ (6.7)

(d) Isotropic body

For an isotropic material body only two elastic constants exist and the second order tensor β_{ij} must also be isotropic. In this case, the stress-strain-temperature relation in matrix form further simplifies to

$$\begin{bmatrix}\sigma_{11}\\ \sigma_{22}\\ \sigma_{33}\\ \sigma_{23}\\ \sigma_{31}\\ \sigma_{12}\end{bmatrix} = \begin{bmatrix} c_{11} & c_{12} & c_{12} & 0 & 0 & 0\\ c_{12} & c_{11} & c_{12} & 0 & 0 & 0\\ c_{12} & c_{12} & c_{11} & 0 & 0 & 0\\ 0 & 0 & 0 & (c_{11}-c_{12}) & 0 & 0\\ 0 & 0 & 0 & 0 & (c_{11}-c_{12}) & 0\\ 0 & 0 & 0 & 0 & 0 & (c_{11}-c_{12})\end{bmatrix}\begin{bmatrix}e_{11}\\ e_{22}\\ e_{33}\\ e_{23}\\ e_{31}\\ e_{12}\end{bmatrix} - \beta\,\theta \tag{6.8}$$

Writing $c_{12} = \lambda$ and $c_{11} = \lambda+2\mu$, the stress-strain-temperature relations for an isotropic body can be expressed in a very simple form

$$\sigma_{ij} = \lambda e_{kk}\delta_{ij} + 2\mu e_{ij} - \beta\theta\,\delta_{ij} \tag{6.9}$$

Here λ and μ are Lame's constants measured at constant temperature θ_0 and $\beta = (3\lambda + 2\mu)\,\alpha$, α being the coefficient of linear thermal expansion. For linear thermoviscoelastic solids the constitutive relations (6.1) may be generated as

$$\sigma_{ij}(x_I, t) = f(D)[c_{ijkl}\ e_{kl}(x_i, t)] - f(D)(\beta_{ij}\ \theta) \quad (6.10)$$

where f (D) is a function of the time operator D $\equiv \frac{\partial}{\partial t}$, which depends upon the choice of the viscoelastic model.

7. Basic Equations of Thermoelasticity

(a) Heat conduction equation

Let us consider a homogeneous isotropic and thermally conducting solid elastic medium. When there is a change in the amount of heat in an elemental volume of the medium, the strains and stresses are produced. Again a mechanical loading and the corresponding strain produce a variation of the temperature field and some heat flow occurs. When the thermomechanical coupling effects are taken into account, the temperature θ satisfies the equation, called the heat conduction equation given by

$$\theta_{,ij} - \frac{1}{\chi}\dot{\theta} - \eta\dot{e} = -\frac{Q}{\chi} = -\frac{W}{\kappa} \quad (7.1)$$

in which

$$\chi = \frac{K}{\rho\ C_E}, \qquad \eta = \frac{\nu\theta_0}{\kappa} = \frac{(3\lambda + 2\mu)\alpha_t\theta_0}{\kappa}, \qquad e = e_{kk}$$

in which θ_0 is the initial temperature, α_t the coefficient of linear thermal expansion, χ the thermal diffusivity, κ the thermal conductivity, ρ the material density and C_E the specific heat at constant deformation. Q is the intensity of the heat source and W represents the rate of heat generation of the source measured per unit volume per unit time.

The equation (7.1) is known as the coupled heat conduction equation and is obtained by introducing the term $\eta\dot{e}$ (corresponding to the energy of deformation) into the well-known Fourier heat conduction equation

$$\theta_{,ii} - \frac{1}{\chi}\dot{\theta} = -\frac{Q}{\chi} \quad (7.2)$$

The term $\eta\dot{e}$ of the equation (7.1) is known as the mechanical coupling term as it is solely responsible for the coupling of the temperature and strain fields.

(b) Equations of motion

$$\sigma_{ij,j} + \rho F_i = \rho \ddot{u}_i \ , (i, j = 1, 2, 3) \tag{7.3}$$

where F_i are the components of the body forces per unit mass and u_i the components of the displacement vector.

(c) Stress-strain-temperature relations

Using the strain-displacement relations, we have from (6.9) the Duhamel-Neumann relations for an isotropic body as

$$\sigma_{ij} = \lambda u_{k,k} \ \delta_{ij} \ + \mu \ (u_{i,j} + u_{j,i}) - \gamma\theta\delta_{ij} \tag{7.4}$$

Substituting (7.4) into (7.3) one gets the displacement formulation of the equations of motion,

$$(\lambda+\mu) \ u_{j,ji} + \mu u_{i,jj} \ + \rho F_i \ - \gamma\theta_{,i} = \rho \ddot{u}_i \ . \tag{7.5}$$

In the equations given above, summation convention over repeated subscripts is implied and a dot over a function represents its time derivative. The equations (7.1) and (7.3) constitute the system of thermoelasticity equations which together with the prescribed initial and boundary conditions enables us to determine the field values, displacement and temperature.

8.(a) Effect of the coupling of temperature and strain fields

The variation of strain within a body caused by an external mechanical agency is, in general, accompanied by changes in temperature and as a consequence a flow of heat occurs. The entropy increases and this leads to an increase in energy in a mechanically irrecoverable manner. This phenomenon is known as thermoelastic dissipation. To study the dissipative process it becomes necessary to consider the coupled heat equation and so the mechanical coupling term shall have to be retained in the heat equation. However, if the heat produced by external loads be very small, the resulting deformation and stress may be reasonably calculated without taking account of the thermal expansion. Similarly strain produced in a body by a non-uniform temperature distribution generally has an exceedingly small influence over the temperature itself.

Thus the coupling term in the heat equation can be dropped in the case of all problems where thermoelastic dissipation is not of primary interest. The effect of coupling is important from the viewpoint of physics and electronics, but may be ignored in case of problems of thermal stresses occurring in machines and structures.

(b) Effect of Inertia

The problem in connection with the inclusion of the inertia term $\rho \ddot{u}_i$, in the equations of motion was first considered by Duhamel [37, 38]. He observed that these terms $\rho \ddot{u}_i$, are not at all significant when the rate of change of temperature is very slow. Later, different research workers studied in detail the influence of the inertia forces for different material configurations. It has been observed that the error due to the neglect of inertia terms is insignificant for a wide class of problems. Thermoelastic problems, when the inertia forces are neglected are known as quasi-static problems and those with the acceleration terms retained are referred to as dynamical problems. Moreover, one may think of a state where all the quantities involved are independent of time; we call such a state a steady state and the problems concerned are known as static problems. Thus thermoelastic or thermoviscoelastic problems can be classified into two groups: coupled and uncoupled. Coupled problems may be dynamic or quasi-static whereas an uncoupled problem may be a dynamic, quasi-static or static one. In the case of a non-homogenous material body, the mechanical and thermal parameters are functions of spatial coordinates, while for a homogenous material they are constants.

9. Generalised Thermoelasticity

The classical uncoupled theory of thermoelasticity has two main shortcomings. The first and most obvious one is that elastic changes have no effect on the temperature. The second one is that when an elastic solid is subjected to thermal disturbances, the effect is instantaneous at a location far away from the source. This means that the thermal wave propagates with an infinite speed which is physically unrealistic. In 1956 Biot [8] derived the equations of the theory of coupled thermoelasticity, which eliminates the first shortcoming. However, in the classical theory of coupled thermoelasticity heat conduction equation still remains parabolic. It predicts infinite speed of propagation for thermal waves. Generalized thermoelasticity theories have been proposed subsequently to eliminate the paradox of infinite velocity of thermal wave propagation and also to correct the classical coupled thermoelasticity theory (CTE). It is based on the assumption that thermal wave propagates with a finite speed. Lord and Shulman [89] formulated a generalized dynamical theory of thermoelasticity for an isotropic body using a form of the heat transport equation which includes the time needed for acceleration of the heat flow. This theory, called as thermoelasticity theory with one relaxation time, takes into account the coupling effect between temperature and strain rates; the resulting coupled equations are hyperbolic. The paradox of an infinite velocity of propagation, inherent in the classical coupled theory of thermoelasticity, is thus eliminated.

Green and Lindsay [68], by introducing two relaxation times in the thermoelastic process and thus correcting the energy equation and Duhamel-Neumann relations, developed a temperature-rate dependent thermoelasticity theory, which does not violate the classical Fourier's law of heat conduction when the body under consideration has a centre of symmetry. The theory also predicts a finite speed for heat propagation. The significant differences between the L-S and G-L theories are as follows:

(i) The energy equation of L-S model depends on both the strain velocity and acceleration, whereas the corresponding equation of G-L model depends only on the strain velocity.

(ii) The equation of motion and the stress-strain-temperature relations for classical coupled thermoelasticity (CTE) and generalized thermoelasticity of L-S theory remain unchanged whereas in G-L theory these are modified.

The fundamental equations of classical coupled thermoelasticity (CTE) and generalized thermoelasticity due to L-S and G-L models for a linear isotropic homogeneous elastic solid are summarized below.

(a) **Coupled Thermoelasticity (CTE)**

Stess-strain-temperature relation:

$$\sigma_{ij} = \mu(u_{i,j} + u_{j,i}) + \lambda u_{k,k}\delta_{ij} - \gamma\theta\delta_{ij} \tag{9.1}$$

Heat conduction equation:

$$K\theta_{,ii} = \rho C\dot{\theta} + \gamma\theta_0 \, \dot{u}_{i,i} - \rho R \tag{9.2}$$

Equation of motion:

$$\mu u_{i,jj} + (\lambda + \mu)\, u_{j,ij} + \rho F_i - \gamma\theta_{,i} = \rho\, \ddot{u}_i \tag{9.3}$$

(b) **Lord-Shulman (L-S) model**

Here the equations (9.1) and (9.3) remain unchanged whereas equation (9.2) is modified as

$$K\theta_{,ii} = \left(1 + \tau_0 \frac{\partial}{\partial t}\right)\left(\rho C\dot{\theta} + \gamma\theta_0\dot{u}_{i,i} - \rho R\right) \tag{9.4}$$

(c) **Green-Lindsay (G-L) model**

In this case, the equations (9.1) - (9.3) assume the forms

$$\sigma_{ij} = \lambda u_{k,k}\delta_{ij} + \mu\,(u_{i,j} + u_{j,i}) - \gamma\left(1+\tau_1\frac{\partial}{\partial t}\right)\theta\delta_{ij} \tag{9.5}$$

$$K\theta_{,ii} = \rho C\left(1+\tau_2\frac{\partial}{\partial t}\right)\dot{\theta} + \gamma\theta_0\dot{u}_{i,i} - \rho R \tag{9.6}$$

$$\mu u_{i,jj} + (\lambda+\mu)u_{j,ij} + \rho F_i - \gamma\left(1+\tau_1\frac{\partial}{\partial t}\right)\theta_{,i} = \rho\ddot{u}_i \tag{9.7}$$

where τ_0 , τ_1 and τ_2 are suitable constants representing thermal relaxation time parameters.

Unified field equations

The field equations (9.1) - (9.7) can be represented in a unified form as

$$\sigma_{ij} = \mu\,(u_{i,j} + u_{j,i}) + \lambda u_{k,k}\delta_{ij} - \gamma\left(1+\tau_1\frac{\partial}{\partial t}\right)\theta\delta_{ij} \tag{9.8}$$

$$K\theta_{,ii} = \rho C\left(1+\tau_2\frac{\partial}{\partial t}\right)\dot{\theta} + \left(1+\tau_3\frac{\partial}{\partial t}\right)(\theta_0\gamma\dot{u}_{i,i} - \rho R) \tag{9.9}$$

$$\mu u_{i,jj} + (\lambda+\mu)\,u_{j,ij} + \rho F_i - \gamma\left(1+\tau_1\frac{\partial}{\partial t}\right)\theta_{,i} = \rho\ddot{u}_i \tag{9.10}$$

Substituting $\tau_1 = \tau_2 = \tau_3 = 0$ in these equations we get equations for conventional coupled thermoelasticity theory (CTE); setting $\tau_1 = 0$, $\tau_2 = \tau_3 = \tau$ (say) the equations reduce to L-S model and for $\tau_3 = 0$ and $\tau_1, \tau_2 \neq 0$ equations for the G-L model are obtained.

Another type of generalized thermoelastic model has been suggested by Ghaleb and Mohamedein [65] in which temperature is governed by a third order differential equation which includes three relaxation times, showing that the resulting coupled equations are more hyperbolic than the equations obtained in L-S and G-L models.

10. New Approach to the Generalized Thermoelasticity Theory

In recent years further theoretical developments on the subject have taken place basing upon the theory given by Green and Naghdi [66, 67, 69] which provide sufficient basic modifications in the constitutive equations that permit treatment of a much wider class of heat flow problems. The characterization of material response for such thermal phenomena explained in [66, 67, 69], is based on three types of constitutive response functions, labelled as Types I, II and III. The nature of these three types of constitutive equations is such that when the respective theories are linearized, Type I is the same as the classical heat conduction theory (based on Fourier's law), while the linearized versions of both Type II and III theories permit propagation of thermal waves at finite speed. The governing field equations for a homogeneous isotropic elastic body in the context of generalized theory of thermoelasticity of Type II and Type III developed by Green and Naghdi are given below.

Type II

Stress-strain-temperature relations:

$$\sigma_{ij} = \lambda u_{i,i}\delta_{ij} + \mu (u_{i,j} + u_{j,i}) - \gamma\theta\delta_{ij} \tag{10.1}$$

Heat conduction equation

$$K^*\theta_{,ii} = C\ddot{\theta} + \gamma\theta_0\ddot{u}_{i,i} - \rho\dot{R} \tag{10.2}$$

Equation of motion

$$\mu u_{i,jj} + (\lambda + \mu) u_{j,ij} + \rho F_i - \gamma\theta_{,i} = \rho\ddot{u}_i \tag{10.3}$$

Type III

The equations (10.1) and (10.3) remain unchanged while the heat conduction equation (10.2) is modified to the form

$$K\dot{\theta}_{,ii} + K^*\theta_{,ii} = \rho C\ddot{\theta} + \gamma\theta_0\ddot{u}_{i,i} - \rho\dot{R}\ , \tag{10.4}$$

where K* is a material constant characteristic of the theory and other symbols have the same meanings as stated earlier.

11. Micropolar Thermoelasticity

A theory of micropolar elasticity was proposed by Eringen and Suhubi [49, 200] to describe the continuum behaviour of materials possessing microstructure. Basically the difference between the classical continuum theory and the micropolar theory is that the latter admits independent rotations of the substructure of the material; that is, the local intrinsic rotations, called microrotations (Φ_k) are taken to be kinematically independent of the linear displacements (u_k). It is believed that such a theory is applicable in the treatment of granular and fibrous composite materials.

Basic Equations of Micropolar Elasticity:

The theories of micropolar solids and fluids developed by Eringen [49] are based on the following balance laws :

$$\frac{\partial \rho}{\partial t} + (\rho v_k)_{,k} = 0 \qquad \text{(mass)}$$

$$t_{kl,k} + \rho f_l = \rho \ddot{u}_l \qquad \text{(momentum)}$$

$$m_{rk,r} + \varepsilon_{klm} t_{lm} + \rho l_k = \rho J \ddot{\varphi}_k \qquad \text{(moment of momentum)}$$

$$\rho \dot{\varepsilon} = t_{kl} \dot{\varepsilon}_{lk} + m_{kl} \dot{\varphi}_{l,k} + q_{k,k} + \rho h \qquad \text{(energy)}$$

$$\rho \Gamma \equiv \rho \dot{\eta} - \frac{\rho h}{\theta} - \left(\frac{q_k}{\theta} \right) \geq 0 \qquad \text{(Clausius – Duhem inequality)}$$

and kinematic relations

$$\varepsilon_{kl} = e_{kl} - \varepsilon_{klm} (r_m - \varphi_m)$$

$$e_{kl} = \frac{1}{2} (u_{k,l} + u_{l,k})$$

$$r_m = \frac{1}{2} \varepsilon_{mnp} u_{p,n}$$

where ρ is the mass density, v_k the velocity vector, t_{kl} the stress tensor, f_l the body force vector, u_l the displacement vector and m_{kl} the couple stress tensor. ε_{klm} is the permutation symbol. l_k represents the body couple vector, J the micro-inertia coefficient, φ_k micro-rotation vector, ε the internal energy density, ε_{kl} the micro-strain tensor and q_k the heat flux vector (directed outward). h, η and θ stand respectively for the heat source density, the entropy density and the temperature. ε_{kl} represents the strain tensor and r_m the rotation vector.

Basic Equations of Linear Micropolar Thermoelasticity:

The linear theory of micropolar thermoelasticity was developed by extending the theory of micropolar continua to include thermal effects (Tauchert et al., [204]). In this case, the basic equations can be listed as follows :

Balance of momentum and moment of momentum:

$$t_{kl,k} + \rho f_l = \rho \ddot{u}_l ,$$

$$m_{rk,r} + \varepsilon_{klm} t_{lm} + \rho l_k = \rho J \ddot{\Phi}_k .$$

Constitutive equations:

$$t_{kl} = \lambda e_{mm} \delta_{kl} + (2\mu + \chi) e_{kl} + \chi \varepsilon_{klm} (r_m - \Phi_m) - \alpha_T (3\lambda + 2\mu + \chi) T \delta_{kl} ,$$

$$m_{kl} = \alpha \Phi_{m,m} \delta_{kl} + \beta \Phi_{k,l} + \gamma \Phi_{l,k} .$$

Coupled heat conduction equation:

$$\kappa T_{,mm} = \rho c_E \dot{T} + (3\lambda + 2\mu + \chi) \alpha_T T_0 \dot{e}_{mm} .$$

and the kinematic relations:

$$e_{kl} = \frac{1}{2}(u_{k,l} + u_{l,k}) ,$$

$$r_m = \frac{1}{2} \varepsilon_{mnp} u_{p,n} ,$$

in which α_T represents the co-efficient of linear thermal expansion, κ the thermal conductivity, T the temperature change, T_0 the temperature for which the material is stress-free, c_E the specific heat at constant deformation, u_k the displacement vector, e_{kl} the strain tensor; $\lambda, \mu, \chi, \alpha, \beta$ and γ are material constants. Other symbols carry the same meanings as above.

The independent variables defined above are referred to a rectangular co-ordinate system (x_1, x_2, x_3). A superposed dot indicates a time derivative and a comma denotes partial differentiation with respect to the space variable x_k. Repeated indices are summed over the range (1,2,3).

12. Mathematical Analyses of Various Problems of Thermoelasticity

As mentioned earlier, the subject of thermoelasticity originated with the work of Duhamel [37, 38] and it dates back to 1837 when his famous memoir containing stress-strain-temperature relations for non-uniformly heated elastic solids appeared. A few years later, Neumann [162] independently formulated the stress-strain-temperature relations similar to those of Duhamel. The equations they derived are known as Duhamel-Neumann's relations in the literature. Sufficient attention was not paid to the field in later years, till the recent developments in nuclear engineering, aeronautics and astronautics took place. Only during the last five decades the theory of thermoelasticity has been a subject of active research with systematic investigations made by mathematicians and engineers covering it in its theoretical as well as experimental aspects. The theory of thermoelasticity for elastic as well as viscoelastic materials is now sufficiently grown up.

In reality even an instantaneous application of mechanical load to solids leads to delayed response. To account for this delayed response, the theory of viscoelasticity has been developed. The linear theory of viscoelascitity based on the assumption of small strain and negligible temperature rise due to dissipative effects has been described in detail in the works of several authors including those of Bland [9] and Christensen [30]. The literature on various problems of thermal stress for elastic and viscoelastic solids is quite extensive and it is very difficult to prepare a complete list of all the investigations here. One may, however, take as general references, the monographs authored by Nowacki [165, 166], Nowinski [167], Parkus [169], Boley and Weiner [12] and Love [90], where thermal stresses developed by non-uniform temperature distribution have been thoroughly discussed; reviews of earlier works on specific problems of thermoelasticity together with lists of references are also given in these monographs. Hetnarski [73, 74] presented the available theories on Thermoelasticity and reviewed a number of specific problems of classical as well as generalised thermoelasticity. Eringen [50, 52], Suhubi [201], Maugin [93], Fung [62] and Lekhnitskii [85] published some related books which may also be used for reference purpose. Method of determining temperature distribution in thermoelastic or thermoviscoelastic materials arising out of surface heating and heat source, based on Fourier heat conduction theory has been explained by Carslaw and Jaeger [13].

The thermoelastic potential technique has been used by many research workers to solve different problems of thermoelasticity. But due to some disadvantages of this potential function technique, Bahar and Hetnarski [5] introduced the state-space approach using the method of direct integration by means of matrix exponential for solving coupled thermoelastic problems. The idea behind the application of the state-space approach in thermoelasticity has been the decoupling of the governing system of partial differential equations through the use of integral transformation. Thereby a new system of first order ordinary differential equations can be formed whose solution is found in terms of a transfer matrix involving a combination of exponential functions of growing and decaying nature as a function of the space variable. In this method the problem is reformulated in terms of the state-space variables, namely, the temperature, the displacement and their gradients. From these the solution of the problem in the Laplace transform domain can be easily derived and the actual physical variables are determined by a corresponding inverse Laplace transformation. Connection between thermoelastic potential technique and the state space approach has been discussed by Bahar and Hetnarski [6]. The state-space approach was adopted by El-Karamany and Ezzat [44] for the solution of a one-dimensional problem of plane distribution of heat sources. The jump discontinuities were calculated and the kinematic conditions of compatibility were derived. Another possible way of solving thermoelastic problems is the eigen value approach.

Given below is a review of some of the mathematical analyses carried out by different investigators.

Classical Thermoelasticity Problems:

Using Fourier / modified Fourier heat conduction equation many problems of classical thermoelasticity have been studied by different research workers considering various material configurations subject to different initial and boundary conditions.

Chen and Chu [28] investigated the transient thermal stress distribution in a finite composite hollow cylinder which is heated by line sources moving periodically on its inner surface and cooled convectively on the exterior surface. The study was based upon the consideration that the axisymmetric heat source moves along the axis of the cylinder with a constant speed.

A three-dimensional analysis of transient thermal stresses in functionally graded plates was put forward by Vel and Ratra [209]. Qian and Batra [170] studied transient thermoelastic deformations of a thick functionally graded plate with edges held at a uniform temperature and either simply supported or clamped. Stresses and deformations induced due to the simultaneous application of the transient thermal and mechanical loads were computed.

Yang et al. [214] studied a one-dimensional quasi-static coupled thermoelastic problem of an infinitely long annular cylinder with axial symmetry. The material considered by them is elastic, homogeneous and isotropic. Laplace transform technique together with the application of Fourier series and matrix operation is used to obtain the general solution of the governing equations. Trehani and Eslami [208] considered the boundary element formulation for the analysis of coupled thermoelastic problems in a finite domain and studied the coupling coefficient and relaxation time effects on thermal and elastic wave propagation. Tran and Singh [207] analyzed wave propagation in coupled non-linear thermoelastic materials. Locations of the leading wave front and the lagging wave front were determined.

Sternberg and Chakraborty [198] solved the problem of an unbounded thermoelastic body having a spherical cavity whose stress-free boundary is subjected to a thermal shock.

Takeuti and Furukawa [202] studied a thermal shock problem for a plate; it was concluded that the coupling effect is more significant than the inertia effect, in the case of ordinary materials. Sosinski [197] studied one-dimensional shock waves in solids with internal state variables.

Thermal stresses produced in a hollow anisotropic cylinder whose outer surface is insulated and the inner surface is subjected to γ-ray heating were studied by Bagchi [4].

Thermoelastic stresses generated due to an internal source of heat in a solid elastic hemisphere was analyzed mathematically by Misra [145].The plane base of the hemispherical solid is considered to rest on a smooth rigid insulating surface, the curved surface being exposed to heat radiation. Deflection of a thin anisotropic rectangular plate having supported edges and with different temperature distributions on the two plane faces was investigated by Misra [139]. The same author [143] made a theoretical study on the bending of a heated solid circular plate of elastic material due to a nucleus of heat concentrated at the centre of the upper face of the plate. The study corresponds to a situation where the lower surface is thermally insulated and the edge of the plate is fixed. Thermal stresses in a long cylindrical solid body made of an anisotropic material have been studied by Misra [144]. The cylindrical body is considered to have a heat source situated at a point on its axis and the free curved surface is radiating heat into the surrounding medium at zero ambient temperature. Misra [126] determined a theoretical estimate of the thermal stresses in an annular circular disc of varying thickness made of a cylindrically aeolotropic material. The disc is considered to be rotating about the central axis, while its inner boundary is rigidly attached to a concentric hub rotating with disc. A thermoelastic problem concerned with a non-homogeneous anisotropic circular disc with a central hole was treated by Misra and Achari [105]. The disc considered is such that the thickness, the thermal conductivity and the elastic constants of the material vary according to different power laws of the distance from the centre. The problem studied pertains to a situation where the disc rotates with a constant angular velocity about its central axis. The inner boundary of the disc is considered to be rigidly welded to a rotating hub while the outer one is stress-free. The problem is numerically analysed further in order to obtain the distribution of hoop stress along the radial direction.

Misra [127-130,140-142, 152-154] performed some more theoretical analyses of a series of problems on thermally stressed solid media with different geometrical configurations subjected to various types of uneven distribution of temperature.

Misra and Achari [108] studied the distribution of thermal stresses in an infinite elastic disc that receives heat through radiation from a point heat source situated at a finite distance. The same authors [104] also investigated the flexural deformation of a thin simply supported rectangular plate due to a point source of heat moving on one of its surfaces. They made an interesting observation that the dynamic solution oscillates about the quasi-static one.

Misra and Samanta [119] carried out an investigation on the thermal shock due to sudden heating of the bounding surface of a viscoelastic half-space by taking into account the coupling of temperature and strain fields as well as the effects of thermal relaxation on the shock propagation. They used the linear standard model to represent the viscoelastic behaviour of the solid in shear which is elastic in bulk deformation. The field equations that they have taken are as follows :

Equation of motion :

$$\sigma_{ij,j} = \rho \ddot{u}_i$$

Local energy balance equation :

$$-h_{i,i} = \rho C_v \dot{\theta} + 3K\alpha T \dot{e}$$

Thermoelastic constitutive relations :

$$S_{i,j} = 2\,G(D)e_{ij}$$

$$\sigma = 3K\,(e - 3\alpha\theta)$$

Constitutive equation for heat conduction with thermal relaxation :

$$h_i + \tau_0 \dot{h}_i = -\chi\,\theta, i$$

in which σij are the stress components, e is the dilatation, $\sigma = \sigma_{kk}$, S_{ij} and e_{ij} are deviatoric stress and strain components, u_i the displacement components, θ the temperature excess over initial temperature T at which the system is free from any disturbance. h_i is the heat flux vector, G(D) is the time –dependent viscoelastic shear modulus , $D \equiv \frac{\partial}{\partial t}$.

K is the bulk modulus of elasticity, α the co-efficient of linear thermal expansion, χ the thermal conductivity and τ_0 the thermal relaxation time. ρ is the volume density of mass and C_v the specific heat at constant volume. The boundary and initial conditions for the problem are expressed mathematically as

$\sigma_{xx}(0, t) = 0$ $\qquad$ $\sigma_{xx}(x, t) = 0$ $\qquad$ $x \to \infty, t > 0$

$\theta(0, t) = \theta_0$ $\qquad$ $\theta(x, t) = 0$ $\qquad$ $x \to \infty, t > 0$

and for $x \geq 0$,

$$u_x(x, 0) = \dot{u}_x(x, 0) = 0$$

$$\sigma_{xx}(x, 0) = \dot{\sigma}_{xx}(x, 0) = 0$$

$$\theta(x, 0) = \dot{\theta}(x, 0) = 0$$

The analysis is valid for small times. The study shows that the stress and temperature fields have discontinuities at two points at a given instant of time. They made an observation that the attenuation takes space more rapidly in the viscoelastic case than in the corresponding case of an elastic continuum. They also found that when the material behaviour changes from elastic to viscoelastic, the velocity of the propagation for the first shock does not undergo any appreciable change but the attenuation factor increases significantly. In the case of the second shock, however, the situation is quite different. If the time for thermal relaxation is not taken into account, they reported that there is a drastic magnification in the velocity of propagation as well as in the attenuation factor of the second shock, in comparison to a case in which the relaxation time is taken to have a non-zero value. This observation led them to suggest that for studying the response at points close to be plane of heating, the consideration of thermal relaxation is of particular importance in the small time range. It was concluded that the second sound can be observed at very low temperatures, at which the relaxation process plays an important role. This study is believed to have applications in cryogenic engineering, where the mechanical response continua exposed to very low temperatures is of particular concern. The results of this study have also been found to possess useful applications in the design of an experiment aimed at the determination of the relaxation time from the measurements of the attenuation and the speed of the second shock front. The asymptotic behaviur of solutions in linear two- or three- dimensional thermoelasticy with second – sound was discussed by Racke [172].

Misra and Samanta [122] also studied the thermal stresses in rotating discs of thermorheologically simple solids – the viscoelastic materials that obey the time-temperature equivalence hypothesis. Assuming that the temperature field is axi-symmetric and that it does not vary along the thickness of the disc, the uncoupled heat conduction equation, in cylindrical polar coordinates (r, ϕ, z) was taken as

$$\eta\frac{\partial\theta}{\partial t}=\frac{\partial^2\theta}{\partial t^2}+\frac{1}{r}\frac{\partial\theta}{\partial r},$$

in which η is the thermal diffusivity assumed to be independent of temperature and θ is the excess temperature at a point (r, ϕ, z) of the disc at time t, over the uniform temperature before heating; all the subsequent temperatures are measured taking this initial temperature as the reference. This investigation was based upon the use of the following initial and boundary conditions for the temperature field :

$$\theta\,(r, t) = \frac{\partial\theta}{\partial t}\,(r,t) = 0, \qquad -\infty < t < 0$$

$$\theta\,(r_1, t) = \theta_1\,H\,(t) \quad \text{and} \quad \theta\,(r_2, t) = \theta_2\,H\,(t),$$

r_1 , r_2 being the internal and external radii of the disc and H(t) the Heaviside function.

Bending of a rectangular viscoelastic plate resting on an elastic foundation and subjected to an uneven distribution of temperature, was investigated by Misra and Samanta [120]. The time dependence of the plate deflection was estimated for a plate made of a Kelvin-Voigt material. Misra and Achari [107] studied the quasi-static stress-field generated in an orthotropic elastic circular disc by a heat source rotating on a concentric circular path. An exact solution was obtained in terms of doubly infinite series, using a stress-function approach. Denoting by K_1, K_2 and K_3 the conductivity co-efficients of the disc having a heat source of constant instenty Q_0, rotating along a circular path $r = r_o$ with constant angular velocity ω, they used the heat conduction equation in plane polar coordinates (r, θ) in the form

$$\beta^2 \alpha^{-2} \left(\frac{\partial^2 T}{\partial r^2} + \frac{1}{r}\frac{\partial T}{\partial r} \right) + \beta^2 r^{-2} \frac{\partial^2 T}{\partial \theta^2} = \frac{\partial T}{\partial t} = \frac{Q_0 \delta \left(r - r_0\right) \, \delta \left(\theta - \omega t\right)}{r}$$

where $\alpha^{-2} = K_1/K_2$, $0 < r \leq a$, $0 < t \leq 2\pi/\omega$, β^2 being the thermal diffusivity in the θ-direction, a the radius of the disc and δ (...) the Dirac's delta function.

The initial and boundary conditions for the temperature T are taken as

$$T(r, \theta; 0) = 0$$

and $$\frac{\partial T}{\partial t} + hT = 0, \qquad r = a$$

in which $h = H/K_1$, H being the surface conductance.

The same authors [109] studied the thermoelastic stresses in a hollow anisotropic finite cylinder arising from axisymmetric temperature variations at the plane ends, where the inner and outer curved surfaces are in contact with smooth rigid insulators. Dynamic thermal stresses in a thin semi-infinite visoelastic rod of Bolzmann model were studied analytically by Misra and Achari [106] . They considered three types of temperature variations : linear, exponential and periodic, at one end of the rod. For the same type of temperature variations, thermoelastic stresses in an elastic non-homogeneous semi-infinite rod were studied by Misra et al. [151]. Thermal shock in a non-homogeneous elastic rod having one of its ends maintained at a constant temperature was studied by Misra et al. [112].

Misra and Samanta [117] presented a theoretical analysis for creep in materials at elevated temperatures. The specimen used by them is a thick orthotropic cylindrical shell subjected to a considerably high pressure and temperature. Using cylindrical polar co-ordinates (r, ϕ, z) and assuming plane strain and axisymmetry whereby all the quantities are independent of ϕ and z, the only non-trivial equation of equilibrium for this study has been taken in the form

$$r\frac{\partial \sigma_r}{\partial r} = (\sigma_\phi - \sigma_r) \exp(e_r - e_\phi)$$

Considering incompressibility of the material of the cylinder, the constitutive equations were suitably modified by them in order to take into account the generated temperature field. Taking the principal axes of stress to be coincident with the axes of symmetry (which are also taken as the axes of co-ordinates), the equations have been derived in the form

$$e_r = \frac{e}{2\sigma}\,[(G + H)\,\sigma_r\ - H\sigma_\phi\ - G\sigma_z\,] + \alpha_1 T$$

$$e_\phi = \frac{e}{2\sigma}\,[(H + F)\,\sigma_\phi\ - F\sigma_z\ - H\sigma_r\,] + \alpha_2 T$$

$$e_z = \frac{e}{2\sigma}\,[(F + G)\,\sigma_z\ - G\sigma_r\ - F\sigma_\phi\,] + \alpha_3 T$$

so that $e_r + e_\phi + e_z = (\alpha_1 + \alpha_2 + \alpha_3\,)T.$

Thus $e_z = -\,\frac{2\sigma}{e}\ \ \alpha_3 T + G\sigma_r + F\sigma_\phi / (\,F + G\,).$

They concluded that the temperature variation has a significant effect on the strain and the strain-rate, particularly when the molecular anisotropy is taken into account.

All the investigations referred to above are concerned with homogeneous media. But similar studies for non-homogeneous media are of considerable interest because many non-homogeneous materials are being extensively used in modern industries. Keeping this in view, Misra et al [148] investigated the effects of mechanical and thermal relaxations on the stresses in a heated viscoelastic continuum with a cylindrical hole. They [149] also performed an analytical study for the stresses generated in a viscoelastic solid due to a heat source in the vicinity of a spherical cavity. The effect of non-homogeneity in the microscopic sense, on the distribution of stress, displacement and temperature in an elastic solid medium having a spherical cavity was examined by Misra et al. [150] by taking the elastic moduli and material density as functions of position. The same authors [147] examined the role of anisotropicity in a viscoelastic plate resting on an elastic foundation, when the plate is subjected to an uneven temperature distribution. They [124] also dealt with the problem of magnetothermoelastic interaction in an infinite continuum with a cylindrical hole, subjected to ramp-type heating.

Hetnarski [74] discussed coupled thermoelastic problems for a half-space and also one-dimensional thermal shock problems [73] for small time approximation. Problems concerning the propagation of thermoelastic waves in elastic media also received attention of several investigators.

Chadwick and Seet [14] investigated the thermoelastic wave propagation in a transversely isotropic heat conducting elastic material. They established three types of elastic waves for such media, namely quasi-longitudinal (QL), quasi-transverse (QT) and a purely transverse (SH) waves of which only two types QL and QT were found to be influenced by thermoelastic coupling effects.

Nowacki [165] discussed the propagation of longitudinal waves in an unbounded thermoelastic medium in the context of coupled thermoelasticity theory. Chadwick and Windle [15] studied the effects of heat conduction upon the propagation of Rayleigh waves in a semi-infinite elastic solid under specific conditions. A boundary element method was used for the study of anisotropic coupled thermoelasticity by Kogl and Gaul [83].

Propagation of thermal shock waves in porous material undergoing thermal loading was studied by Sluzalec [194]. Iovane and Passarella [79] dealt with a study of porous materials with a memory effect for the intrinsic equilibrated body forces. An appropriate time-weighted volume measure was considered to be associated with the backward-in-time thermoelastic processes. The same authors [80] treated a problem for a linear thermoelastic porous material with a constitutive equation for heat flux with memory. An approximate theory of thermodynamics was also presented for this model and a maximum pseudo free energy was determined. The response of saturated porous media to cyclic thermal loading was examined by Blond et al. [10]. Eshelby's formula for an ellipsoidal elastic inclusion was applied by Levin and Alvarez – Tostado [86] for the study of anisotropic poroelasticity and thermoelasticity. Thermal stresses in plane strain in completely separate terms were studied by Iesan and Nappa [76].

Studies with Generalized Thermoelasticity Theories

As the concept of instantaneous propagation of thermal waves needs modification for the cause of finite velocity of thermal waves (determined also from experiments conducted at low temperature), generalized thermoelasticity theories were developed by different research workers [65-69, 89] following different lines of reasoning. The effects of temperature on the relaxation and coupling constants were discussed by Lord and Shulman [89]. The exact solution obtained by them leads to the following important conclusions:

(i) At a very low temperature the coupling constant becomes zero. Consequently, the coupling between temperature and strain becomes negligible at temperature near absolute zero.

(ii) At high temperatures the relaxation constant of the generalized theory becomes very small. Thus, at high temperatures the generalized theory reduces to the conventional coupled theory of thermoelasticity.

(iii) The solution obtained by applying the generalized theory to a particular boundary value problem is generally similar to that obtained by using the conventional coupled theory of elasticity but is different in one important aspect, namely, the generalized theory predicts a thermal wave front having a finite propagation velocity. Hence it is expected that when applied to other problems, the generalized theory should yield solutions not very different in form from those predicted by conventional coupled theory except where the propagation of thermal energy is predominant.

(iv) Finally the generalized theory presented by Lord and Shulman [89] serves to eliminate the paradox of an infinite propagation velocity inherent in the conventional coupled theory of thermoelasticity and indicates the effect of neglecting higher order terms in the heat conduction equation.

During the last two decades, mathematical analyses of different thermoelastic as well as thermoviscoelastic problems in solid continua were carried out by several investigators using generalized theories of thermoelasticity developed by Lord and Shulman [89], Green and Lindsay [68] as well as Green and Naghdi [66, 67, 69]. Sherief and Dhaliwal [185] extended the theory of Lord and Shulman [89] for anisotropic material media and also studied the problem in the presence of a heat source. A one-dimensional problem for an infinitely long hollow cylinder was studied by Sherief et al. [188] in the context of generalized thermoelasticity with one relaxation time.

Integral transform techniques have been used by most of the investigators in order to obtain the solution of generalized thermoelastic problems but the difficulties arise during the exact inversion process. An approximate analysis using asymptotic expansion valid for small values of time was performed in order to avoid the complexity of the nature of inversions. Generalized thermoelasticity for an unbounded solid medium having a spherical/cylindrical cavity was employed by different research workers for finding the solution, using small-time approximations.

In another study on the propagation and stability of harmonically time-dependent thermoelastic plane waves of assigned frequency using G-L model of generalized thermoelasticity, it was shown that the transverse (shear) waves remain unaffected.

Misra et al. [158] conducted another theoretical study of the magneto-thermoelastic stress fields generated in an aeolotropic solid cylindrical medium subjected to a ramp-type heating on its surface, when the cylindrical body placed in vacuum is under the action of a uniform axial magnetic field. The perturbed magnetic field induced in the free space surrounding the cylinder has also been analysed by them. The linearized equations that they used as the equations governing the interactions between the mechanical, thermal and magnetic fields are given below.

Equation of motion

$$\sigma_{ij,j} = \rho \frac{\partial^2 u_i}{\partial t^2}$$

in which the stress components σ_{ij} and the displacement components u_i include the thermal and electromagnetic contributions, so that

$$\sigma_{ij} = C_{ijkl}e_{kl} - \beta_{ij}\,\theta + \mu\,(h_i H_j + h_j H_i - h_k H_k \delta_{ij})$$

where ρ is the material density, u_i the displacement vector, θ the excess temperature over the ambient one, C_{ijkl} stiffness tensor, β_{ij} dilatation tensor, μ the magnetic permeability and H_i the magnetic field vector given by

$$H_i = H_{oi} + h_i \ ,$$

H_{oi} being the applied magnetic field and h_i its perturbed value; the strain tensor e_{kl} satisfies the kinematic relation

$$e_{kl} = \frac{1}{2}\left(\frac{\partial u_k}{\partial x_l} + \frac{\partial u_l}{\partial x_k}\right),$$

x_k being the co-ordinates of a representative point.

Energy equation

$$\rho C_v \frac{\partial \theta}{\partial t} - \frac{\partial}{\partial x_i}\left(K_{ij}\frac{\partial \theta}{\partial x_j}\right) = 0$$

where K_{ij} is the thermal conductivity tensor and C_v the specific heat at constant strain.

Electrodynamic constitutive equations

$$\nabla \times \mathbf{h} = \mathbf{J} + \mathbf{D}$$

$$\nabla \times \mathbf{E} = - \frac{\partial \mathbf{B}}{\partial t}$$

$$\nabla . \mathbf{B} = 0$$

$$\nabla . \mathbf{D} = \rho_e$$

where **E**, **J**, **D** and **B** are the electric field, current density, electric displacement and magnetic induction vectors respectively and ρ_e is the charge density.

Electromagnetic constitutive relations

$$\mathbf{B} = \mu \, \mathbf{H}$$

$$\mathbf{D} = \varepsilon \left[\mathbf{E} + \left(\mu - \frac{1}{\varepsilon c^2} \right) \left(\frac{\partial \mathbf{u}}{\partial t} \times \mathbf{H} \right) \right]$$

$$\mathbf{J} = \eta_0 \left[\mathbf{E} + \mu \left(\frac{\partial \mathbf{u}}{\partial t} \times \mathbf{H} \right) \right] - \tau \nabla \theta + \rho_e \frac{\partial \mathbf{u}}{\partial t}$$

with ε as the dielectric constant, η_0 the electrical conductivity and τ the Thompson coefficient.

For this problem, the ramp-type heating on the curved surface of the cylinder has been mathematically defined as

$$\theta = \begin{cases} 0, & t \le 0 \\ \theta_0 \dfrac{t}{t_0}, & 0 < t < t_0 \\ \theta_0, & t \ge t_0 \end{cases}$$

where θ is the excess temperature at time t, θ_0 is the steady temperature and t_0 is the time of rise of temperature of the cylindrical surface.

On the basis of the study, they made a conjecture that wave-like disturbances arise out of the hyperbolic nature of the displacement equations – one of these corresponds to the beginning of the heating ($t = 0$) and the other to the attainment of the steady state of the temperature of the curved surface (at $t = t_0$). Diffusion-like disturbances have their origin in the differential nature of the heat conduction equation. Moreover, the magnetic disturbance consists of two parts each of which is wave-like and propagates with the velocity of light. This is due to the hyperbolic nature of the equation governing the perturbed magnetic field. One of the constituents corresponds to the onset of the shock ($t= 0$) and the other to its termination on the cylindrical surface.

Misra et al. [157] also studied magnetothermomechanical interaction in an aeolotropic viscoelastic cylinder permeated by a magnetic field and subjected to a periodic loading. They assumed that the viscoelastic nature of the material is described by the Voigt model of linear viscoelasticity and considered

$$c_{ij}(D) = c_{ij}\left(1 + t_0 \frac{\partial}{\partial t}\right),$$

where $D = \dfrac{\partial}{\partial t}$ and t_0 is the relaxation time assumed to be the same for each c_{ij}.

The problem of a half-space under the influence of an external primary magnetic field and an elevated temperature field arising out of a ramp-type heating of the surface was investigated by Misra et al. [159]. In the mathematical analysis, due attention has been paid to the viscous property of the medium as well as the finite time of rise in temperature. A quantitative estimate is made for the stress-field developed in the medium and the perturbation magnetic field in the adjoining free-space on the surface of the separation at different instants of time for a Voigt-type solid. They made an observation that the stress distribution and the secondary magnetic field are almost independent of the thermal relaxation time but are significantly dependent on the mechanical relaxation time.

A theoretical estimate of the distribution of temperature, displacement as well as the radial, axial and hoop stresses in a circular cylindrical shell of perfectly conducting material placed in a uniform magnetic field was put forward by Misra et al. [123]. The situation considered by them is one in which the inner surface of the shell is maintained at a constant temperature while the outer surface is exposed to convective heat transfer. The initial and boundary conditions used in this problem for the temperature of the shell have been mathematically represented as

$$T(r, 0) = 0, \quad r_1 \le r \le r_2$$

$$T(r_1, t) = T_0\eta(t), \quad t>0$$

$$\left[\frac{\partial T}{\partial r}+gT\right]_{r_2}=0, \quad t>0$$

in which r_1 and r_2 are respectively the internal and external radial of the shell, $\eta(t)$ is the Heaviside function and g is the convection constant.

The influence of the thermomechanical coupling on the transient thermoelastic deformation of a cylindrical solid medium in the presence of a magnetic field was also investigated by Misra et al. [131].

The problem of thermally induced vibration in an infinite anisotropic viscoelastic solid with a cylindrical cavity in the presence of a uniform axial magnetic field was also analyzed by Misra et al. [132]. The study corresponds to a situation where a harmonic temperature field is applied at the surface of the cavity so that

$$T = T_0\, e^{i\omega t} \quad \text{on } r = a$$

where a stands for the radius of the cylindrical cavity, while T_0 and ω are constants. The displacement equation of the resulting magnetothermoelastic wave propagtion has been considered in the form

$$\frac{\partial^2 u}{\partial r^2}+\frac{1}{r}\frac{\partial u}{\partial r}-\frac{u}{r^2}-\frac{\beta_{11}(1+i\omega t_0)}{c_{11}+\mu H_0^2+i\omega t_0 c_{11}}=\frac{\rho}{c_{11}+\mu H_0^2+i\omega t_0 c_{11}}\frac{\partial^2 u}{\partial t^2}$$

This is supplemented by the heat conduction equation

$$\frac{\partial^2 T}{\partial r^2}+\frac{1}{r}\frac{\partial T}{\partial r}-\frac{1}{\kappa}\frac{\partial}{\partial t}\left(1+t_1\frac{\partial}{\partial t}\right)T=\eta(1+i\omega t_0)\frac{\partial}{\partial t}\left(1+t_1\frac{\partial}{\partial t}\right)\left(\frac{\partial u}{\partial r}+\frac{u}{r}\right)$$

where the Lord - Shulman model has been used to account for the finite velocity of heat propagation and

$$\eta=\frac{\beta_{11}T_0'}{K_{11}}, \qquad \beta_{11}=(c_{11}+c_{12})\,\alpha_{11}+c_{13}\,\alpha_{33}$$

α's being the linear thermal expension coefficients. K_{11} and κ are, respectively, the thermal conductivity and diffusivity, ρ is the material density, T'_0 the reference temperature, t_1 the thermal relaxation time and μ the magnetic permeability of the material medium. In the above equations, r is the radial coordinate, t the time variable, t_0 the mechanical relaxation time and c_{ij} the elastic constants.

Ezzat [57] used the equations of generalized magneto-thermoelasticity with two relaxation times for a perfectly conducting isotropic medium to solve the problem of determining stress and temperature distributions with continuous line source of heat in an infinite elastic body, permeated by an axial uniform magnetic field. Propagation of electro-magneto-thermoelastic waves in rotating media with thermal relaxation was studied by Choudhury [29]. A thermal-shock problem in magneto – thermoelasticity with thermal relaxation was studied by Sherief and Ezzat [191]. Generalized magneto-thermoelasticity was used to analyze another problem concerned with an infinitely long annular cylinder. The problem was solved analytically by the same authors [192]. Ezzat et al. [60] discussed the propagation of electro-magneto – thermoelastic plane waves with thermal relaxation in a medium of perfect conductivity. The uniqueness and reciprocity theorems for linear thermo-electro-magneto-elasticity were studied by Li [87].

Sosa and Bahar [196] studied a one-dimensional thermoelastic problem of a homogeneous isotropic body bounded by two planes using state-space approach to deal with coupled thermoelasticity. Two completely independent formulations were taken for solving the problem. One-dimensional problems of thermoelasticity were treated in non-dimensional form to reduce the excessive number of parameters appearing in their formulations. In the first part of their work thermal coupling and inertia effect were tied together on the basis of the normalization or change of variables and in the second part, the analysis was treated by a different approach. An alternative state-space formulation for magneto-electric thermoelasticity for solids with transverse isotropy were giving by Chen and Lee [25]. They applied their analysis to study the bending of non-homogeneous plates. Magneto-thermo-elastic instability of ferromagnetic plates in thermal and magnetic fields was investigated by Wang et al.[211]. Sharma et al. [181] studied the thermoelastic interaction in an infinite Kelvin-Voigt type viscoelastic, thermally conducting plate. They derived the secular equations for the plate in closed form and isolated mathematical conditions for symmetric and skew-symmetric wave mode propagation in completely separate terms.

Sherief and Anwar [183] considered a one-dimensional thermoelastic problem for a long cylindrical shell made of two different materials with axial symmetry. The problem was solved under the framework of L-S model of generalized thermoelasticity using potential function approach and Laplace transform technique. Two-dimensional thermoelastic problems under axisymmetric temperature distributions were considered by Sherief and Hamza [186] in the context of generalized thermoelasticity with one relaxation time (L-S model). The general solution was obtained by using Laplace transformation and by resorting to a direct approach instead of using potential functions. The resulting formulation was used to solve two problems – one concerning a solid sphere and another an infinite space with a spherical cavity.

Using L-S model of generalized thermoelasticity, Sherief [189] also treated the problem of determining stress and temperature distributions in an infinite homogeneous isotropic thermoelastic solid having a continuous point heat source situated at the origin. A similar problem for a continuous line source of heat located along an axis was studied by Sherief and Anwar [184]. The state–space approach developed by Bahar and Hetnarski [5-6] was used by them [183] for finding the solution of another coupled thermoelastic problem, under the purview of generalized thermoelasticity with one relaxation time. Suhubi and Erbay [199] studied a longitudinal wave propagation problem for an infinite circular cylinder in the context of generalized thermoelasticity, where the dispersion relation was obtained for the particular case in which the temperature is kept constant on the surface of the cylinder.

An eigenvalue approach, together with Laplace and Hankel transform techniques was employed by Sharma and Kumar [177] to study a two-dimensional axisymmetric problem for a homogeneous transversely isotropic medium. With an aim to illustrate the applicability of the approach, an example of infinite space with concentrated force at the origin was considered by them. The same authors [178], employing an eigenvalue approach together with the techniques of Laplace and Fourier transforms, investigated plane strain problems of a homogeneous transversely isotropic medium under the purview of generalized thermoelasticity theory. A problem concerning a half-space having a step unit of temperature was solved by them in order to demonstrate the application of the approach.

An alternative state-space approach to generalized thermoelasticity (both L-S and G-L models) was reformulated by Sharma [180] in which an attempt was made to establish that the solutions for the generalized thermoelastic fields can be formed by an entirely different approach as was illustrated by Sosa and Bahar [196] for coupled thermoelastic problems. Using G-L model of thermoelasticity the reflection and refraction of thermoelastic waves at an interface of two semi-infinite media in contact with each other was studied by Sinha and Elsibai [193].

Several analyses on thermal shock problem and one-dimensional wave propagation problem for isotropic elastic media are available in the literature. The analyses have been made by using generalized thermoelasticity theory of type II and III developed by Green and Naghdi [66, 67, 69]. Exact solutions in closed form for displacement, temperature, strain and stress fields have been presented.

Thermoelastic stress waves in a spherically aeolotropic homogeneous solid medium having a spherical cavity were studied by Misra et al. [135]. The cavity surface was considered to be maintained at a fixed temperature. The infinite medium was supposed to have a heat source distributed exponentially and originated by the absorption of radioactive and electromagnetic radiations occurring in it. The analytical expressions for temperature displacement and stress were obtained by using integral transform techniques. The same authors (Misra et al. [136]) also examined the thermoelastic interactions in an elastic half-space at an elevated temperature field arising out of ramp-type heating at the traction – free bounding surface. The governing equations were taken in a unified system from which the field equations for coupled thermoelasticity as well as generalized thermoelasticity can be easily obtained as particular cases. The problem has been solved analytically by using state-space approach.

The distributions of temperature, stress and displacement in a viscoelastic solid medium having a spherical cavity were studied analytically by Misra et al. [137], where the material medium is considered non-homogeneous and non-isotropic and is considered to receive heat from a distributed heat source. The type of non-homogeneity is such that the elastic constants and the coefficient of linear thermal expansion are proportional to the nth power of the radial distance. The viscoelastic nature of the material is characterized by a model where the elastic constants are replaced by appropriate differential operators. The solution of a problem for a non-homogeneous orthotropic cylinderical shell for axisymmetric plane strain dynamic thermoelastic problem was obtained by Ding, et al. [36].

The generalized thermoelasticity theory based on Lord-Shulman (L-S) model was used by Misra et al. [134] to account for the finite velocity of heat conduction involving relaxation time in the thermoelastic process. Under the purview of this theory, propagation of thermoviscoelastic waves in an infinite viscoelastic aeolotropic body with a circular cylindrical cavity has been studied. A constant heat flux was considered to be prescribed on the entire cavity surface in such a manner that the displacement at the surface of the cavity is constrained. The problem was solved analytically by using integral transform techniques.

For this problem which is treated as one-dimensional, the differential equation governing the conduction of heat in the medium considered is given by

$$\frac{\partial^2 T}{\partial r^2}+\frac{1}{r}\frac{\partial T}{\partial r}-\frac{1}{\chi}\left(t_0\frac{\partial^2 T}{\partial r^2}+\frac{\partial T}{\partial t}\right)=\eta\frac{\partial}{\partial r}\left[\frac{\partial u}{\partial r}+\frac{u}{r}+t_0\frac{\partial}{\partial t}\left(\frac{\partial u}{\partial r}+\frac{u}{r}\right)\right],$$

where

$$\eta = \frac{T_0}{K}[(c'_{11}+c'_{12})\,\alpha_{11} + c'_{13}\,\alpha_{33}],$$

u being the displacement, T the temperature excess over the initial temperature T_0, at any point r, at any time t and t_0 the thermal relaxation time. χ denotes the thermal diffusivity of the material body given by

$$\chi = \frac{K}{\rho_1 c_v}$$

ρ_1, K and c_v being respectively the density, thermal conductivity and specific heat of the body and

$$c'_{ij} = c_{ij}\left(\frac{1+t_1 D}{1+t_2 D}\right), \qquad D \equiv \frac{\partial}{\partial t}$$

t_1, t_2 and c_{ij} being the material parameters.

The initial and boundary conditions are taken as follows :

$$\text{when } t = 0, \quad u = \frac{\partial u}{\partial t} = 0, \qquad T = \frac{\partial T}{\partial t} = 0$$

$$\text{when } r = a, \qquad u = 0, \qquad \frac{\partial T}{\partial r} = -\frac{q_0}{K} H(t)$$

where a is the radius of the circular cylindrical hole and H(t) is the Heaviside unit step function.

The study shows that in both the elastic and viscoelastic cases, the wave propagation velocity of the thermal wave front is greater than that of the stress wave front.

Thermal stresses generated in a long hollow viscoelastic cylinder made of an inhomogeneous transversely isotropic material were investigated by Misra et al. [138]. They also studied the thermoelastic interactions of an isotropic homogeneous elastic half-space. The mathematical analysis was carried out by employing the generalized theory of thermoelasticity of type III developed by using Green and Naghdi model. Of specific concern here is the propagation of waves owing to ramp-type heating of the boundary surface which is either (i) free of tractions or (ii) held rigidly fixed whereby the displacement is constrained.

One-dimensional thermoelastic disturbances are investigated in a half-space due to (i) suddenly applied constant temperature on the boundary which is rigidly fixed, and (ii) time-dependent "exponential-type" temperature change applied to the stress-free boundary, which includes (a) sudden heating to constant temperature, (b) smooth heating for all times, (c) smooth heating followed by sudden cooling and (d) smooth heating that becomes constant after a given lapse of time among its particular cases. Employing the Laplace transform technique, exact expression in closed form, for the displacement, temperaturc and stress field are obtained in each case. The solutions for the "step-type" as well as "ramp-type" temperature change on the boundary are deduced as particular cases.

Chandrasekharaiah and Srinath [18] studied the problem of one-dimensional wave propagation in a thermoelastic half-space without energy dissipation. They [20] also dealt with the propagation of thermoelastic plane waves without energy dissipation in a half-space due to time-dependent heating of the boundary. Thermoelastic interactions without energy dissipation due to a point heat source were considered by them [22]. Chandrasekharaiah and Srinath further analyzed the propagation of thermoelastic plane waves without energy dissipation in a rotating body [21]. Quintanilla [171] studied thermoelasticity without energy dissipation of non-simple materials.

Axisymmetric thermoelastic interactions without energy dissipation in an unbounded body with a cylindrical cavity was considered by Chandrasekharaiah and Srinath [21]. The disturbances were assumed to be due to a constant step in radial stress or temperature applied to the boundary of the cavity, which is maintained at a constant temperature or zero radial stress. Solutions for small time are obtained using Laplace transform method. The same authors [17] also studied the problem for the case when thermoelastic solid is subjected to a continuous line heat source. Emplying Laplace and Hankel transform, exact expressions were obtained for the temperature and stress fields.

The spherically symmetric thermoelastic interactions in an unbounded thermoelastic solid with a spherical cavity due to a constant step in temperature applied to the stress-free boundary of the cavity were studied by Chandrasekharaiah and Srinath [19]. Solutions for small values of time were obtained by employing Laplace transformation technique.

Bagri and Eslami [11] analysed a one-dimensional generalized thermoelasticity model of a disc based on L-S theory. The dynamic thermoelastic response of the disc under axisymmetric thermal shock loading was studied by using Galerkin finite element method. Bakshi et al. [7] dealt with thermoelastic wave propagation in a rotating medium with relaxation effects. Using the joint Laplace and Fourier transforms, the governing equations are transformed into a vector-matrix differential equation which is them solved by the eigen value approach. Sharma et al. [182] put forward an exact analysis of the free vibrations of a simply supported, homogeneous, transversely isotropic cylindrical pannel, by using three dimensional thermoelasticity theory. It was noted that a purely transverse mode is independent of piezoelectric and pyroelectric effects and gets decoupled from the rest of the motion. Schulz et al [175] analyzed the behaviour of piezoelectric materials at elevated temperature. Another three dimensional vibration problem for a homogeneous, transversely isotropic thermoelastic simply supported and insulated cylindrical panel in the context of various generalized theories of thermoelasticity was analyzed by Sharma and Sharma [179]. The uniqueness and reciprocity theorems for generalized thermoviscoelasticity for anisotropic media were discussed by Ezzat and El-Karamany [55]. Ezzat et al. [56] dwelt on the boundary integral formulation of thermoviscoelasticity theory.

Al-Huniti et al. [2] studied the transient variations of thermal stresses and the resulting residual stresses within a thin plate during welding processes. The temperature was determined analytically using Greens' functions. Propagation of the transient generalized magneto thermoelastic waves in a rotating half-space was analyzed by Chand et al. [16], while generalized magnetothermoelastic waves in an infinite elastic solid with a cylindrical cavity were studied by Dhaliwal and Saxena [35]. Zhou [216] discussed the magnetothermoelastic theory of Type II superconductors in mixed state. The semically driven magnetothermoelastic effect was investigated by Novik [163]. Ezzat [58] gave a mathematical analysis for a problem concerned with the generation of generalized magnetothermoelastic waves by thermal shock in a perfectly conducting half-space. The magnetothermodynamic stress and perturbation of the magnetic field vector were discussed by Wang and Dai [210] for an orthotropic thermoelastic cylinder.

A thermal shock problem in generalized thermo-viscoelasticity under four theories was discussed by El-Karamany and Ezzat [40]. In this paper a model of the equations of generalized thermoviscoelasticity for isotropic media is given. The formulation is applied to four generalized thermoelasticity theories, viz. L-S model, G-L model, Chandrasekharaiah and Tzou as well as the two dynamic coupled theories. Hetnarski and Ignaczak [72] examined five generalizations to the coupled theory and obtained a number of important results. The model of the equations of generalized thermoviscoelasticity ignoring relaxation effects of the volume, with one relaxation time and two relaxation times was established respectively by Ezzat et al. [56] and Ezzat et al. [59]. Jordan and Puri [81] studied the thermal stresses in a spherical shell under three thermoelastic models. They calculated the jump discontinuities for short time, of the solution in the spherical shell problem under L-S, G-L and the coupled theory.

A two-dimensional electromagneto-thermoelastic problem was dealt with by Sherief and Helmy [187] for a finitely conducting half-space by using Laplace and Fourier transforms. He et al [71] studied a thermo-elastic-piezoelectric coupled two-dimensional generalized thermal shock problem of a thick piezoelectric plate of infinite extent by the hybrid Laplace transform – finite element method. Ashida and Tauchert [3] discussed the transient response of piezothermoelastic plate subjected to axisymmetric surface heating. Kapuria et al. [82] presented a coupled third-order consistent theory for thermal load on hybrid piezoelectric beams. The potential thermal fields are discretized layer-wise as piecewise linear. The axial and transverse electric fields are considered. Analytical solutions are obtained for simply supported beams. The theory is assessed by comparison with the exact piezothermoelasticity solution and coupled first-order shear deformation theory solution for two thermal loads.

Segall [176] investigated thermal stresses in an infinite slab under an arbitrary thermal shock. Tianhu et al. [206] studied the electromagneto-thermoelastic interactions in a semi-infinite perfectly conducting solid subjected to a thermal shock on its surface, when the solid and its adjoining vacuum is subjected to a uniform magnetic field. They used the theory of generalized thermoelasticity, based on the theory of Lord and Shulman with one relaxation time.

Chen et al. [24] investigated thermoelastic stress waves in a solid medium generated by subpicosecond – pulse laser heating, using three different approaches : ultrafast thermoelasticity, Lord-Shulman theory and the classical thermoelasticity. They considered both the conditions of uniaxial strain and uniaxial stress. A combnined finite difference / finite element algorithm was developed for solving the coupled nonlinear, transient differential equations. It was shown that the thermomechanical results obtained from the ultrafast thermoelasticity theory are significantly different from those obtained from L-S theory and the classical thermoelasticity theory. El-Maghraby [45] dealt with a two-dimensional problem in generalized thermoelasticity with one relaxation time, with heat sources.

Yu et al [215] made an attempt to prove the evidences of thermal stress wave induced in deeply frozen biomaterials by a strong and instantaneous heating. The study was made both experimentally and theoretically. A thermal stress applied to characterize the wave feature of the shock stress induced in the strong freezing and heating process.

Crack Problems in Mathematical Thermoelasticity

For studying the fracture strength of solids under various situations, many problems on elastic continua containing cracks of different shapes have been considered over the past couple of decades. A theory of crack problems for two dimensional elastic continua seems to have been first developed Griffith [70]. Stress distribution in the vicinity of a Griffith crack was subsequently studied by various investigators, e.g. Sneddon and Elliott [195] and Lowengrub [91,92] for isotropic solids and Willmore [213] for anisotropic media.

Pressure necessary to open a Griffith crack to a given shape in an orthotropic solid was investigated by Misra et al. [101a]. The distribution of stresses in the neighbourhood of cracks opened in a long anisotropic elastic strip of finite width was investigated by Misra et al. [94] for a situation where the cracks are situated in the direction of the length of the strip and are arrayed periodically. The problem is reduced to solving a pair of dual integral equations which are further reduced to a Fredholm integral equation of the second kind. The Fredholm equation is treated by employing appropriate numerical techniques. Expressions for the stress intensity factor, the deformation of the crack surface and the strain energy are derived. A theoretical estimate of the highly intensified stress field was obtained for a strip made of boron-epoxy (anisotropic) and the results are compared with those for an identical strip made of iron (isotropic). Two different cases have been studied viz, (a) when the edges of the strip are in contact with smooth and rigid planes and (b) when the edges are stress-free.

In order to examine the fracture strength of materials subjected to differential heating, a couple of thermoelastic crack problems for cylindrical bodies were analysed by Das [32, 33], while the problem of growth and stability of thermaly induced equally spaced parallel cracks in an elastic semi-space of a brittle material was discussed by Nemat-Nasser et al. [161]. Misra et al. [95,96,98,99] carried out a series of investigations for various problems concerned with solids of finite and infinite extent having internal / external cracks, by formulating the problems as Mixed Boundary Value Problems. These problems were, however, restricted to the configuration of a single crack in a given solid. In a practical situation a solid medium may have a number of cracks which are responsible for the weak regions of the medium. Solids containing an infinite row of parallel cracks were considered by England and Green [46] as well as by Watanabe and Atsumi [212] for isothermal conditions. The problems of thermally stressed long cylinders having an infinite row of parallel penny-shaped cracks were studied by Misra and Maiti [97]. Misra [102] also performed a mathematical analysis for the study of the thermoelastic stresses in the vicinity of a row of cracks in an elastic strip subjected to temperature variations and having an infinite row of parallel Griffith cracks which are arrayed periodically and situated along the length of the strip, the distribution of temperature on the crack surfaces being assumed to be identical for all the cracks. Denoting by 2a the length of each crack, by 2b the finite width of the strip, by 2l the distance between any to consecutive cracks and by $f(\frac{x}{a})$ the prescribed inner temperature on the crack surface, the thermal and mechanical boundary conditions have been expressed mathematically as

$$T = f\left(\frac{x}{a}\right), \quad 0 \leq x < a, \; y = 0$$

$$\frac{\partial T}{\partial y} = 0, \quad a < x \leq b, \; y = 0$$

$$\frac{\partial T}{\partial y} = 0, \quad 0 \leq x \leq b, \; y = l$$

$$T = 0, \quad 0 \leq y \leq l, \quad x = \pm b$$

$$\sigma_{yy} = 0, \quad 0 \leq x < a, \; y = 0$$

$$u_y = 0, \quad a < x \leq b, \quad y = 0$$

$$\sigma_{xy} = 0, \; 0 \leq x \leq b, \quad y = 0$$

$$\sigma_{xy} = 0, \; 0 \leq x \leq b, \quad y = l$$

$$u_y = 0, \; 0 \leq x \leq b, \quad y = l$$
$$\sigma_{xy} = 0, \; 0 \leq y \leq l, \quad x = \pm b$$

$$u_x = 0, \; 0 < y \leq l, \quad x = \pm b$$

In the above, T denotes the temperature, u_i the displacement and σ_{ij} the stress at any point.

The heat conduction problem is first reduced to that of solving a pair of dual integral equations with trigonometric kernels. The solution is then reduced to a Fredholm integral equation of second kind. Next by the use of Neuber-Papkovitch potential functions, the thermoelastic problem is reduced to that of solving another

Fredholm equation of the same kind. The expressions for the stress intensity factor and the crack profile are derived.

Misra and Maiti [97] derived the integral equations for studying a thermoelastic crack problem concerned with the determination of thermal stresses in a long circular cylinder containing an infinite row of parallel penny-shaped cracks arrayed periodically in the direction of the axis of the cylinder and opened due to some prescribed temperature. Denoting by b the radius of the cylinder, by a the radius of each penny-shaped crack and by 2l the spacing between any two consecutive cracks and by taking the line of centres of the cracks as the axis of symmetry of the configuration (coinciding with the axis of the cylinder), the locations of the cracks are defined by

$$z = 2nl, \quad 0 \le r < a, \quad n = 0, \ \pm 1, \ \pm 2, \ldots$$

Under these assumptions, they considered that the infinite row of parallel cracks forms the image system of a short circular cylinder bounded by the planes $z = \pm l$ having a single crack at its centre. Thus the problem reduces to the determination of the stress-field in a finite circular cylinder of length l enclosed between $z = 0, l$. Denoting by T, u_i, σ_{ij} respectively the temperature, the displacement components and the stress components at any point and assuming that the curved surface of the cylinder is maintained at zero ambient temperature and further that it is in contact with a smooth rigid insulating cover they write the boundary conditions as

$$T = 0 = u_r = \sigma_{rz} \quad \text{on } r = b$$

Denoting by $f\left(\frac{r}{a}\right)$ the prescribed temperature, the thermal boundary conditions on the plane ends are written as

$$T = f(r/a), \quad 0 \le r < a, \; z = 0$$

$$\frac{\partial T}{\partial z} = 0, \quad a < r \le b, \quad z = 0$$

and $$\frac{\partial T}{\partial z} = 0, \quad 0 \le r \le b, \quad z = l;$$

while the mechanical boundary conditions are

$$\sigma_{rz} = 0, \quad 0 \le r \le b, \qquad z = 0, l$$

$u_z = 0, \quad a < r \le b, \qquad z = 0$

$u_z = 0, 0 \le r \le b, \qquad z = 1$

$\sigma_{zz} = 0, \quad 0 \le r \le a, \qquad z = 0$

An analytical study for the determination of the distribution of thermal stresses in a non-isotropic infinite plate having a penny-shaped crack was performed by Misra and Maiti [114]. The crack is assumed to be opened due to the application of a prescribed normal load and a prescribed flux of heat or a prescribed temperature. They consider an elastic plate of finite thickness 2d but of infinite radius containing a penny-shaped crack of radius b defined by

$0 \le r \le b, \quad z = 0$

where (r, θ, z) are the cylindrical co-ordinates of any point and the axis of the crack as the z-axis (which is also taken to be the axis of elastic symmetry).

Treating the problem as that for the elastic layer $0 \le z \le d$, they assumed that the plate is constrained by two rigid walls which permit no normal displacement and that the contact between the elastic layer and its rigid casing is frictionless so that the shearing stress on the planes z = 0, and z = d is zero. Mathematically the boundary conditions have been described as

$\sigma_{rz} = 0 \qquad \text{on } z = 0$

$\sigma_{rz} = 0 \text{ and } w = 0 \qquad \text{on } z = d$

In a situation where the crack is opened by a heat flux and a prescribed a normal pressure applied at its surfaces, the boundary conditions on z = 0 are taken as

$$\frac{\partial T}{\partial z} = \begin{cases} Q(\rho) & 0 \le \rho < 1 \\ 0, & \rho > 1 \end{cases}$$

$\sigma_{zz} = -p(\rho), \qquad 0 \le \rho < 1$

$w = 0, \qquad \rho > 1$

where w is the normal displacement, σ_{zz} the stress component, ρ (= r/a) is the non-dimensional radial distance, while Q(ρ) and P(ρ) are given functions of ρ.

In case, the crack is opened only by the application of a prescribed temperature on its surfaces, the boundary conditions on z = 0 assume the form as below:

$$T = T_0 = \text{const}, \; 0 \le \rho < 1, \;\; T_0 \text{ being a constant}$$

$$\frac{\partial T}{\partial z} = 0, \quad \rho > 1$$

$$\sigma_{zz} = 0, \quad 0 \le \rho < 1$$

$$w = 0, \; \rho > 1$$

The problem has been solved by using Hankel Transformation and employing dual integral equations technique. By using numerical methods, the derived analytical expressions have been computed to examine the magnitude of the normal stress generated in the plane of the crack, at some points in the neighbourhood of the crack.

The distribution of surface temperature necessary to maintain a crack in a prescribed shape on the surface of the crack was investigated analytically by Misra and Maiti [113]. The problem was formulated as a mixed boundary value problem of the form:

$$T(r,0) = -f(r/a), \quad 0 \le r < a$$

$$\frac{\partial T}{\partial z}(r,0) = 0, \quad r > a$$

$$\sigma_{rz}(r,0) = 0, \quad 0 \le r < \infty$$

$$\sigma_{zz}(r,0) = 0, \quad 0 \le r < a$$

$$u_z(r,0) = 0, \quad r > a$$

in which $f(r/a)$ is a given function of r and T(r,z) denotes the temperature at any point (r,z), (r,z) being the radial and axial coordinates. σ_{rz} and σ_{zz} are the stress components, while u_z represents the longitudinal displacement.

A theoretical estimate of the intensified stress-field generated in a non-isotropic long circular cylinder having an infinite row of parallel penny-shaped cracks was reported by Misra et al. [101]. The cracks are considered to be located periodically along the cylinder-axis. All the cracks are assumed to be opened by the same distribution of internal pressure on their surfaces. The problem is treated mathematically through the use of integral equation approach.

An analysis was carried out by Misra et al. [125] for studying the stress-field in a sandwich structure consisting of a core material having cylindrical configuration surrounded by a dissimilar material, where the core material is assumed to contain a number of penny-shaped cracks situated periodically along the axis of the cylinder.

An analytical solution for a generalized two-dimensional problem of an interface crack between two dissimilar magnetoelectroelastic materials under uniform heat flow was presented by Gao and Noda [64]. The crack is assumed to be electrically permeable. The thermal induced electric-magnetic fields within the crack and the stress intensity factor at the crack tip are obtained in closed form.

Liu et al. [88] studied, by deriving complex potentials, electromagnetic, thermoelastic solids with an elliptical cavity or a crack under uniform heat flow. Gao et al. [63] put forward a fracture analysis of electromagnetic thermoelastic solids. Chen et al. [26] extended the potential theory method to the mixed boundary value problems of magneto – electro-thermoelastic materials and thereby solved the problem of a penny-shaped crack in an infinite space subjected to symmetric mechanical, electric, magnetic as well as temperature loads at the crack surfaces.

Rokne et al. [174] studied a mixed boundary value problem in a non-homogeneous medium under steady distribution of temperature. Ozturk and Erdogan [168] studied an axisymmetric crack problem for a non-homogeneous medium. Erdogan [48] discussed the facture mechanics of functionally graded materials. The surface crack problem for a plate with functionally graded properties was discussed by Erdogan and Wu [47]. Interface cracking of functionally graded material coating under steady-state heat flow was investigated by Lee and Erdogan [84].

El-Borgi et al. [39] put forward a mathematical analysis for the study of a partially insulated embedded crack in an infinite functionally graded medium under thermo-mechanical loading. A crack-closure algorithm is developed to handle the problem of having negative mode I stress intensity factors. The principal object of the study has been to examine the effect of non-homogeneity parameters, crack-closure and partial crack surface insulation on the crack tip stress intensity factors for the purpose of gaining better understanding of the thermo-mechanical behaviour of the graded materials.

Chen [27] solved the problem of a circular arc-shaped crack in an infinite plate under the action of a heat source or a doublet by using the complex variable method.

Theoretical Estimates of Thermal Stresses Generated in Human Body

Temperature distributions in the human head due to certain environmental temperatures were measured experimentally by Thron [205]. Through a study on a model of the human head, the problem of heat conduction in the head was examined by Richardson and Whitelaw [173]. It was pointed out by Colin et al. [31] that the temperature of the environment has a significant effect on the conduction of heat inside the cranium. Flesh [61] mentioned that the temperature –distribution in the human head can be calculated corresponding to an environmental temperature of 26°C if the amount of heat produced at a point is assumed to be proportional to the square of the radial distance of the point and further that if the temperature of the environment ranges between 10°C and 20°C, the amount of heat produced at a point is proportional to the cube of the radial distance. He calculated the temperature-distributions by solving the heat conduction equation on the assumption that the temperature –field is a function of the radial coordinate only. This assumption is, however, an idealized one.

Misra [155] put forward a mathematical analysis for the study of the stress-field in the human head generated due to thermogenesis. By considering a spherical model of the human head, the differential equation governing the temperature – field in the cranium has been taken as

$$Kr^{-2}\left[\frac{\partial}{\partial r}\left(r^{2}\frac{\partial T}{\partial r}\right)+\operatorname{cosec}\theta\frac{\partial}{\partial\theta}\left(\sin\theta\frac{\partial T}{\partial\theta}\right)\right]+Q_{1}r^{m}+R_{1}=0,$$

in which $Q_1 r^m + R_1$ represents the amount of heat generated in the head per unit volume in unit time (Q_1 and R_1 are constants), K is the average thermal conductivity of the head, m is an arbitrary constant whose value is 2 when the temperature of the environment is 26°C and 3 when the environmental temperature ranges between 10°C and 20°C, as mentioned earlier; r and θ are respectively the radial and angular coordinates of any representative point of the head.

The effect of solar radiation on the human head was studied analytically by Misra and Chakravarty [111]. The sun's rays touching the head are assumed to be parallel to each other and normal to the surface. The cranium is considered as a two-layered transversely isotropic elastic shell, the outer layer being the representative of the scalp and the inner layer that of the skull. The study is believed to be useful in the treatment of diseases like sun-stroke.

A theoretical estimate of the stresses in a tubular bone exposed to heat radiation from a distant heat source (e.g. solar radiation) was also presented by Misra [103]. In this study he developed a mathematical model by considering the two-layered structure (spongy and cortical) of bone. Denoting by a and b the endosteal and periosteal radii of the bone specimen and by r_1 the radius of the interface between the two layers, the thermal conditions were represented mathematically as

$$-\kappa^{(2)}\frac{\partial T^{(2)}}{\partial r} = -\tfrac{1}{2}E^{(2)}\sin\theta + hT^{(2)},\ 0 \le \theta \le \pi \qquad \text{on } r = b$$

$$= hT^{(2)}, \qquad \pi < \theta \le 2\pi$$

$$\left.\begin{aligned} -\kappa^{(1)}\frac{\partial T^{(1)}}{\partial r} &= \kappa^{(2)}\frac{\partial T^{(2)}}{\partial r} \\ T^{(1)} &= T^{(2)} \end{aligned}\right\} \qquad \text{on } r = r_1$$

$$\kappa^{(1)}\frac{\partial T^{(1)}}{\partial r} + h^{(1)}T^{(1)} = 0, \qquad \text{on } r = a$$

where κ denotes the coefficient of thermal conductivity, E the emissivity of the cortical surface of the bone specimen and h the coefficient of heat radiation; r denotes the radial distance, θ the angular distance and T the temperature at any point. The superscripts (2), (1) represent the quantity for the second (outer cortical) layer and first (inner spongy) layer respectively.

The study is valuable particularly to the people of tropical and sub-tropical countries where the human head is often is exposed to burning rays of the sun in summer during daily activities of life.

Micropolar Thermoelasticity

The plane problem of micropolar thermoelasticity was discussed by Nowacki [164]. Thermal stresses and couple stresses in a micropolar elastic solid containing a spherical inclusion were studied by Aköz and Tauchert [1]. Two general solutions to the linear theory of thermoelasticity were presented by Tauchert [203]; one representation involves deformation potentials, while the other employs stress-functions. The two representations were used by him to analyse the thermal stresses and thermal couple stresses in the vicinity of a cavity and a rigid inclusion during uniform heat flow.

In a series of papers Eringen [51,53,54] established the general theory *micromorphic continua* for predicting the behaviour of materials with inner structure. This theory is a generalization of the theory of micropolar elasticity. In [51], Eringen extended the theory of *microstretch elastic solids* to include heat conduction. The linear theory of thermoelastic microstretch solids has been studied in various papers [34,77,78,160]. In [77] the thermoelastic deformation of a tube and thermal stresses in a cylinder subjected to a uniform temperature gradient have been dealt with.

El-Karamany [42] putforward a boundary integral equation formulation for the generalized micro-polar thermoelasticity. The initial mixed boundary value problem is considered as an example illustrating the BIE formulation. It has been reported that the results are applicable to different generalized theories, viz. L-S model with one relaxation time, G-L. model with two relaxation times and G-N theory.

The uniqueness and reciprocity theorems in generalized linear micropolar thermoviscoelasticity were discussed by El-Karamany [43].

Acknowledgement : The author wishes to acknowledge the help received from Dr. N. C. Chattopadhyay in the preparation of a portion of the manuscript of this chapter.

References :

1. Aköz , A. Y. and Tauchert, T. R., Bull. D.E. L'Acad. Pol. Des. Sci., Ser. Des Sci. Tech., **18**, 349 (1970).
2. Al-Huniti, N. S., Al-Nimr, M.A. and Da'as, M. A., J. Thermal Stresses, **27**, 671-689 (2004).
3. Ashida, F. and Tauchert, T. R., J. Thermal Stresses, **27**, 513-536 (2004).
4. Bagchi,M.C.,J. Sci.Engng.Res., **9**, Part 1,171-178 (1965).
5. Bahar,L.Y. and Hetnarski, R.B., J. Thermal Stresses, **1**, 135-145 (1978).
6. Bahar,L.Y. and Hetnarski, R.B., J. Thermal Stresses, **2**, 285-290 (1979).
7. Bakshi, A., Bera, R. K. and Debnath, L., Int. J. Engng. Sci., **42**, 1573-1585 (2004).
8. Biot, M.C. ,J.Appl. Phys., **27**, 240-253 (1956).
9. Bland,D.R.,The theory of linear viscoelasticity, Pergamon Press, London (1960).

10. Blond, E., Schmitt, N. and Hild, F., Int. J. Numer. Anal. Meth., **27**, 883-904 (2003).

11. Bogri, A. and Eslami, M. R., J. Thermal Stresses, **27**, 691-704 (2004).

12. Boley, B.A. and Weiner,J.H., Theory of Thermal Stresses, John Wiley and Sons, New York (1967).

13. Carslaw,H.S. and Jaeger,J.C.,Conduction of heat in solids, Clarendon Press, Oxford (1959).

14. Chadwick ,P. and Seet,L.T.C,Mathematika, **17**, 255-274 (1970).

15. Chadwick ,P., and Windle,D.W.,Proc.Roy.Soc.Amer., **280**, 47-71 (1964).

16. Chand, D., Sharma, J. N. and Sud, S. P., Int. J. Engng. Sci., **28**, 547-556 (1990).

17. Chandrasekharaiah, D. S. and Srinath, K. S., Acta Mechanica, **128,** 243-251 (1998).

18. Chandrasekharaiah, D. S. and Srinath, K. S., Int. J. Engng. Sci., **34**, 1447-1455 (1996).

19. Chandrasekharaiah, D. S. and Srinath, K. S., International Journal of Mathematics and Mathematical Sciences, **22** (1999).

20. Chandrasekharaiah, D. S. and Srinath, K. S., J. Thermal Stresses, **20**, 659-676 (1997).

21. Chandrasekharaiah, D. S. and Srinath, K. S., Journal of Elasticity, **46,** 19-31 (1997).

22. Chandrasekharaiah, D. S. and Srinath, K. S., Journal of Elasticity, **50,** 97-108 (1998).

23. Chandrasekharaiah, D. S. and Srinath, K. S., Mech. Res. Comm., **24**, 551-560 (1997).

24. Chen, J. K., Beraun, J. U. and Tham, C. L., Int. J. Engng. Sci., **42**, 793-807 (2004).

25. Chen, W. Q. and Lee, K. Y., Int. J. Solids Struct., **40**, 5689-5706 (2003).

26. Chen, W. Q., Lee, K. Y. and Ding, H. J., Int. J. Engng. Sci., **42**, 1361-1379 (2004).

27. Chen, Y. Z., J. Thermal Stresses, **27**, 137-150 (2004).

28. Chen,L.S. and Chu, H.S., J. Thermal Stresses, **13**, 131-149 (1990).

29. Choudhuri, S., Int. J. Engng. **22**, 519-530 (1984).

30. Christensen, R. M, Theory of viscoelasticity :an introduction, Academic Press, New York (1971).

31. Colin, J., Boutelier, C., Houdas, Y., J. Physiol Paris, **58**, 500 (1966).

32. Das, B. R., Int. J. Engng. Sci., **6** (1968).

33. Das, B. R., Int. J. Engng. Sci., **7** (1989).

34. De Cicco, S and Nappa, L., J. Maths. Mechs. Solids, **5**, 467-484 (2000).

35. Dhaliwal, S. and Saxena, H. S., J. Thermal Stresses, **14**, 353-369 (1991).

36. Ding, H. J., Wang, H. M. and Chen, Y. Q., J. Sound Vibration, **263**, 815-830 (2003).

37. Duhamel, J.M.C., Memores par divers savans, **5**, 440-498 (1838).

38. Duhamel, J.M.C.,J. de I'Ecole Polytechnique, **15**, 1-57 (1837).

39. El-Borgi, S., Erdogan, F. and Hidri, L., Int. J. Engng. Sci., **42**, 371-393 (2004).

40. El-Karamany, A. S. and Ezzat, M. A., Int. J. Engng. Sci., **42**, 649-671 (2004).

41. El-Karamany, A. S., Int. J. Engng. Sci., **40**, 2097-2117 (2002).

42. El-Karamany, A. S., Int. J. Engng. Sci., **42**, 157-186 (2004).

43. El-Karamany, A.S. and Ezzat, M. A., Int. J. Engng. Sci., **40**, 1943-1956 (2002).

44. El-Karamany, A.S. and Ezzat, M. A., J. Thermal Stresses, **27**, 1187-1212 (2004).

45. El-Maghraby, N. M., J. Thermal Stresses, **27**, 227-239 (2004).

46. England, A. H. and Green, A. E., Proc. Camb. Phil. Soc., **59**, (1963).

47. Erdogan, F. and Wu, B. H., ASME. J. Appl. Mech., **64**, 449-456 (1997).

48. Erdogan, F., Compos. Engng., **5**, 753-770 (1995).

49. Eringen, A. C., J. Math. Mech., **15**, 909 (1966).

50. Eringen, A. C., Mechanics of continua,2nd Edition., Krieger, Huntington, New York (1980).

51. Eringen, A. C., Theory of Thermo-microstretch elastic solids, Int. Engng. Sci, **28**, 1291-1301 (1990).

52. Eringen, A.C and Maugin, G.A., Electrodynamics of continua,Vols-I,II ,Springer, New York (1989).

53. Eringen, A.C. and Kafadar, C. B., Polar Field Theories, in : A. C. Eringen (Ed.), Continuum Physics,Vol. IV, Academic Press, New York, 1-73, (1976).

54. Eringen, A.C., Microcontinuum Field Theories I : Foundations and Solids, Springer-Verlag, New York, Berlin, Heidelberg (1999).

55. Ezzat, M. A. and El-Karamany, A. S., J. Thermal Stresses, **25**, 507-522 (2002).

56. Ezzat, M. A., El-Karamany, A. S. and Samaan, A. A., J. Thermal Stresses, **24**, 823-846 (2001).

57. Ezzat, M. A., Int. J. Engng. **42**, 1503-1519 (2004).

58. Ezzat, M. A., J. Thermal Stresses, **22**, 617-633 (1997).

59. Ezzat, M. A., Othman, M.I. and El-Karamany, A. S., Int. J. Engng. Sci., **40**, 1251-1274 (2002).

60. Ezzat, M., Othman, M. and El-Karamany, A., J. Thermal Stresses, **24**, 411-432 (2001).

61. Flesh, U., J. Theor Biol., **54**, 285 (1975).

62. Fung, Y.C., Foundations of Solid Mechanics, Prentice-Hall, Englewood Cliffs, New York (1965)

63. Gao, C. F., Kessler, H. and Balke, H., Eur. J. Mech. A – Solid, **22**, 433-442 (2003).

64. Gao, Cun-Fa and Noda, N., Int. J. Engng. Sci., **42**, 1347-1360 (2004).

65. Ghaleb, A.F. and Mohamedein ,M.S.E.,Int. J.Engng.Sci., **27**, 1367-1377 (1989).

66. Green, A.E and Naghdi, P.M., Proc. Roy. Soc, London, SerA, **432**, 171-194 (1991).

67. Green, A.E and Naghdi,P.M., J. Elasticity,**31**,189-208 (1993).

68. Green, A.E. and Lindsay, K.A. , J. Elasticity, **2**, 1-7 (1972).

69. Green, A.E. and Naghdi, P.M., J. Thermal Stresses, **15**,253-264 (1992).

70. Griffith, A.A., Phil.Trans. A. **221**, 163 (1920).

71. He, T. H., Tian, X. G. and Shen, Y. P., Int. J. Engng. Sci., **40**, 2249-2264 (2002).

72. Hetnarski, R. B. and Ignaczak, J., J. Thermal Stresses, **22**, 451-476 (1999).

73. Hetnarski, R.B. (Ed.),Thermal Stresses, Elsevier Science Publishers, B.V., Vol. .I (1986).

74. Hetnarski, R.B. (Ed.),Thermal Stresses, Elsevier Science Publishers, B.V., Vol. .II (1987).

75. Hetnarski, R.B. (Ed.),Thermal Stresses, Elsevier Science Publishers, B.V., Vol. .III (1989).

76. Iesan, D. and Nappa, L., Meccanica, **39**, 125-138 (2004).

77. Iesan, D. and Quintanilla, R., Int. J. Engng. Sci., **42**, 1957-1972 (2004).

78. Iesan, D., Acta Mechanica, **11,** 141-144 (1971).

79. Iovane, G. and Passarella, F., J. Thermal Stresses, **27**, 97-109 (2004).

80. Iovane, G. and Passarella, F., J. Thermal Stresses, **27**, 983-999 (2004).

81. Jordan, P. M. and Puri, P., J. Thermal Stresses, **24**, 47-70 (2001).

82. Kapuria, S. Ahmed, A. and Dumir, P. C., J. Thermal Stresses, **27**, 405-424 (2004).

83. Kogl, M. and Gaul, L., Arch. Appl. Mech., **73**, 377-398 (2003).

84. Lee, Y. –D. and Erdogan, F., Engng. Fract. Mech., **59**, 361-380 (1998).

85. Lekhnitskii, S.G., Theory of elasticity of anisotropic body, Mir Publishers, Moscow (1981).

86. Levin, V. M. and Alvarez-Tostado, J. M., Int. J. Fract., **199**, 79-82 (2003).

87. Li, J. Y., Quart. J. Mech. Appl. Math., **56**, 35-44 (2003).

88. Liu, X. L., Liu, X. F. and Liu, J. X., Key Engng. Mater., **183**, 139-144 (2000).

89. Lord, H.W and Shulman, Y., J. Mech Phys.Solids,**15**,299-309 (1967).

90. Love, A.E.H., A Treatise on the Mathematical Theory of Elasticity, Oxford (1927).

91. Lowengrub, M., Int. J. Engg. Sci. **4**, 289 (1966).

92. Lowengrub, M., Int. J. Engg. Sci. **4**, 69 (1966).

93. Maugin, G.A., Continuum Mechanics of Electromagnetic Solids, North-Holland, Amsterdam (1988).

94. Mishra, M. and Misra, J. C., International Journal of Engineering Science (England) **21**, 187-198 (1983).

95. Misra, J. C. and Maiti, S. N., Bull Acad. Pol. Sci. Ser. Sci. Tech., **22**, 567-578 (1974).

96. Misra, J. C. and Maiti, S. N., Bull. Cal. Math. Soc., **70,** 63-71 (1978).

97. Misra, J. C. and Maiti, S. N., J. Math. Phys. Sci., **13**, 311-323 (1979).

98. Misra, J. C. and Maiti, S. N., Rev. Fac. Sc. Univ. Istanbul. Ser. A. **40**, 17-28 (1975).

99. Misra, J. C. and Maiti, S. N., Rev. Roum. Sci. Tech. Mech. Appl. (Bucarest), **21**, 339-351 (1976).

100. Misra, J. C. and Mishra, M., Engineering Fracture Mechanics (England), **19**, 101-112 (1984).

101. Misra, J. C., Mishra, M. and Maiti, S. N., Engng. Fracture Mech. (England), **17**, 471-489 (1983)

101a. Misra, J. C., Mishra, M. and Maiti, S. N., Intl. Journal of Fracture, **17**, 477-482 (1981)

102. Misra, J. C., Rev. Roum. Sci. Tech. Mec. Appl., **29**, 403-413 (1984).

103. Misra, J. C., Rheological Acta (Germany), **24**, 520-533 (1985).

104. Misra, J.C. and Achari, R.M., Der Stahlbau (Germany), **49**, 23-25 (1980).

105. Misra, J.C. and Achari, R.M., Forschung im Ingenieurwesen (Germany), **45**, 80-82 (1979).

106. Misra, J.C. and Achari, R.M., J. Math. Phys. Sci., **15**, 123-131 (1981).

107. Misra, J.C. and Achari, R.M., J. Thermal Stresses, **6**, 115-123 (1983).

108. Misra, J.C. and Achari, R.M., J. Thermal Stresses (USA), **3**, 57-66 (1980).

109. Misra, J.C. and Achari, R.M., J. Thermal Stresses (USA), **3**, 509-520 (1980).

110. Misra, J.C. and Achari, R.M., Rev. Mec. Appl., **22**, 687-694 (1997).

111. Misra, J.C. and Chakravarty, S., International J. Engg. Sci. (England), **20**, 445-454 (1982).

112. Misra, J.C., Khankari, A.K. and Kar, S.B., Modelling Simulation and Control, B, (France), **41**, 9-21 (1992).

113. Misra, J.C. and Maiti, S.N., Forschung im Ingenieurwesen (Germany), **47**, 82-84 (1981).

114. Misra, J.C. and Maiti, S.N., Forschung im Ingenieurwesen (Germany), **52**, 33-36 (1986).

115. Misra, J.C. and Maiti, S.N., Rev. Roum. Sci. Tech. Mec. Appl. (Bucarest), **2**, 531-543 (1981).

116. Misra, J.C. and Mishra, M., Engg. Fracture Mechanics (England), **22**, 65-76 (1985).

117. Misra, J.C. and Samanta, S., Acta Mechanica (Wien-New York), **41**, 149-155 (1981).

118. Misra, J.C. and Samanta, S., Beton und Stahlbetonbau (Germany), **76**, 143-145 (1981).

119. Misra, J.C. and Samanta, S., J. Thermal Stresses (USA), **5**, 365-375 (1982).

120. Misra, J.C. and Samanta, S., Journal of Non-Equilibrium Thermodynamics (Germany), **5**, 137-142 (1980).

121. Misra, J.C. and Samanta, S., Math. Forum, **3**, 23-30 (1980).

122. Misra, J.C. and Samanta, S., Solid Mechanics Archives (Oxford), **9**, 259-267 (1984).

123. Misra, J.C. and Samanta, S.C., Computers and Structures (UK), **46**, 851-857 (1993).

124. Misra, J.C. and Samanta, S.C., International J. of Engg. Sci. (UK), **29**, 1505-1514 (1991).

125. Misra, J.C. and Samantarai, J. C., International Journal of Fracture (The Netherlands) **24**, 239-254 (1984).

126. Misra, J.C., Arch. Bud. Masz. (Poland), **22**, 138-144 (1975).

127. Misra, J.C., Arch. Bud. Masz. (Poland), **27**, 219-225 (1980).

128. Misra, J.C., Bul. Inst. Pol. Timisora, Ser. Mechanica, **23**, 163-168 (1978).

129. Misra, J.C., Bull. Cal. Math. Soc., **63**, 135-141 (1971).

130. Misra, J.C., Bull. Cal. Math. Soc., **63**, 183-190 (1971).

131. Misra, J.C., Chakraborty, A., Samanta, S.C. and Misra, S.C., Mechanics Reseach Communications (USA), **19**, 415-426 (1992).

132. Misra, J.C., Chakraborty, A.K. and Misra, S.C., International J. of Engg. Sci. (UK), **31**, 1047-1052 (1993).

133. Misra, J.C., Chattopadhyay, N.C. and Chakravorty, A., J. Thermal Stresses (USA), **23**, 327-351 (2000).

134. Misra, J.C., Chattopadhyay, N.C. and Samanta, S.C., Computers and Structures (UK), **52**, 705-717 (1994).

135. Misra, J.C., Chattopadhyay, N.C. and Samanta, S.C., International J. Engg. Sci. (UK), **32**, 1769-1789 (1994).

136. Misra, J.C., Chattopadhyay, N.C. and Samanta, S.C., International J. Engg. Sci. (UK), **34**, 579-596 (1996).

137. Misra, J.C., Chattopadhyay, N.C. and Samanta, S.C., J. of Thermal Stresses (USA), **23**, 169-195 (2000).

138. Misra, J.C., Chattopadhyay, N.C. and Samanta, S.C., Modelling Simulation and Control, B, (France), **42**, 49-64 (1991).

139. Misra, J.C., Ind. J. Mech. Maths., **11**, 33-36 (1973).

140. Misra, J.C., Ind. J. Mech. Maths., **12**, 14-19 (1974).

141. Misra, J.C., Ind. J. Th. Phys., **19**, 57-64 (1971).

142. Misra, J.C., Ind. J. Th. Phys., **19**, 91-96 (1971).

143. Misra, J.C., Ind. J. Th. Phys., **21**, 35-39 (1973).

144. Misra, J.C., Ind. Jour. Pure and Appl. Maths., **2**, 143-153 (1971).

145. Misra, J.C., Ind. Jour. Pure and Appl. Maths., **3**, 612-621 (1972).

146. Misra, J.C., Kar, S.B. and Samanta, S., Arch. Mech. (Poland), **36**, 299-306 (1984).

147. Misra, J.C., Kar, S.B. and Samanta, S., CSME Transactions (Canada), **9**, 228-232 (1985).

148. Misra, J.C., Kar, S.B. and Samanta, S., CSME Transactions (Canada), **11**, 151-159 (1987).

149. Misra, J.C., Kar, S.B. and Samanta, S., CSME Transactions (Canada), **11**, 87-94 (1987).

150. Misra, J.C., Kar, S.B. and Samanta, S., Solid Mechanics Archives (Oxford), **11**, 103-112 (1986).

151. Misra, J.C., Kar, S.B. and Samanta, S., Solid Mechanics Archives (Oxford), **12**, 423-438 (1987).

152. Misra, J.C., Pure and Applied Geophysics (Switzerland), **112**, 83-89 (1974).

153. Misra, J.C., Pure and Applied Geophysics (Switzerland), **112**, 92-95 (1974).

154. Misra, J.C., Rev. Mec. Appl., **17**, 1165-1173 (1972).

155. Misra, J.C., Rheologica Acta (Germany), **25**, 201-205 (1986).

156. Misra, J.C., Rheologica Acta (Germany), **25**, 485-490 (1986).

157. Misra, J.C., Samanta, S.C. and Chakraborty, A.K., International J. of Engg. Sci. (UK), **29**, 1209-1216 (1991).

158. Misra, J.C., Samanta, S.C. and Chakraborty, A.K., International J. of Engg. Sci. (UK), **29**, 1065-1075 (1991).

159. Misra, S. C., Samanta, S. C. and Chakrabarti, A. K., Computers & Structures, **43**, 951-957 (1992).

160. Nappa , L., J. Thermal Stresses, **18,** 537-550 (1995).

161. Nemat-Nasser, S., Keer, L.M. and Parihar, K. S., Int. J. Solids Structures, **14** (1978).

162. Neumann, F.E., Abhandl. der Königlich Akad. der Wiss., Berlin (1841).

163. Novik, O.B., Trans.(Doklady) Russian Acad. Sci. : Earth Sci. Sec., **341A**, 5-11 (1996).

164. Nowacki, W., Arch. Mech. Stos. **22**, 1-26 (1970).

165. Nowacki, W., Dynamical Problems of Thermoelasticity, Noordhoff, Leyden (1975).

166. Nowacki, W., Thermoelasticity, Pergamon Press(1962).

167. Nowinski, J.l., Theory of Thermoelasticity with Applications, Sijthoff and Noordhoff, Alpen aan den Rijn (1978).

168. Ozturk, M. and Erdogan, F., ASME J. Appl. Mech., **44**, 631-636 (1993).

169. Parkus, H. Thermoelasticity, Blaisdell Pub. Comp., London (1968).

170. Qian, L. F. and Batra, R. C., J. Thermal Stresses, **27**, 705-740 (2004).

171. Quintanilla, R., Z. Angew. Math. Mech., **83**, 172-180 (2003).

172. Racke, R., Quart. Appl. Math., **61**, 315-325 (2003).

173. Richardson, P.D. and Whitelaw, J. H., J. Franklin Inst., **286**, 169 (1968).

174. Rokne, J., Singh, B. M., Dhaliwal, R. S. and Vrbik, J., Meccanica, **39**, 113-124 (2004).

175. Schulz, M. J.,Sundaresan, M. J., McMichael, J., Clayton, D., Sandler, R. and Nagel, B., J. Intell. Mater. Sys, Struct., **14**, 693-706 (2003).

176. Segall, A. E., J. Appl. Mech., **70**, 779-782 (2003).

177. Sharma, J. N. and Kumar, V., J. Thermal Stresses, **19**, 781-794 (1996).

178. Sharma, J. N. and Kumar, V., J. Thermal Stresses, **20**, 467-475 (1997).

179. Sharma, J. N. and Sharma, P.K., Journal of Thermal Stresses, **26**, 169-182 (2002).

180. Sharma, J. N., J. Thermal Stresses, **20**, 115-127 (1997).

181. Sharma, J. N., Pal, M. and Chand, D., Int. J. Engng. Sci., **42**, 1655-1673 (2004).

182. Sharma, J. N., Singh, D. and Kumar, R., J. Thermal Stresses, **27**, 645-668 (2004).

183. Sherief, H. H. and Anwar, M. N., J. Thermal Stresses, **9**, 165-181 (1986).

184. Sherief, H. H. and Anwar, M.N., J. Thermal Stresses, **11**, 352-365 (1988).

185. Sherief, H. H. and Dhaliwal, R. S., J. Thermal Stresses, **3**, 223-231 (1980).

186. Sherief, H. H. and Hamza, A.F., J. Thermal Stresses, **19**, 55-76 (1996).

187. Sherief, H. H. and Helmy, K. A., Int. J. Engng. Sci., **40**, 587-604 (2002).

188. Sherief, H. H., Elmisiery, A. E. M. and Elhagary, M. A., J. Thermal Stresses, **27**, 885-902 (2004).

189. Sherief, H. H., J. Thermal Stresses, **9**, 151-164 (1986).

190. Sherief, H.H. and Dhaliwal, R. S., J. Thermal Stresses, **3**, 1-8 (1980).

191. Sherif, H. and Ezzat, M., Int. J. Solids Struct., **33**, 4449-4459 (1996).

192. Sherif, H. and Ezzat, M., J. Engng. Math., **34**, 387-402 (1998).

193. Sinha, S. B., and Elsibai, K. A., J. Thermal Stresses, **20**, 129-145 (1997).

194. Sluzalec, A., Int. J. Heat and Mass Transfer, **46**, 1607-1612 (2003).

195. Sneddon, I. N., and Elliot, H. A. Quart. Appl. Math. **4**, 262 (1946).

196. Sosa, H. A. and Bahar, L. Y., J. Thermal Stresses, **16**, 421-436 (1993).

197. Sosinski, W., Arch.Mech., **27**, 445-458 (1975).

198. Sternberg, E. and Chakravorty, J. G., Quart. Appl. Math., **17**, 205-218 (1959).

199. Suhubi, E. S. and Erbay, S., J. Thermal Stresses, **9**, 279-295 (1986).

200. Suhubi, E.S. and Eringen, A. C., Int. J. Engng. Sci., **2**, 289 (1964).

201. Suhubi, E.S., Thermoelastic solids, In: Continuum Physics (ed. A. C. Eringen), Academic Press, New York (1975).

202. Takeuti, Y. and Furukawa, T., J. Appl. Mech. ASME., **48**, 113-118 (1981).

203. Tauchert, T. R., Acta Mechanica, **11**, 155 (1971).

204. Tauchert, T. R., Claus, W. D., Jr. and Ariman, T., Int. J. Engng. Sci., **6**, 37 (1968).

205. Thron, H.L., Pflügers Arch. Geo. Physiol., **263**, 107 (1956).

206. Tianhu, H., Yapeng, S. and Xiaogeng, T., Int. J. Engng. Sci., **42**, 809-823 (2004).

207. Tran, D.V.D. and Singh, M. C., J. Thermal Stresses, **27**, 741-777 (2004).

208. Trehani, P. H. and Eslami, M. R., AIAA J., **38**, 534-541 (2004).

209. Vel, S. S. and Ratra, R. C., Int. J. Solids Struct., **40**, 7181-7196 (2003).

210. Wang, X. and Dai, H. L., Int. J. Engng. Sci., **42**, 539-556 (2004).

211. Wang, X. Z., Lee, J. S. and Zheng, X. J., Int. J. Solids Struct., **40**, 6125-6142 (2003).

212. Watanabe, K. and Atsumi, A., Int. J. Engg. Sci., **10,** 173 (1972).

213. Willmore, T. J., Quart. J. Mech. Appl. Math., **2**, 53 (1949).

214. Yang, Y., Wang, T. and Chen, C., J. Thermal Stresses, **9**, 19-30 (1986).

215. Yu, T. H., Liu, J. and Zhou, Y. X., J. Thermal Stresses, **27**, 1089-1100 (2004).

216. Zhou, S. A., Int. J. Appl. Electromagnetics mater., **53**, 211-222 (1994).

Modern Applied Mathematics
Editor: J.C. Misra

Some Problems in Newtonian and Non-Newtonian Fluid Dynamics Having Physiological Applications

J.C. Misra and B. Pal

Department of Mathematics, Indian Institute of Technology, Kharagpur-721302, India

This chapter gives a brief review of some flow problems in the dynamics of Newtonian and non-Newtonian fluids having the promise of applications to the study of Physiological Fluid Dynamics, particularly in studies pertaining to the haemodynamic flow of blood. For the convenience of readers who are not very familiar with this interesting emerging field, some relevant information on the circulatory system, common diseases of the circulatory system, flow properties of blood and basic equations used in the study of Newtonian and non-Newtonian fluid flows are also included.

1. Introduction

As fluids are one of the major components of a living organism, it is rational to think that fluid mechanics plays a vital role in analyzing and simulating the fluid flow problems associated with physiological processes. One of the common and challenging fluid mechanics problems in the physiological system is the flow of blood through arteries. This is an interesting problem. However, the problem is quite complex owing to the fact that the rheological characters of blood are of very complex nature and that the conduits through which blood flows are of non-uniform cross-sections or are constricted under various pathological situations.

A Newtonian fluid, by definition, is one in which the co-efficient of viscosity is constant at all rates of shear. At a sufficiently high shear rate blood behaves like a Newtonian fluid. Also, it has been observed by various investigators that flow of blood in large arteries, can be modelled as a Newtonian viscous fluid.

It is now well established that whole blood and suspensions of normal red cells in various suspending media exhibit non-Newtonian viscoelastic behaviour. For example, the steady state viscosity decreases with increasing shear rate. Many experimental studies have revealed that the major factors determining the non-

Newtonian viscosity of blood are red cell aggregation and its deformation. The human red cell has a biconcave discoid shape at rest with a major diameter of about 8μm. Aggregation of red cells into rod-like formations called rouleaux due to increasing shear rate is considered to be a cause of the non-Newtonian behaviour of blood. The existence of yield stress, non-linear variation of shear stress versus rate-of-strain and non-parabolic velocity profiles are the different experimental observations for blood, which are significantly different from Newtonian behaviour. Several non-Newtonian aspects of fluid flow were discussed by different investigators in the past.

The laminar flow in channels or tubes with constriction has received serious attention of researchers in view of its important applications to flow in arteriosclerotic blood vessels. With an aim to understand arteriosclerosis, the blood flow in a circular cylindrical tube with a local constriction has been analyzed by Lee and Fung (1970). It has been pointed out by researchers [Misra and Chakravarty (1986)] that the deposits of cholesterol, pearly substance on the arterial wall and proliferation of connective tissues in an arterial wall form plaques which grow inward and restrict the flow of blood. The flow pattern of blood in such a pathological state is deviated from its normal course and the pressure field is greatly disturbed. The collapsibility of blood vessels in the lungs is also of prime concern and has been a subject of extensive investigations [Fung (1981)].

Pulsatility as well non-linearity are also the factors which complicate the flow of blood through arteries. At any given location of the arterial system, blood pressure and velocity change periodically in accordance with the pumping action of the heart. In most cases the equations that govern the flow problems are non-linear. Various aspects (effects of keeping or omitting non-linear terms in the governing equations, effects of varying the dimensionless parameters of the problem, representation of the higher-frequency components of the flow, stability of the computations) of the problem of computation of non-linear pulsatile blood flow in large arteries have been investigated extensively by Imaeda and Goodman (1980). It may also be mentioned that under different physiological conditions (e.g. during pumping action of heart) some blood vessels (coronary arteries) pulsate in a definite manner. As a result, flow in the vessel can be modified abruptly.

The applications of magnetohydrodynamic (MHD) principles to biology and medicine has been receiving considerable attention of physiologists, medical practitioners and fluid dynamicists owing to their importance in biomedical engineering in general and in the treatment of various pathological states in particular. Since blood is electrically conducting, its flow in a cardiovascular system is likely to be influenced by the application of an external magnetic field. In fact the Lorentz force that owes its origin to the flow across magnetic lines of force acts on the constituent particles of blood and alters the haemodynamic indicators of blood flow.

Another kind of studies such as flow past a continuously moving surface are applicable in biomechanics where the flexible surfaces of biological conduits, cells and membranes in living systems are typically lined or surrounded by fluids and are stretched constantly.

2. The Circulatory System

The concept of blood was first established by William Harvey in 1615. Before Harvey no evidence of any connection between arteries and veins was established and it was also believed that the arterial and venous blood were of different look. It was only Harvey who showed first that blood can only leave the ventricle of the heart in one direction. He measured the capacity of the heart and observed it to be two ounces. The heart beats 72 times a minute, so that in one hour it throws into the system $2 \times 72 \times 60$ ounces which is about 540 pounds. This led him to come to the conclusion that the existence of the circulation is a necessary condition for the functioning of the heart.

The central pump, heart along with different blood vessels (aorta, arteries, arterioles and capillaries) constitute the circulatory system. The primary objects of this system are to transport oxygen and nutrients (glucose, amino acids, polypeptides, fats, vitamins, mineral substances, water etc.) to the tissues and to carry several waste-products (ammonia, urea, uric acid etc.) of metabolism and carbon dioxide which are then eliminated from the body by kidneys, sweat glands, lungs and intestines. Thus the transport medium blood is circulated throughout the body in the network system. The circulatory system may be divided into two parts viz., the **systemic** and **pulmonary** circulations. In the systemic circulation blood is supplied to all the body tissues except the lungs, whereas the lungs are supplied by the pulmonary circulation.

3. The Circulation of Blood

The pumping action i.e., the contraction (systole) and relaxation (diastole) of the heart causes a constant unidirectional flow of blood through the vessels. During the systole heart sends arterial (oxygenated) blood to all parts of the body through aorta, arteries and capillaries (except pulmonary artery) and during diastole heart receives venous (de-oxygenated) blood from all parts of the body (except pulmonary vein) through the veins and venous trunk.

At first venous blood which comes from the upper part of the body through superior vena cava and from the lower part of the body through inferior vena cava to the right atrium (RA) of the heart. The RA also receives reduced blood from the heart itself through the coronary sinus. From RA the venous blood passes into the right ventricle (RV) by opening the tricuspid valve (atrioventricular valve). After filling, RV expels the venous blood to the lungs by pulmonary artery through an atrioventricular orfice. In the lungs pulmonary trunk divides into numerous capillaries network. As the blood circulates through the pulmonary (lungs) capillaries, the blood is brought into close relationship with the inspired air in lungs alveoli. It gives off some of its carbon dioxide and acquires a fresh supply of oxygen. This is commonly known as oxygenation of blood.

After oxygenation the venous blood converts into the arterial blood. This blood then returns to the left atrium (LA) through four pulmonary veins (two arising from each lung). From LA blood passes into the left ventricle (LV) by opening the bicuspid (mitral) valve. During contraction of the ventricles, the oxygenated blood from LV flows into aorta by opening aortic (semilunar) valve. The aorta divides and subdivides to form a network of extremely fine vessels called capillaries. The arterial blood while flowing through capillaries supplies oxygen and nutrients to the tissues and takes waste product and carbon dioxide from the tissues.

4. Importance of Studies of Mechanics of the Circulatory System

Studies related to the mechanics of the circulatory system, in particular, dynamics of the cardiovascular system are very important because the number of deaths due to various cardiovascular diseases has increased quite remarkably. Of these atherogenesis, atherosclerosis, cardiovascular neurosis, angina pectoris, myocardial infarction and heart failure deserve special mention. These are closely related to the structure and the mechanical properties of vascular walls and also to mechanical and hydrodynamical conditions imposed on them. The relevant information is very much helpful to medical persons for the treatment of the vascular diseases as well as to those bio-engineers, who are engaged in the design and construction of improved artificial organs. Thus studies on the haemodynamical flow of blood are of prime importance to clinicians.

Extensive studies are being carried out with the purpose of knowing the cause and development of the diseases and obtaining optimum conditions for their treatment. On the basis of mathematical laws, which specify the normal function of living organism and organs, their force and motion constitutive relations have been derived for illustrating their application to different situations. These constitutive relations which lead to an interaction with mechanics and physiology are of fundamental importance in studies pertaining to the mechanical behaviour of biological tissues.

5. Some Common Diseases of the Circulatory System

(a) Hypertension

High arterial pressure is usually referred to as the principal indication of hypertension. The increased blood pressure can be due to many causes and hence hypertension should be considered as a symptom rather than a disease.

(b) Arteriosclerosis and Atherosclerosis

Arteriosclerosis is the generic term for three patterns of vascular diseases, all of which cause thickening and inelasticity of arteries. The dominant pattern is called atherosclerosis, which is hardening of the arterial walls due to deposition of substances which are normally not found in them, such as calcium, cholesterol and hyaline (a substance of protein nature). When large arteries become hardened, they lose their elastic properties, can not be easily distended, do not absorb any of the pressure developed by the ventricles; thereby systolic blood pressure rises very high. Atherosclerosis does not usually develop in the vessels evenly. In some cases it affects mainly aorta and vessels of the brain etc. The occurrence of these diseases in cerebral arteries gives the patient a giddy feeling and headache. It may even cause brain disorder. If the sclerosis i.e., hardening of vessel wall material occurs in the aorta and involves also the aortic valves, the aortic valves shrink and no longer ensure complete closure of the mouth of the aorta during cardiac dilation. This results in aortic insufficiency, which may sometimes lead to heart failure. Atherosclerosis is also of common occurrence in coronary arteries of the heart. This may enhance the attacks of the heart diseases like angina pectoris and may lead to myocardial infarction. Angina pectoris occurs when a portion of the heart muscle temporarily receives inadequate blood supply but no damage to the tissue occurs. It is sometimes referred to as heart attack.

(c) Myocardial Infarction

The word 'infarction' is a general term used to describe the situation when death of tissues occurs as a result of oxygen deficiency. Myocardial infarction is caused by obstruction in the lumen of a coronary artery due to thrombi (which is due to blood clot or some other reasons) which are responsible for the supply of blood to some portion of the blood muscle (particularly, the myocardium). The portion of the muscle, which is deprived of its blood supply gradually softens and is then replaced by a scar. In severe infarctions, the scar tissue gets multiplied over large portions of the muscle with a limited capacity of contraction of the heart.

One can thus realize the relevance and importance of the haemodynamical flow of blood in pathology.

6. Composition of Blood

Blood is a wonderful fluid containing several enzymes and hormones. It transports oxygen and carbon dioxide between the lungs and the cells of the tissues. Thus blood plays the most vital role in nurturing life.

Blood is a body fluid, which is a suspension of several formed elements (i.e. cells) in an aqueous solution, called plasma. It is comprised of erythrocytes (red cells or RBC), leukocytes (white cells or WBC) and platelets. Plasma consists of about 90% water, about 7% plasma protein and some organic and inorganic substances. The red cells occupy the major portion of total cellular volume and half of the blood volume. The number of red cells is about 5-million/cu. mm. While the white cells occupy only about 1/600 th of the total cellular content (about 8000/cu. mm. in count). On the other hand the platelets cover less than 1/800 th of the cellular volume (about 250 000 – 300 000/ cu. mm.) The normal human red cells are usually disk-shaped with diameter and thickness about 7.6μm and 2.8μm respectively. White cells are quite rounded, while platelets are much smaller in size having diameter about 2.5μm [cf. Fung (1981)].

7. Flow Properties of Blood

The red cells of blood (RBC) play a major role in determining the mechanical properties of blood. Hence it is important to know the volume concentration of RBC. To represent this volume concentration, the term hematocrit is used, which is defined as a percentage volume concentration (which normally varies from 42 to 45). If blood is allowed to stand in a container, the corpuscles settle progressively to the bottom leaving clear plasma above. The rate at which this takes place is constant in health and is known as the erythrocyte sedimentation rate (ESR). The cause of settling corpuscles is that their density is greater than that of the plasma. During illness, the ESR usually increases abruptly due to the formation of red cell aggregation. The white cells (WBC) perform the main role in fighting against diseases. As already mentioned, they are relatively fewer in number compared to the RBC, and so their contribution to the circulatory characteristic parameters is, in general, not that significant. The platelets are numerous, but they are also very small in size, so they are not given much importance as far as the flow of blood is concerned.

Human blood is generally considered to posses the behaviour of a non-Newtonian fluid. Experimental studies have, however, revealed that at a high shear rate and when it flows through larger vessels, blood behaves more or less like a Newtonian fluid with a constant co-efficient of viscosity.

8. Constitutive Equations

Many physiological phenomena, e.g. contraction of the heart, pulsation of the artery etc. occur at the continuum level. To have a better understanding of these phenomena, one must have a thorough knowledge of the forces that cause or are caused by the motion. It is important to know the description of the motion, the deformation of the body and the forces that are transmitted into the body. The mechanical properties of the materials are described by constitutive equations, which are simply the relation between different stress and rate-of-strain components.

For an isotropic incompressible Newtonian fluid the constitutive equation may be written as

$$\tau_{ij} = -p\delta_{ij} + 2\mu e_{ij},$$

in which $$e_{ij} = \frac{1}{2}\left(\frac{\partial u_i}{\partial x_j} + \frac{\partial u_j}{\partial x_i}\right),$$

$$e_{ii} = 0$$

τ_{ij} being the stress tensor, e_{ij} the rate of –strain tensor, u_i the velocity components, δ_{ij} the isotropic tensor (the Kronecker delta), p the hydrostatic pressure and μ the co-efficient of viscosity.

9. Basic Equations for a Newtonian Fluid

Two Dimensional Flow

Here the flow is the same in all planes parallel to a plane (the *xy*-plane, say). In this case it may be considered as the volume of fluid of unit thickness moving parallel to the *xy*-plane. The equations governing the flow of an incompressible fluid in the Cartesian coordinate system (*x, y)* may be expressed as

$$\rho\left(\frac{\partial u}{\partial t}+u\frac{\partial u}{\partial x}+v\frac{\partial u}{\partial y}\right)=\rho F_x-\frac{\partial p}{\partial x}+\mu\left(\frac{\partial^2 u}{\partial x^2}+\frac{\partial^2 u}{\partial y^2}\right)$$

$$\rho\left(\frac{\partial v}{\partial t}+u\frac{\partial v}{\partial x}+v\frac{\partial v}{\partial y}\right)=\rho F_y-\frac{\partial p}{\partial x}+\mu\left(\frac{\partial^2 v}{\partial x^2}+\frac{\partial^2 v}{\partial y^2}\right)$$

while the equation of continuity reads

$$\frac{\partial v}{\partial x}+\frac{\partial v}{\partial y}=0$$

where u and v are the respectively velocity components of fluid along x and y directions respectively, ρ is the density, μ the co-efficient of viscosity, F_x and F_y the body forces in the x and y directions, p the pressure and t the time variable.

Axi-symmetric Flow

When the flow is symmetric about an axis, the cylindrical polar coordinates $(r,\ \theta,\ z)$ may be used, where the axis of symmetry is along the z-axis. Let v_r be the velocity component of the fluid along the radius vector perpendicular to the axis, v_θ that along the perpendicular to the axis and the radius vector and v_z parallel to the z-axis. In the axi-symmetric case $v_\theta = 0$ and all the flow characteristics are independent of θ. Hence the equations of momentum and the equation of continuity for an incompressible fluid in the absence of external body forces may be expressed (in the cylindrical system of coordinates) as

$$\rho\left(\frac{\partial v_r}{\partial t}+\frac{\partial v_r}{\partial r}+v_z\frac{\partial v_r}{\partial z}\right)=-\frac{\partial p}{\partial r}+\mu\left(\frac{\partial^2 v_r}{\partial r^2}+\frac{1}{r}\frac{\partial v_r}{\partial r}-\frac{v_r}{r^2}+\frac{\partial^2 v_r}{\partial z^2}\right)$$

$$\rho\left(\frac{\partial v_z}{\partial t}+\frac{\partial v_z}{\partial r}+v_z\frac{\partial v_z}{\partial z}\right)=-\frac{\partial p}{\partial z}+\mu\left(\frac{\partial^2 v_z}{\partial r^2}+\frac{1}{r}\frac{\partial v_z}{\partial r}+\frac{\partial^2 v_z}{\partial z^2}\right)$$

$$\frac{1}{r}\frac{\partial}{\partial r}(rv_r)+\frac{\partial}{\partial z}v_z=0$$

10. Non-Newtonian Fluids

A non-Newtonian fluid, by definition, is one which exhibits a non-linear relationship between the shear stress and the rate-of-strain. Thick liquids generally belong to this category. It is now a well established fact that particularly in the low shear rate region, blood behaves like a non-Newtonian fluid. The deviations in the non-linear relationship between stress and deformation rate, existence of a yield stress and the variation of viscosity according to shear rate may sometimes appear. Such anomalous behaviour of blood is often referred to as non-Newtonian properties of blood [cf. Murata (1983)]. These properties are of two types. Firstly, at low shear rates, the apparent viscosity increases markedly, sometimes even a certain yield stress is required for the onset of the flow. Secondly, at high shear rates the apparent viscosity is lesser in narrow vessels than it is in larger vessels. These two types of anomalies are often referred to as 'low shear' and 'high shear' effects respectively.

Classification of Non-Newtonian Fluids

(a) Time –Independent Fluids

They are the simplest type of non-Newtonian fluids. The rate-of-strain, which is a non-linear function of the shear stress is independent of shearing time as well as previous strain-rate history. The Casson fluid, power-law fluids, Herschel-Bulkley fluid, Bingham fluids and biviscosity fluids are the examples of this class of non-Newtonian fluids.

(b) Time-Dependent Fluids

For this class of fluids rate-of strain is a multivalued function of the shear stress. Because here strain-rates depend on the shearing time and on the past strain-rate histories. If the shear stress increases with time, for a fluid when it is sheared, the fluid is called rheopectic fluid. On the other hand if the shear stress decreases with time, the concerned fluid is called a thixotropic fluid.

(c) Viscoelastic Fluids

A viscoelastic fluid possesses both viscous and elastic properties. It has normal stress difference. It is well known that the whole blood and suspensions of normal red cells in different suspending media exhibit non-Newtonian viscoelastic behaviour. For example, the steady state viscosity decreases with increase in shear rate. Many experimental studies suggest that the major factors determining non-Newtonian viscosity of blood are red cell aggregation and deformation.

Equations Governing the Flow of Non-Newtonian Fluids

For an isotropic, incompressible and non-Newtonian fluid, the equations of motion and continuity in cartesian coordinate system (*x, y*) (for – dimensional flow) may be taken as

$$\rho\frac{Du}{Dt}=\frac{\partial}{\partial x}\tau_{xx}+\frac{\partial}{\partial y}\tau_{xy},$$

$$\rho\frac{Dv}{Dt}=\frac{\partial}{\partial x}\tau_{xy}+\frac{\partial}{\partial y}\tau_{yy},$$

and $$\frac{\partial u}{\partial x}+\frac{\partial v}{\partial y}=0$$

and in cylindrical polar coordinate system (r, θ, z) (for axi-symmetric flow) the equations of motion and continuity take the following form.

$$\rho\frac{D\upsilon_r}{Dt}=\frac{\partial}{\partial r}\tau_{rr}+\frac{\partial}{\partial z}\tau_{rz}+\frac{\tau_{rr}-\tau_{\theta\theta}}{r},$$

$$\rho\frac{Dv_z}{Dt}=\frac{\partial}{\partial r}\tau_{rz}+\frac{\partial}{\partial z}\tau_{zz}+\frac{\tau_{rz}}{r}$$

and $$\frac{\partial v_r}{\partial r}+\frac{\partial v_z}{\partial z}+\frac{v_r}{r}=0$$

where $\frac{D}{Dt}$ stands for material derivative, τ_{xx}, τ_{xy}, τ_{yy} (Cartesian system) and τ_{rr}, $\tau_{\theta\theta}$, τ_{zz}, τ_{rz}, (cylindrical system) are the principal components of the stress tensor τ_{ij}, whose various forms (in the constitutive relations for various non-Newtonian fluids) have been presented below. All other symbols carry the came meaning as mentioned earlier.

(a) Casson Fluid

$$\tau_{12}^{1/2}=\eta_c\dot{\gamma}^{1/2}+\tau_y^{1/2},\quad |\tau_{12}|\geq\tau_y$$

$$\dot{\gamma}=0, |\tau_{12}|<\tau_y\ ,$$

where τ_{12}, γ, η_c, and τ_y denote the shear stress, the shear rate, Casson viscosity, and yield stress respectively.

(b) Power-Law Fluid

$$\tau_{12} = K\left|\dot{\gamma}\right|^{n-1}\dot{\gamma}$$

where K is the consistency parameter and *n* is the power-law index.

(c) Bingham Fluid

$$\tau_{12} = \mu_B\,\dot{\gamma} + \tau_y, \quad \tau_{12} \geq \tau_y,$$

$$\dot{\gamma} = 0, \qquad \tau_{12} < \tau_y,$$

where μ_B is the plastic viscosity.

(d) Herschel-Bulkley Fluid

$$\tau_{12} = \mu_B\,\dot{\gamma}^{\,n} + \tau_y, \quad \tau_{12} \geq \tau_y,$$

$$\dot{\gamma} = 0, \qquad \tau_{12} < \tau_y,$$

(e) Bingham Plastic Fluid

$$\tau_{ij} = 2\left(\mu_B + \frac{py}{\sqrt{2P}}\right)e_{ij}, \quad \frac{1}{2}\tau_{ij}\tau_{ji} < p_y^2,$$

$$e_{ij} = 0, \quad \frac{1}{2}\tau_{ij}\tau_{ji} \geq p_y^2,$$

τ_{ij} being the stress tensor, p_y the yield stress, e_{ij} the rate –of-strain tensor and $P = e_{ij}.\, e_{ji}$.

(f) Biviscosity Fluid

$$\tau_{ij} = \begin{cases} 2\left(\mu_B + \dfrac{p_y}{\sqrt{2P}}\right)e_{ij}, & P \geq P_c, \\ 2\left(\mu_B + \dfrac{p_y}{\sqrt{2P_c}}\right)e_{ij}, & P < P_c, \end{cases}$$

in which P_c is a constant so that $\beta\ (= \mu_B \sqrt{2P_c} / p_y)$ is a non-dimensional quantity.

g) Viscoelastic Fluid

$$\tau_{ij} = - p\delta_{ij} + \eta_0 A_{(1)}$$

$$\tau_{ij} = p\delta_{ij} + \eta_0 A_{(1)_{ij}} + \beta_0 A_{(1)_{i\alpha}} + \gamma_0 A_{(2)_{ij}},$$

where p is an indeterminate pressure and η_0, β_0 and γ_0 are material constants. The rate-or-strain tensor $A_{(1)_{ij}}$ and the acceleration tensor $A_{(2)_{ij}}$ are defined by

$$A_{(1)_{ij}} = v_{i,j} + v_{j,i}$$

$$A_{(2)_{ij}} = a_{i,j} + a_{j,i} + 2v_{m,i}\, v_{m,j},$$

where a_i 's are the acceleration components given by $\partial v_i / \partial t + v_j v_{i,j}$

Electrically Conducting Fluid under Magnetic Field

An electrically conducting fluid is a fluid which conducts due to the presence of free charges (ions/electrons) that can move indefinitely. Due to the motion of an electrically conducting fluid in the presence of a magnetic field, electric current is induced in the fluid which produces its own magnetic field, called induced magnetic field and thus perturbs the original magnetic field. In addition to this the induced current interacts with the magnetic field to produce electromagnetic force perturbing the original motion. Thus two important basic effects of magnetohydrodynamics (MHD) may be summarized as follows :

(a) the motion of the fluid affects the magnetic field and (b) the magnetic field affects the motion of the fluid. Hence there is a two-fold interaction between the flow and the electromagnetic field.

Net Electromagnetic Force and Governing MHD Equations

Let us consider an electrically conducting fluid moving with velocity $\vec{q}$ in the presence of an applied magnetic field $\vec{B}_0$. The motion of the fluid across the imposed magnetic field $\vec{B}_0$ gives rise to an induced e.m.f. in a direction normal to the direction of both the field and the flow. Since the fluid is electrically conducting, induced current will flow along the direction of the induced e.m.f. This current gives rise to induced magnetic field $\vec{b}$ which distorts the imposed magnetic field $\vec{B}_0$. The current $\vec{j}$ given by

$$\vec{j} = \sigma(\vec{E} + \mu_e \vec{q} \times \vec{B})$$

(where $\vec{B} = \vec{B}_0 + \vec{b}$) interacts with the magnetic field to give rise to the electromagnetic force $\vec{j} \times \vec{B}$ per unit volume of the fluid. Here σ, $\vec{E}$ and μ_e denote respectively the electrical conductivity, the electric field and the magnetic permeability. This force affects the flow field. The Navier-Stokes equation expressing the conservation of momentum gets modified to include this electromagnetic force.

The equations of continuity, momentum along with Maxwell's equations for steady flow of a viscous incompressible and electrically conducting fluid are

$\nabla.\vec{q} = 0$, (continuity equation)

$(\vec{q}.\nabla)\vec{q} = -(1/\rho)\nabla p + \nu\nabla^2\vec{q} + \frac{1}{\rho}\left(\vec{j} \times \vec{B}\right)$, (equation of momentum)

$\nabla \times \vec{B} = \mu_e \vec{j}$, (Amp'ere's law)

$\nabla \times \vec{E} = 0$, (Faraday's law)

$\nabla.\vec{B} = 0$,

$\vec{j} = \sigma\left(\vec{E} + \vec{q} \times \vec{B} - \frac{1}{en_e}\vec{j} \times \vec{B} + \frac{1}{en_e}\nabla p_e\right)$, (generalized Ohm's law)

in which ρ, ν, e, n_e and p_e are respectively the density, the kinematic viscosity, the electric charge, the number density of electrons and the electron pressure. It is to be noted that in writing the last equation above, the ion slip and thermoelectric effects have been ignored. Further in writing Maxwell's equations, the terms due to convection current and the displacement current are ignored along with the assumption that the flow velocity is much smaller than the velocity of light.

11. Effect of Magnetic Field on Blood Flow

In the medical literature, it is often noted that the prevention and rational therapy of arterial hypertension are intimately connected with closer study of the machanisms of its origin and development. However for the prevention and rational therapy of the arterial hypertension, studies pertaining to the hydrodynamic changes associated with the action of certain external forces are of great importance. Since blood is an electrically conducting fluid, one such force may be the magnetic field. The Lorentz force will act on the constituent particles of blood. Being a body force, the magnetic force must alter the haemodynamic indicators of the flow of blood. It must therefore be supposed that prolonged exposure to a magnetic field and the associated haemodynamic shifts will lead to disturbances in the normal functioning of the cardiovascular system. Experimental investigations on exposure of the living body [cf. Karmilov et al. (1948)] have also shown that the magnetic field exerts a very important influence on the vascular system. However, the mechanism of a magnetic field on the cardiovascular system remains obscure. An important step in elucidating this problem is to study the patterns of the effect of the magnetic field on the movement of blood.

12. Various Studies in Fluid Mechanics Having Relevance to the Haemodynamic Flow of Blood

(i) Studies on the Use of Different Constitutive Equations of Blood

The viscoelastic fluid behaviour of whole blood is associated with the elastic properties of the red-cell membrane and the viscosity of the internal and external fluids. Detailed discussions on the rheological properties of blood and its constituents can be found in the reports of Skalak (1976), Chien (1979), Chien et al. (1984), Cokelet (1987), Hochmuth (1987) and Schmid-Schonbein (1987). The viscoelastic properties of blood in transient flow in a coaxial cylinder viscometer were examined by Stoltz et al. (1980). Kaibara and Fukada (1981), Thurston (1981), Murata (1983), Liepsch (1986) carried out different studies on non-Newtonian behaviour of blood. The steady and unsteady flows of blood through an axi-symmetric stenosis was studied by Nakamura and Sawada (1988, 1990), treating blood as a biviscosity fluid. Luo and Kuang (1992) proposed and studied a three-parameter constitutive equation for whole human blood. Recently, Dutta and Tarbell (1996) discussed the use of Casson's model and power-law model, as the constitutive equation of blood, as well as the validity of these models in describing the flow behaviour of blood.

(ii) General Studies on Blood Flow

A numerical solution was given by O'Brien et al. (1976) for non-linear pulsatile flow of a Newtonian fluid in a channel having a symmetrical bifurcation. Secomb (1978) studied theoretically the problems of both two-dimensional and axisymmetric flow fields of an incompressible Newtoninan viscous fluid, where the walls of the vessels pulsate in a prescribed way under different physiological conditions. Different analyses on non-linear pulsatile blood flow in arteries were investigated by Ling and Atabek (1972), Imaeda and Goodman (1980) as well as by Misra and Singh (1987). Mark et al. (1985) studied the pulsatile flow through the proximal portion of the left coronary artery of man, that exhibits quasi-steady characteristics. The flow in curved tubes simulating blood flow in curved arteries was analysed by several investigators (Chandran et al., 1979 as well as Chang and Tarbell, 1985 for aortic arch); (Perktold et al., 1987, 1988 for carotid siphon) and (Sabbah et al., 1984; Altobelli and Nerem, 1985; Chang and Tarbell, 1988; Asakura and Karino, 1990 for coronary arteries). Liepsch and Moravec (1984) studied the pulsatile flow of non-Newtonian fluid in distensible models of human arteries. In this investigation, emphasis was given on the difference between rigid and elastic models and also on the Newtonian and non-Newtonian flow behaviours. Mcdonald (1986) studied the pursatile flow of blood in a catheterized artery. It was asserted by Charara et al. (1985) that the transient rheological behaviour of blood which involves non-Newtonian viscosity, elasticity and thixotropy can be modelized with a Maxwell rheological state equation which depends on a structural parameter having dimension of shear rate. Axial velocities in an enlarged, two-dimensional, rigid tube model of the carotid artery bifurcation by means of a Laser Doppler Anemometer, under both steady and unsteady flow conditions were measured by Rindt et al. (1987). Sung and Yoganathan (1990) investigated experimentally a model which addressed the hypothesis that altered velocity patterns relate to the severity of various diseases in the pulmonary artery. A finite element approximation of the steady flow in a rigid three-dimensional model of the caotid artery bifurcation was presented by Rindt et al. (1990). Sud and Sekhon (1990) presented a finite element model for the analysis of the steady flow of a viscous fluid through a connected system of elastic tubes with the aim to simulate the conditions of blood flow through the human arterial system. The flow pattern in the left main coronary artery was studied by Perktold et al. (1991) by considering an idealized geometry and by numerically solving the full Navier-Stokes equations for a Newtonian fluid. On the other hand a two –dimensional numerical model was developed by Lou and Yang (1993) to determine the effect of the non-Newtonian behaviour of blood on a pulsatile flow at the aortic bifurcation. Taylor and Yamaguchi (1994) simulated the three-dimensional blood flow in an abdominal aortic aneurysm to predict how the irregularities of blood flow, pressure and shear stress occur there. They also compared steady and unsteady flow predictions. The mathematical models of the flow in the basilar artery and in the vertebro-basilar junction were developed by Krijger et al. (1989, 1992) and Ravensbergen et al. (1995). Based on cam-piston –valve arrangement, a mechanical pulsatile flow generator was designed by Shi et al. (1997) to investigate sinusoidal flow and other types of pulsatile flows in straight rigid tubes.

(iii) On the Entry Flow Problems

The oscillatory entry-flow problem in a circular pipe by applying the Oseen approximation was investigated by Atabek and Chang (1961). They also computed the entrance length as a function of time. Wang and Longwell (1964) solved the flow problem in the inlet region by employing the full Navier-Stokes equations. The governing partial differential equations were expressed in finite difference form and solution was obtained by an iterative relaxation technique. The entry flow problem of blood vessels at arbitrary Reynolds number was studied by Lew and Fung (1970). Morihara and Cheng (1973) investigated the steady flow in the entrance region of a channel between two semi-infinite parallel plates by the method of quasilinearization. Caro et al. (1978) obtained a rough estimate of the unsteady entry length by neglecting the interaction between oscillatory and steady flow. Creff et al. (1985) computed developing laminar pulsatile flow in a tube by using finite difference method. Krijger et al. (1991) studied the pulsating entry flow in a channel by solving numerically the unsteady Navier-Stokes equations and obtained estimates for the time-dependent entry length. On the other hand the unsteady entrance flow development in a straight tube to quantify the effect of variations in Womersley parameter was investigated by He and Ku (1994).

Misra et al. (2001) studied the unsteady flow of an incompressible viscous fluid in the entrance region of a parallel-sided channel whose walls pulsate in a prescribed manner. The flux of fluid entering the channel oscillates cosinusoidally with time so that there is always inflow at the entrance. The time dependent Natier-Stokes equations are solved numerically using the finite difference method. Axial velocity profiles at various distances from the entrance of the channel are found at different instants of time during one period oscillation of the flux of fluid entering the channel. It is shown that for large values of frequency of pulsation f_0 of the channel walls oscillating sinusoidally with time, back flow occurs near the channel walls at a certain location far downstream from the entrance section at a certain instant of time. It has also been shown that at a subsequent time there is flow reversal at the same location across the entire cross section of the channel. For sufficiently small values of f_0, such back flow does not occur. The effect of variation of the phase difference in the oscillation of entry flux of fluid and the channel wall oscillation on the time intervals in which wall shear stress does not vanish has also been determined by them.

This study by Misra et al. (2001) has been appreciated in the book "Theory and Applications of Viscous Fluid Flows" authored : by R. K. Zeytounian (France) and published by Springer in 2003. On p. 203, the author writes, "... It is interesting to note that the motivation behind studying this problem is related to physiological flow problems (e.g., blood flow through arteries, ...)."

(iv) Flow under Externally Imposed Body Accelerations

The flow of blood in large arteries under the influence of linear periodic acceleration was studied by Sud et al. (1983). A mathematical model of flow of blood treated as a Newtonian fluid in a single artery subjected to a pulsating pressure gradient as well as a body acceleration was discussed by Sud and Sekhon (1985). Subsequently the effects of externally imposed body accelerations on blood in a branched system of arteries by using a finite element model of flow in the arterial system were studied by same authors (1986). Belardinelli et al. (1989) studied theoretically the pressure variations when the body was subjected to shock accelerations. By considering blood as a non-Newtonian fluid Misra and Kar (1990) investigated the unsteady flow of blood through arteries in vibration environments. They observed that the rate of flow, shear stress, velocity and acceleration are significantly affected when the amplitude of vibration is considerably high. It was noted that the mass flow magnitude is significantly influenced by the microrotation of blood cells. Chaturani and Palanisami (1990a, 1990b) considered respectively Casson and power-law fluid models for pulsatile flow of blood under periodic body acceleration. An analytical estimate of the flow-field in a porous stenosed artery subject to a body acceleration was presented by Chakravarty and Sannigrahi (1998).

Misra and Pal (1999) developed a mathematical model for the study of pulsating flow of blood under an externally imposed body acceleration, by treating blood as a non-Newtonian fluid. Use has been made of the biviscosity model for blood flow. Of specific concern here is the flow of blood in an artery subject to both the pulsatile pressure gradient due to normal heart action and a periodic body acceleration. The governing equations are solved by using finite-difference technique. Closed from solutions are also obtained by using Laplace transformation technique for the low pulse frequency region. Computational results are presented for the effects of periodic body acceleration on the flow of blood passing through a straight cylindrical vessel.

(v) Flow on the Stretching Surface

Crane (1970) investigated the steady two-dimensional boundary layer flow of a Newtonian viscous fluid caused by a stretching sheet. The sheet moves in its plane with a velocity varying linearly with distance from a fixed point due to application of a uniform stress. Gupta and Gupta (1977) investigated the heat and mass transfer in the flow when the sheet is subject to uniform suction or blowing. Brady and Acrivos (1981) gave an exact solution to the steady-state Navier-Stokes equations for incompressible viscous flow in an infinitely long channel or tube, when the surface velocity of the channel or tube varies linearly with the streamwise coordinate. Rajagopal et al. (1984) investigated the steady flow over a stretching surface for a certain class of non-Newtonian fluids known as incompressible second-order fluids which are viscoelastic in nature. Dandapat and Gupta (1989) studied the same problem with heat transfer and gave an exact analytical solution of the non-linear equations governing the self-similar flow and heat transfer.

(vi) Flow in Non-uniform Vessels

Young (1968) investigated the problems of steady flow of blood through an arterial segment having a collar-like mild stenosis. He presented an analysis on the effect of an axially-symmetric time-dependent growth into the lumen of a blood vessel of constant cross-section. In his analysis the convective acceleration terms in the Navier-Stokes equations were neglected and the effect of stenosis on the pressure distribution and wall shear stress was discussed along with possible biological implications. Lee and Fung (1970) studied blood flow through a circular cylindrical tube with a local constriction. They presented numerical results for the streamlines and distributions of velocity, pressure, vorticity and shear stress at the low Reynolds number. An asymptotic series solution was obtained by Manton (1971) for the low Reynolds number flow through an axi-symmetric tube where radius varies slowly in the axial-direction. Chow and Soda (1972) studied laminar flow in a tube with constriction by assuming that the spread of surface roughness is large in comparison with the mean radius of the tube. The unsteady viscous flow in a pipe of slowly varying cross-section was studied by Hall (1974). A mathematical model was proposed by Morgan and Young (1974) for the analysis of blood flow in arterial stenosis, by making use of the momentum integral method. They simulated the axi-symmetric constriction by a geometry such that solution obtained was applicable to both mild and severe stenosis for Reynolds number below transition. Deshpande et al. (1976) investigated the steady laminar flow through modelled vascular stenoses.

Back et al. (1977) examined the rheological effects of multiple non-obstructive plaques in main coronary arteries of man. Wild et al. (1977) studied the viscous flow in collapsible tubes of slowly varying elliptical cross section. Khalifa and Giddnes (1978) performed measurements of blood velocity in the descending thoracic aorta of dogs with a controlled degree of stenosis. Young (1979) studied in detail the causes and effects of stenoses from fluid mechanical point of view. A technique for solving the approximate equations governing steady flow through an axi-symmetric tube of slowly varying cross-section was discussed Mcdonald (1979). Bestman (1982) studied analytically the low Reynolds number non-Newtonian flow in slowly varying axi-symmetric tubes from physiological aspects. Fukushima et al. (1982) solved numerically the problem of flow of a Newtonian fluid. Cho et al. (1983) investigated experimentally the pulsatile and steady flows through a smooth tube and an atherosclerotic coronary artery casting a man. O'Brien and Ehrlich (1985) analyzed a simple pulsatile flow in an artery with constriction by using finite-difference techniques. Bitoun and Bellet (1986) carried out an experimental study by considering a model of two-component pulsating fluid flow through a cylindrical duct having a mild stenosis. Their model is applicable to flow of blood through small diameter stenosed vessels. A review on collapsible-tube flow with various aspects was presented by Kamm (1987). The unsteady flow of blood through vessels whose endothelial surface is non-uniform was studied by Misra and Sahu (1989). A mathematical model of both two –dimensional and axi-symmetric flows through an irregular stenosis was developed by Johnston and Kilpatrick (1991). Johnson et al. (1992) developed a model of pulsatile flow in a uniform deformable vessel by employing finite difference technique. while Tu et al. (1992) gave a finite element simulation of pulsatile flow of blood through arterial stenosis. treating blood as a Newtonian fluid. On the other hand Tutty (1992) studied the pulsatile flow of a Newtonian fluid in a constricted channel. A non-Newtonian fluid model for blood flow through arteries under stenotic conditions was investigated by Misra et al. (1993). The steady laminar flow in a vascular tube with double constrictions of realistic bell shaped Gaussian normal distribution profiles was studied by Lee (1994). Tu and Deville (1996) developed the pulsatile flow of different non-Newtonian fluids through arterial stenosis by Galerkin finite element method. Jenson (1998) developed an asymptotic model of viscous flow limitation in a highly collapsed channel. On the other hand Luo and Pedley (1998) investigated the effect of wall inertia on the self-excited oscillations in a collapsible channel flow by solving the full coupled two-dimensional membrane-flow equations. Dash et al (1999) investigated the flow in a catheterized curved artery with stenosis.

(vii) Flow in the Presence of Magnetic Fields

The first attempt at a theoretical evaluation of the effect of a magnetic field on blood flow was made by Belousova (1965) as well as by Korchevskii and Marochnik (1965). By considering a blood vessel in the form of a cylindrical tube Vardanyan (1973) investigated the problem of the effect of a uniform transverse magnetic field on the flow of a viscous electrically conducting fluid (e.g. blood). McMichael and Deutsch (1984) investigated the steady MHD flow of an imcompressible viscous fluid in a circular tube of variable cross-section under the influence of an axial magnetic field. A physiological-type flow in a straight circular tube in the presence of a transverse magnetic field was analyzed by Rao and Deshikachar (1988). They solved the governing equations by using the method of Laplace transforms. It was found that the effect of a magnetic field is to reduce the magnitude of velocity near the central region of the tube. The influence of a uniform transverse magnetic field on the motion of an electrically conducting fluid past a stretching surface was investigated by Kumari et al (1990), Andersson (1992) as well as by Andersson et al. (1992). Hall effects of MHD boundary layer flow over a continuous semi-infinite flat plate moving with a uniform velocity in its own plane in an incompressible viscous and electrically conducting fluid (e.g. blood) in the presence of a uniform transverse magnetic field were investigated by Watanabe and Pop (1995). Ram and Takhar (1994) studied the unsteady MHD flow with a suspension of spherical particles through a square channel from physiological points of view.

Gupta and Misra et al. (2000) studied the Hall effects on magnetohydrodynamic boundary layer flow over a stretching surface in an incompressible viscous electrically conducting fluid, when the fluid is permeated by a uniform transverse magnetic field. Such a study is applicable to the elongation of the bubbles and in bioengineering where the flexible surfaces of the biological conduits, cells and membrances in living systems are typically lined or surrounded with fluids which are electrically inducting (e.g. blood) and are being stretched constantly. A similarity solution for the velocity distribution in the flow is obtained when the surface is stretched with a velocity proportional to the distance from a fixed point. The velocity profiles and skin friction on the surface are computed and studied in detail for various values of the Hall and the magnetic parameters. A novel result of this analysis is that Hall currents may induce flow reversal at a certain height above the stretching surface. A similarity solution in the flow under consideration is found and the rate of heat transfer from the surface held at uniform temperature has been calculated. The study is believed to have some implications in physiological fluid dynamics.

The steady flow of an incompressible viscoelastic and electrically conducting fluid in a parallel plate channel in the presence of a uniform transverse magnetic field, was investigated by Pal, Misra and Gupta (1999). The motivation for this problem arises in studying the flow of blood in arteries whose walls are stretchable. It was observed by Fukada and Kaibara (1980), Thurston (1972) and Stolz and Lucius (1981) that under certain conditions blood exhibits viscoelastic behaviour which may be attributed to the viscoelastic properties of a individual red cells and the internal structures fromed by cellular interactions. Further since blood is also electrically conducting, its flow in a cardiovascular system (e.g. arteries) is like to be influenced by a magnetic field. The aforesaid study reveals that the reverse flow arising out of the streamwise velocity changing sign in the transverse direction in the absence of magnetic field (M=0) can be almost eliminated by the application of a strong magnetic field. This interesting result is likely to have an important bearing on the physiological problem of controlling blood flow in arteries (with elastic walls) by applying magnetic field.

An asymptotic series solution for steady flow of an incompressible viscous and electrically conducting fluid in a channel permeated by a uniform transverse magnetic field was put forward by Pal, Misra and Gupta (1996). The depth of the channel varies slowly in the axial direction. Analytical expressions are derived for the wall shear stress and also the pressure drop along the channel. It is found that for a fixed value of the Reynolds number R, the wall shear stress increases with an increase in the magnetic parameter K. Computational results have been obtained for some specific slowly varying channels. The results show that for fixed K, flow separation occurs for a value of R which is higher than that for the non-magnetic case (K=0). The results are likely to have important bearing on the haemodynamical flow of blood in constricted arteries giving rise to the pathological condition of arteriosclerosis.

Misra et al (1998) gave an asymptotic series solution for steady flow of an incompressible, second –grade electrically conducting fluid in a channel permeated by a uniform transverse magnetic field. The depth of the channel is assumed to vary slowly in the axial direction. Analytical expressions are derived for the velocity, wall shear stress and pressure drop along the channel. It is found that for fixed values of the Reynolds number R and the non-Newtonian parameter K_1, the wall shear stress increases with an increase in the value of the magnetic parameter M. Numerical computations carried out for a specific slowly varying channel show that flow separation occurs for both second-grade ($K_1 < 0$) and second order ($K_1 > 0$) fluids when $|K_1| < 0.15$. The study also reveals the interesting result that while flow separation takes place for a second order fluid for $K_1 \geq 0.15$, no separation occurs at all for $K_1 \geq 0.15$ for a second –grade fluid. The motivation behind the study is to find the extent of influence a uniform transverse magnetic field exerts on blood flow through constricted arteries, although the channel geometry in this analysis is rather a crude model for an actual arteriosclerotic blood vessel. Nevertheless, the results of this study throw some light on the qualitative features of blood flow in constricted arteries and would serve as an important step towards understanding of the magnetohydrodynamic flow of blood in an actual arteriosclerotic artery.

References:

1. Altobelli, S. A. and Nerem, R. M. : Trans. ASME J. Biomech. Engng. **107**, 16-23 (1985).
2. Andersson, H. I. : Acta Mech. **95**, 227-230 (1992).
3. Andersson, H. I., Bech, K. H. and Dandapat, B. S. : Int. J. Non-Linear Mech. **27**, 929-936 (1992).
4. Asakura, T. and Karino, T. : Circ. Res. **66**, 1045-1066 (1990).
5. Atabek, H. B. and Chang, C. C. : Z. Angew. Math. Phys. **12**, 185-201 (1961).
6. Back, L. D., Radbill, J. R. and Crawford, D. W. : J. Biomech. **10**, 339-353, (1977).
7. Belardinelli, E., Ursino, M. and Lemmi, E. : Trans. ASME J. Biomech. Engng. **111**, 233-240 (1989).
8. Belousova, L. Ye. : Biofizika, **10**, 365 (1965).
9. Bestman, A. R. : Acta Mech. **44**, 107-119 (1982).
10. Bitoun, J. P. and Bellet, D. : Biorheol. **23**, 51-61 (1986).
11. Brady, J. F. and Acrivos, A. : J. Fluid Mech. **112**, 127-150 (1981).
12. Caro, C. R., Pedley, T. J., Schroter, R. C. and Seed, W. A. : The Mechanics of the Circulation, Oxford University Press (1978).
13. Chakravarty, S. and Sannigrahi, A. K. : Int. J. Engg. Sci. **36**, 1083-1102 (1998).
14. Chandran, K. B., Yearwood, T. L. and Wietnig, D. W. : J. Biomech. **12**, 793-805 (1979).
15. Chang, L. J. and Tarbell, J. M. : J. Biomech. **21**, 927-937 (1988).

16. Chang, L. J. and Tarbell, J. M. : J. Fluid Mech. **161**, 175-198 (1985).

17. Charara, J., Aurengo, A., Lelievre, J. C. and Lacombe, C. : Biorheology **22**, 509-520 (1985).

18. Chaturani, P. and Palanisami, V. : Biorheology **27**, 619-630 (1990a).

19. Chaturani, P. and Palanisami, V. : Biorheology **27**, 747-758 (1990b).

20. Chien, S., Usami, S. and Skalak, R.: Blood Flow in Small Tubes. In Handbook of Physiology : The Cardiovascular System, IV, Renkin, E. M. and Michel, C. C. (eds.), Md. Am. Physiol. Soc., 217-249 (1984).

21. Chien, S.: Blood Rheology. In : Quantitative Cardiovascular Studies : Clinical and Research Applications of Engineering Principles, Hwang, N. H. C. , Gross, D. R. and Patel, D. J.(eds.), Baltimore. University Park Press, 241-287 (1979).

22. Cho, Y. I., Back, L. H., Crawford, D. W. and Cuffel, R. F. : J. Biomech. **16**, 933-946 (1983).

23. Chow, J. C. F. and Soda, K. : Phys. Fluids. **15**, 1700-1706 (1972).

24. Cokelet, G. R. : The Rheology of Human Blood. Biomechanics – Its Foundations and Objectives, Fung, Y. C., Perrone, N. and Aniliker, M. (eds.), Prentice Hall, 63-103 (1981).

25. Crane, L. J. : Z. Angew. Math. Phys. **21**, 645-647 (1970).

26. Creff, R., Andre, P. and Batina, J. : Int. J. Num. Methods in Fluids. **5**, 745-760 (1985).

27. Dandapat, B. S. and Gupta, A. S. : Int. J. Non-linear Mech. **24**, 215-219 (1989).

28. Dash, R. K., Jayaraman, G. and Mehta, K. N. : J. Biomech. **32**, 49-61 (1999).

29. Deshpande, M. D., Giddens, D. P. and Mabon, R. F. : J. Biomech. **9**, 165-174 (1976).

30. Dutta, M. and Tarbell, J. M. : Trans. ASME J. Biomech. Engng. **118**, 111-119 (1996).

31. Fukada, E. and Kaibara, M. : Biorheol. **17**, 177-182 (1980).

32. Fukushima, T., Azuma, T. and Matsuzawa, T. : Trans. ASME J. Biomech. Engng. **104**, 143-147 (1982).

33. Fung, Y.C. : Mechanical Properties of Living Tissues, Springer-Verlag (1981).

34. Gupta, A. S. and Pal, A., Misra, J. C., Pal, B. and Takhar, H. S. : Hall effects on magnetohydrodynamic flow and heat transfer over a stretching surface, In : Mathematics and its Application to Industry and New Emerging Areas, 49-62, Special Volume (Ed. S. K. Malik) INSA, New Delhi, 2000.

35. Gupta, P. S. and Gupta, A. S. : Can. J. Chem. Engng. **55**, 744-746 (1977).

36. Hall, P. : J. Fluid Mech. **64**, 209-226 (1974).

37. He, X. and Ku, D. N. : Trans. ASME J. Biomech. Engng. **116**, 355-360 (1994).

38. Hochmuth, R. H. : Properties of Red Blood Cells. In : Handbook of Bioengineering, Skalak, R. and Chien, S. (eds.), New York. Mcgraw-Hill, 12.1-12.7 (1987).

39. Imeada, K. and Goodman, F. O. : J. Biomech. **13**, 1007-1021 (1980).

40. Jensen, O. E. : Trans. ASME J. Biomech. Engng. **120,** 544-546 (1998).

41. Johnson, G. A., Borovetz, H. S. and Anderson, J. L. : J. Biomech. **25**, 91-100 (1992).

42. Johnson, P.R. and Kilpatrick, D. : J. Biomech. **24**, 1069-1077 (1991).

43. Kaibara, M. and Fuckada, E. : Biorheology **18**, 405-413 (1981).

44. Kamm, R. D. : Flow through Collapsible Tubes. In : Handbook of Bioengineering, Skalak, R. and Chien, S. (eds.) McGraw-Hill, New York. 23.1-23.19 (1987).

45. Karmilov, V.I. (ed.) : Collection : Biological and Therapeutic Action of a Magnetic Field and Strictly Periodic Vibration (Experimental and Clinical Material) (In Russian) (1948).

46. Khalifa, A. M. A. and Giddens, D. P. : J. Biomech. **11**, 129-141 (1978).

47. Korchevskii, E. M. and Marochnik, L. S. : Biofizika, **10**, 371-373 (1965).

48. Krijger, J. K. B., Hillen, B. and Hoogstraten, H. W. : J. Biomech. **22**, 1192-1202 (1989).

49. Krijger, J. K. B., Hillen, B. and Hoogstraten, H. W. : J. Biomech. **25**, 1451-1465 (1992).

50. Krijger, J. K. B., Hillen, B. and Hoogstraten, H. W. : Z. Angew. Math. Phys. **42**, 139-153 (1991).

51. Kumari, M., Takhar, H. S. and Nath, G. : Warmestoffübertrag, **25**, 331-336 (1990).

52. Lee, J. S. and Fung, Y. C. : Trans. ASME J. Appl. Mech. **37**, 9-16 (1970).

53. Lee, T. S. : Trans. ASME J. Biomech. Engng. **116**, 66-71 (1994).

54. Lew, H. S. and Fung, Y. C. : J. Biomech **3**, 23-38 (1970).

55. Liepsch, D. and Moravec, St. : Biorheology **21**, 571-586 (1984).

56. Liepsch, D. W. : Biorhelogy **23**, 395-433 (1986).

57. Ling, S. C. and Atabek, H. B. : J. Fluid Mech. **55**, 492-511 (1972).

58. Lou, Z. and Yang, W. J. : J. Biomech. **26**, 37-49 (1993).

59. Luo, X. Y. and Kuang, Z. B. : J. Biomech. **25**, 929-934 (1992).

60. Luo, X. Y. and Pedley, T. J. : J. Fluid Mech. **363**, 253-280 (1998).

61. Manton, M. J. : J. Fluid Mech. **49**, 451-459 (1971).

62. Mark, F. F., Bargeron, C. B., Deters, O. J. and Friedman, M. H. : Trans. ASME J. Biomech. Engng. **107**, 24-28 (1985).

63. McDonald, D. A. : J. Biomech. **12**, 13-20 (1979).

64. McDonald, D. A. : J. Biomech. **19**, 239-249 (1986).

65. McMichael, J. M. and Deutsch, S. : Phys. Fluids, **27**, 110-118 (1984).

66. Misra, J. C. and Chakravarty, S. : J. Biomech. (Oxford), **19**, 907-916 (1986).

67. Misra, J. C. and Kar, B. K,.: Mathematical and Computer Modelling (UK), **13**, 7-17 (1990).

68. Misra, J. C. and Pal, B. : Mathematical and Computer Modelling (USA), **29**, 89-106 (1999).

69. Misra, J. C. and Sahu, B. K. : Modelling, Simulation and Control, C, **15**, 13-64 (1989).

70. Misra, J. C. and Singh, S. I. : Bull. Math. Biol. **49**, 257-277 (1987).

71. Misra, J. C., Pal, B. and Gupta, A. S. : Mathematical Models and Methods in Applied Sciences, (Italy), **8**, 1323-1342 (1998).

72. Misra, J. C., Pal, B., Pal, A. and Gupta, A. S. : International Journal of Non-linear Mechanics (U.K.), **36**, 731-741 (2001).

73. Misra, J. C., Patra, M. K. and Misra, S. C. : J. Biomech. **26**, 1129-1141 (1993).

74. Morgan, B. E. and Young, D. F. : Bull. Math. Biol. **36**, 39-53 (1974).

75. Morihara, H. and Cheng, R. T. : J. Comp. Phys. **11**, 550-572 (1973).

76. Murata, T. : Biorheology **20**, 471-483 (1983).

77. Nakamura, M. and Sawada, T. : Trans. ASME J. Biomech. Engng. **110**, 137-143 (1988).

78. Nakamura, M. and Sawada, T. : Trans. ASME J. Biomech. Engng. **112**, 100-103 (1990).

79. O' Brien, V., Ehrlich, L. W. and Friedman, M. H. : J. Fluid Mech. **75**, 315-336 (1976).

80. O'Brien, V. and Ehrlich, L. W. : J. Biomech. **18**, 117-127 (1985).

81. Pal, B., Misra, J. C. and Gupta, A. S. : Proc. Nat. Acad. Sci. India. **66(A)**, 247-262 (1996).

82. Pal, B., Misra, J. C., Pal, A. and Gupta, A. S. : Ind. Journal of Mathematics, **41**, 231-247 (1999).

83. Perktold, K., Florian, H. and Hilbert, D. : J. Biomed. Engng. **9**, 46-53 (1987).

84. Perktold, K., Florian, H., Hilbert, D. and Peter, R. O. : J. Biomech. **21**, 663-671 (1988).

85. Perktold, K., Nerem, R. M. and Peter, R. O. : J. Biomech. **24**, 175-189 (1991).

86. Rajagopal, K. R., Na, T. Y. and Gupta, A. S. : Rheol. Acta. **23**, 213-215 (1984).

87. Ram, P. C. and Takhar, H. S. : Int. J. Non-Linear Mech. **29**, 775-780 (1994).

88. Rao, A. R. and Deshikachar, K. S. : J. Indian Inst. Sci. **68**, 247-260 (1988).

89. Ravensbergen, J., Krijger, J. K. B., Hillen, B. and Hoogstraten, H. W. : J. Fluid Mech. **304**, 119-141 (1995).

90. Rindt, C. C. M., Steenhoven, A. A. V., Janssen, J. D. and Reneman, R. S. : J. Biomech. **23**, 461-473 (1990).

91. Rindt, C. C. M., Voose, F. N., Steenhoven, A. A. V. and Janssen, J. D. : J. Biomech. **20**, 499-509 (1987).

92. Sabbah, H. N., Walburn, F. J. and Stein, P. D. : Trans. ASME J. Biomech. Engng. **106**, 272-279 (1984).

93. Schmid-Schoenbein, G. W. : Rheology of Leukocytes. In : Handbook of Bio-engineering, Skalak, R. and Chien, S. (eds.), New York. Mcgraw-Hill, 13.1-13.25 (1987).

94. Secomb, T. W. : J. Fluid Mech. **88**, 273-288 (1978).

95. Shi, Z. D., Winoto, S. H. and Lee, T. S. : Trans. ASME J. Biomech. Engng. **119**, 213-216 (1997).

96. Skalak, R. : Rheology of Red Blood Cell Membrane. In : Microcirculation, Grayson, J. and Zingg, W. (eds.), New York, Plenum. **1**, 53-70 (1976).

97. Stoltz, J. F. and Lucius, M. : Biorheol. **18,** 453-473 (1981).

98. Stoltz, J. F., Gaillard, S. and Lucius, M. : J. Biomech. **13**, 341-346 (1980).

99. Sud, V. K. and Sekhon, G. S. : Bull. Math. Biol. **47**, 35-52 (1985).

100. Sud, V. K. and Sekhon, G. S. : J. Biomech. **19**, 929-941 (1986).

101. Sud, V. K. and Sekhon, G. S. : J. Biomech. **23**, 513-527 (1990).

102. Sud, V. K., von Gierks, H. E., Kaleps, I. and Oestreicher, H. L. : Med. Biol. Engng. Comput. **21**, 446-452 (1983).

103. Sung, H. W. and Yoganathan, A. P. : J. Biomech. **23**, 201-214 (1990).

104. Taylor, T. W. and Yamaguchi, T. : Trans. ASME J. Biomech. Engng. **116**, 89-97 (1994).

105. Thurston, G. B. : Biophysical J. **12**, 1205-1217 (1972).

106. Thurston, G. B. : Significance and Methods of Measurements of Viscoelastic Behaviour of Blood. In : The Rheology of Blood, Blood Vessels and Associated Tissues, Gross, D. R. and Hwang, N. H. C. (eds.), Alphen aan den Sijthoff, Noordhoff, 236-256 (1981).

107. Tu, C. and Deville, M. : J. Biomech. **29**, 899-908 (1996).

108. Tu, C., Deville, M., Dheur, L. and Vanderschuren, L. : J. Biomech. **25**, 1141-1152 (1992).

109. Tutty, O. R. : Trans. ASME J. Biomech. Engng. **114**, 50-54 (1992).

110. Vardanyan, V. A. : Biofizika, **18**, 491-496 (1973).

111. Wang, Y. C. and Longwell, P. A. : AIChE J. **10**, 323 (1964).

112. Watanabe, T. and Pop, I. : Acta Mech. **108**, 35-47 (1995).

113. Wild, R., Pedley, T. J. and Riley, D. S. : J. Fluid Mech. **81**, 273-291 (1977).

114. Young, D. F. : Trans. ASME J. Biomech. Engng. **101**, 157-175 (1979).

115. Young, D. F. : Trans. ASME J. Engng. Ind. **90**, 248-254 (1968).

Modern Applied Mathematics
Editor: J.C. Misra

Methods of Constrained and Unconstrained Approximation for Mappings in Probability Spaces*

Anatoli Torokhti[1,2], Phil Howlett[1] and Charles Pearce[2]

[1]Centre for Industrial and Applicable Mathematics, University of South Australia, SA 5093, Australia
[2]Applied Mathematics Department, University of Adelaide, SA 5005, Australia

Abstract

In this chapter we propose new approaches to the best constructive approximation of nonlinear operators in probability spaces. It is assumed that the only available information on the operator is given by certain covariance matrices. The proposed approaches are based on the approximant produced by a polynomial operator. The approximant minimizes the mean-squared error between a desired image and the image of the approximating operator.

Keywords: Best approximation; nonlinear operator; signal processing

1 Introduction

A general theory for best approximation in normed linear spaces has been developed for a long time and a number of deep theoretical results established relating to the existence, uniqueness and characterization of best-approximation elements (see, for example, [16], [20] and the references therein). However the theory of best *con-*

*This work was supported by the Australian Research Council under the ARC Large Grants Scheme.

structive approximation of nonlinear operators is rather less developed and pioneering achievements in this area, such as those in [3], are very recent.

The problem of the constructive approximation of nonlinear operators has been a topic of intensive research for some decades. In 1970 Prenter [15] generalized the Weierstrass theorem to nonlinear operator approximation. Since then, a number of fundamental papers have appeared establishing further significant advances in the approximation of nonlinear operators. Some relevant references can be found in [2], [5]–[10], [12], [15], [17]–[19], [21]–[29].

Known results [2], [7]–[10], [12], [15], [17]–[19], [21]–[25] for the constructive approximation of nonlinear operators *with any preassigned accuracy* have been concerned mainly with proving the existence and uniqueness of approximating operators and with justifying error bounds arising from approximation methods. The assumptions used are that the spaces of images and preimages are deterministic and that elements of these spaces can be represented in an analytical form, that is, by equations.

It is well–known that the problem of constructive operator approximations arises in various applications, in particular, the modelling of nonlinear systems. A number of relevant papers, including work by Weierstrass, Volterra and Wiener, can be found in the bibliographies to [5]–[10], [17]–[19], [21]–[29]. In many applications, the spaces of images and preimages are probabilistic and their elements cannot be expressed analytically. Nevertheless, it is possible to describe the elements of these spaces in terms of their numerical characteristics, such as estimates of mathematical expectation and covariance matrices. A typical example is image processing where a digitized image, represented by a matrix, is often interpreted as a sample of a stochastic signal.

In this chapter we propose new approaches to the best constructive approximation of nonlinear operators in probability spaces. It is assumed that the only available information about the operator is given by certain covariance matrices. The proposed methods are based on the approximant produced by a polynomial operator. The approximant minimizes the mean–squared error between a desired image and the image of the approximating operator. Standard software packages, such as MATLAB, can easily be used to implement the methods.

The problem of the best unconstrained approximation of nonlinear operators is considered in Section 2. The extension of this problem and its solution to the case of the best successive approximation is given in Section 3. In Section 4 we consider the best constrained approximation for a random vector. The theoretical results obtained are illustrated by applications to the modelling of signal and image processing systems.

2 Best unconstrained approximation of mappings in probability spaces

2.1 Statement of the problem

Let (Ω, Σ, μ) be a probability space, where $\Omega = \{\omega\}$ is the set of outcomes, Σ a σ–field of measurable subsets of Ω and $\mu : \Sigma \mapsto [0, 1]$ an associated probability measure on Σ with $\mu(\Omega) = 1$. Suppose that $x \in L^2(\Omega, \mathbb{R}^p)$ and $y \in L^2(\Omega, \mathbb{R}^n)$ are random vectors with realisations $x(\omega) \in \mathbb{R}^p$, $y(\omega) \in \mathbb{R}^n$.

Each operator $F : \mathbb{R}^p \to \mathbb{R}^m$ defines an associated operator $\mathcal{F}_F : L^2(\Omega, \mathbb{R}^p) \to L^2(\Omega, \mathbb{R}^m)$ *via* the equation

$$[\mathcal{F}_F(x)](\omega) = F[x(\omega)] \tag{1}$$

for each $\omega \in \Omega$.

We can interpret F as a mathematical model of a nonlinear system transforming input signals x into output signals $F(x)$. As in [21], we interpret x as a given "idealized" input signal (without any distortion) and y as an observed input signal. In particular, y can be interpreted as x contaminated with noise so that no specific relationships between signal and noise are assumed to be known. It is customary to write $F(x)$ rather than $\mathcal{F}_F(x)$, since then we have $[F(x)](\omega) = F[x(\omega)]$ for each $\omega \in \Omega$. It is also convenient to write x instead of $x(\omega)$, y instead of $y(\omega)$, *etc.*

Let $P_r : \mathbb{R}^n \to \mathbb{R}^m$ be the operator associated with the input-output map $\mathcal{P}_{P_r} : L^2(\Omega, \mathbb{R}^n) \to L^2(\Omega, \mathbb{R}^m)$ according to an equation similar to (1), and let P_r be given by

$$P_r(y) = A^{(0)} + \sum_{q=1}^{r} A^{(q)}(y^q). \tag{2}$$

Here $A^{(0)} \in \mathbb{R}^m$, $r \le n$, $A^{(q)} : (\mathbb{R}^n)^q \to \mathbb{R}^m$ is a $q-$linear operator and

$$y^q = \underbrace{(y, \ldots, y)}_{q} \in (\mathbb{R}^n)^q.$$

Thus the operator P_r is completely defined by $A^{(q)}$ $(q = 0, 1, \ldots, r)$. We call P_r a (generalized) polynomial operator of the rth degree. Hereafter the terms "system," "operator" and "input-output map" will be identified.

The problem is to find $\tilde{A}^{(q)}$ $(q = 0, 1, \ldots, r)$ such that

$$J(\tilde{A}^{(0)}, \tilde{A}^{(1)}, \ldots, \tilde{A}^{(r)}) = \min_{A^{(0)}, A^{(1)}, \ldots, A^{(r)}} J(A^{(0)}, A^{(1)}, \ldots, A^{(r)}), \tag{3}$$

where

$$J(A^{(0)}, A^{(1)}, \ldots, A^{(r)}) = E[\|F(x) - P_r(y)\|^2], \tag{4}$$

with E the expectation operator and $\|\cdot\|$ the Frobenius norm.

In other words, we wish to find the mathematical model $\tilde{P}_r$ (defined by $\tilde{A}^{(0)}$, $\tilde{A}^{(1)}$, ..., $\tilde{A}^{(r)}$) of the system F which is the best in the sense of (3) in the class of (generalized) polynomial operators (2).

The problem considered in [21] is a particular case of (2)–(4) when $r = 2$ in (2) and $A^{(2)}(y^2) = A^{(2)}\tilde{y}$, where $\tilde{y}$ is defined by the Hadamard product, so that $\tilde{y} = (y_1^2, \ldots, y_n^2)^T$, where $y = (y_1, \ldots, y_n)^T \in \mathbb{R}^n$.

2.2 Auxiliary results

We write $z_{j_1,\ldots,j_{q-1}} = y_{j_1} \cdots y_{j_{q-1}} y$, where $j_1, \ldots, j_{q-1} = 1, \ldots, n$ and $q = 2, \ldots, r$. For our purposes, it is more convenient to use a representation of the operator P_r in matrix terms. This is established in the following lemma.

Lemma 1 *There exist matrices* $A^{(q)}_{j_1,\ldots,j_{q-1}} \in \mathbb{R}^{m \times n}$ *such that*

$$P_r(y) = A^{(0)} + A^{(1)}y + \sum_{q=2}^{r} \left[\sum_{q-1 \le \sigma_{q-1} \le (q-1)n} A^{(q)}_{j_1,\ldots,j_{q-1}} z_{j_1,\ldots,j_{q-1}} \right], \tag{5}$$

where $\sigma_{q-1} = j_1 + \ldots + j_{q-1}$ *and the inner sum is extended for all summands with subscripts satisfying the inequality* $q - 1 \le \sigma_{q-1} \le (q-1)n$.

Proof. Let $\{e_1, \ldots, e_n\}$ be the standard basis in $\mathbb{R}^n$. Then

$$\begin{aligned} A^{(q)}(y^q) &= A^{(q)}\left(\sum_{j_1=1}^{n} y_{j_1} e_{j_1}, \ldots, \sum_{j_q=1}^{n} y_{j_q} e_{j_q}\right) \\ &= \sum_{j_1=1}^{n} \cdots \sum_{j_q=1}^{n} A^{(q)}(e_{j_1}, \ldots, e_{j_q}) y_{j_1} \ldots, y_{j_q}. \end{aligned}$$

If we write

$$A^{(q)}(e_{j_1}, \ldots, e_{j_q}) = a^{(q)}_{j_1,\ldots,j_q} \in \mathbb{R}^m,$$

then we can define the matrix $A^{(q)}_{j_1,\ldots,j_{q-1}} \in \mathbb{R}^{m \times n}$ by the formula

$$A^{(q)}_{j_1,\ldots,j_{q-1}} u = \sum_{j_q=1}^{n} a^{(q)}_{j_1,\ldots,j_q} u_{j_q},$$

where $u = (u_1, \ldots u_n)^T \in \mathbb{R}^n$. Therefore

$$\begin{aligned} A^{(q)}(y^q) &= \sum_{j_1=1}^{n} \cdots \sum_{j_{q-1}=1}^{n} y_{j_1} \cdots y_{j_{q-1}} \left(\sum_{j_q=1}^{n} a^{(q)}_{j_1,\ldots,j_q} y_{j_q}\right) \\ &= \sum_{q-1 \le \sigma_{q-1} \le (q-1)n} A^{(q)}_{j_1,\ldots,j_{q-1}} z_{j_1,\ldots,j_{q-1}} \end{aligned}$$

and the lemma is proved. □

Example 1 *For $r = 2$, formula (5) takes the form*

$$P_2(y) = A^{(0)} + A^{(1)} y + \sum_{j_1=1}^{n} A^{(2)}_{j_1} z_{j_1}, \tag{6}$$

where

$$A^{(2)}_1 = \{a^{(2)}_{i,1,j_2}\}_{i,j_2=1}^{i=m,j_2=n} \in \mathbb{R}^{m \times n}, \ldots, A^{(2)}_n = \{a^{(2)}_{i,n,j_2}\}_{i,j_2=1}^{i=m,j_2=n} \in \mathbb{R}^{m \times n}$$

or

$$A^{(2)}_1 = \{a^{(2)}_{i,j_1,1}\}_{i,j_1=1}^{i=m,j_1=n} \in \mathbb{R}^{m \times n}, \ldots, A^{(2)}_n = \{a^{(2)}_{i,j_1,n}\}_{i,j_1=1}^{i=m,j_1=n} \in \mathbb{R}^{m \times n}$$

and $a^{(2)}_{i,j_1,j_2}$ are entries of the $m \times n \times n$ tensor $A^{(2)}$.

Let us now reduce (5) to a more compact form similar to (6) but for $r > 2$.

If we consider the matrices $A^{(q)}_{j_1,\ldots,j_{q-1}}$ for each value of $q = 2, \ldots r$ and each combination of $j_1, \ldots, j_{q-1}$, where $j_k = 1, \ldots n$ for $k = 1, \ldots, q-1$, then for each order q we obtain n^{q-1} matrices. There are $N = n + n^2 + \ldots + n^{r-1}$ matrices altogether which we denote by $C_1, \ldots, C_N$ with the corresponding operands denoted by $u_1, \ldots, u_N \in \mathbb{R}^n$. We can then write

$$\sum_{q-1 \le \sigma_{q-1} \le (q-1)n} A^{(q)}_{j_1,\ldots,j_{q-1}} z_{j_1,\ldots,j_{q-1}} = \sum_{j=1}^{N} C_j u_j.$$

Therefore the following is true.

Corollary 1 *The polynomial operator P_r can be written as*

$$P_r(y) = A^{(0)} + A^{(1)} y + \sum_{j=1}^{N} C_j u_j. \qquad (7)$$

The operator P_r is represented diagrammatically in Figure 1.

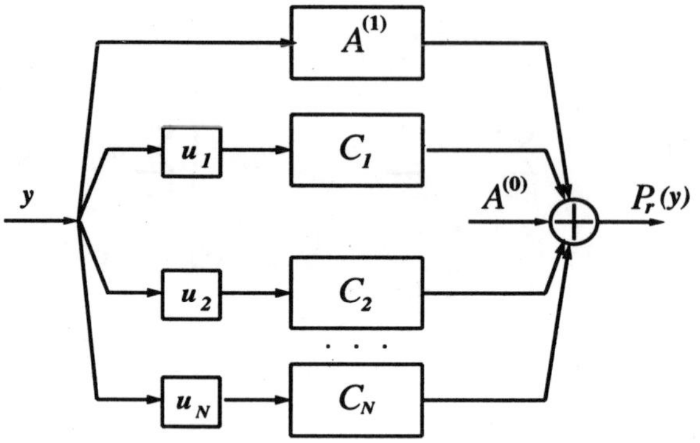

Figure 1: Representation of the operator P_r.

The results in the next lemmas will be used in the proof of Theorems 1 and 2.

We denote by $\mathcal{N}(M)$ and $\mathcal{R}(M)$ the null space and range space respectively of a matrix M and by $M^\dagger$ the Moore-Penrose pseudo-inverse of M. Note (see [14]) that

$$\mathcal{N}(M) = \mathcal{R}(M^T)^\perp \quad \text{and} \quad \mathcal{N}(M^T) = \mathcal{R}(M)^\perp.$$

Lemma 2 *Let $P \in \mathbb{R}^{n \times m}$ and $Q \in \mathbb{R}^{n \times m}$. Then*

$$\mathcal{N}(P) \subseteq \mathcal{N}(Q) \qquad (8)$$

implies

$$QP^\dagger P = Q.$$

Proof. If $q_{\mathcal{N}} \in \mathcal{N}(P)$, then $QP^{\dagger}Pq_{\mathcal{N}} = 0$ and since (8) shows that $q_{\mathcal{N}} \in \mathcal{N}(Q)$, we have also that $Qq_{\mathcal{N}} = 0$. Hence

$$Q(I - P^{\dagger}P)q_{\mathcal{N}} = 0.$$

On the other hand, if $q_{\mathcal{R}} \in \mathcal{N}(P)^{\perp} = \mathcal{R}(P^T)$, then there exists $p \in \mathbb{R}^n$ such that $q_{\mathcal{R}} = P^{\dagger}p$ (see [14]). Hence

$$QP^{\dagger}Pq_{\mathcal{R}} = QP^{\dagger}PP^{\dagger}p = Qq_{\mathcal{R}}$$

and therefore

$$Q(I - P^{\dagger}P)q_{\mathcal{R}} = 0.$$

The desired result follows from the unique representation of any vector $q \in \mathbb{R}^n$ in the form $q = q_{\mathcal{N}} + q_{\mathcal{R}}$. □

For any w, $v \in \mathbb{R}^n$, we denote

$$\mathcal{E}_{wv} = E[wv^T] - E[w]E[v^T].$$

Lemma 3 *We have the equations*

$$\mathcal{E}_{xy}\mathcal{E}_{yy}^{\dagger}\mathcal{E}_{yy} = \mathcal{E}_{xy}, \quad \mathcal{E}_{u_jy}\mathcal{E}_{yy}^{\dagger}\mathcal{E}_{yy} = \mathcal{E}_{u_jy} \quad \text{and} \quad \mathcal{E}_{xu_j}\mathcal{E}_{u_ju_j}^{\dagger}\mathcal{E}_{u_ju_j} = \mathcal{E}_{xu_j}. \tag{9}$$

Proof. Let h and r be random vectors with realizations in $\mathbb{R}^p$ and $\mathbb{R}^n$, respectively. If $u \in \mathcal{N}(E[rr^T])$, then

$$u^T E[rr^T]u = 0$$

and hence

$$E[(r^T u)^2] = 0.$$

But for each $w \in \mathbb{R}^p$ we have

$$\begin{aligned} |w^T E[hr^T]u| &= |E[(w^T h)(r^T u)]| \\ &\le (E[(w^T h)^2])^{1/2}(E[(r^T u)^2])^{1/2} \\ &= 0. \end{aligned}$$

Therefore $E[hr^T]u = 0$ and hence $u \in \mathcal{N}(E[hr^T])$. This means that

$$\mathcal{N}(E[rr^T]) \subseteq \mathcal{N}(E[hr^T]). \tag{10}$$

Now let $h = \tilde{x}$ and $r = \tilde{y}$, where

$$\tilde{x} = x - E[x] \quad \text{and} \quad \tilde{y} = y - E[y].$$

Then

$$\mathcal{E}_{xy} = E[hr^T] \quad \text{and} \quad \mathcal{E}_{yy} = E[rr^T],$$

and by Lemma 1 the first equality in (9) follows from (10). Two other equalities in (9) are proved similarly. □

Lemma 4 *Let* $F(x) = v$,

$$D = \begin{bmatrix} D_{11} & \dots & D_{1N} \\ D_{21} & \dots & D_{2N} \\ \dots & \dots & \dots \\ D_{N1} & \dots & D_{NN} \end{bmatrix}$$

and

$$G = [G_1 G_2 \dots G_N],$$

where

$$D_{ij} = \mathcal{E}_{u_i u_j} - \mathcal{E}_{u_i y}\mathcal{E}_{yy}^{\dagger}\mathcal{E}_{yu_j} \quad \text{and} \quad G_j = \mathcal{E}_{vu_j} - \mathcal{E}_{vy}\mathcal{E}_{yy}^{\dagger}\mathcal{E}_{yu_j}$$

for $i, j = 1, 2, \dots N$. *Then*

$$GD^{\dagger}D = G. \tag{11}$$

Proof. We write $u = (u_1^T, \dots, u_N^T)^T$. The matrices D and G can be written as

$$D = \mathcal{E}_{uu} - \mathcal{E}_{uy}\mathcal{E}_{yy}^{\dagger}\mathcal{E}_{yu}$$

and

$$G = \mathcal{E}_{vu} - \mathcal{E}_{vy}\mathcal{E}_{yy}^{\dagger}\mathcal{E}_{yu}.$$

Then

$$D = E[(\tilde{u} - \mathcal{E}_{uy}\mathcal{E}_{yy}^{\dagger}\tilde{y})(\tilde{u}^T - \tilde{y}^T\mathcal{E}_{yy}^{\dagger}\mathcal{E}_{yu})]$$

and

$$G = E[(\tilde{v} - \mathcal{E}_{vy}\mathcal{E}_{yy}^{\dagger}\tilde{y})(\tilde{u}^T - \tilde{y}^T\mathcal{E}_{yy}^{\dagger}\mathcal{E}_{yu})].$$

Let $w \in \mathcal{N}(D)$, where $w \in \mathbb{R}^N$. Then $w^T D w = 0$ and

$$w^T E[(\tilde{u} - \mathcal{E}_{vy}\mathcal{E}_{yy}^\dagger \tilde{y})(\tilde{u}^T - \tilde{y}^T \mathcal{E}_{yy}^\dagger \mathcal{E}_{yu})]w = 0.$$

Therefore

$$E[\{(\tilde{u} - \mathcal{E}_{uy}\mathcal{E}_{yy}^\dagger \tilde{y})^T w\}^2] = 0.$$

Next, for all $z \in \mathbb{R}^m$ we have

$$\begin{aligned}(z^T G w)^2 &= (E[z^T(\tilde{v} - \mathcal{E}_{vy}\mathcal{E}_{yy}^\dagger \tilde{y})(\tilde{u}^T - \tilde{y}^T \mathcal{E}_{yy}^\dagger \mathcal{E}_{yu})w])^2 \\ &\le E[\{z^T(\tilde{v} - \mathcal{E}_{vy}\mathcal{E}_{yy}^\dagger \tilde{y})\}^2] E[\{(\tilde{u} - \mathcal{E}_{uy}\mathcal{E}_{yy}^\dagger \tilde{y})^T w\}^2] \\ &= 0,\end{aligned}$$

that is,

$$Gw = 0 \quad \text{or} \quad w \in \mathcal{N}(G).$$

Hence $\mathcal{N}(D) \subseteq \mathcal{N}(G)$ and (11) follows from Lemma 1. □

2.3 Main results

2.3.1 General solution

By Corollary 1, the problem formulated in Section 2.1 reduces to finding $\tilde{A}^{(0)} \in \mathbb{R}^m$, $\tilde{A}^{(1)} \in \mathbb{R}^{m\times n}$ and $\tilde{C}_j \in \mathbb{R}^{m\times n}$ $(j = 1, 2, \ldots, N)$ which minimize the functional (4), where $P_r(y)$ is of the form (7). Thus

$$J(\tilde{A}^{(0)}, \tilde{A}^{(1)}, \tilde{C}_1, \ldots, \tilde{C}_N) = \min_{A^{(0)}, A^{(1)}, C_1, \ldots, C_N} J(A^{(0)}, A^{(1)}, C_1, \ldots, C_N) \tag{12}$$

where

$$J(A^{(0)}, A^{(1)}, C_1, \ldots, C_N) = E\left[\left\| F(x) - \left(A^{(0)} + A^{(1)}y + \sum_{j=1}^{N} C_j u_j\right)\right\|^2\right]. \tag{13}$$

As before, we write $F(x) = v$.

Theorem 1 *The solution to the problem (12), (13) is given by*

$$\tilde{A}^{(0)} = E[v] - \tilde{A}^{(1)}E[y] - \sum_{j=1}^{N} \tilde{C}_j E[u_j], \tag{14}$$

$$\tilde{A}^{(1)} = \left(\mathcal{E}_{vy} - \sum_{j=1}^{N} \tilde{C}_j \mathcal{E}_{u_j y}\right) \mathcal{E}_{yy}^{\dagger} + K_1 \left[I - \mathcal{E}_{yy}^{1/2} (\mathcal{E}_{yy}^{1/2})^{\dagger}\right], \tag{15}$$

$$[\tilde{C}_1 \, \tilde{C}_2 \, \ldots \, \tilde{C}_N] = G D^{\dagger} + K_2 [I - D^{1/2} (D^{1/2})^{\dagger}], \tag{16}$$

where K_1 and K_2 are arbitrary matrices.

Proof. Note that each matrix $\tilde{C}_j \in \mathbb{R}^{m \times n}$ in (14)–(16) is defined as a correponding $m \times n$ submatrix of the matrix $GD^{\dagger} + K_2[I - D^{1/2}(D^{1/2})^{\dagger}]$.

We write

$$D^{\dagger} = \begin{bmatrix} Q_{11} & \ldots & Q_{1N} \\ Q_{21} & \ldots & Q_{2N} \\ \ldots & \ldots & \ldots \\ Q_{N1} & \ldots & Q_{NN} \end{bmatrix},$$

where $Q_{ij} \in \mathbb{R}^{n \times n}$ for $i, j = 1, \ldots, N$.

Let us show that $J(A^{(0)}, A^{(1)}, C_1, \ldots, C_N)$, given by (13), can be represented as

$$\begin{aligned} J(A^{(0)}, A^{(1)}, C_1, \ldots, C_N) &= J_0 + J_1(A^{(0)}, A^{(1)}, C_1, \ldots, C_N) \\ &\quad + J_2(A^{(1)}, C_1, \ldots, C_N) + J_3(C_1, \ldots, C_N), \end{aligned} \tag{17}$$

where

$$J_0 = \operatorname{tr}\{\mathcal{E}_{vv}\} - \|\mathcal{E}_{vy}(\mathcal{E}_{yy}^{\dagger})^{1/2}\|^2 - \sum_{i=1}^{N} \|G_i Q_{ii}^{1/2}\|^2 - \sum_{j,k=1,\ldots,N,\; j \neq k} \operatorname{tr}\{G_j Q_{jk} G_k^T\},$$

$$J_1(A^{(0)}, A^{(1)}, C_1, \ldots, C_N) = \left\| A^{(0)} - E[v] + A^{(1)} E[y] + \sum_{j=1}^{N} C_j E[u_j] \right\|^2, \tag{18}$$

$$J_2(A^{(1)}, C_1, \ldots, C_N) = \left\| \left[A^{(1)} - \left(\mathcal{E}_{vy} - \sum_{j=1}^{N} C_j \mathcal{E}_{u_j y} \right) E_{yy}^{\dagger} \right] \mathcal{E}_{yy}^{1/2} \right\|^2 \tag{19}$$

and

$$J_3(C_1, \ldots, C_N) = \|([C_1, \ldots, C_N] - GD^{\dagger}) D^{1/2}\|^2. \tag{20}$$

We have

$$J_0 = \operatorname{tr}\left\{ \mathcal{E}_{vv} - \mathcal{E}_{vy} \mathcal{E}_{yy}^{\dagger} \mathcal{E}_{yv} - \sum_{j,k=1,\ldots,N} C_j Q_{jk} C_k^T \right\}, \tag{21}$$

$$
\begin{aligned}
&J_1(A^{(0)}, A^{(1)}, C_1, \ldots, C_N) \\
&= \operatorname{tr}\left\{A^{(0)}A^{(0)T} - A^{(0)}E[v^T] + A^{(0)}E[y^T]A^{(1)}\right. \\
&\quad + A^{(0)}\sum_{k=1}^{N} E[u_k]C_k^T - E[v^T]A^{(0)} + E[v]E[v^T] - E[v]E[y^T]A^{(1)T} \\
&\quad + (A^{(1)}E[y] - E[v])\sum_{k=1}^{N} E[u_k^T]C_k^T + A^{(1)}E[y]A^{(0)T} - A^{(1)}E[y]E[v^T] \\
&\quad + A^{(1)}E[y]E[y^T]A^{(1)T} + \sum_{k=1}^{N} C_k E[u_k]\left(A^{(0)T} - E[v^T] + E[y^T]A^{(1)T}\right) \\
&\quad \left. + \sum_{j,k=1,\ldots,N} C_j E[u_j]E[u_k]C_k^T\right\},
\end{aligned} \tag{22}
$$

and

$$
\begin{aligned}
&J_2(A^{(1)}, C_1, \ldots, C_N) + J_3(C_1, \ldots, C_N) \\
&= \operatorname{tr}\left\{A^{(1)}\mathcal{E}_{yy}A^{(1)T} - A^{(1)}\left(\mathcal{E}_{yv} - \sum_{j=1}^{N}\mathcal{E}_{yu_j}C_j^T\right) - \mathcal{E}_{vy}A^{(1)T}\right. \\
&\quad + \sum_{j=1}^{N} C_j\mathcal{E}_{u_jy}A^{(1)T} + \mathcal{E}_{vy}\mathcal{E}_{yy}^{\dagger}\mathcal{E}_{yv} + \sum_{j,k=1,\ldots,N} C_j\mathcal{E}_{u_ju_k}C_k^T \\
&\quad \left. - \sum_{j=1}^{N} C_j\mathcal{E}_{u_jv} - \sum_{j=1}^{N}\mathcal{E}_{vu_j}C_j^T + \sum_{j,k=1,\ldots,N} C_jQ_{jk}C_k^T\right\}.
\end{aligned} \tag{23}
$$

Using Lemmas 2–4, we may combine (21)–(23) to obtain

$$
\begin{aligned}
&J_0 + J_1(A^{(0)}, A^{(1)}, C_1, \ldots, C_N) + J_2(A^{(1)}, C_1, \ldots, C_N) + J_3(C_1, \ldots, C_N) \\
&= \operatorname{tr}\left\{E[vv^T] - E[v]A^{(0)T} - E[vy^T]A^{(1)T} - \sum_{k=1}^{N} E[vu_k^T]C_k^T - A^{(0)}E[v^T]\right. \\
&\quad + A^{(0)}A^{(0)T} + A^{(0)}E[y^T]A^{(1)T} + A^{(0)}\sum_{k=1}^{N} E[u_k^T]C_k^T - A^{(1)}E[yv^T] + A^{(1)}E[y]A^{(0)T} \\
&\quad + A^{(1)}E[yy^T]A^{(1)T} + A^{(1)}\sum_{k=1}^{N} E[yu_k^T]C_k^T - \sum_{k=1}^{N} C_kE[u_kv^T] + \sum_{k=1}^{N} C_kE[u_k]A^{(0)T} \\
&\quad \left. + \sum_{k=1}^{N} C_kE[u_ky^T]A^{(1)T} + \sum_{j,k=1,\ldots,N} C_jE[u_ju_k^T]C_k^T\right\} \\
&= J(A^{(0)}, A^{(1)}, C_1, \ldots, C_N),
\end{aligned}
$$

giving (17).

It follows from (17)–(20) that $J(A^{(0)}, A^{(1)}, C_1, \ldots, C_N)$ is minimized when

$$A^{(0)} = \tilde{A}^{(0)},$$

$$A^{(1)}\mathcal{E}_{yy}^{1/2} = \left(\mathcal{E}_{vy} - \sum_{j=1}^{N} C_j \mathcal{E}_{u_j y}\right) (\mathcal{E}_{yy}^{1/2})^\dagger, \tag{24}$$

$$[C_1 C_2 \dots C_n] D^{1/2} = G D^\dagger D^{1/2}, \tag{25}$$

since

$$\mathcal{E}_{yy}^\dagger \mathcal{E}_{yy}^{1/2} = ([\mathcal{E}_{yy}^{1/2}]^T \mathcal{E}_{yy}^{1/2})^\dagger [\mathcal{E}_{yy}^{1/2}]^T = (\mathcal{E}_{yy}^{1/2})^\dagger = (\mathcal{E}_{yy}^\dagger)^{1/2}.$$

From Lemmas 2 and 4, the necessary and sufficient conditions given in [1] for (24) and (25) to have solutions are clearly satisfied and the corresponding solutions are $A^{(1)} = \tilde{A}^{(1)}$ and $[C_1\ C_2\ \dots\ C_N] = [\tilde{C}_1\ \tilde{C}_2\ \dots\ \tilde{C}_N]$. □

Possible and natural choices for K_1 and K_2 in (15) and (16) are $K_1 = \mathbb{O}$ and $K_2 = \mathbb{O}$, where as throughout this chapter $\mathbb{O}$ denotes the zero matrix.

We note also that the best polynomial operator $\tilde{P}_r$ defined by (14)–(16) requires knowledge of the matrices $E[v]$, $E[y]$, $E[u_k]$, $\mathcal{E}_{vy}$, $\mathcal{E}_{vu_k}$, $\mathcal{E}_{yy}$, $\mathcal{E}_{yu_k}$ and $\mathcal{E}_{u_j u_k}$. Methods for the estimation of these can be found, for example, in [13] and [30].

Theorem 2 *The error of approximation by the best polynomial operator $\tilde{P}_r$ defined by (14)–(16) is*

$$E[\|F(x) - \tilde{P}_r(y)\|^2]$$
$$= \operatorname{tr}\{\mathcal{E}_{vv}\} - \|\mathcal{E}_{vy}(\mathcal{E}_{yy}^\dagger)^{1/2}\|^2 - \sum_{i=1}^{N} \|G_i Q_{ii}^{1/2}\|^2 - \sum_{j,k=1,\dots,N,\ j\neq k} \operatorname{tr}\{G_j Q_{jk} G_k^T\}.$$

Proof. The proof follows directly from (14)–(17). □

Comparison of this expression with the error associated with the method in [21] (see Theorem 2 in [21]) demonstrates a clear advantage for our proposed method.

2.3.2 Alternative solution

By Theorem 1, the matrices $A_2^0, \dots, A_{n+1}^0$ are determined with a computation of the pseudo-inverse $D^\dagger$ which is normally realized by a singular–value decomposition ([4, page 256]). This procedure can encounter computational difficulties when n is large. In this subsection, we consider a method which reduces the computation of $D^\dagger$ to computations of pseudo–inverses for smaller matrices.

To this end, we note that

$$[C_1C_2\dots C_N]D^{1/2}D^{1/2}(D^{1/2})^{\dagger} = [C_1C_2\dots C_N]DD^{1/2} = [C_1C_2\dots C_N]D^{1/2}$$

and therefore (25) can equivalently be represented as

$$G = [C_1C_2\dots C_N]D + K[I - (D^{1/2})^{\dagger}D^{1/2}], \tag{26}$$

where $K \in \mathbb{R}^{m\times n}$ is arbitrary. Let us rewrite (26) in the form

$$D\mathcal{A} = \mathcal{G}, \tag{27}$$

where

$$\mathcal{A} = [C_1C_2\dots C_N]^T \quad \text{and} \quad \mathcal{G} = G^T + [I - D^{1/2}(D^{1/2})^{\dagger}]K^T.$$

We note that D is symmetric.

By Lemma 3, the system of matrix equations (27) is consistent. A device for its solution is as follows.

We partition D into $m_i \times n_j$ submatrices $\mathcal{D}_{ij}$ so that

$$D = \begin{bmatrix} \mathcal{D}_{11} & \dots & \mathcal{D}_{1\mu} \\ \mathcal{D}_{21} & \dots & \mathcal{D}_{2\mu} \\ \dots & \dots & \dots \\ \mathcal{D}_{p1} & \dots & \mathcal{D}_{\mu\mu} \end{bmatrix},$$

where $i = 1,\dots,\mu$, $j = 1,\dots,\mu$, $m_1 + \dots + m_\mu = N$ and $n_1 + \dots + n_\mu = N$. Next, let

$$\mathcal{N}_k = I_N - W_k Z_k^T,$$

where I_N is the $N \times N$ identity matrix,

$$W_k^T = [W_{1k}^T, W_{2k}^T, \dots, W_{\mu k}^T],$$

$$Z_k^T = [\mathbb{O}, \dots, \mathbb{O}, I, \mathbb{O}, \dots, \mathbb{O}]$$

with $W_{1k}, W_{2k}, \dots, W_{\mu k} \in \mathbb{R}^{m_i\times m_i}$, $\mathbb{O}$ is the $m_i \times m_i$ zero matrix and I is the $m_i \times m_i$ identity matrix which is treated as the k-th component of Z_k.

We say that $\mathcal{N}_k$ is a transform matrix of index k if

$$W_k^T = [\mathbb{O}, \dots, \mathbb{O}, W_{k+1,k}^T, W_{k+2,k}^T, \dots, W_{\mu k}^T]. \tag{28}$$

Then $\mathcal{N}_k$ has the form

$$\mathcal{N}_k = \begin{bmatrix} I & \mathbb{O} & \dots & \mathbb{O} & \dots & \mathbb{O} \\ \mathbb{O} & I & \dots & \mathbb{O} & \dots & \mathbb{O} \\ \dots & \dots & \dots & \dots & \dots & \dots \\ \mathbb{O} & \mathbb{O} & \dots & I & \dots & \mathbb{O} \\ \mathbb{O} & \mathbb{O} & \dots & -W_{k+1,k} & \dots & \mathbb{O} \\ \dots & \dots & \dots & \dots & \dots & \dots \\ \mathbb{O} & \mathbb{O} & \dots & -W_{\mu k} & \dots & I \end{bmatrix},$$

that is, $\mathcal{N}_k$ is block–identity lower–triangular.

Lemma 5 *Let $V_k = [V_{1k}^T, V_{2k}^T, \dots, V_{\mu k}^T]^T$ be a matrix such that*

$$V_{jk} V_{kk}^{\dagger} V_{kk} = V_{jk}$$

for $k = 1, \dots, \mu$ and $j = k+1, \dots, \mu$, where $V_{jk} \in \mathbb{R}^{m_i \times n_j}$ for all $i, k = 1, \dots, \mu$. Then there is a unique transform matrix N_k of index k such that

$$\mathcal{N}_k V_k = [V_{1k}^T, \dots, V_{kk}^T, \mathbb{O}, \dots, \mathbb{O}]^T \tag{29}$$

with matrices $W_{k+1,k}, \dots, W_{nk}$ determined by

$$W_{jk} = V_{jk} V_{kk}^{\dagger} + \mathcal{M}_{jk}(I - V_{kk} V_{kk}^{\dagger}) \quad \textit{for} \quad j = k+1, \dots, \mu, \tag{30}$$

where $\mathcal{M}_{jk} \in \mathbb{R}^{m_i \times m_i}$ is arbitrary.

Proof. Since $\mathcal{N}_k$ is to be of index k, condition (28) must be satisfied. Next,

$$\begin{aligned} \mathcal{N}_k V_k &= (I_N - W_k Z_k^T) V_k \\ &= V - W[\mathbb{O}, \dots, \mathbb{O}, I_n, \mathbb{O}, \dots, \mathbb{O}]V \\ &= [V_{1k}^T \dots V_{kk}^T \dots V_{\mu k}^T]^T - [W_{1k}^T \dots W_{kk}^T \dots W_{\mu k}^T]^T V_{kk}. \end{aligned}$$

The last $n - k$ components of $\mathcal{N}_k V$ are to be zero matrices, so we must have

$$W_{jk} V_{kk} = V_{jk} \quad \text{for} \quad j = k+1, \dots, \mu.$$

By assumption, the necessary and sufficient condition $V_{jk} V_{kk}^{\dagger} V_{kk} = V_{jk}$ for this equation to have a solution is satisfied and therefore the solution is given by (30) [1].

Hence $\mathcal{N}_k$ exists and uniquely defined by (28) and (30). On the other hand, if W_k is determined by (28) and (30), then the matrix $\mathcal{N}_k = I_N - W_k Z_k^T$ satisfies (29). $\square$

Let us now consider a matrix $D^{(s)}$ defined recursively for $s = 1, 2, \ldots, \mu - 1$ by

$$D^{(s)} = \mathcal{N}_s D^{(s-1)}, \tag{31}$$

where

$$D^{(s-1)} = \begin{bmatrix} \mathcal{D}_{11} & \mathcal{D}_{12} & \mathcal{D}_{13} & \ldots & \mathcal{D}_{1,s-2} & \mathcal{D}_{1,s-1} & \mathcal{D}_{1,s} & \ldots & \mathcal{D}_{1\mu} \\ \mathbb{O} & \mathcal{D}_{22}^{(1)} & \mathcal{D}_{23}^{(1)} & \ldots & \mathcal{D}_{2,s-2}^{(1)} & \mathcal{D}_{2,s-1}^{(1)} & \mathcal{D}_{2,s}^{(1)} & \ldots & \mathcal{D}_{2\mu}^{(1)} \\ \mathbb{O} & \mathbb{O} & \mathcal{D}_{33}^{(2)} & \ldots & \mathcal{D}_{3,s-2}^{(2)} & \mathcal{D}_{3,s-1}^{(2)} & \mathcal{D}_{3,s}^{(2)} & \ldots & \mathcal{D}_{3\mu}^{(2)} \\ \ldots & \ldots & \ldots & \ldots & \ldots & \ldots & \ldots & \ldots & \ldots \\ \mathbb{O} & \mathbb{O} & \mathbb{O} & \ldots & \mathbb{O} & \mathcal{D}_{s-1,s-1}^{(s-2)} & \mathcal{D}_{s-1,s}^{(s-2)} & \ldots & \mathcal{D}_{s-1,\mu}^{(s-2)} \\ \mathbb{O} & \mathbb{O} & \mathbb{O} & \ldots & \mathbb{O} & \mathbb{O} & \mathcal{D}_{s,s}^{(s-1)} & \ldots & \mathcal{D}_{s,\mu}^{(s-1)} \\ \ldots & \ldots & \ldots & \ldots & \ldots & \ldots & \ldots & \ldots & \ldots \\ \mathbb{O} & \mathbb{O} & \mathbb{O} & \ldots & \mathbb{O} & \mathbb{O} & \mathcal{D}_{\mu,s}^{(s-1)} & \ldots & \mathcal{D}_{\mu,\mu}^{(s-1)} \end{bmatrix},$$

$D^{(0)} = D$ and

$$W_{\mu s} = \mathcal{D}_{\mu,s}^{(s-1)}(\mathcal{D}_{s,s}^{(s-1)})^\dagger + \mathcal{M}_{\mu s}(I - \mathcal{D}_{s,s}^{(s-1)}(\mathcal{D}_{s,s}^{(s-1)})^\dagger) \qquad (s = 1, 2, \ldots, \mu - 1).$$

By Lemma 5, it follows from (31) that if

$$\mathcal{D}_{s+1,s}^{(s-1)}(\mathcal{D}_{s,s}^{(s-1)})^\dagger \mathcal{D}_{s,s}^{(s-1)} = \mathcal{D}_{s+1,s}^{(s-1)} \text{ for } s = 1, 2, \ldots, \mu - 1,$$

then $D^{(\mu-1)}$ is block upper–triangular. We write $D^{(\mu-1)} = U$.

It is also clear that the matrix

$$L = \mathcal{N}_1^{-1} \mathcal{N}_2^{-1} \ldots \mathcal{N}_{\mu-1}^{-1},$$

where $\mathcal{N}_k^{-1} = I_N + W Z_k^T$, is *block-identity lower–triangular,* that is,

$$L = \begin{bmatrix} I & \mathbb{O} & \ldots & \mathbb{O} \\ L_{21} & I & \ldots & \mathbb{O} \\ \ldots & \ldots & \ldots & \ldots \\ L_{\mu 1} & L_{\mu 2} & \ldots & I \end{bmatrix}$$

with $L_{ij} \in \mathbb{R}^{m_i \times m_i}$ for all $i, j = 1, \ldots, \mu$. As a result we have

$$D = LU. \tag{32}$$

Next we partition $\mathcal{A}$ and $\mathcal{G}$ so that $\mathcal{A} = [\mathcal{A}_1^T \dots \mathcal{A}_\mu^T]^T$ and $\mathcal{G} = [\mathcal{G}_1^T \dots \mathcal{G}_\mu^T]^T$, where $\mathcal{A}_j \in \mathbb{R}^{n_j \times m}$ and $\mathcal{G}_j \in \mathbb{R}^{n_j \times m}$ for all $j = 1, \dots, \mu$.

The system (27) can be written as $LU\mathcal{A} = \mathcal{G}$. Therefore its solution consists of the following steps. We set

$$U\mathcal{A} = \mathcal{Q}, \tag{33}$$

where $\mathcal{Q} = [\mathcal{Q}_1^T \dots \mathcal{Q}_\mu^T]^T$ with $\mathcal{Q}_j \in \mathbb{R}^{n_j \times m}$ for all $j = 1, \dots, \mu$, and solve $L\mathcal{Q} = \mathcal{G}$ for $\mathcal{Q}_1 \dots \mathcal{Q}_\mu$ by forward substitutions. Next we solve (33) for $\mathcal{A}_1 \dots \mathcal{A}_\mu$ by back substitutions.

The solution to the problem (12), (13) is now given by (14), (15) and by $\tilde{C}_1, \dots, \tilde{C}_N$ determined as submatrices of the solution $[\mathcal{A}_1^T \dots \mathcal{A}_\mu^T]$ obtained from (33).

As a result, instead of a computation of the pseudo-inverse $D^\dagger$ in the general solution given in Theorem 1, we now need to compute pseudo-inverses for smaller $m_i \times n_j$ matrices.

Example 2 *Let $n = m = 10$ and $r = 2$ in (2). Then $D \in \mathbb{R}^{100\times 100}$, $[\tilde{C}_1 \dots \tilde{C}_{10}] \in \mathbb{R}^{10\times 100}$, $G \in \mathbb{R}^{10\times 100}$. Let us set $K_1 = K_2 = \mathcal{M}_{ij} = \mathbb{O}$. We wish to demonstrate that the alternative solution above requires less computational work than a direct computation of the pseudo-inverse matrix (16).*

We computed the pseudo-inverse $D^\dagger$ in (16) with the eigenvalue decomposition of D. The computation of $[\tilde{C}_1 \dots \tilde{C}_{10}]$ by (16) then requires 9×10^6 flops by the software package MATLAB.

To apply the alternative solution above, let us partition D into four submatrices $\mathcal{D}_{ij} \in \mathbb{R}^{50\times 50}$ with $i, j = 1, 2$. Accordingly the matrices $\mathcal{A}$ and $\mathcal{G}$ are also partitioned into submatrices $\mathcal{A}_j \in \mathbb{R}^{50\times 10}$ and $\mathcal{G}_j \in \mathbb{R}^{50\times 10}$, where $j = 1, 2$. Then by (32), (33)

$$\mathcal{A}_1 = (\mathcal{D}_{11} - \mathcal{D}_{12}\mathcal{D}_{22}^\dagger \mathcal{D}_{21})^\dagger (\mathcal{G}_1 - \mathcal{D}_{12}\mathcal{D}_{22}^\dagger \mathcal{G}_2), \tag{34}$$

$$\mathcal{A}_2 = \mathcal{D}_{22}^\dagger (\mathcal{G}_2 - \mathcal{D}_{21}\mathcal{A}_1) \tag{35}$$

and $\tilde{C}_1, \dots, \tilde{C}_{10}$ are the corresponding submatrices of $[\mathcal{A}_1^T \; \mathcal{A}_2^T]$. The computation of $\mathcal{A}_1$ and $\mathcal{A}_2$ by (34), (35), where the pseudo-inverses are computed via the eigenvalue decomposition, requires 0.96×10^6 flops (using MATLAB*).*

Thus (34), (35) requires 9 times less computational work than (16).

(a) Given "idealized" inputs X_{ij}

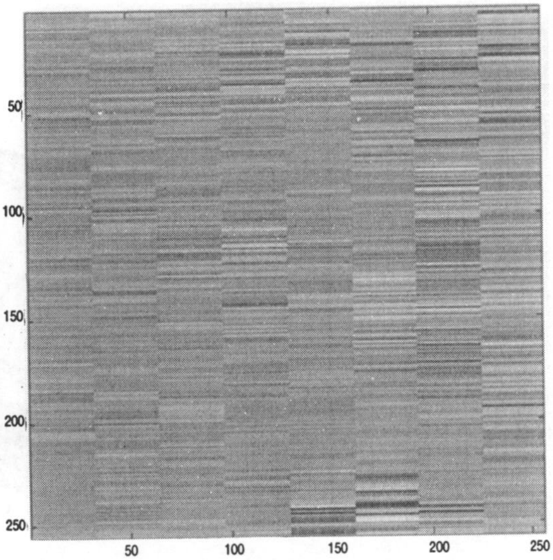

(b) Observed inputs $Y_{ij}^{(1)}$.

(c) Outputs of $\tilde{H}_{ij}^{(k)}$ by [21].

(d) Outputs of $\tilde{P}_{2,ij}^{(1)}$ by (7), (14)-(16).

Figure 2: The performance of the proposed method.

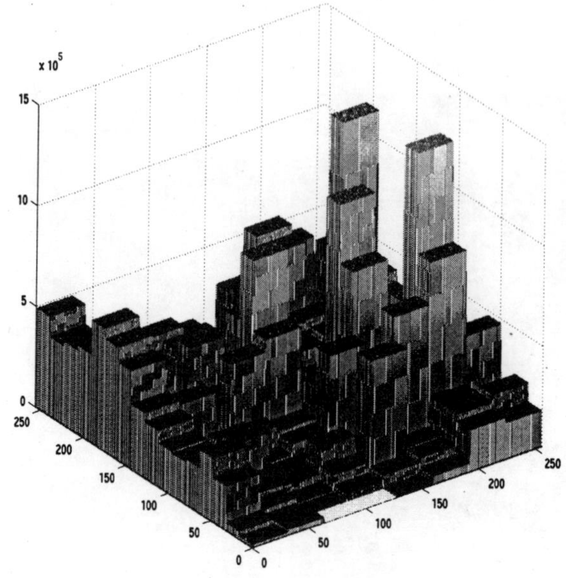

(a) Errors associated with method [21].

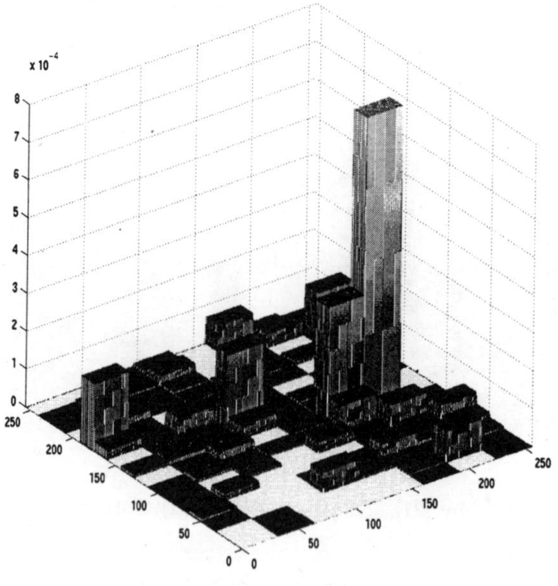

(b) Errors associated with $\tilde{P}^{(1)}_{2,ij}(Y^{(1)}_{ij})$.

Figure 3: The error associated with the proposed method and method [21].

2.4 Simulations

To illustrate the results obtained above, we apply the proposed method to the best simulation of systems transforming digitized images.

Let matrices $X \in \mathbb{R}^{256\times 256}$ and $V \in \mathbb{R}^{256\times 256}$ be counterparts of the image presented in Figure 1 (a) and the known image "Lena," respectively. We partition X and V into 128 submatrices $X_{ij}, V_{ij} \in \mathbb{R}^{16\times 32}$ with $i = 1, \ldots, 16$ and $j = 1, \ldots, 8$ so that $X = \{X_{ij}\}$ and $V = \{V_{ij}\}$. Let $F_{ij} : X_{ij} \to V_{ij}$. Each submatrix X_{ij} has been treated as a set of 32 realizations of a random signal with columns representing realizations. The operator $F_{ij} : X_{ij} \to V_{ij}$ is interpreted as the mathematical model of a nonlinear system, where a column of X_{ij} is the input signal and the corresponding column of V_{ij} the output signal.

TABLE 1.

k	Observed inputs	$\max_{ij} \Delta^{(k)}_{H,ij}$	$\max_{ij} \Delta^{(k)}_{2,ij}$
1	$Y^{(1)}_{ij} = 20\mathcal{R}^{(1)}_{ij} X_{ij} \mathcal{Q}^{(1)}_{ij} + 50\,\mathcal{Q}^{(2)}_{ij}$	1.4074×10^{6}	7.7375×10^{-4}
2	$Y^{(2)}_{ij} = X_{ij}\mathcal{R}^{(2)}_{ij} + 20\,\mathcal{Q}^{(3)}_{ij}$	6.0448×10^{4}	2.4749×10^{-15}
3	$Y^{(3)}_{ij} = \mathcal{Q}^{(4)}_{ij} X_{ij} + X_{ij}\mathcal{R}^{(3)}_{ij}$	2.4671×10^{5}	1.2915×10^{-11}

In practice, input signals are contaminated with noise. We simulated three different types of noisy input signals in the form $Y^{(k)}_{ij}$ with $k = 1, \ldots, 5$ presented in Table 1, where $\mathcal{R}^{(k)}_{ij}$ is a matrix with normally distributed entries with mean 0 and variance 1 and $\mathcal{Q}^{(k)}_{ij}$ a matrix with entries uniformly distributed on the interval $(0, 1)$. In Table 1,

$$\Delta^{(k)}_{H,ij} = \|F_{ij}(X_{ij}) - \tilde{H}^{(k)}_{ij}(Y^{(k)}_{ij})\|^2$$

and

$$\Delta^{(k)}_{2,ij} = \|F_{ij}(X_{ij}) - \tilde{P}^{(k)}_{2,ij}(Y^{(k)}_{ij})\|^2,$$

where $\tilde{H}^{(k)}_{ij}$ is the best approximation for F_{ij} by the method of [21] and $\tilde{P}^{(k)}_{2,ij}$ the best approximation of second degree for F_{ij} constructed from (6), (7), (14)–(16) with $r = 2$.

Here $\tilde{H}^{(k)}_{ij}(Y^{(k)}_{ij})$ and $\tilde{P}^{(k)}_{2,ij}(Y^{(k)}_{ij})$ have been calculated with MATLAB for each k, i and j (that is, the method has been tested 384 times). We put $K_1 = K_2 = \mathbb{O}$ in (15), (16).

Figures (a)–(f) illustrate the performance of the methods for $k = 1$, that is, for $Y^{(1)}_{ij}$ in Table 1. The matrices for the digitized images in Figures (b)–(d) have been composed from submatrices $Y_{ij}^{(1)}$, $\tilde{H}_{ij}^{(1)}(Y_{ij}^{(1)})$ and $\tilde{P}_{2,ij}^{(1)}(Y_{ij}^{(1)})$ correspondingly. The expectations and covariance matrices in (14)–(16) have been estimated from known simple equations [30]. For instance, for each k, i and j we estimated $\mathcal{E}_{vz_q}$ as $V_{ij}(Z_{ij}^{(k)})^T/32 - \hat{V}_{ij}(\hat{Z}_{ij}^{(k)})^T$, where $Z_{ij}^{(k)} = Y_{ij}^{(k)}(\text{diag}Y_{ij}^{(k)}(q,:))$, with $\text{diag}Y_{ij}^{(k)}(q,:))$ is a diagonal matrix with elements from the qth row of $Y_{ij}^{(k)}$ on its diagonal and $\hat{M}$ is a vector formed from means of the rows of a matrix M. These simple estimates have been chosen to illustrate the performance of the proposed method. Special estimates can be found, for example, in [13], [30].

Figures (e), (f) represent the matrices $\{\Delta_{H,ij}^{(k)}\}$ and $\{\Delta_{2,ij}^{(k)}\}$ of errors associated with the best approximations $\tilde{H}_{ij}^{(k)}(Y_{ij})$ by the method of [21] and the best approximations $\tilde{P}_{2,ij}^{(1)}(Y_{ij})$ by (7), (14)–(16) with $r = 2$, respectively.

We see from Table 1 and the figures that the best approximations $\tilde{P}_{2,ij}^{(k)}$ give significant improvements in the accuracy of the F_{ij} approximation compared to the best approximations by [21]: the error associated with the proposed method with $r = 2$ is at least 10^{10} times less than the error associated with the method in [21]. This illustrates Theorem 2.

3 Best successive approximations of mappings in probability spaces

In this section we propose another new approach to the best approximation of a nonlinear operator F in probability spaces. As before, it is assumed that the only available information on F is given by certain covariance matrices formed from the sets of preimages and images.

Some motivation for the proposed approach is as follows. Best–approximation methods have the aim of obtaining the best solution within a certain class and therefore the solution cannot be improved by these techniques in cases when the approximation is not satisfactory. By contrast, iterative methods are normally convergent, but the error associated with each iteration loop of the particular method is not the smallest possible error. As a result, convergence can be quite slow for a rather wide variety of problems. Moreover, in practice only a finite number of iteration loops can be carried out, and therefore the final approximate solution is often unsatisfactorily inaccurate.

A natural idea is to combine the above techniques to exploit their advantageous features. We present a method which realizes this idea. First, a special iterative procedure is proposed with the aim of improving the accuracy of approximation with each consequent iteration loop. Secondly, the best approximation problem is solved providing the smallest associated error within the chosen class of approximants for each iteration loop.

We show that the combination of these techniques allows us to build a computationally efficient and flexible method. In particular, we prove that the error in approximating F by the proposed method decreases with an increase in the number of iterations. An application to the optimal filtering of stochastic signals is given.

3.1 Description of the method

In this section, we present the general structure of the proposed method and its particularities, that is, a solution of the best approximation problem and the scheme of a numerical realization.

3.1.1 General Structure

As before, we write (Ω, Σ, μ) for a probability space, where Ω is the set of outcomes, Σ a σ-field of measurable subsets of Ω and $\mu : \Sigma \to [0, 1]$ an associated probability measure on Σ with $\mu(\Omega) = 1$. Suppose that $x \in L^2(\Omega, \mathbb{R}^m)$, $y \in L^2(\Omega, \mathbb{R}^n)$ are random vectors with realizations $x(\omega) \in \mathbb{R}^m$ and $y(\omega) \in \mathbb{R}^n$.

If $F : \mathbb{R}^m \to \mathbb{R}^p$, then an associated operator is $\mathcal{F}_F : L^2(\Omega, \mathbb{R}^m) \to L^2(\Omega, \mathbb{R}^p)$ defined by

$$[\mathcal{F}_F(x)](\omega) = F[x(\omega)]$$

for each $\omega \in \Omega$. As before, we write $F(x)$ rather than $\mathcal{F}_F(x)$, x instead of $x(\omega)$, y instead of $y(\omega)$, etc. Let

$$F(x) = s.$$

It is supposed that a constructive representation of the operator F is either unknown or hardly computationally realizable.

To find the approximating operator S, we propose the following device.

Let us suppose that $s_0 = y$ and that $s_k \in L^2(\Omega, \mathbb{R}^p)$ is known for $k = 1, 2, \ldots$. We set

$$s_{k+1} = \mathcal{S}_k(y), \tag{36}$$

where the nonlinear operator $\mathcal{S}_k : \mathbb{R}^\nu \to \mathbb{R}^p$, with $\nu = n$ for $k = 0$ and $\nu = p$ for $k = 1, 2, \ldots$, is determined by

$$\mathcal{S}_k(y) = S_k(s_k), \tag{37}$$

where

$$S_k(s_k) = \tilde{A}_k^{(0)} + \sum_{q=1}^{r} \tilde{A}_k^{(q)}(s_k^q) \tag{38}$$

and where $\tilde{A}_k^{(0)} \in \mathbb{R}^p$, $\tilde{A}_k^{(q)} : (\mathbb{R}^\nu)^q \to \mathbb{R}^p$ is the q-linear operator. The vector $\tilde{A}_k^{(0)}$ and operators $\tilde{A}_k^{(q)}$ are determined from a solution to the problem

$$J(\tilde{A}_k^{(0)}, \tilde{A}_k^{(1)}, \ldots, \tilde{A}_k^{(r)}) = \min_{A_k^{(0)}, A_k^{(1)}, \ldots, A_k^{(r)}} J(A_k^{(0)}, A_k^{(1)}, \ldots, A_k^{(r)}) \tag{39}$$

for each $k = 1, 2, \ldots$, where

$$J(A_k^{(0)}, A_k^{(1)}, \ldots, A_k^{(r)}) = E\left[\left\| F(x) - \left(A_k^{(0)} + \sum_{q=1}^{r} A_k^{(q)}(s_k^q) \right) \right\|^2\right]. \tag{40}$$

The operator $\mathcal{S}_k = S$ is the proposed approximation of the mapping F.

In other words, the method of F approximation consists of the suuccessive solutions of the best nonlinear approximation problem (3) with s_{k+1} in the form (36)–(38). The operator S_k represents the best approximant of the rth degree for each loop of the procedure (36)–(40).

Note that (37) can be written as

$$\mathcal{S}_k(y) = S_k \circ S_{k-1} \circ \ldots \circ S_0(s_0).$$

3.1.2 Particularities: Determination of $A_k^{(0)}, A_k^{(1)}, \ldots, A_k^{(r)}$ in (38)

We write $s_k = (s_{k,1}, \ldots, s_{k,\nu})^T \in \mathbb{R}^\nu$, $z_{k\,(j_1,\ldots,j_{q-1})} = s_{kj_1} \cdots s_{kj_{q-1}} s_k$, where $j_1, \ldots, j_{q-1} = 1, \ldots, \nu$ and $q = 2, \ldots, r$.

The operator S_k can be written in terms of matrices similarly to (5): for each value $k = 1, 2, \ldots$, there exist matrices $A^{(q)}_{k\,(j_1,\ldots,j_{q-1})} \in \mathbb{R}^{\nu \times p}$ such that

$$S_k(y) = A_k^{(0)} + A_k^{(1)} s_k + \sum_{q=2}^{r} \left[\sum_{q-1 \le \sigma_{q-1} \le (q-1)n} A^{(q)}_{k\,(j_1,\ldots,j_{q-1})} z_{k\,(j_1,\ldots,j_{q-1})} \right], \tag{41}$$

(a) Given "idealized" input.

(b) Desired output.

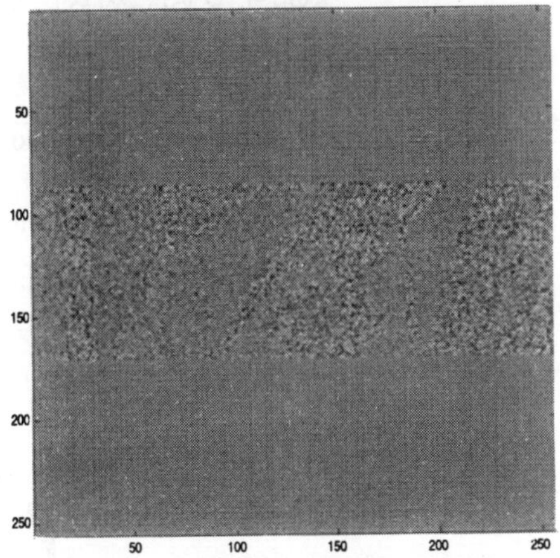

(c) Observed input.

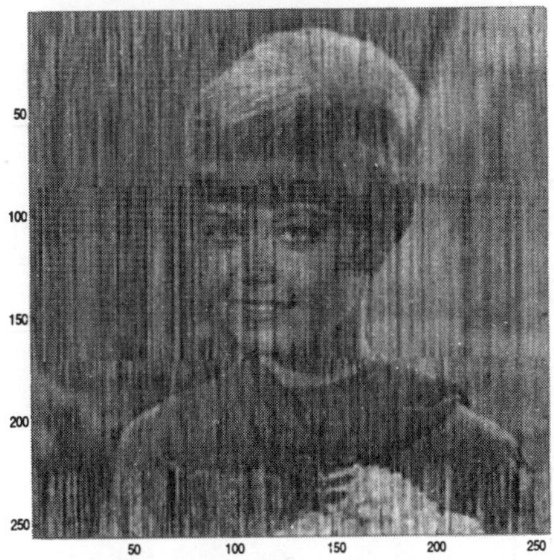

(d) Output of the model S_0.

(e) Output of the model S_1.

(f) Output of the model S_{10}.

Figure 4: The performance of the proposed method.

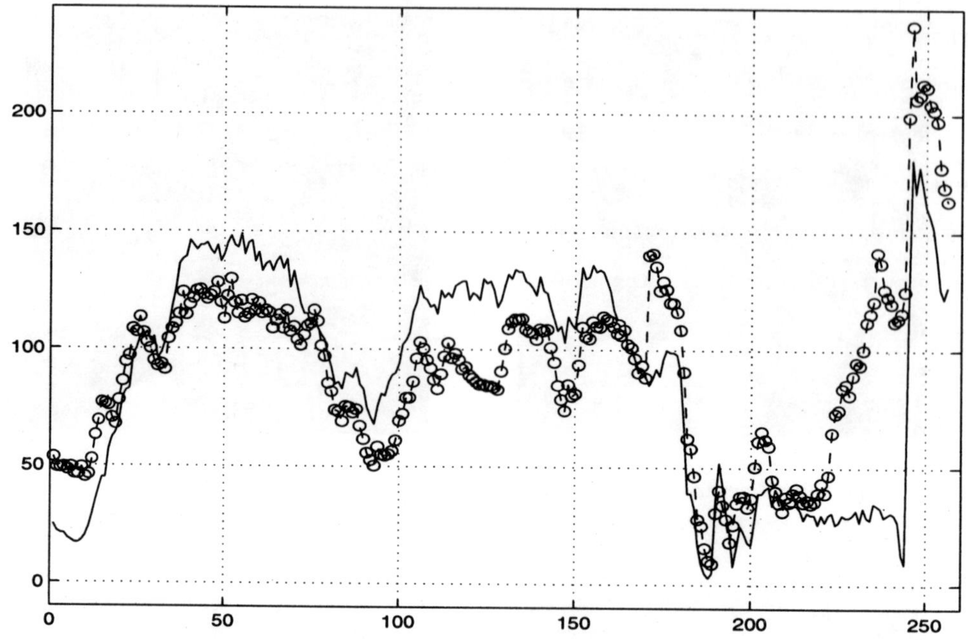

(a) Approximation (dashed line with circles) of the 205-th column (solid line) in matrix V by the procedure (42), (45)–(47) for $k = 0$ and $r = 2$.

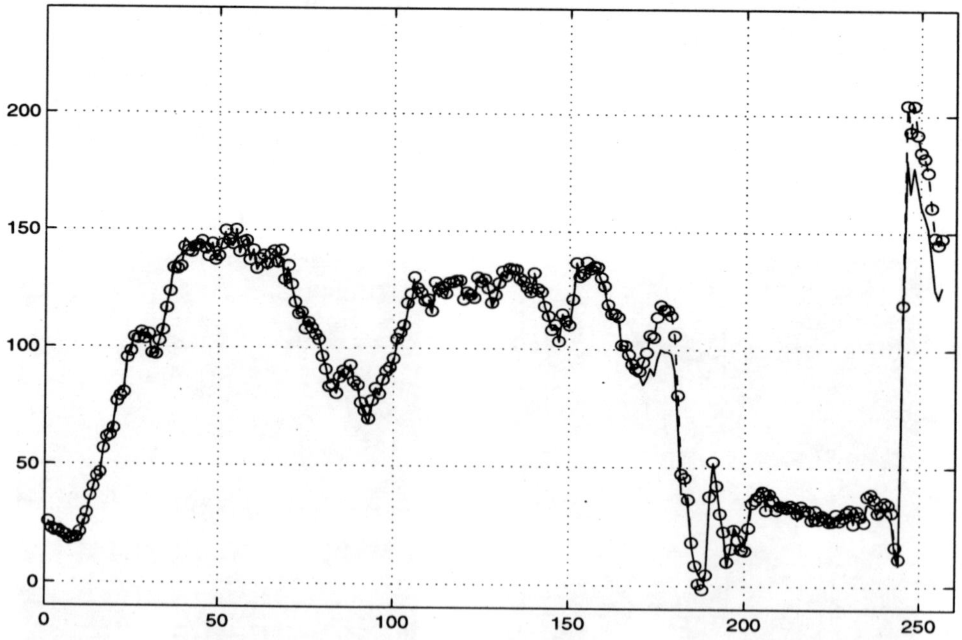

(b) Approximation (dashed line with circles) of the same column (solid line) in matrix V by the procedure (42), (45)–(47) for $k = 10$ and $r = 2$.

Figure 5: Typical example of a column approximation in matrix V (which is the given output) by the procedure (42), (45)–(47)

where $\sigma_{q-1} = j_1 + \ldots + j_{q-1}$ and the inner sum is extended for all summands with subscripts satisfying the inequality $q-1 \le \sigma_{q-1} \le (q-1)\nu$.

Next, if for fixed k, we consider the matrices $A^{(q)}_{k\ j_1,\ldots,j_{q-1}}$ for each value of $q = 2,\ldots r$ and each combination of $j_1,\ldots,j_{q-1}$, where $j_i = 1,\ldots\nu$ for $i = 1,\ldots,q-1$, then for each order q we obtain ν^{q-1} matrices. There are $N = \nu + \nu^2 + \ldots + \nu^{r-1}$ matrices altogether and if we denote them by $C^{(1)}_k,\ldots,C^{(N)}_k$ with the corresponding operands denoted by $u_{k1},\ldots,u_{kN} \in \mathbb{R}^n$ then we can write

$$\sum_{q-1\le\sigma_{q-1}\le(q-1)n} A^{(q)}_{k\ (j_1,\ldots,j_{q-1})} z_{k\ (j_1,\ldots,j_{q-1})} = \sum_{j=1}^{N} C^{(j)}_k u_{kj}$$

and

$$S_k(s_k) = A^{(0)}_k + A^{(1)}_k s_k + \sum_{j=1}^{N} C^{(j)}_k u_{kj}. \tag{42}$$

As a result, the problem (3), (4) is now presented in the following form. For each $k = 1,2,\ldots$, find $\tilde{A}^{(0)}$, $\tilde{A}^{(1)}_k$, $\tilde{C}^{(1)}_k,\ldots,\tilde{C}^{(N)}_k$ such that

$$J(\tilde{A}^{(0)}_k, \tilde{A}^{(1)}_k, \tilde{C}^{(1)}_k,\ldots,\tilde{C}^{(N)}_k) = \min_{A^{(0)}_k, A^{(1)}_k, C^{(1)}_k,\ldots,C^{(N)}_k} J(A^{(0)}_k, A^{(1)}_k, C^{(1)}_k,\ldots,C^{(N)}_k), \tag{43}$$

where

$$J(A^{(0)}_k, A^{(1)}_k, C^{(1)}_k,\ldots,C^{(N)}_k) = E\left[\left\|F(x) - \left(A^{(0)}_k + A^{(1)}_k s_k + \sum_{j=1}^{N} C^{(j)}_k u_{kj}\right)\right\|^2\right]. \tag{44}$$

We write $E_{gh} = E[gh^T]$ and $\mathcal{E}_{gh} = E_{gh} - E[g]E[h^T]$ for any random vectors g and h with realizations in $\mathbb{R}^m$ and $\mathbb{R}^\nu$ respectively.

Let

$$u_k = (u^T_{k1},\ldots,u^T_{kN})^T, \quad \mathcal{D}_k = \mathcal{E}_{u_k u_k} - \mathcal{E}_{u_k s_k}\mathcal{E}^\dagger_{s_k s_k}\mathcal{E}_{s_k u_k} \quad \text{and} \quad \mathcal{G}_k = \mathcal{E}_{s u_k} - \mathcal{E}_{s s_k}\mathcal{E}^\dagger_{s_k s_k}\mathcal{E}_{s_k u_k}.$$

Theorem 3 *For each $k = 1,2,\ldots$, the solution to problem (43), (44) is given by*

$$\tilde{A}^{(0)}_k = E[s] - \tilde{A}^{(1)}_k E[s_k] - \sum_{j=1}^{N} \tilde{C}^{(j)}_k E[u_{kj}], \tag{45}$$

$$\tilde{A}^{(1)}_k = \left(\mathcal{E}_{s s_k} - \sum_{j=1}^{N} \tilde{C}^{(j)}_k \mathcal{E}_{u_{kj} s_k}\right)\mathcal{E}^\dagger_{s_k s_k} + K_{1k}[I - \mathcal{E}^{1/2}_{s_k s_k}(\mathcal{E}^{1/2}_{s_k s_k})^\dagger], \tag{46}$$

$$[\tilde{C}^{(1)}_k \ \ldots\ \tilde{C}^{(N)}_k] = \mathcal{G}_k\mathcal{D}^\dagger_k + K_{2k}[I - \mathcal{D}^{1/2}_k(\mathcal{D}^{1/2}_k)^\dagger], \tag{47}$$

where K_{1k} and K_{2k} are arbitrary matrices.

Proof. The proof is similar to that of Theorem 1.

3.2 Algorithm

The computational procedure for the proposed method in a stylized MATLAB notation is as follows.

Given s_0, $E[s]$, $E[s_0]$, $E[u_0]$, $\mathcal{E}_{ss_0}$, $\mathcal{E}_{s_0s_0}$, $\mathcal{E}_{s_0u_0}$, $\mathcal{E}_{u_0u_0}$, K_{1k} and K_{2k}.

for $k = 0, 1, \ldots$

$$\mathcal{D}_k = \mathcal{E}_{u_ku_k} - \mathcal{E}_{u_ks_k}\mathcal{E}^{\dagger}_{s_ks_k}\mathcal{E}_{s_ku_k};$$

$$\mathcal{G}_k = \mathcal{E}_{su_k} - \mathcal{E}_{ss_k}\mathcal{E}^{\dagger}_{s_ks_k}\mathcal{E}_{s_ku_k};$$

$$\tilde{A}^{(0)}_k = E[s] - \tilde{A}^{(1)}_k E[s_k] - \sum_{j=1}^{N} \tilde{C}^{(j)}_k E[u_{kj}];$$

$$\tilde{A}^{(1)}_k = \left(\mathcal{E}_{ss_k} - \sum_{j=1}^{N} \tilde{C}^{(j)}_k \mathcal{E}_{u_{kj}s_k}\right)\mathcal{E}^{\dagger}_{s_ks_k} + K_{1k}[I - \mathcal{E}^{1/2}_{s_ks_k}(\mathcal{E}^{1/2}_{s_ks_k})^{\dagger}]:$$

$$[\tilde{C}^{(1)}_k \ \ldots \ \tilde{C}^{(N)}_k] = \mathcal{G}_k\mathcal{D}^{\dagger}_k + K_{2k}[I - \mathcal{D}^{1/2}_k(\mathcal{D}^{1/2}_k)^{\dagger}];$$

$$s_k = \tilde{A}^{(0)}_k + \tilde{A}^{(1)}_k s_k + \sum_{j=1}^{N} \tilde{C}^{(j)}_k u_{kj};$$

end

The algorithm above requires knowledge of the matrices

$$E[s], \quad E[s_k], \quad E[u_k], \quad \mathcal{E}_{ss_k}, \quad \mathcal{E}_{s_ks_k}, \quad \mathcal{E}_{s_ku_k} \quad \text{and} \quad \mathcal{E}_{u_ku_k}. \tag{48}$$

The techniques for their estimation can be found, for example, in [13], [30].

3.3 Error analysis associated with the operator $\mathcal{S}_k$

3.3.1 The error equation

Theorem 4 *Let* $\mathcal{G}_k = [\mathcal{G}_{k1} \ldots \mathcal{G}_{kN}]$, $\mathcal{G}_{kj} = \mathcal{E}_{su_{kj}} - \mathcal{E}_{ss_k}\mathcal{E}^{\dagger}_{s_ks_k}\mathcal{E}_{s_ku_{kj}}$ *and*

$$\mathcal{D}^{\dagger}_k = \begin{bmatrix} \mathcal{Q}_{11,k} & \cdots & \mathcal{Q}_{1N,k} \\ \mathcal{Q}_{21,k} & \cdots & \mathcal{Q}_{2N,k} \\ \cdots & \cdots & \cdots \\ \mathcal{Q}_{N1,k} & \cdots & \mathcal{Q}_{NN,k} \end{bmatrix},$$

where $\mathcal{Q}_{ij,k} \in \mathbb{R}^{n\times n}$ *for* $i, j = 1, \ldots, N$, *and let*

$$\Delta(s_k) = \sum_{i=1}^{N} \|\mathcal{G}_{ki}(\mathcal{Q}_{ii,k})^{1/2}\|^2 + \sum_{j,i=1,\ldots,N,\ j\neq i} \mathrm{tr}\{\mathcal{G}_{kj}\mathcal{Q}_{ji,k}\mathcal{G}^T_{ki}\}.$$

The error of F approximation by the operator $\mathcal{S}_k$ is

$$E[\|F(x) - \mathcal{S}_k(y)\|^2] = \text{tr}\{\mathcal{E}_{ss}\} - \|\mathcal{E}_{sy}(\mathcal{E}_{yy}^{\dagger})^{1/2}\|^2 - \sum_{j=0}^{k} \Delta(s_j), \tag{49}$$

where $k = 0, 1, \ldots$.

Proof. It follows from (17)–(20) with $A^{(0)} = \tilde{A}_k^{(0)}$, $A^{(1)} = \tilde{A}_k^{(1)}$ and $C_j = \tilde{C}_k^{(j)}$ that

$$E[\|F(x) - \mathcal{S}_k(y)\|^2] = \text{tr}\{\mathcal{E}_{ss}\} - \|\mathcal{E}_{ss_k}(\mathcal{E}_{s_k s_k}^{\dagger})^{1/2}\|^2 - \Delta(s_k).$$

Next, let us denote $w = s - E[s]$, $w_k = s_k - E[s_k]$, $v_k = u_k - E[u_k]$ and consider the functional

$$J(A_k^{(0)}, A_k^{(1)}, C_k^{(1)}, \ldots, C_k^{(N)}) = E\left[\left\| w - \left(A_k^{(0)} + A_k^{(1)} w_k + \sum_{j=1}^{N} C_k^{(j)} v_{kj}\right)\right\|^2\right],$$

where w_k and v_k are assumed to be known for $k = 0, 1, \ldots$.

The minimum

$$J(\hat{A}_k^{(0)}, \hat{A}_k^{(1)}, \hat{C}_k^{(1)}, \ldots, \hat{C}_k^{(N)}) = \min_{A_k^{(0)}, A_k^{(1)}, C_k^{(1)}, \ldots, C_k^{(N)}} J(A_k^{(0)}, A_k^{(1)}, C_k^{(1)}, \ldots, C_k^{(N)})$$

of this functional is achieved for

$$\hat{A}_k^{(0)} = \mathbb{O}, \tag{50}$$

$$\hat{A}_k^{(1)} = (E_{ww_k} - \sum_{j=1}^{N} \hat{C}_k^{(j)} E_{v_{kj} w_k}) E_{w_k w_k}^{\dagger} + K_{1k}[I - E_{w_k w_k}^{1/2}(E_{w_k w_k}^{1/2})^{\dagger}] \tag{51}$$

and

$$[\hat{C}_k^{(1)} \ \ldots \ \hat{C}_k^{(N)}] = \mathcal{G}_k \mathcal{D}_k^{\dagger} + K_{2k}[I - \mathcal{D}_k^{1/2}(\mathcal{D}_k^{1/2})^{\dagger}]. \tag{52}$$

Equations (50)–(52) follow directly from Theorem 1, since

$$E[w] = E[w_k] = E[v_k] = 0, \ \mathcal{E}_{ss_k} = E_{ww_k}, \ \mathcal{E}_{z_k s_k} = E_{v_k w_k}, \tag{53}$$

$$\mathcal{E}_{su_k} = E_{wv_k} \quad \text{and} \quad \mathcal{E}_{s_k s_k} = E_{w_k w_k}.$$

We write

$$w_{k+1} = \hat{A}_k^{(1)} w_k + \sum_{j+1}^{N} \hat{C}_k^{(j)} v_{kj}.$$

Let us denote

$$W_k = [A_k^{(1)} \ C_k^{(j)} \ \ldots \ C_k^{(j)}], \qquad g_k = [w_k^T v_{k1}^T \ldots v_{kN}^T]^T,$$

$$J_w(W_k) = E[\|w - W_k g_k\|^2]$$

and

$$J_w(A_k^{(1)}, C_k^{(1)}, \ldots, C_k^{(N)}) = E\left[\left\| w - \left(A_k^{(1)} w_k + \sum_{j+1}^{N} C_k^{(j)} v_{kj}\right)\right\|^2\right].$$

Then

$$J_w(W_k) = J_w(A_k^{(1)}, C_k^{(1)}, \ldots, C_k^{(N)})$$

and

$$\min_{W_k} E[\|w - W_k g_k\|^2] = E[\|w - W_k^0 g_k\|^2] = J_w(W_k^0),$$

where

$$W_k^0 = E_{wg_k} E_{g_k g_k}^{\dagger} + K_k[I - E_{g_k g_k}^{1/2}(E_{g_k g_k}^{\dagger})^{1/2}], \tag{54}$$

with an arbitrary matrix K_k. Here $E_{wg_k} = E[wg_k^T]$, $E_{g_k g_k} = E[g_k g_k^T]$.

We note that $\hat{A}_k^{(1)}, \hat{C}_k^{(1)}, \ldots, \hat{C}_k^{(N)}$ defined by (51), (52) respectively, give the minimum to the functional $J_w(A_k^{(1)}, C_k^{(1)}, \ldots, C_k^{(N)})$, so that

$$J_w(\hat{A}_k^{(1)}, \hat{C}_k^{(1)}, \ldots, \hat{C}_k^{(N)}) = \min_{A_k^{(1)}, C_k^{(1)}, \ldots, C_k^{(N)}} J_w(A_k^{(1)}, C_k^{(1)}, \ldots, C_k^{(N)}).$$

Also

$$\hat{A}_k^{(1)} = \tilde{A}_k^{(1)}, \hat{C}_k^{(1)} = \tilde{C}_k^{(1)}, \ldots, \hat{C}_k^{(N)} = \tilde{C}_k^{(N)}.$$

Since $J_w(\hat{A}_k^{(1)}, C_k^{(1)}, \ldots, C_k^{(N)})$ and $J_w(W_k^0)$ are the global minima of the same functional, they coincide, that is,

$$J_w(\hat{A}_k^{(1)}, \hat{C}_k^{(1)}, \ldots, \hat{C}_k^{(N)}) = J_w(W_k^0)$$

and

$$[\tilde{A}_k^{(1)}, \tilde{C}_k^{(1)}, \ldots, \tilde{C}_k^{(N)}] = W_k^0.$$

Hence

$$W_k^0 g_k = \tilde{A}_k^{(1)} w_k + \sum_{j+1}^{N} \tilde{C}_k^{(j)} v_{kj} = w_{k+1}. \tag{55}$$

Next

$$E[\|w - W_k^0 g_k\|^2] = \text{tr}\{E_{ww} - E_{wg_k} E_{g_k g_k}^{\dagger} E_{g_k w}\}$$

and

$$E[\|w - w_{k+1}\|^2] = \text{tr}\{E_{ww} - E_{ww_k} E_{w_k w_k}^{\dagger} E_{w_k w}\} - \tilde{\Delta}(w_k),$$

where

$$\tilde{\Delta}(w_k) = \sum_{i=1}^{N} \|G_{ki}(Q_{ii,k})^{1/2}\|^2 + \sum_{j,i=1,\ldots,N,\ j\neq i} \mathrm{tr}\{G_{kj}Q_{ji,k}G_{ki}^T\},$$

$$G_{ki} = E_{wv_k} - E_{ww_k}E^{\dagger}_{w_kw_k}E_{wv_k}, \quad v_k = [v_{k1}^T \ldots v_{kN}^T]^T$$

and $Q_{ii,k}$ are entries of the matrix

$$D_k^{\dagger} = \begin{bmatrix} Q_{11,k} & \ldots & Q_{1N,k} \\ Q_{21,k} & \ldots & Q_{2N,k} \\ \ldots & \ldots & \ldots \\ Q_{N1,k} & \ldots & Q_{NN,k} \end{bmatrix}$$

with $D_k = E_{v_kv_k} - E_{v_kw_k}E^{\dagger}_{w_kw_k}E_{w_kv_k}$.

Therefore

$$\begin{aligned} E[\|w - w_{k+1}\|^2] &= \mathrm{tr}\{E_{ww} - E_{wg_k}E^{\dagger}_{g_kg_k}E_{g_kw}\} \\ &= \mathrm{tr}\{E_{ww} - E_{ww_k}E^{\dagger}_{w_kw_k}E_{w_kw}\} - \tilde{\Delta}(w_k). \end{aligned} \tag{56}$$

Let us now suppose that (49) is true for $k = i - 1$. Then on the strength of (53) and (56) with $k = i - 1$,

$$\begin{aligned} E[\|s - s_i\|^2] &= E[\|w - w_i\|^2] \\ &= \mathrm{tr}\{E_{ww} - E_{wg_{i-1}}E^{\dagger}_{g_{i-1}g_{i-1}}E_{g_{i-1}w}\} \\ &= \mathrm{tr}\{E_{ww}\} - \|E_{ww_0}(E^{\dagger}_{w_0w_0})^{1/2}\|^2 - \sum_{j=0}^{i-1}\tilde{\Delta}(w_j), \end{aligned} \tag{57}$$

where

$$E_{ww} = \mathcal{E}_{ss}. \tag{58}$$

Next, (54) and (55) imply that

$$\begin{aligned} E_{ww_i} &= E[wg_{i-1}^T(W_{i-1}^0)^T] \\ &= E_{wg_{i-1}}E^{\dagger}_{g_{i-1}g_{i-1}}E_{g_{i-1}w} + E_{wg_{i-1}}\{I - (E^{\dagger}_{g_{i-1}g_{i-1}})^{1/2}E^{1/2}_{g_{i-1}g_{i-1}}\}K_{i-1}^T)] \\ &= E_{wg_{i-1}}E^{\dagger}_{g_{i-1}g_{i-1}}E_{g_{i-1}w} \end{aligned}$$

by Lemma 3. For the same reason

$$E_{w_iw_i} = E[W_{i-1}^0 g_{i-1}g_{i-1}^T(W_{i-1}^0)^T] = E_{wg_{i-1}}E^{\dagger}_{g_{i-1}g_{i-1}}E_{g_{i-1}u}.$$

Therefore

$$E_{ww_i}E_{w_iw_i}^{\dagger}E_{w_iw} = E_{wg_{i-1}}E_{g_{i-1}g_{i-1}}^{\dagger}E_{g_{i-1}w}. \tag{59}$$

Hence, by virtue of (56) and (59),

$$\begin{aligned} E[\|w - w_{i+1}\|^2] &= \text{tr}\{E_{ww} - E_{ww_i}E_{w_iw_i}^{\dagger}E_{w_iw}\} - \tilde{\Delta}(w_i) \\ &= \text{tr}\{E_{ww} - E_{wg_{i-1}}E_{g_{i-1}g_{i-1}}^{\dagger}E_{g_{i-1}w}\} - \tilde{\Delta}(w_i) \end{aligned}$$

and (57) implies

$$\begin{aligned} E[\|w - w_{i+1}\|^2] &= E[\|s - s_{i+1}\|^2] \\ &= \text{tr}\{E_{ww}\} - \|E_{ww_0}(E_{w_0w_0}^{\dagger})^{1/2}\|^2 - \sum_{j=0}^{i-1}\tilde{\Delta}(w_j) - \tilde{\Delta}(w_i). \end{aligned} \tag{60}$$

Taking into account (53) and (58), we now see that (49) with $k = i$ follows from (60) and the theorem is proved. □

Corollary 2 *The error in the operator F approximation by the method (36)–(47) decreases with an increase in the number of iterations k due to the term $\sum_{j=0}^{k}\Delta(s_j)$ in (49).*

3.4 Simulations

In our simulations, the performance of the proposed approach is illustrated with an example of its application to image processing.

We suppose that the signals to be transformed by a system are given by digitized images presented by matrices. A column of the matrix can be considered as a realization of a stochastic signal. The known digitized images (see subfigures (a) and (b) in Figure 2) have been chosen to represent the sets of input and output signals. We denote by $U \in \mathbb{R}^{256\times256}$ and $V \in \mathbb{R}^{256\times256}$ matrices which are numerical counterparts of the images in Figure 4(a) and (b), respectively.

To illustrate the performance of the presented technique for different types of signals, the matrices U and V have been partitioned into submatrices $U^{(1)} = U(1:85,:)$, $U^{(2)} = U(86:170,:)$, $U^{(3)} = U(171:256,:)$ and $V^{(1)} = V(1:85,:)$, $V^{(2)} = V(86:170,:)$, $V^{(3)} = V(171:256,:)$, where $U(n_1:n_2,:)$ is a matrix formed by $n_2 - n_1 + 1$

consequent rows of U beginning with the n_1-th row. Then each submatrix $U^{(i)}$ has been distorted to the matrix $W^{(i)}$ so that $W^{(1)} = R_1. * \cos(U^{(i)})$, $W^{(2)} = 100R_2. * U^{(2)}$ and $W^{(3)} = \sin(U^{(3)})$, where R_1 is a matrix with uniformly distributed entries in the interval $(0, 1)$, R_2 is a matrix with normally distributed entries with mean 0 and variance 1, $\cos(U^{(i)}) = \cos(u_{kj}^{(i)})$, $\sin(U^{(i)}) = \sin(u_{kj}^{(i)})$, $u_{kj}^{(i)}$ is the entry of $U^{(i)}$ and the symbol $.*$ denotes the Hadamard product of matrices.

The proposed method has been applied to each pair $W^{(i)}$, $V^{(i)}$ separately to find the best approximation to the operator $S = S_i : U^{(i)} \to V^{(i)}$ in the form (42), (45)–(47) for $r = 2$. The input of the system S_i is a column of the matrix $U^{(i)}$ and the input of the approximating system is a column of the matrix $W^{(i)}$. The matrices (48) have been estimated by the known sample estimates formed by the matrices $W^{(i)}$, $V^{(i)}$ and $Z^{(i)} = V^{(i)}. * V^{(i)}$ for each $i = 1, 2, 3$. For example, for $i = 1$ we calculated E_{sy} as $V^{(1)}W^{(1)T}/85$ etc. [1]

We denote by $V_{k+1}^{(i)}$ the best approximation of $V^{(i)}$ (in the sense of (43)) obtained in the $k-$th iteration of the procedure (37), (38) with s and s_k substituted by $V^{(i)}$ and $V_k^{(i)}$ respectively.

Figures 4(b) and (c) have been created from the matrices $[V^{(1)^T}\ V^{(2)^T}\ V^{(3)^T}]^T$ and $[W^{(1)^T} W^{(2)^T}\ W^{(3)^T}]^T$ correspondingly.

Figures 4(d)–(f) are digitized images created from the matrices $[V_1^{(1)^T}\ V_1^{(2)^T}\ V_1^{(3)^T}]^T$, $[V_2^{(1)^T}\ V_2^{(2)^T}\ V_2^{(3)^T}]^T$ and $[V_{10}^{(1)^T}\ V_{10}^{(2)^T}\ V_{10}^{(3)^T}]^T$ respectively.

To illustrate the same performance of the proposed method in a more conspicuous way, we represent in Figures 5(a) and (b) a plot of the 205-th column in the matrix V and plots of the 205-th column in the matrices $[V_1^{(1)^T}\ V_1^{(2)^T}\ V_1^{(3)^T}]^T$ and $[V_{10}^{(1)^T}\ V_{10}^{(2)^T}\ V_{10}^{(3)^T}]^T$.

These illustrative results clearly demonstrate the efficiency of the method presented above.

[1] This estimation technique is an area of special study and is not a subject of this paper. The simple estimation method used in our simulations has been chosen for the purpose of illustration only.

4 Constrained best approximation of mappings in probability spaces

4.1 Statement of the Problem

Optimal transforms of random vectors have been succesfully applied to many problems in signal processing including, for example, filtering and compression of random signals and classification and clustering of signals [11].

The known transforms are mainly based on the linear model. The Karhunen–Loève transform (KLT) is perhaps the most popular linear transform, since it achieves the smallest associated error among all linear transforms of given rank. Recently Hua and Liu [11] generalized the KLT for the case where no relationship is assumed between a stochastic signal and noise. The transform in [11] is the best among all fixed rank linear transforms and therefore the associated error cannot be reduced by any other linear transform with the same rank. Nevertheless in many practical problems its performance is still not satisfactory. See the simulations in Subsection 4.8 in this connection.

We present a new *nonlinear* transform with a substantially better performance than that of the generalized KLT [11]. In particular, we show that for the same fixed rank, the transform given possesses a much smaller associated error than that associated with the method in [11].

The idea behind the proposed method is based on the best constrained approximation of a stochastic signal by the approximant generated by a polynomial operator of the second degree. This technique is essentially based on the development of the primary concept presented in [21]–[23].

We begin with a rigorous statement of the problem.

Suppose that $x \in L^2(\Omega, \mathbb{R}^m)$, $y \in L^2(\Omega, \mathbb{R}^n)$ are random vectors with realizations $x(\omega) \in \mathbb{R}^m$, $y(\omega) \in \mathbb{R}^n$. We interpret x as a given "idealized" signal (without any distortion), and y as an observed signal. In particular, y can be interpreted as x contaminated with noise so that no specific relationships between signal and noise are assumed to be known. For instance, noise can be additive, multiplicative or a combination of these.

Let $T : \mathbb{R}^n \to \mathbb{R}^m$ be the operator associated with the map $\mathcal{T}_T : L^2(\Omega, \mathbb{R}^n) \to$

$L^2(\Omega, \mathbb{R}^m)$ from an equation similar to (1). Let T be given by

$$T(y) = A + \sum_{j=0}^{n} B_j z_j, \tag{61}$$

where $A \in \mathbb{R}^m$, $B_j \in \mathbb{R}^{m\times n}$ for $j = 0, 1, \ldots n$, $z_0 = y$, $y = (y_1, \ldots, y_n)^T \in \mathbb{R}^n$ and $z_j = y_j y$ for $j = 1, \ldots n$. Hence the operator T is completely defined by A and B_j with $j = 0, 1, \ldots n$.

The problem is to find vector A^0 and matrices B_j^0 such that

$$J(A^0, B_0^0, \ldots B_n^0) = \min_{A, B_0, \ldots, B_n} J(A, B_0, \ldots, B_n) \tag{62}$$

subject to

$$\operatorname{rank}[A\ B_0\ B_1 \ldots B_n] = r \tag{63}$$

with $r \le m$, where

$$J(A, B_0, \ldots, B_n) = E\left[\left\| x - \left(A + \sum_{j=0}^{n} B_j z_j\right)\right\|^2\right] \tag{64}$$

with E the expectation operator, $\|\cdot\|$ the Frobenius norm, $[A\ B_0\ B_1 \ldots B_n] \in \mathbb{R}^{m\times q}$ and $q = n^2 + n + 1$.

4.2 Motivation of the Statement of the Problem

The equations (3)–(4) represent the best constrained approximation problem. It is well–known that a nonlinear approximant normally possesses a smaller associated error than a linear approximation. Therefore it is natural to seek for a suitable nonlinear form of the approximant.

Let us consider the nonlinear operator T given by

$$T(y) = A + B_0 y + C(y, y), \tag{65}$$

where $C : \mathbb{R}^n \times \mathbb{R}^n \to \mathbb{R}^m$ is a bilinear operator, that is, $(y, y) \in \mathbb{R}^n \times \mathbb{R}^n$. The operator C is a $(m \times n \times n)$–tensor, $C = \{c_{ijk}\} \in \mathbb{R}^{m\times n\times n}$. Therefore the vector $C(y, y)$ can be presented as a product of the tensor C with the vector y and then as a product of the matrix Cy with the vector y. As a result, we have

$$C(y, y) = (Cy)y = B_1 y_1 y + \ldots + B_n y_n y,$$

where $B_1 = \{c_{i1k}\} \in \mathbb{R}^{m\times n}$, ..., $B_n = \{c_{ink}\} \in \mathbb{R}^{m\times n}$. Alternatively, $B_1 = \{c_{ij1}\} \in \mathbb{R}^{m\times n}$, ..., $B_n = \{c_{ijn}\} \in \mathbb{R}^{m\times n}$. Hence (65) coincides with (61).

In the following sections we show that the transform T^0 produced by the nonlinear approximant

$$T^0(y) = A^0 + \sum_{j=0}^{n} B_j^0 z_j \tag{66}$$

possesses a much smaller associated error than the generalized KLT [11].

4.3 Preliminaries

For any u, $w \in \mathbb{R}^n$, we write $E_{uw} = E[uw^T]$ and $\mathcal{E}_{uw} = E_{uw} - E[u]E[w^T]$.

Lemma 6 *Let*

$$P_{11} = 1 - P_{12}E[y] - P_{13}E[z], \quad P_{12} = P_{21}^T, \quad P_{13} = -E[y^T]P_{23} - E[z^T]P_{33},$$

$$P_{21} = -P_{22}E[y] - P_{23}E[z], \; P_{22} = \mathcal{E}_{yy}^{\dagger} - P_{23}\mathcal{E}_{zy}\mathcal{E}_{yy}^{\dagger}, \quad P_{23} = P_{32}^T,$$

$$P_{31} = -P_{33}E[z] - P_{32}E[y], \quad P_{32} = -P_{33}\mathcal{E}_{zy}\mathcal{E}_{yy}^{\dagger}, \quad P_{33} = \mathcal{D}^{\dagger}.$$

Then

$$E_{ss}^{\dagger} = \begin{bmatrix} P_{11} & P_{12} & P_{13} \\ P_{21} & P_{22} & P_{23} \\ P_{13} & P_{32} & P_{33} \end{bmatrix}. \tag{67}$$

Proof. Let

$$t = \begin{bmatrix} 1 \\ y \end{bmatrix}, \quad S_{11} = 1 - S_{12}E[y],$$

$$S_{12} = -E[y^T]S_{22}, \quad S_{21} = S_{12}^T, \quad S_{22} = \mathcal{E}_{yy}^{\dagger}.$$

We show first that

$$E_{tt}^{\dagger} = \begin{bmatrix} S_{11} & S_{12} \\ S_{21} & S_{22} \end{bmatrix}. \tag{68}$$

We have

$$E_{tt} \begin{bmatrix} S_{11} & S_{12} \\ S_{21} & S_{22} \end{bmatrix} E_{tt} = \begin{bmatrix} Q_{11} & Q_{12} \\ Q_{21} & Q_{22} \end{bmatrix},$$

where

$$Q_{11} = S_{11} + E[y^T]S_{21} + S_{12}E[y] + E[y^T]S_{22}E[y] = 1,$$

$$Q_{12} = S_{11}E[y^T] + E[y^T]S_{21}E[y^T] + S_{12}E_{yy} + E[y^T]S_{22}E_{yy} = E[y^T],$$

$$Q_{21} = E[y]S_{11} + E_{yy}S_{21} + E[y]S_{12}E[y] + E_{yy}S_{22}E[y] = E[y]$$

and

$$Q_{22} = E[y]S_{11}E[y^T] + E_{yy}S_{21}E[y^T] + E[y]S_{12}E_{yy} + E_{yy}S_{22}E_{yy} = E_{yy}.$$

Hence $E_{tt} \begin{bmatrix} S_{11} & S_{12} \\ S_{21} & S_{22} \end{bmatrix} E_{tt} = E_{tt}$, that is, the first Moore-Penrose condition is satisfied. The remaining Moore-Penrose conditions for $E_{tt}^{\dagger}$, defined by (68), are easily verified as well.

Next, let

$$R_{11} = E_{tt}^{\dagger} - R_{12}E_{zt}E_{tt}^{\dagger}, \quad R_{12} = R_{21}^{T}, \tag{69}$$

$$R_{21} = -R_{22}E_{zt}E_{tt}^{\dagger} \quad \text{and} \quad R_{22} = D_{zt}^{\dagger}, \tag{70}$$

where $D_{zt} = E_{zz} - E_{zt}E_{tt}^{\dagger}E_{tz} = \mathcal{D}$.

As above, it can be shown by Lemmas 2 and 3 that

$$E_{ss}^{\dagger} = \begin{bmatrix} R_{11} & R_{12} \\ R_{21} & R_{22} \end{bmatrix}. \tag{71}$$

Then (67) follows from (71) by virtue of (68)–(70). □

4.4 Main results

We denote by $U\Sigma V^T$ the singular–value decomposition (SVD) of $E_{xs}(E_{ss}^{1/2})^{\dagger}$, that is,

$$U\Sigma V^T = E_{xs}(E_{ss}^{1/2})^{\dagger}, \tag{72}$$

where

$$U = (u_1, \ldots, u_q) \in \mathbb{R}^{m\times q}, \quad V = (v_1, \ldots, v_q) \in \mathbb{R}^{q\times q}$$

are orthogonal matrices and

$$\Sigma = \text{diag}(\sigma_1, \ldots, \sigma_q) \in \mathbb{R}^{q\times q}$$

is a diagonal matrix with $\sigma_1 \geq \cdots \geq \sigma_k > 0$ and $\sigma_{k+1} = \cdots = \sigma_q = 0$. Put

$$U_r = (u_1, \ldots, u_r), \quad V_r = (v_1, \ldots, v_r), \quad \Sigma_r = \text{diag}(\sigma_1, \ldots, \sigma_r)$$

and define

$$\Theta_r = \Theta_{r(x,s)} = U_r \Sigma_r V_r^T.$$

Set

$$\Phi = [A\ B_0 \dots B_n] \in \mathbb{R}^{m \times q}$$

and let $\Phi(:, \eta : \tau)$ be a matrix formed by the $\tau - \eta + 1$ sequential columns of Φ beginning from the η-th column.

The optimal transform T^0, introduced by (66), is defined by the following theorem.

Theorem 5 *The solution to problem (62)-(64) is given by*

$$A^0 = \Phi^0(:, 1 : 1), \quad B_j^0 = \Phi^0(:, jn + 2 : jn + n + 1) \tag{73}$$

with $j = 0, 1, \dots n$, *where*

$$\Phi^0 = \Theta_r (E_{ss}^{1/2})^\dagger + M_r [I - E_{ss}^{1/2} (E_{ss}^{1/2})^\dagger] \tag{74}$$

and $M_r \in \mathbb{R}^{m \times q}$ *any matrix such that rank* $\Phi^0 \leq r < m$.

Proof. We have

$$J(A, B_0, \dots, B_n) = E[\|x - \Phi s\|^2].$$

By Lemma 1,

$$\begin{aligned} J(A, B_0, \dots, B_n) &= \operatorname{tr}\{E_{xx} - E_{xs} E_{ss}^\dagger E_{sx}\} + \operatorname{tr}\{(\Phi - E_{xs} E_{ss}^\dagger) E_{ss} (\Phi - E_{xs} E_s^\dagger)^T\} \\ &= \operatorname{tr}\{E_{xx} - E_{xs} E_{ss}^\dagger E_{sx}\} + \|(\Phi - E_{xs} E_{ss}^\dagger) E_{ss}^{1/2}\|^2. \end{aligned} \tag{75}$$

The minimum of this functional subject to constraint (63) is achieved (see [4]) if

$$\Phi E_{ss}^{1/2} = \Theta_r, \tag{76}$$

where we have used

$$E_{ss}^\dagger E_{ss}^{1/2} = ([E_{ss}^{1/2}]^T E_{ss}^{1/2})^\dagger [E_{ss}^{1/2}]^T = (E_{ss}^{1/2})^\dagger = (E_{ss}^\dagger)^{1/2}.$$

It is easy to see that the necessary and sufficient condition [1] for (76) to have a solution is satisfied and therefore the solution is given [1] by $\Phi = \Phi^0$. The theorem is proved. $\square$

Remark 1 *The proof above is essentially based on Lemmas 2–4. In this connection, note that the first equation in Lemma 3 has been presented in [11] but without a proof.*

Theorem 6 *Let*

$$\Delta = \|(\mathcal{E}_{xz} - \mathcal{E}_{xy}\mathcal{E}_{yy}^{\dagger}\mathcal{E}_{yz})(\mathcal{D}^{\dagger})^{1/2}\|^2.$$

The error associated with the transform T^0 is

$$E[\|x - T^0(y)\|^2] = \mathrm{tr}\{\mathcal{E}_{xx}\} + \sum_{i=r+1}^{k} \sigma_i^2 - \|\mathcal{E}_{xy}(\mathcal{E}_{yy}^{\dagger})^{1/2}\|^2 - \Delta.$$

Proof. By Lemma 3, it follows from (75) and (76) that

$$E[\|x - T^0(y)\|^2] = \mathrm{tr}\{\mathcal{E}_{xx} - \mathcal{E}_{xy}\mathcal{E}_{yy}^{\dagger}\mathcal{E}_{yx}\} - \|(\mathcal{E}_{xz} - \mathcal{E}_{xy}\mathcal{E}_{yy}^{\dagger}\mathcal{E}_{yz})(\mathcal{D}^{\dagger})^{1/2}\|^2 + \|U\Sigma V^T - \Theta_r\|^2,$$

where

$$\|U\Sigma V^T - \Theta_r\|^2 = \sum_{j=r+1}^{k} \sigma_j^2$$

(see [4]). This proves the theorem. □

4.5 Comparative analysis of the transform T^0 and the generalized KLT

The generalized KLT [11] is a particular case of the proposed transform T^0 when $A^0 = \mathbb{O}$ and all $B_j^0 = \mathbb{O}$ in (73), with $\mathbb{O}$ the corresponding zero vector or zero matrix. To compare the transform T^0 and the generalized KLT, we also put $A^0 = \mathbb{O}$ in (73). Then the vector s in (72) is written as $s = \tilde{s} = [y^T\ z^T]^T$. We denote by $\tilde{\sigma}_j$ the eigenvalues in (72) for $s = \tilde{s}$ and by $\tilde{T}^0$ the transform which follows from (66), (73) with $A^0 = \mathbb{O}$ and $s = \tilde{s}$.

We denote the generalized KLT [11] by H.

Theorem 7 *Let $\vartheta_1, \ldots, \vartheta_l$ be l nonzero singular values of the matrix $E_{xy}(E_{yy}^{1/2})^{\dagger}$, rank $H = p \le l$ and $D = E_{zz} - E_{zy}\mathcal{E}_{yy}^{\dagger}E_{yz}$. If*

$$\sum_{j=r+1}^{k} \tilde{\sigma}_j^2 < \|(E_{xz} - E_{xy}E_{yy}^{\dagger}E_{yz})(D^{\dagger})^{1/2}\|^2 + \sum_{i=p+1}^{l} \vartheta_i^2, \tag{77}$$

then the error associated with the transform $\tilde{T}^0$ is less than the error $E[\|x - Hy\|]^2$ associated with the generalized KLT H [11], that is,

$$E[\|x - \tilde{T}^0(y)\|]^2 < E[\|x - Hy\|]^2.$$

Proof. It is easy to show that

$$E[\|x - Hy\|]^2 = \mathrm{tr}\{E_{xx} - E_{xy}E_{yy}^{\dagger}E_{yx}\} + \sum_{i=p+1}^{l} \vartheta_i^2.$$

Then

$$E[\|x - Hy\|]^2 - E[\|x - \tilde{T}^0(y)\|]^2$$
$$= \|(E_{xz} - E_{xy}E_{yy}^{\dagger}E_{yz})(D^{\dagger})^{1/2}\|^2 + \sum_{i=p+1}^{l} \vartheta_i^2 - \sum_{j=r+1}^{k} \tilde{\sigma}_j^2$$

and the theorem is proved. □

Note that condition (77) is not restrictive and is normally satisfied in practice. In this connection, see the results of our simulations in Section 4.7.

4.6 Special case: solution of the unconstrained minimization problem (62)

Solution of the minimization problem (3) without constraint (63) is important in its own right. The solution is a special form of the transform T^0 and represents a model of the optimal *nonlinear* filter with x an actual signal and y the observed data.

Let $\mathcal{P} = \begin{bmatrix} \mathcal{P}_{11} & \dots & \mathcal{P}_{1n} \\ \mathcal{P}_{21} & \dots & \mathcal{P}_{2n} \\ \dots & \dots & \dots \\ \mathcal{P}_{n1} & \dots & \mathcal{P}_{nn} \end{bmatrix}$ and $G = [\mathcal{G}_1\mathcal{G}_2 \dots \mathcal{G}_n]$, where $\mathcal{D}_{ij} = \mathcal{E}_{z_iz_j} - \mathcal{E}_{z_iy}\mathcal{E}_{yy}^{\dagger}\mathcal{E}_{yz_j} \in \mathbb{R}^{n\times n}$ for $i, j = 1, \dots, n$ and $\mathcal{G}_j = \mathcal{E}_{xz_j} - \mathcal{E}_{xy}\mathcal{E}_{yy}^{\dagger}\mathcal{E}_{yz_j} \in \mathbb{R}^{m\times n}$ for $i = 1, \dots, n$.

We denote a solution to the unconstrained problem (3) by the same symbols as before, that is, by A^0 and B_j^0 with $j = 0, \cdots, n$.

Theorem 8 *The solution to the problem (3) is given by*

$$A^0 = E[x] - B_0^0E[y] - \sum_{k=1}^{n} B_k^0E[z_k],$$

$$B_0^0 = (\mathcal{E}_{xy} - \sum_{k=1}^{n} B_k^0\mathcal{E}_{z_ky})\mathcal{E}_{yy}^{\dagger} + M_1[I - \mathcal{E}_{yy}^{1/2}(\mathcal{E}_{yy}^{1/2})^{\dagger}],$$
$$[B_1^0B_2^0 \dots B_n^0] = G\mathcal{P}^{\dagger} + M_2[I - \mathcal{P}^{1/2}(\mathcal{P}^{1/2})^{\dagger}],$$

where B_1^0, B_2^0, …, $B_n^0 \in \mathbb{R}^{m\times n}$, $M_1 \in \mathbb{R}^{m\times n}$ *and* $M_2 \in \mathbb{R}^{m\times n^2}$ *are arbitrary matrices.*

4.7 Applications and further modifications and extensions

Applications of the proposed technique are abundant and include, for example, simultaneous filtering and compression of noisy stochastic signals, feature selection in pattern recognition, a blind channel equalization and the optimal rejection to colored noise in some neural systems. Background for these applications can be found, for example, in [11].

Note that the efficiency of a fixed-rank transform is mainly characterized by the two parameters – the compression ratio (see [11]) and the accuracy of the signal restoration.

The signal compression is realized through the following device. Let p be the rank of the transform H [11]. Then H can be represented in the form $H = H_1 H_2$, where the matrix $H_2 \in \mathbb{R}^{p\times n}$ implies compression of the signal (see [11]) and the matrix $H_1 \in \mathbb{R}^{m\times p}$ implies its reconstruction. The compression ratio of the transform [11] is given by $c_H = p/m$. Similarly, the transform T^0 can be represented in the form $\Phi^0 = C_1 C_2$, where, for example, $C_1 = U_r \in \mathbb{R}^{m\times r}$ and $C_2 = \Sigma_r V_r^T (E_{ss}^{1/2})^\dagger \in \mathbb{R}^{r\times q}$, so that the matrix C_2 implies compression and the matrix C_1 implies the signal reconstruction. Then the compression ratio of the transform T^0 is given by $c_T = r/m$.

Modifications of the proposed method are mostly motivated by a desire to reduce computations to estimate the SVD of $E_{xs}(E_{ss}^{1/2})^\dagger$. In our further work we shall do this, first by exploiting the representation (73)–(74) in such a way that the matrices $B_1^0, \cdots, B_n^0$ in (73) are estimated by a scheme similar to the Gaussian elimination scheme in linear algebra. Then the rank restriction can be imposed on the matrices $B_1^0, \cdots, B_n^0$ that will allow reduced computational work in finding certain pseudo-inverse matrices.

Extensions of the technique presented in the preceding sections will be developed in the following directions. First, we intend to combine the proposed method with a special iterative procedure to improve the associated accuracy of the signal estimation. Secondly, an attractive extension is based on the representation of the operator T (65) in the form

$$T(y) = A_0 + A_1 y + A_2(y, y) + \cdots + A_k(y, \ldots, y),$$

where $A_k : (\mathbb{R}^n)^k \to \mathbb{R}^m$ is a $k-$linear operator.

(a) Given signals.

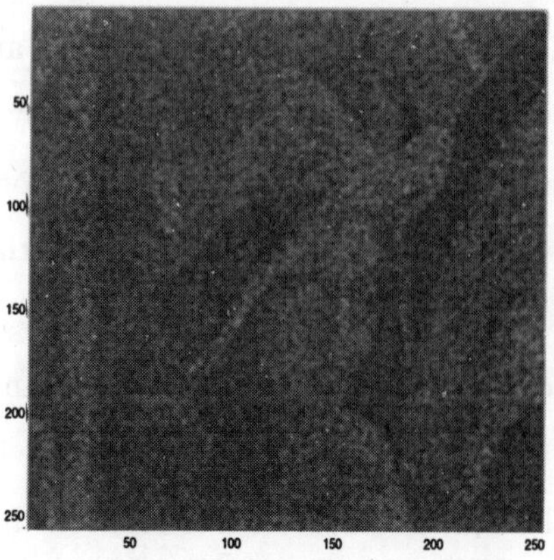

(b) Observed signals.

(c) Reconstruction after filtering and compression by the generalized KLT (GKLT).

(d) Reconstruction after filtering and compression by our transform with the same rank as that of GKLT.

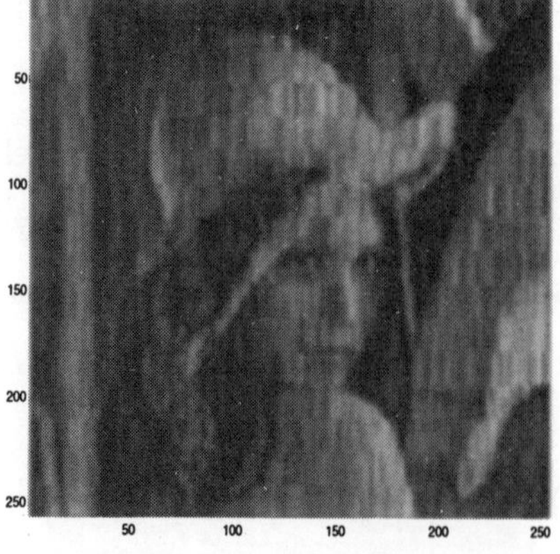

(e) Estimates by the filter $H^{(1)}$.

(f) Estimates by the filter $T^{(1)}$.

Figure 6: Illustration of the performance of the proposed method.

Table 2: Ratios ρ_{ij} of errors associated with our transform T^0 and the generalized KLT H [11] with the same compression ratios

$\downarrow i \parallel j \rightarrow$	1	2	3	4	5	6	7	8
1	5268.3	3880.6	1864.5	1094.7	2605.4	2878.0	4591.6	1052.7
2	2168.4	995.1	1499.7	338.6	1015.1	3324.0	2440.5	336.1
3	2269.3	803.5	158.4	136.4	66.7	2545.4	1227.1	326.6
4	1394.3	716.2	173.7	62.9	451.4	721.6	227.8	691.6
5	3352.4	1970.1	98.9	192.8	390.0	92.8	680.4	3196.8
6	1781.5	758.6	93.6	79.3	59.8	223.2	110.5	2580.8
7	2077.4	1526.0	67.4	30.3	172.5	70.3	1024.4	4749.3
8	3137.2	901.2	27.1	38.5	475.3	445.6	1363.2	2917.5
9	2313.2	117.0	18.0	39.3	180.6	251.0	1500.4	2074.2
10	1476.0	31.5	35.7	119.3	859.3	883.5	2843.1	3270.6
11	1836.7	35.3	36.4	1015.5	460.6	487.0	2843.1	8902.3
12	1808.5	74.5	38.2	419.0	428.0	387.2	2616.9	8895.3
13	1849.1	17.6	30.3	492.4	1175.5	135.8	1441.9	1649.2
14	2123.6	54.9	38.6	302.0	1310.5	2193.8	2681.5	1347.9
15	1295.1	136.3	31.8	711.1	2561.7	5999.2	550.7	996.0
16	2125.5	114.9	31.5	732.3	2258.2	5999.2	550.7	427.1

4.8 Simulations

The aim of our simulations is to demonstrate the advantages of the proposed technique over the best–known linear transform, the generalized KLT H [11]. To this end we use the standard digitized image "Lena" presented by a 256×256 matrix X.

To compare our transform T^0 and transform H [11] for different noisy signals, we partition matrix X into 128 submatrices $X_{ij} \in \mathbb{R}^{16\times 32}$ with $i = 1, \ldots, 16$ and $j = 1, \ldots, 8$, and treat each X_{ij} as a set of 32 realizations of a random vector so that a column of X_{ij} represented the vector realization.

Observed data have been simulated in the form

$$Y_{ij} = 10\,\mathcal{R}_{ij}^{(1)} .* X_{ij} + 500\,\mathcal{R}_{ij}^{(2)}$$

with $i = 1, \ldots, 16$ and $j = 1, \ldots, 8$, where each $\mathcal{R}_{ij}^{(1)}$ is a matrix with uniformly distributed entries in the interval $(0, 1)$ and each $\mathcal{R}_{ij}^{(2)}$ is a matrix with normally distributed entries with mean 0 and variance 1.

The transforms T^0 and H [11] have been applied to each pair X_{ij}, Y_{ij} with the same rank $r = 8$, that is, with the same compression ratio. The corresponding covariance matrices have been estimated from the samples X_{ij} and Y_{ij}. Special methods for their estimation can be found, for example, in [13], [30].

Table 2 presents values of the ratios

$$\rho_{ij} = \|X_{ij} - H(Y_{ij})\|^2 / \|X_{ij} - T^0(Y_{ij})\|^2$$

for each $i = 1, \ldots, 16$ and $j = 1, \ldots, 8$, where $\|X_{ij} - H(Y_{ij})\|^2$ and $\|X_{ij} - T^0(Y_{ij})\|^2$ are the errors associated with the transforms H [11] and T^0, respectively. The value ρ_{ij} is placed in the cell situated at the intersection of the i-th row and the j-th column.

From Table 2 we see that, for the same compression ratio, the proposed transform T^0 has associated error which is less than that of the transform H [11] from 17.6 to 8,895.3 times.

We also applied our filter $T^{(1)}$ (constructed from Theorem 8) and the optimal linear filter $H^{(1)} = E_{xy}E_{yy}^{\dagger}$ [11] to the same signals and data as above, that is, to each pair X_{ij}, Y_{ij} with $i = 1, \cdots, 16$ and $j = 1, \cdots, 8$.

The error associated with filters $T^{(1)}$ and $H^{(1)}$ are

$$\|X - X_T\|^2 = 1.4 \times 10^{-12} \quad \text{and} \quad \|X - X_H\|^2 = 3.9 \times 10^7,$$

where the matrices X_T and X_H have been constructed from submatrices $X_{Tij} \in \mathbb{R}^{16\times 32}$ and $X_{Hij} \in \mathbb{R}^{16\times 32}$ correspondingly, $X_T = \{X_{Tij}\} \in \mathbb{R}^{256\times 256}$ and $X_H = \{X_{Hij}\} \in \mathbb{R}^{256\times 256}$ with $X_{Tij} = T^{(1)}(Y_{ij})$ the estimate of X_{ij} by the filter $T^{(1)}$ and $X_{Hij} = H^{(1)}Y_{ij}$

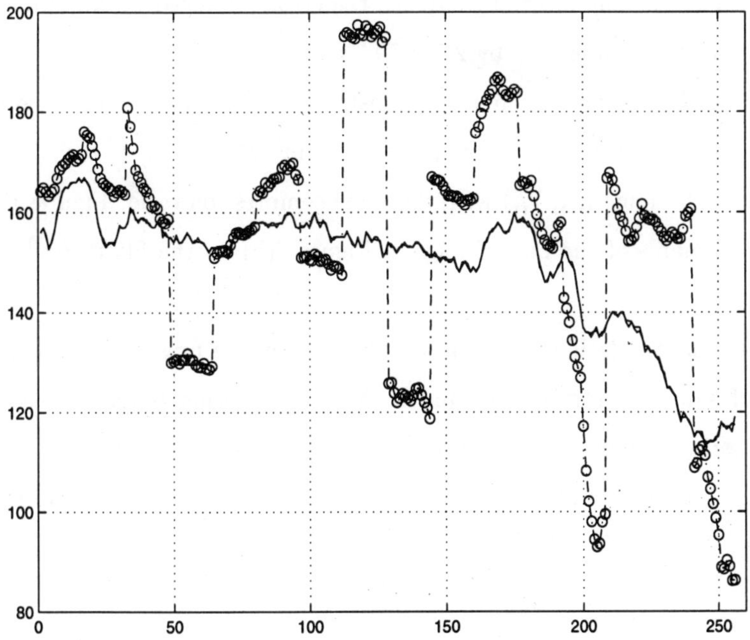

(a) Estimates of the $18th$ column in matrix X.

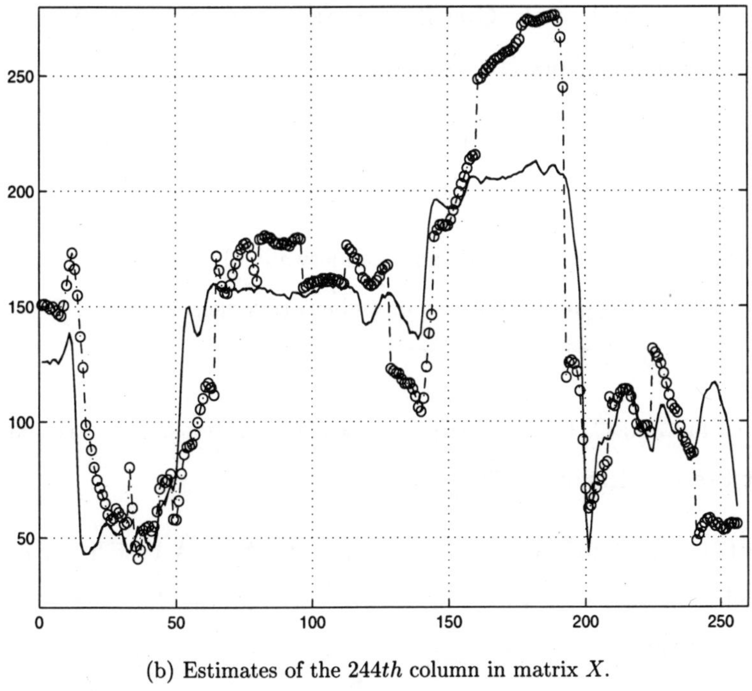

(b) Estimates of the $244th$ column in matrix X.

Figure 7: Typical examples of a column reconstruction in matrix X (image "Lena") after filtering and compression of the observed noisy image (Fig. 7(b)) by transforms H (line with circles) and T^0 (solid line) with the same ranks. In both subfigures, the plot of the column (solid line) virtually coincides with the plot of the estimate by transform T^0.

the es timate of X_{ij} by the filter $H^{(1)}$. Hence the error produced by the filter $H^{(1)}$ is greater than that of the filter $T^{(1)}$ by 2.7×10^{19} times.

Figures 6(c) and (d) represent images reconstructed after filtering and compression of the noisy image in Figure 6(b) by the transforms H and T^0 (which have been applied to each sub-image X_{ij}, Y_{ij}) with the same compression ratio. Figures 6(e) and (f) represent estimates of the noisy image in Figure 6(b) by the filters $H^{(1)}$ [11] and $T^{(1)}$, respectively.

To illustrate the simulation results in a different way, we present typical examples of the plots of a column estimate in matrix X by transforms H and T^0. Note that differences of estimation by T^0 from the column plot are almost invisible.

Table 2 and Figures 6 and 7 demonstrate clearly the advantages of the proposed technique.

References

[1] A. Ben-Israel and T. N. E. Greville, Generalized Inverses: Theory and Applications, Wiley, New York (1974).

[2] Bruno, V. I. An approximate Weierstrass theorem in topological vector space. J. Approx. Theory 42 1–3 (1984),

[3] De Lathauwer, L., De Moor, B. and Vanderwalle, J. On the best Rank-1 and Rank-(R_1, ..., R_N) approximation of higher-order tensors. SIAM J. on Matrix Anal. and Appl. 21 1324–1342 (2000).

[4] G. H. Golub and C. F. van Loan, Matrix Computations, Johns Hopkins University Press, Baltimore (1996).

[5] Howlett, P. G., Pearce, C. E. M. and Torokhti, A. P. A best linear estimator for random vectors with values in Hilbert space. Maths. Res. (St. Petersburg) 4 99–107 (1999).

[6] Howlett, P. G., Pearce, C. E. M. and Torokhti, A. P. Best estimators of second degree for data analysis. Appl. Stoch. Models and Data Anal., Proc. 10th Int.

Symp. June 12–15, 2001. Université de Technologie de Compiègne, France. 2 (2) 549–560 (2001).

[7] Howlett, P. G., Pearce, C. E. M. and Torokhti, A. P. On nonlinear operator approximation with preassigned accuracy. J. Comput. Anal. and Applic. (to appear).

[8] Howlett, P. G. and Torokhti, A. P. A methodology for the constructive approximation of nonlinear operators defined on noncompact sets. Numer. Funct. Anal. and Optimiz. 18 (3) 343–365 (1997).

[9] Howlett, P. G. and Torokhti, A. P. Weak interpolation and approximation of non–linear operators on the space $\mathcal{C}([0,1])$. Numer. Funct. Anal. and Optimiz. 19 (9) 1025–1043 (1998).

[10] Howlett, P. G., Torokhti, A. P. and Pearce, C. E. M. The modelling and numerical simulation of causal non–linear systems. Nonlin. Anal. 47 (8) 5559–5572 (2001).

[11] Hua, Y. and Liu, W. Q. Generalized Karhunen–Loève Transform. IEEE Sig. Proc. Letters. 5, (6) 141–143 (1998).

[12] Istratescu, V. I. A Weierstrass theorem for real Banach spaces. J. Approx. Theory. 19 118-122 (1977).

[13] E. I. Lehmann, Testing Statistical Hypotheses, Wiley, New York (1986).

[14] D. G. Luenberger, Optimization by Vector Space Methods, Wiley, New York (1969).

[15] Prenter, P. M. A Weierstrass theorem for real, separable Hilbert spaces. J. Approx. Theory. 3 341–357 (1970).

[16] Rubinov, A. M. and Singer, I. Best approximation by normal and conormal sets. J. Approx. Theory. 107 212–243 (2000).

[17] Sandberg, I. W. Criteria for uniform approximation using dynamical neural networks. Plenary address, Proc. 4th IEEE Mediterranean Symp. Control Automation. Crete, June 1996, 1–5 (1996).

[18] Sandberg, I. W. Separation conditions and approximation of continuous–time approximately finite memory systems. IEEE Trans. on Circuits and Systems. Part 1, Fundamental theory and applications. 46 (1) 820–829 (1999).

[19] Sandberg, I. W. Time–delay polynomial networks and quality of approximation. IEEE Trans. on Circuits and Systems. Part 1, Fundamental theory and applications. 47 (1) 40–49 (2000).

[20] I. Singer, Best Approximation in Normed Linear Spaces by Elements of Linear Subspaces, Springer, New York (1970).

[21] Torokhti, A. P. On polynomial synthesis of non–linear systems. Electronic Modelling J. of the USSR Acad. of Sci. 11 28–34 (1989). (in Russian)

[22] Torokhti, A. P. Modelling of non-linear dynamical systems. Electronic Modelling J. of the USSR Acad. of Sci. 6 10–16 (1990). (in Russian)

[23] Torokhti, A. P. On constructive restoration of continuous maps of Banach Spaces. Izv. VUZ. Math. 1 86–89 (1990). (in Russian)

[24] Torokhti, A. P. On constructive approximation of non-linear operators. Izv. VUZ. Math. 7 24–28 (1991). (in Russian)

[25] Torokhti, A. P. and Howlett, P. G. On the constructive approximation of non-linear operators in the modelling of dynamical systems. J. Austral. Math. Soc. Ser. B 39 1–27 (1997).

[26] Torokhti, A. P. and Howlett, P. G. On the best quadratic approximation of nonlinear systems. IEEE Trans. on Circuits and Systems. Part I, Fundamental theory and applications 48 595–602 (2001).

[27] Torokhti, A. P. and Howlett, P. G. An optimal filter of the second order. IEEE Trans. on Signal Proc. 49 1044–1048 (2001).

[28] Torokhti, A. P. and Howlett, P. G. Optimal fixed rank transform of the second degree. IEEE Trans. on Circuits & Systems. Part II, Analog & Digital Signal Proc. 48 pp. 309–315 (2001).

[29] Torokhti, A. P., Howlett P. G. and Pearce, C. E. M. Method of best successive approximations for nonlinear operators. J. of Comput. Anal. and Applic. (to appear).

[30] V. N. Vapnik, Estimation of dependences based on empirical data, Springer–Verlag, New York (1982).

Modern Applied Mathematics
Editor: J.C. Misra

Mathematical Modelling of Seismic Sources

Sarva Jit Singh

Emeritus Scientist (CSIR), Department of Mathematics, Maharshi Dayanand University
Rohtak-124001, India

ABSTRACT

In this study we review the work done in the recent past in the direction of modelling of seismic sources. The elasticity theory of dislocations has proved to be a very useful tool in modelling earthquake displacement fields. Dislocations have been commonly used as models of seismic sources. As a mathematical representation of faulting, a displacement dislocation surface, across which the displacement vector is discontinuous, is assumed. A more general approach is the use of a second order tensor, known as moment tensor, for the representation of seismic sources. The moment tensor representation is elegant and very general. The seismic moment tensor describes the forces acting at a seismic source. In the point source approximation, the seismic moment tensor is symmetric, possessing six independent elements, three real eigenvalues and mutually orthogonal eigenvectors. It represents a shear dislocation if the trace and one of the eigenvalues of the moment tensor vanish.

Keywords: Dislocation; Modelling; Moment tensor; Seismic source; Shear fault.

1. Introduction

A major research interest in seismology is the description of the physics of seismic sources. A common approach is the approximation of seismic sources by a model of equivalent forces that correspond to the linear wave equation neglecting non-linear effects in the near-source regions. Equivalent forces are defined as body forces producing displacements at the Earth's surface that are identical to those from the actual forces of the physical process at the source.

The elasticity theory of dislocations, developed by Steketee [19, 20], Maruyama [12-14] and others, has been used very extensively for modelling seismic sources [16,17]. The mathematical foundation for the calculation of the near-field ground motion was laid in the early 1960s by Maruyama [12] and Burridge and Knopoff [2]. They showed that the point source equivalent to fault-slip is a double-couple. However, recent studies have shown that a large number of earthquake sources cannot be represented by a double-couple source mechanism [3, 4, 9, 15, 18]. Therefore, it was felt that we need a more general representation of the seismic source than the double-couple. Gilbert [5,6] introduced moment tensors for calculating the displacement at the free surface which can be expressed as a sum of moment tensor elements times the corresponding Green's function. An elastodynamic Green's function is a displacement field due to a unidirectional unit impulse, i.e. the Green's function is the impulse response of the medium between source and receiver. The response of the medium to any other time function is the convolution of that time function with the impulse response.

The reason that the moment tensors are important is that they completely describe, in a first order approximation, the equivalent forces of general seismic point sources. The equivalent forces can be correlated to physical source models such as sudden relative displacement at a fault surface, rapidly propagating metastable phase transitions, sudden volume collapse due to phase transitions, or sudden volume increase due to explosions. The equivalent forces representing a sudden displacement on a fault plane form the familiar double-couple. A seismic moment tensor is a general concept, describing a variety of seismic source models, the shear dislocation (double-couple source) being just one of them.

The equivalent forces can be determined from an analysis of the eigenvalues and eigenvectors of the moment tensor. The sum of the eigenvalues of the moment tensor describes the volume change in the source (isotropic component of the moment tensor). If the sum is positive, the isotropic component is due to an explosion. The source has an implosive component if the sum is negative. If the sum of the eigenvalues vanishes, then the moment tensor has only deviatoric components. The deviatoric moment tensor represents a pure double-couple source if one eigenvalue equals zero. If none of the eigenvalues vanishes and their sum still equals zero, the moment tensor can be decomposed into a major and a minor double-couple, or a double-couple and a compensated linear vector dipole (CLVD). A CLVD describes a seismic source which has no volume change, net force, or net moment.

2. General Elastodynamic Source

By using the representation theorem for seismic sources [7], the observed displacement u_n at an arbitrary position **x** at the time t due to a distribution of equivalent body force density, f_k, in a source region is

$$u_n(\mathbf{x}, t) = \int_{-\infty}^{\infty} \int_V G_{nk}(\mathbf{x}, t; \mathbf{r}, \bar{t})\, f_k(\mathbf{r}, \bar{t}) dV(\mathbf{r}) d\bar{t}, \qquad (1)$$

where G_{nk} are the components of the Green's function containing the propagation effects and V is the source volume where f_k are non-zero. We assume the summation convention for repeated indices; the x-axis points towards north, the y-axis towards east, and the z-axis down (Figure 1).

By assuming that the Green's functions vary smoothly within the source volume, the Green's functions can be expanded into a Taylor series around the centroid $\mathbf{r} = \xi$ to facilitate the spatial integration in equation (1). The physical source region is characterized by the existence of the equivalent forces. These forces arise due to difference between the model stress and the actual physical stress (stress glut). Outside the source region, the stress glut vanishes as do the equivalent forces. The Taylor series expansion around the centroid is [8]

$$G_{nk}(\mathbf{x}, t; \mathbf{r}, \bar{t}) = G_{nk}(\mathbf{x}, t; \xi, \bar{t}) + (r_j - \xi_j) G_{nk,j}(\mathbf{x}, t; \xi, \bar{t}), \qquad (2)$$

neglecting higher order terms, where

$$G_{nk,j} = \partial/\partial\xi_j\, G_{nk}(\mathbf{x}, t; \xi, \bar{t}).$$

Therefore,

$$\int_V G_{nk}(\mathbf{x}, t; \mathbf{r}, \bar{t})\, f_k(\mathbf{r}, \bar{t}) dV(\mathbf{r})$$

$$= G_{nk}(\mathbf{x}, t; \xi, \bar{t}) \int_V f_k(\mathbf{r}, \bar{t}) dV(\mathbf{r}) + G_{nk,j}(\mathbf{x}, t; \xi, \bar{t}) \int_V (r_j - \xi_j) f_k(\mathbf{r}, \bar{t}) dV(\mathbf{r})$$

$$= 0 + G_{nk,j}(\mathbf{x}, t; \xi, \bar{t})\, M_{kj}(\xi, \bar{t}), \qquad (3)$$

where

$$M_{kj}(\xi, \bar{t}) = \int_V (r_j - \xi_j) f_k(\mathbf{r}, \bar{t})\, dV(\mathbf{r}) \qquad (4)$$

is known as the force moment tensor. In deriving (3), we have assumed that for an indigeneous source,

$$\int_V f_k(\mathbf{r}, \bar{t}) dV(\mathbf{r}) = 0, \qquad (5)$$

which follows from the principle of conservation of linear momentum. The total force, linear and angular momenta must vanish for the equivalent forces of an indigeneous source. The conservation of angular momentum for the equivalent forces leads to the symmetry of the seismic moment tensor. From (1) and (3), we find

$$u_n(\mathbf{x}, t) = \int_{-\infty}^{\infty} G_{nk,j}(\mathbf{x},t;\xi, \bar{t}) M_{kj}(\xi, \bar{t})\, d\bar{t} \qquad (6)$$

$$= \int_{-\infty}^{\infty} G_{nk,j}(\mathbf{x}, t-\bar{t}; \xi, 0) M_{kj}(\xi, \bar{t}) d\bar{t} \qquad (7)$$

$$= G_{nk,j}(\mathbf{x}, t; \xi, 0)* M_{kj}(\xi, t), \qquad (8)$$

where * denotes the temporal convolution :

$$f(t)* g(t)$$

$$= \int_{-\infty}^{\infty} f(\bar{t}) g(t-\bar{t})\, d\bar{t} = \int_{-\infty}^{\infty} f(t-\bar{t}) g(\bar{t})\, d\bar{t}.$$

Equation (7) is obtained from equation (6) on shifting the time origin in G_{nk} since it depends on t and $\bar{t}$ via the combination $t-\bar{t}$.

Assuming that all components of the time dependent seismic moment tensor in (8) have the same time dependence s(t), we have

$$u_n(x, t) = M_{kj}(\xi) [G_{nk,j} (x, t; \xi, 0)* s(t)]. \qquad (9)$$

M_{kj} are the components of the second order seismic moment tensor M, usually termed the moment tensor. Note that the displacement u_n is a linear function of the moment tensor elements and the terms in the square brackets in (9). If the source time function s(t) is a delta function δ(t), then

$$G_{nk,j} (x, t; \xi, o)* s(t) = G_{nk,j} (x, t; \xi, 0). \qquad (10)$$

The entities $G_{nk,j}$ describe nine generalized couples. If k = j, $G_{nk,j}$ represents a vector dipole in the k-direction. If k ≠ j, $G_{nk,j}$ represents a single couple with the forces in the k-direction and the arm in the j-direction.

3. Double-Couple Sources: Shear Faults

Faults represent zones of crustal weakness. Seismic events have been, are and will continue to be related to them. The mass of the rock below an inclined fault plane is known as the foot wall, and the mass of the rock above it as the hanging wall. Figure 1 shows the surface of the foot wall. The slip $\mathbf{u}_0$ is taken as the motion of the hanging wall relative to the foot wall. The line of intersection of the fault plane with the surface of the Earth is known as the strike direction or the strike of the fault. We take the positive direction of the strike to the right of an observer facing the foot wall.

In the case of a shear fault, the slip $\mathbf{u}_0$ is parallel to the fault. Let $\mathbf{u}_0 = U\mathbf{e}$. Then U is the slip (or dislocation) and the unit vector **e** is the slip vector. Let **n** be a unit vector normal to the fault plane and $\mathbf{b} = \mathbf{n} \times \mathbf{e}$. Then the unit vector **b** is known as the null vector. We choose **n** in such a manner that the angle between **n** and the vertical (up) is acute. The vectors (**e, b, n**) form a right – handed system. The angle (λ) between **e** and the strike of the fault is known as the slip angle ($0 \leq \lambda < 2\pi$). It is measured anticlockwise when viewed from the hanging wall side of the fault plane. The angle (δ) which the fault plane makes with the horizontal plane is known as the dip angle ($0 \leq \delta \leq \pi/2$).

Considering the direction of crustal block movements, there are two types of shear faults:

1. **Strike – slip** in which movement is parallel to strike of the fault ($\lambda = 0^\circ$ or 180°).
2. **Dip – slip** in which movement along the fault is perpendicular to the strike of the fault ($\lambda = 90^\circ$ or 270°).

By using elastodynamic representation theorem it can be shown that a shear fault, in point source approximation, is equivalent to a double – couple of moment $M_0 = \mu UA$, where μ is the rigidity of the surrounding medium, U is the magnitude of the relative displacement (slip) and A is the fault area. A shear fault is not a double – couple but represents physically relative tangential movement across a break in the medium. The double – couple serves merely as an alternative mathematical representation of the seismic source. As required by the principles of conservation of linear and angular momenta, the double – couple has zero net force and zero net moment.

A shear fault can be specified completely by two unit vectors, **n** and **e**, and a single scalar, $M_0 = \mu$ UA. **n** is a unit vector normal to the fault (Figure 1) and describe the **fault geometry**. **e** is the unit vector parallel to the direction of the relative displacement $\mathbf{u}_0 = U\,\mathbf{e}$ of the two sidesof the fault and describes the **slip geometry**. M_0 is called seismic moment and describes the **size of the source**. The directions of the forces of the equivalent double – couple are parallel to the directions of the unit vectors **e** and **n** of the shear fault.

The seismic moment determines the intensity of the emitted seismic radiation and is therefore a good measure of the size of an earthquake, at least as far as the elastic radiation is concerned. It is a more logical indication of the size of an event than the earthquake magnitude, which is based on an arbitrary measure of the radiated elastic energy (either body waves or surface waves).

In 1776, Coulomb postulated that a brittle material under stress fractures along a plane of greatest tangential stress. Let $\tau_1, \tau_2, \tau_3,$ where $\tau_1 > \tau_2 > \tau_3$, be the principle stresses just before a fracture. From the theory of elasticity, the maximum shearing stress is equal to one – half the difference between the largest and the smallest principle stresses and acts on a plane that bisects the angle between the directions of these principle stresses. Therefore, from the Coulomb postulate, the plane of fracture passes through the direction of τ_2 and bisects the angle between the directions of τ_1 and τ_3, thus making an angle of $\pm 45^\circ$ with τ_1; the magnitude of the greatest shear stress being $(\tau_1 - \tau_2)/2$. Since there is no tangential stress at a liquid – solid boundary, at the surface of the Earth or at the ocean bottom, one of the principle stresses can be taken to be vertical. Then there is dip – slip faulting if the vertical stress is the largest or the smallest of the three principle stresses and strike – slip faulting if the vertical stress is intermediate.

Let **p** be a unit vector in the direction of $\tau_1$and **t** be a unit vector in the direction of τ_3. Let **b** be the unit vector chosen in such a manner that **(p, b, t)** form a right – handed system. This system can be obtained from the **(e, b, n)** system by a rotation about the **b**-axis through 45°. **b** is known as the null vector. The directions of the vectors **p** and **t** are known as the pressure (P) axis and the tension (T) axis, respectively.

It has been known almost since the beginning of instrumental seismology that during an earthquake certain stations record a P wave impulse upward and away from the epicenter (which is called a compression), whereas other stations record an impulse downward and toward the epicenter (a dilatation or rarefaction). For a shear fault, the quadrant in the **e n** – plane in which **t** lies will yield compression and the quadrant in which **p** lies will yield dilatation. Opposite quadrants have similar patterns. This property is of use in source mechanism studies using first motion observations. The boundaries that separate those stations which record compressional impulses from those which record dilatational impulses are the conjugate planes. One of these planes is the fault plane; the other plane is known as the auxiliary plane. There is an ambiguity between the fault plane and the auxiliary plane.

The moment tensor components in an isotropic medium corresponding to a shear fault are given by [1],

$$M_{kj} = \mu\, AU\, (e_k\, n_j + e_j\, n_k). \qquad (11)$$

From the symmetry of M, we note that the roles of the vectors **e** and **n** could be interchanged. This is the source of the fault plane – auxiliary plane ambiguity which cannot be resolved from the seismic radiation of a point source. Hence studies of locations of aftershocks, surface faulting, rupture directivity or static final displacements need to be done in order to resolve this ambiguity.

The moment tensor components describing shear faulting in an isotropic medium are given by [21]

$$M_{kj} = AU\, C_{kjpq}\, e_p\, n_q, \tag{11a}$$

where C_{kjpq} are the elastic constants, satisfying the symmetry relations

$$C_{kjpq} = C_{jkpq} = C_{pqkj}.$$

For an isotropic medium

$$C_{kjpq} = \lambda\delta_{kj}\,\delta_{pq} + \mu\,(\delta_{kp}\,\delta_{jq} + \delta_{kq}\,\delta_{jp}).$$

Since the vectors **e** and **n** are orthogonal, $e_k n_k = 0$, and (11a) reduces to (11).

The entities $e_k n_j + e_j n_k$ in (11) form a tensor **D** describing a double-couple;

$$\mathbf{D} = \mathbf{e\,n} + \mathbf{n\,e} \equiv \begin{pmatrix} 0 & 0 & 1 \\ 0 & 0 & 0 \\ 1 & 0 & 0 \end{pmatrix}, \tag{12}$$

in the (**e, b, n**) system. This tensor is real and symmetric, having real eigenvalues and orthogonal eigenvector. The eigenvalues are (−1, 0, 1). Hence, the characteristic properties of a moment tensor representing a double-couple (point shear fault) are:

(i) one eigenvalue of the moment tensor vanishes,

(ii) the sum of the eigenvalues vanishes (the trace of the moment tensor is zero), i.e. the non-zero eigenvalues are equal in magnitude but opposite in sign.

As mentioned earlier, the (**p, b, t**) system is obtained from the (**e, b, n**) system by a rotation about the **b**–axis through 45°. Therefore,

$$\mathbf{t} = (\mathbf{e} + \mathbf{n})/\sqrt{2},\ \mathbf{p} = (\mathbf{e} - \mathbf{n})/\sqrt{2}, \tag{13a}$$

$$\mathbf{e} = (\mathbf{t} + \mathbf{p})/\sqrt{2},\ \mathbf{n} = (\mathbf{t} - \mathbf{p})/\sqrt{2}, \tag{13b}$$

$$\mathbf{D} = \mathbf{e\,n} + \mathbf{n\,e}$$

$$= \mathbf{t\,t} - \mathbf{pp}$$

$$\equiv \begin{pmatrix} -1 & 0 & 0 \\ 0 & 0 & 0 \\ 0 & 0 & 1 \end{pmatrix}, \tag{14}$$

in the (**p, b, t**) system. Thus **p**, **b** and **t** are the orthogonal eigenvectors corresponding to the eigenvalues – 1, 0 and 1, respectively.

The strike of the fault plane, Φ ($0 \le \Phi < 2\pi$), is measured clockwise from north, with the fault plane dipping to the right when looking along the strike direction, i.e. the hanging wall is to the right (Figure 1). The (**e, b, n**) system is obtained from the eqicentral system ($\mathbf{e}_x$, $\mathbf{e}_y$, $\mathbf{e}_z$) through the following transformations (Figure 1):

(i) rotation about the z-axis through an angle Φ ; followed by

(ii) rotation about the new x-axis through an angel $\pi + \delta$; followed by

(iii) rotation about the new z-axis through an angle λ .

We therefore have

$$\begin{pmatrix} \mathbf{e} \\ \mathbf{b} \\ \mathbf{n} \end{pmatrix} = \begin{pmatrix} \cos\lambda & \sin\lambda & 0 \\ -\sin\lambda & \cos\lambda & 0 \\ 0 & 0 & 1 \end{pmatrix} \begin{pmatrix} 1 & 0 & 0 \\ 0 & -\cos\delta & -\sin\delta \\ 0 & \sin\delta & -\cos\delta \end{pmatrix} \begin{pmatrix} \cos\Phi & \sin\Phi & 0 \\ -\sin\Phi & \cos\Phi & 0 \\ 0 & 0 & 1 \end{pmatrix} \begin{pmatrix} \mathbf{e}_x \\ \mathbf{e}_y \\ \mathbf{e}_z \end{pmatrix} . \quad (15)$$

This implies

	$\mathbf{e}_x$	$\mathbf{e}_y$	$\mathbf{e}_z$
e	$\cos\lambda\cos\Phi + \sin\lambda\cos\delta\sin\Phi$	$\cos\lambda\sin\Phi - \sin\lambda\cos\delta\cos\Phi$	$-\sin\lambda\sin\delta$
b	$-\sin\lambda\cos\Phi + \cos\lambda\cos\delta\sin\Phi$	$-\sin\lambda\sin\Phi - \cos\lambda\cos\delta\cos\Phi$	$-\cos\lambda\sin\delta$
n	$-\sin\delta\sin\Phi$	$\sin\delta\cos\Phi$	$-\cos\delta$

(16)

Equations (11) and (16) yield the Cartesian components of the moment tensor in terms of the strike (Φ), dip (δ) and slip (λ) angles:

$M_{xx} = -M_0 (\sin\delta \cos\lambda \sin2\Phi + \sin2\delta \sin\lambda \sin^2\Phi)$,

$M_{yy} = M_0 (\sin\delta \cos\lambda \sin2\Phi - \sin2\delta \sin\lambda \cos^2\Phi)$,

$M_{zz} = M_0 (\sin2\delta \sin\lambda)$,

$M_{xy} = M_{yx} = M_0 (\sin\delta \cos\lambda \cos2\Phi + 1/2 \sin2\delta \sin\lambda \sin2\Phi)$,

$M_{zx} = M_{xz} = -M_0 (\cos\delta \cos\lambda \cos\Phi + \cos2\delta \sin\lambda \sin\Phi)$,

$M_{yz} = M_{zy} = -M_0 (\cos\delta \cos\lambda \sin\Phi - \cos2\delta \sin\lambda \cos\Phi)$, (17)

where

$$M_0 = \mu U A \quad (18)$$

is the scalar seismic moment.

The seismic radiation from a dislocation on a plane dipping at an arbitrary angle can be expressed as a linear combination of three simple focal mechanisms: vertical strike-slip, vertical dip-slip and 45° dip-slip [1].

3.1. Vertical strike-slip

Assuming

$$\Phi = 0^0, \delta = 90^o, \lambda = 0^o, \quad (19)$$

equation (17) reveals that the moment tensor is given by

$$\mathbf{M} \equiv M_0 \begin{bmatrix} 0 & 1 & 0 \\ 1 & 0 & 0 \\ 0 & 0 & 0 \end{bmatrix}. \quad (20)$$

The eigenvalues are M_0, $-M_0$ and 0; the corresponding eigenvectors are $(1/\sqrt{2}, 1/\sqrt{2}, 0)$, $(1/\sqrt{2}, -1/\sqrt{2}, 0)$ and $(0, 0, -1)$ which give the tension axis (T), the pressure axis (P) and the null axis, respectively. The directions of the axes are so chosen as to make the (**p, b, t**) system right-handed, i.e. $\mathbf{t} \times p = \mathbf{b}$.

3.2. Vertical dip-slip

Assuming

$$\Phi = 0^0, \delta = 90^o, \lambda = 270^o, \quad (21)$$

equation (17) reveals that the moment tensor for a vertical dip-slip fault is given by

$$\mathbf{M} \equiv M_0 \begin{bmatrix} 0 & 0 & 0 \\ 0 & 0 & 1 \\ 0 & 1 & 0 \end{bmatrix}. \quad (22)$$

The eigenvalues are M_0, $-M_0$ and 0; the corresponding eigenvectors are $(0, 1/\sqrt{2}, 1/\sqrt{2})$, $(0, -1/\sqrt{2}, 1/\sqrt{2})$ and $(1, 0, 0)$ which give the tension axis (T), the pressure axis (P) and the null axis, respectively.

3.3. 45° dip-slip

Assuming

$$\Phi = 0^0, \delta = 45^o, \lambda = 270^o, \quad (23)$$

the moment tensor for a 45° dip-slip fault is

$$\mathbf{M} \equiv M_0 \begin{bmatrix} 0 & 0 & 0 \\ 0 & 1 & 0 \\ 0 & 0 & -1 \end{bmatrix} . \tag{24}$$

The eigenvalues are M_0, - M_0 and 0; the corresponding eigenvectors are (0, 1, 0), (0, 0, 1) and (1, 0 ,0) which give the tension axis (T), the pressure axis (P) and the null axis, respectively.

4. General Seismic Point Source

Let us now consider a seismic point source which cannot be represented by a pure double-couple. Let m_i be the eigenvalues and a_i (a_{i1}, a_{i2}, a_{i3}) be the corresponding unit eigenvectors. Since the moment tensor is symmetric, m_i are real and the eigenvectors constitute an orthonormal set:

$$\mathbf{a_i} . \mathbf{a_j} = \delta_{ij},\ \mathbf{a}_1 \times \mathbf{a}_2 = \mathbf{a}_3. \tag{25}$$

Let Q denote the matrix

$$Q = \begin{bmatrix} a_{11} & a_{21} & a_{31} \\ a_{12} & a_{22} & a_{32} \\ a_{13} & a_{23} & a_{33} \end{bmatrix} . \tag{26}$$

Then the tensor **M** can be diagonalized through the principal axes transformation [11]

$$\text{diag}\ (m_1, m_2, m_3) = Q^T M Q, \tag{27}$$

$$M = Q\ \text{diag}(m_1, m_2, m_3)\ Q^T, \tag{28}$$

where Q^T denotes the transpose of the matrix Q. From (28), we can express the components of the moment tensor in terms of its eigenvalues:

$$M_{xx} = m_1\, a_{11}^2 + m_2 a_{21}^2 + m_3 a_{31}^2,$$

$$M_{yy} = m_1 a_{12}^2 + m_2 a_{22}^2 + m_3 a_{32}^2,$$

$$M_{zz} = m_1 a_{13}^2 + m_2 a_{23}^2 + m_3 a_{33}^2,$$

$$M_{xy} = M_{yx} = m_1\, a_{11} a_{12} + m_2 a_{21} a_{22} + m_3 a_{31} a_{32},$$

$$M_{yz} = M_{zy} = m_1\, a_{12} a_{13} + m_2 a_{22} a_{23} + m_3 a_{32} a_{33},$$

$$M_{xz} = M_{zx} = m_1\, a_{11} a_{13} + m_2 a_{21} a_{23} + m_3 a_{31} a_{33}. \tag{29}$$

We next decompose the diagonal moment tensor $\mathbf{m} \equiv$ diag (m_1, m_2, m_3) into its isotropic and deviatoric parts:

$$\mathbf{m} \equiv \text{diag}(m_1, m_2, m_3)$$

$$= 1/3\ \text{tr}(M)I + \text{diag}(m_1^*, m_2^*, m_3^*), \quad (30)$$

where

$$\text{tr}(M) = M_{xx} + M_{yy} + M_{zz} = m_1 + m_2 + m_3,$$

$$I = \text{diag}(1, 1, 1),$$

$$m_i^* = m_i - 1/3\ \text{tr}(M). \quad (31)$$

Obviously

$$m_1^* = 1/3(2m_1 - m_2 - m_3),$$

$$m_2^* = 1/3(2m_2 - m_3 - m_1),$$

$$m_3^* = 1/3(2m_3 - m_1 - m_2). \quad (32)$$

Equation (30) may be written in the dyadic notation for easy interpretation:

$$\mathbf{m} = m_1\, \mathbf{a_1}\, \mathbf{a_1} + m_2\, \mathbf{a_2}\, \mathbf{a_2} + m_3\, \mathbf{a_3}\, \mathbf{a_3}$$

$$= 1/3\ \text{tr}(M)\, \mathbf{I} + (m_1^*\, \mathbf{a_1}\, \mathbf{a_1} + m_2^*\, \mathbf{a_2}\, \mathbf{a_2} + m_3^*\, \mathbf{a_3}\, \mathbf{a_3}), \quad (33)$$

where

$$\mathbf{I} = \mathbf{e_x e_x} + \mathbf{e_y e_y} + \mathbf{e_z e_z}$$

$$= \mathbf{a_1}\, \mathbf{a_1} + \mathbf{a_2}\, \mathbf{a_2} + \mathbf{a_3}\, \mathbf{a_3} \quad (34)$$

is the unit dyadic.

The first term on the right – hand side of (30) or (33) is the isotropic part of the moment tensor and describes a center of expansion. A center of expansion is associated with volume change in the source. The second term is the deviatoric part with which no volume change is associated. Obviously, the deviatoric part can be represented by three vector dipoles m_i^* along the three eigenvectors $\mathbf{a_i}$ of the moment tensor.

Inserting the values of m_i from (32) into (33), we have

$$\mathbf{m} = 1/3\ \text{tr}(M)\, \mathbf{I} + 1/3(2m_1 - m_2 - m_3)\mathbf{a_1 a_1} + 1/3(2m_2 - m_3 - m_1)\mathbf{a_2 a_2}$$

$$+ 1/3(2m_3 - m_1 - m_2)\mathbf{a_3 a_3} \quad (35)$$

$$= 1/3\ \text{tr}(M)\, \mathbf{I} + 1/3(2m_1 - m_2 - m_3)(\mathbf{a_1 a_1} - \mathbf{a_2 a_2})$$

$$+ 1/3(2m_3 - m_1 - m_2)(\mathbf{a_3 a_3} - \mathbf{a_2 a_2}) \quad (36)$$

$$= 1/3\ \text{tr}(M)\, \mathbf{I} + 1/3(m_1 - m_2)(\mathbf{2a_1 a_1} - \mathbf{a_2 a_2} - \mathbf{a_3 a_3})$$

$$+ 1/3(m_3 - m_2)(\mathbf{2a_3 a_3} - \mathbf{a_1 a_1} - \mathbf{a_2 a_2}). \quad (37)$$

From equation (14) we note that, for example, $\mathbf{a_1 a_1} - \mathbf{a_2 a_2}$ represents a double couple in the $\mathbf{a_1 a_2}$-plane. Therefore, from (36), deviatoric part of the moment tensor can be considered as a superposition of two double –couples in two mutually orthogonal planes. If one eigenvalue equal the mean of the other two eigenvalues, the deviatorc part can be represented by just one double–couple. The combination $2\mathbf{a}_1\mathbf{a}_1 - \mathbf{a}_2\mathbf{a}_2 - \mathbf{a}_3\mathbf{a}_3$ corresponds to three mutually orthogonal dipoles of strengths 2, –1 and –1 in the directions of the vectors $\mathbf{a}_1$, $\mathbf{a}_2$ and $\mathbf{a}_3$, respectively. This combination is known as compensated linear vector diple (CLVD). Therefore, from (37), the deviatoric part of the moment tensor can be regarded as due to two CLVDs. If two eigenvalues are equal, the deviatoric part can be represented by a single CLVD.

Noting that $m_1^* + m_2^* + m_3^* = 0$, we may write (33) in the form

$$\mathbf{m} = 1/3 \text{ tr}(M)\, \mathbf{I} + m_1^* (\mathbf{a}_1\mathbf{a}_1 - \mathbf{a}_2\mathbf{a}_2) + \mathbf{m}_3^* (\mathbf{a}_3\mathbf{a}_3 - \mathbf{a}_2\mathbf{a}_2). \quad (38)$$

Therefore, the deviatoric part of the moment tensor can be considered as due to two double-couples. Assuming

$$|m_3|^* \geq |m_2^*| \geq |m_1^*|, \quad (39)$$

m_3^* is called major double-couple and m_1^* minor double-couple. The major double-couple seems to be a good approximation of the diviatoric part of the moment tensor because the directions of the principal axes remain unchanged. The eigenvector of the smallest eigenvalue (in the absolute sense) m_1^* is the null axis of the major double – couple.

Using the relation $m_1^* + m_2^* + m_3^* = 0$ once again, we may write [18],

$$m_1^* = -\in m_3^*\,,\ m_2^* = -(1-\in)\, m_3^*, \quad (40)$$

where

$$0 \leq \in \leq ½. \quad (41)$$

The relation (41) follows from (39). From (33) and (40), we have

$$\mathbf{m} = 1/3 \text{ tr}(M)\, \mathbf{I} + (1 - 2\in) m_3^* (\mathbf{a}_3\mathbf{a}_3 - \mathbf{a}_2\mathbf{a}_2)$$
$$+ \in m_3^* (2\mathbf{a}_3\mathbf{a}_3 - \mathbf{a}_1\mathbf{a}_1 - \mathbf{a}_2\mathbf{a}_2). \quad (42)$$

Consequently, the deviatoric part of the moment tensor is equivalent to a double – couple and a CLVD. The parameter $\in$ gives an estimate of the deviation of the seismic source from the double – couple mechanism. For a pure double – couple, $\in = 0$; for a CLVD, $\in = ½$. The deviatoric part of the moment tensor can be decomposed into a double – couple and a CLVD in a large number of ways. In decomposition expressed by (42), we have made the physically plausible assumption that the maximum principle stress producing the radiation pattern of the double – couple is parallel to that producing the radiation pattern of the CLVD. We thus conclude that:

(i) If all the three eigenvalues of **M** are equal, it represents a center of explosion.

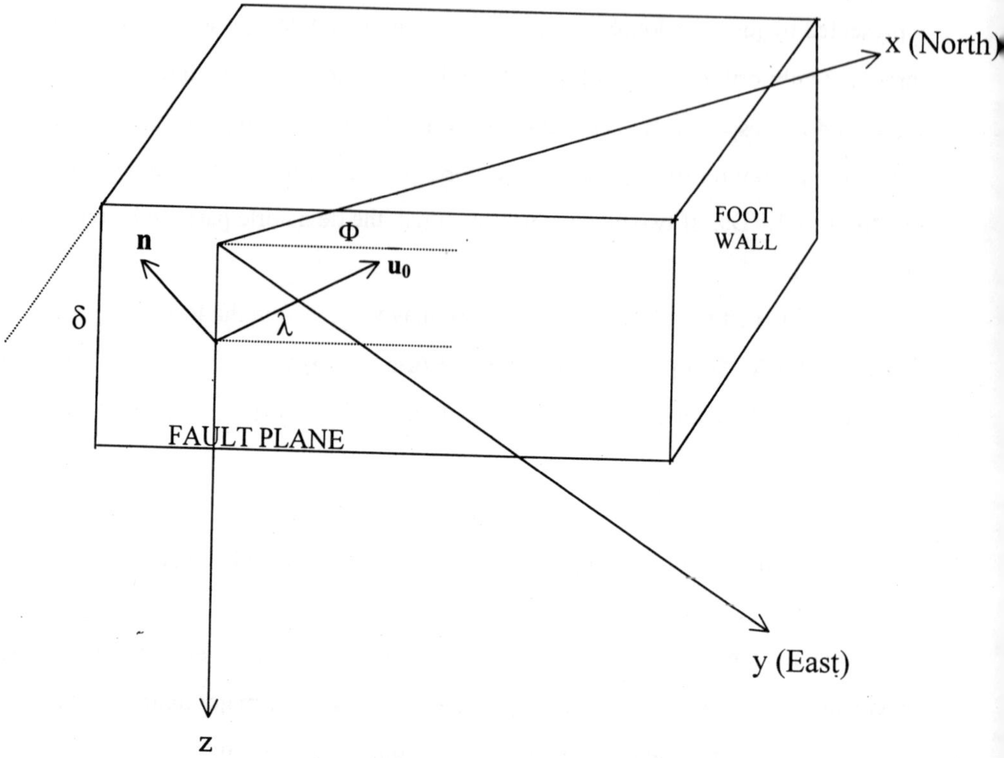

Figure 1. Definition of the epicentral coordinate system (x,y,z). The origin is at the epicentre on the surface of the Earth, strike-angle (Φ) is measured clockwise from north, dip angle (δ) from horizontal down and slip angle (λ) anti-clockwise from horizontal.

(ii) If two eigenvalues of **M** are equal, it represents a center of explosion plus a CLVD.

(iii) If one eigenvalue of **M** is equal to the average of the other two eigenvalues, it represents a center of explosion plus a double – couple. In this case, one eigenvalue of the deviatoric part of **M** is zero. The eigenvector corresponding to this eigenvalue is the null vector.

Acknowledgement

The author is grateful to the Council of Scientific and Industrial Research, New Delhi for financial support through the Emeritus Scientist Scheme.

References

1. A. Ben-Menahem and S.J. Singh, *Seismic Waves and Sources*, 2nd edition, Dover Publications, New York (2000).
2. Burridge R. and Knopoff L., Body force equivalents for seismic dislocations. *Bull. Seism. Soc. Am.* **54,** 1875-1888 (1964).
3. Dziewonski A.M., Chou T.A. and Woodhouse J.H. Determination of earthquake source parameters from waveform data for global and regional seismicity. *J. Geophys. Res.* **86,** 2825-2852 (1981).
4. Frohlich C., Earthquakes with non-double-couple mechanism. *Science* **264,** 804-809 (1994).
5. Gilbert, F., Excitation of the normal modes of the earth by earthquake sources, *Geophys. J. Roy. Astr. Soc.* **22,** 223-226 (1970).
6. Gilbert F., Derivation of source parameters from low-frequency spectra. *Phil. Trans. R. Soc. Lond. A.* **274,** 369-371 (1973).
7. D. Gubbins, *Seismology and Plate Tectonics*, Cambridge University Press, London (1990).
8. Jost M.L. and Herrmann R.B., A student's guide to and review of moment tensors, *Seism. Res. Lett.* **60,** 37-57 (1989).
9. Julian B.R., Miller A.D. and Foulger G.R., Non-double-couple earthquakes 1, Theory. *Rev. Geophys.* **36,** 525-549 (1998).
10. Knopoff L. and Randall M.J., The compensated linear-vector dipole: a possible mechanism for deep earthquakes. *J. Geophys. Res.* **75,** 4957-4963 (1970).
11. A.K. Mal and S.J. Singh, *Deformation of Elastic Solids*, Prentice-Hall, New Jersey (1991).
12. Maruyama T., On the force equivalents of dynamical elastic dislocations with reference to the earthquake mechanism. *Bull. Earthq. Res. Inst. (Tokyo Univ.)* **41,** 467-486 (1963).
13. Maruyama T., Statical elastic dislocations in an infinite and semi-infinite medium. *Bull. Earthq. Res. Inst. (Tokyo Univ.)* **42,** 289-368 (1964).
14. Maruyama T., On two-dimensional elastic dislocation in an infinite and semi-infinite medium. *Bull. Earthq. Res. Inst. (Tokyo Univ.)* **44,** 811-871 (1966).

15. Ross A., Fougler G.R. and Julian B.R., Non-double-couple earthquake mechanisms at the Geysers geothermal area, California. *Geophys. Res. Lett.* **23**, 877-880 (1996).

16. Rybicki K., Dislocations and their geophysical applications. In *Continuum Theories in Solid Earth Physics*, (Edited by R. Tiesseyre), pp. 18-186, Elsevier, Amsterdam (1986).

17. Savage J.C., Dislocations in seismology. In *Dislocations in Solids, Vol. 3, Moving Dislocations,* (Edited by F.R. N. Nabarro), pp. 251-339, Elsevier, Amsterdam (1980).

18. Singh S.J. and Ben-Menahem A., Seismic waves from non-double-couple sources. *Phys. Earth Planet. Int.* **50,** 272-279 (1988).

19. Stekete J.A., On Volterra's dislocations in a semi-infinite medium. *Can. J. Phys.* **36,** 192-205 (1958a).

20. Steketee J.A., Some geophysical applications of the elasticity theory of dislocations. *Can J. Phys.* **36**, 1168-1198 (1958b).

21. A. Udias, *Principles of Seismology*, Cambrdige Univ. Press, Cambridge (1999).

Modern Applied Mathematics
Editor: J.C. Misra

Study of Indian Climate Variability using Predictability Indices

Dhananjay A. Sant[1] and Govindan Rangarajan[2]

[1]Department of Geology, Faculty of Science, M.S. University of Baroda, Vadodara 390 002, India
[2]Department of Mathematics and Centre for Theoretical Studies, Indian Institute of Science
Bangalore 560 012, India

ABSTRACT

In this paper, we investigate local climatic dynamics of 31 stations spread over the Indian subcontinent. Our study is focused on the classic monsoon region which is known for monsoonal fluctuations. Time series data of historical climatological variables (viz. temperature, pressure and precipitation) for about a century is analysed using the Climate Predictability Index proposed by us in a previous paper. The index is a collection of three interrelated quantities, derived from a fractal dimensional analysis of the climatological time series. We study climate variability at two temporal resolutions – for a season as a whole and for individual months. We also investigate variability both at a local level (for individual stations) and at a regional level (for groups of stations). Our studies suggest that regional climatic models typically would not be able to predict local climate since they deal with averaged quantities. Each station has its own characteristic dynamics regulated by the local interplay of climatic variables. We do, however, find that precipitation during the south-west monsoon is affected by temperature during the preceding winter. This may be similar to the well-known effect of the amount of snowfall over the Tibetan plateau in the preceding winter on south-west monsoon precipitation. Finally, we identify regional and temporal patterns in the predictability indices.

I. INTRODUCTION

Variability is an intrinsic property of the earth's climate system. In this paper, we study climate variability of the Indian subcontinent because it lies at the heart of the classic monsoon region and is the area most sensitive to monsoon fluctuations. In a previous paper [1], we had introduced the concept of a climate predictability index as a tool for studying climate dynamics and briefly explored its applications. In this paper, we undertake a detailed analysis of the Indian subcontinental climate using the predictability indices as the main tool.

Predictability indices are calculated for 31 stations spread throughout India. These indices are computed using temperature, pressure and precipitation time series data for these stations available from the Global Historical Climatology Network (GHCN) dataset [2]. We calculate these indices both for a season as a whole and for individual months. This enables us to study the climate dynamics at two temporal resolutions. Next, we study individual stations as well as groups of stations. This allows us to identify both local dynamics as well as regional dynamics.

II. PRELIMINARIES

The regional climatic dynamics of the south-west monsoon is based on warming up of Tibetan plateau [3]. The Tibetan plateau warms rapidly relative to the Indian Ocean resulting in low pressure over Asia and high pressure over ocean. This pressure difference regulates the atmospheric circulation, giving rise to strong low level atmospheric pressure gradient that in turn generates the south-west monsoon. The rise/fall in temperature increases/decreases the rate of evaporation over the ocean leading to more/less precipitation averaged over the Indian subcontinent. But the precipitation does not increase/decrease by a constant amount everywhere over the subcontinent in a given year. The actual distribution of precipitation is determined by local parameters (geographic location etc.), temperature and local atmospheric circulation patterns caused by the pressure difference. The above dynamics suggests that temperature, pressure and precipitation constitute the important aspects of the Indian climate system at a given location.

We observed the following gross features of the Indian climate upon studying the mean temperature, pressure and precipitation data (obtained from GHCN) for 31 stations. The mean monthly pressure for all stations has a characteristic trough invariably during the months of June and July. An example of this phenomenon is illustrated in Figure 1 for a particular station (Nagpur). This is irrespective of the amount of rainfall the station receives and the season in which it receives the rainfall. The pressure trough therefore is a regional characteristic and a part of the south-west monsoon climatic dynamics. It is know to be caused by insolation and the surface albedo of Tibetan plateau [4,3].

The mean temperature over the subcontinent has a high during the summer months and a shallow dip during the south-west monsoon season due to the rainfall. The amount of dip in temperature can also be used as proxy to the amount of total rainfall. Figure 2 shows the temperature variabilty for Nagpur. The north-east region which has a long rainy season (starting well before the onset of the south-west monsoon) does not show this dip in temperature.

When the monthly rainfall distribution across the subcontinent is studied, two distinct regions are observed. The central and north Indian regions have peak rainfall concentrated in the months of July and August. On the other hand, the north-eastern and peninsular regions have a more widespread monthly rainfall distribution.

III. CLIMATE PREDICTABILITY INDEX

We study the time series for the three climatic variables (temperature, pressure and precipitation) which have already been shown [5–9] to correspond to a fractional Brownian noise. Consider a discrete time series given by $(t_i, x(t_i))$ $(i = 1, 2, \ldots, N)$. Here t denotes the time and x the amplitude of the variable under consideration (either temperature, pressure or precipitation in our case). For a fractional Brownian motion, the amplitude increments $x(t_j) - x(t_i)$ which give the fractional Brownian noise have a Gaussian distribution with variance [10]:

$$< [x(t_j) - x(t_i)]^2 > \sim (t_j - t_i)^{2H},$$

where the brackets $<>$ denote the average over many samples of $x(t)$. The parameter H is called the Hurst exponent and takes values between 0 and 1. If $H = 0.5$, we obtain the usual Brownian motion. The Hurst exponent is related to the fractal dimension D of the time series curve by the formula [10]

$$D = 2 - H.$$

If the fractal dimension D for the time series is 1.5, we again get the usual Brownian motion. In this case, there is no correlation between amplitude changes corresponding to two successive time intervals. Therefore, no trend in amplitude can be discerned from the time series and hence the process is unpredictable. However, as the fractal dimension decreases to 1, the process becomes more and more predictable as it exhibits "persistence". That is, the future trend is more and more likely to follow an established trend [8]. As the fractal dimension increases from 1.5 to 2, the process exhibits "anti-persistence". That is, a decrease in the amplitude of the process is more likely to lead to an increase in the future. Hence, the predictability again increases. However, we will be concerned only with persistence behaviour since all geophysical time record analysed till date [5–9] exhibit this behaviour.

We obtain the fractal dimensions of the time series corresponding to temperature, pressure and precipitation for a given location. Thc fractal dimensions are denoted by D_T, D_P and D_R respectively. These are obtained using the rescaled range (R/S) analysis [6] described briefly below: Again, consider the discrete time series given by $(t_i, x(t_i))$ $(i = 1, 2, \ldots, N)$. Define $X^*(t_i)$ as follows:

$$X^*(t_i) = \sum_{j=1}^{i} x(t_i).$$

Then the average of amplitudes in the time period $[t_i, t_i + \tau]$ is given by

$$\overline{X}(t_i, \tau) = [X^*(t_i + \tau) - X^*(t_i)]/\tau.$$

Let the sample variance for this time period be $S^2(t_i, \tau)$. Next, we define the quantity $R(t_i, \tau)$ as

$$R(t_i, \tau) = \max_{0 \le t \le \tau} [X^*(t_i + t) - X^*(t_i) - t\overline{X}(t_i, \tau)] - \min_{0 \le t \le \tau} [X^*(t_i + t) - X^*(t_i) - t\overline{X}(t_i, \tau)].$$

The ratio $R(t_i, \tau)/S(t_i, \tau)$ is called the rescaled range [6]. For each t_i, we get a different ratio. All these values are averaged to get the average ratio R/S for each value of τ. The slope of the plot of of $\log(\tau)$ vs. log(R/S) gives the Hurst exponent. This in turn is related to the fractal dimension D by the formula $D = 2 - H$. The R/S analysis is used merely.

because it has been the conventional technique used for geophysical time records [5–9]. Any other method would be equally adequate.

Predictability indices (denoted by PI_T, PI_P and PI_R respectively) for temperature, pressure and precipitation are defined as follows [1]:

$$\mathrm{PI}_T = 2|D_T - 1.5|; \quad \mathrm{PI}_P = 2|D_P - 1.5|; \quad \mathrm{PI}_R = 2|D_R - 1.5|.$$

Here $|D|$ denotes the absolute value of the number D. We use absolute values since predictability increases in both the following cases – when the fractal dimension becomes less that 1.5 and when it becomes greater than 1.5. In the former case, we have correlation (persistence) behaviour and in the latter case, anti-correlation (anti-persistance) behaviour. However, in either case, the process becomes more predictable. Thus, use of absolute values ensures that a process with $D = 1.3$ has the same predictability index as a process with $D = 1.7$.

The climate predictability index (PI_C) is defined as the collection of the above three indices:

$$\mathrm{PI}_C = (\mathrm{PI}_T, \mathrm{PI}_P, \mathrm{PI}_R).$$

If one of these indices is close to zero, then the corresponding process approximates the usual Brownian motion and is therefore unpredictable. If it is close to one, the process is very predictable. Note that the PI_R value is not related to the amount of precipitation, but to how precipitation changes from year to year. The rationale for introducing the climate predictability index is as follows. In this paper, we are interested in studying the inter-relationships between the three climatic components from the viewpoint of fractal dimensions. Hence, it is useful to have all three of them represented in a single index. Then it is easier to see how the three sub-indices change in relation to one another as the seasons change. Further, by introducing predictability indices instead of fractal dimensions, we focus on how predictable the process is. This is especially useful for precipitation.

Data from thirty one measuring stations spread throughout India were studied. The temperature, pressure and precipitation time series for these stations were obtained from the Global Historical Climatology Network (GHCN) dataset [2]. The dataset gives monthly mean values for the above three climatic variables. Mean values for various seasons were obtained by averaging over the months constituting the season under study. These time series data were then studied using R/S analysis and predictability indices calculated using the formulas given above.

Table 1 gives the predictability indices for temperature, pressure and precipitation respectively for all 31 stations. Predictability indices for individual months in the year and for two seasons — south-west monsoon, north-east monsoon — are listed.

IV. ANALYSIS OF THE DATA

We now study how the predictability indices change with the seasons and within the seasons. We concentrate on the following seasons — pre-south-west monsoon season (March – May), south-west monsoon season (June – September), and north-east monsoon season (October and November). Each of these seasons is studied separately below. Four stations exhibit a consistent predictability pattern across the seasons. We list them first before going into a season-wise analysis. Calcutta has a predictable temperature whereas Darjeeling, Hyderabad and Kodaikanal have a predictable pressure throughout the year.

A. Pre South-West Monsoon Season

First, we consider predictability indices calculated for this season as a whole. Figure 3 gives the scatter plot of PI_T vs. PI_P for this season. We see that the stations are equally distributed across all regions of the graph. Roughly, one-third of the stations have unpredictable PI_T and PI_P. Even if one considers only those stations (concentrated mainly in the North-Eastern region) where it rains significantly during this season, they do not exhibit any consistent behaviour as far as PI_T and PI_P are concerned. This shows that local factors play an important role and further detailed studies are required. Predictability index for precipitation was not studied since a majority of the stations do not get significant rainfall during this season.

Next, we consider how the predictability indices change with the months (viz. March, April and May) constituting this season. From Table 1, we observe that all coastal stations (with the exception of Dwaraka) have predictable temperature during each month. However, pressure is not predictable for all months for these stations. For central Indian stations, temperature is not predicatble. For north-eastern stations, pressure is not predictable.

B. South-West Monsoon Season

If one looks at the seasonal values of the predictability indices, the following regional patterns are observed (see Figure 4). The coastal and north-eastern stations have predictable temperature (again with the exception of Dwaraka). Stations situated in the core of the Indian monsoon (centered around Central India) have unpredictable temperature whereas pressure is predictable. For these stations, the variability in precipitation is linked to the variability in temperature.

As for PI_R, nine stations have reasonably high predictability (greater than or equal to 0.5) if one looks at the South-West monsoon season (June-September) as a whole [cf. Table 1]. However, if one looks at PI_R for the individual months constituting the season, no station has high predictability for all these months. This anamoly can be explained as follows. It is quite possible that the total rainfall for the season follows a more consistent pattern across the years as compared to the rainfall in individual months. The monthly rainfall could be more unpredictable because of delays in onset of monsoon for that station etc whereas even a delayed monsoon can give the same amount of rainfall for the entire season as compared to a normal monsoon. The difference between the seasonal PI_R and monthly PI_R quantifies this for each station. Also note that PI_R is not directly related to the amount of rainfall. It only measures the predictability of the rainfall.

Next, we looked for correlations between the south-west monsoon rainfall and the various monthly predictability indices [cf. Table 1]. Correlations were calculated using data from all stations except those which have a long rainy season (stretching from March/April to October/November). Significant positive correlations of the south-west monsoon rainfall (significant at the 5% level) were observed with both February PI_T and March PI_P. This suggests that the temperature/pressure dynamics during the February-March period (the transition period between winter and summer) prior to the monsoon could play a role in the south-west monsoon rainfall over the Indian subcontinent.

Next, we looked for characteristic patterns in the temperature and pressure predictability graphs (i.e. graphs of PI_T and PI_P as a function of months) for a given station prior to the onset of the south-west monsoon. An example of such a graph is given in Figure 5 for Nagpur. No pattern was observed in the pressure predictability graphs. On the other hand, the temperature predictability graphs for stations getting significant rainfall (greater than 10 cms monthly) during the south-west monsoon exhibited a characteristic dip prior to the onset of the monsoon. However, there were a few exceptions. Stations at high elevations did not exhibit this pattern which is understandable since such stations typically have a quite different climatic dynamics. The other exceptions were Veraval (situated along the West coast) and New Delhi where other local/geographical factors probably play a role.

C. North-Eastern Monsoon Season

The regional distribution of PI_T and PI_P for the North-East monsoon season (October – November) is given in Figure 6. We first consider stations in the southern and eastern India where it rains significantly during the North-East monsoon. The north-eastern region where it rains practically throughout the year appears to have a different dynamics dictated by local geographical factors. However, other stations exhibit the following patterns in the predictability indices calculated for this season. Rainfall is unpredictable for all stations [cf. Table 1]. Except for three stations (considered separately below), temperature is predictable whereas pressure is unpredictable. Therefore, it is the variability in pressure which causes the rainfall unpredictability.

The two high elevation stations (Kodaikanal and Darjeeling) follow a different pattern. Both their temperature and pressure are predictable whereas rainfall is unpredictable. Here, their elevation seems to play a role. The other exception is Madras. Here temperature is unpredictable whereas pressure is predictable. This is in contrast to what is observed for other stations. One interesting fact (which may be related to this contrary behaviour) is that Madras alone of all these stations gets its maximum rainfall during the month of November rather than October.

V. DISCUSSIONS

Most climatic studies deal with a region or a season as a whole. Even predictions for the Indian monsoon deal with the whole subcontinent. However, for (say) a farmer at a specific location, the only thing which matters is whether rainfall is good at her/his location in a given month. In our paper, we have therefore studied predictability of the climate at two temporal scales – for the season as a whole and for individual months within the season. We find that even if the seasonal rainfall is predictable at a given location, the monthly rainfall is not. For individual stations spread throughout the Indian subcontinent, our studies quantify the discrepancy between predictability of seasonal vs. monthly rainfall and consequently the limitations of gross climatic models.

We have also investigated whether regional patterns exist for the predictability indices. We have found some patterns (detailed in the previous section). But these patterns exist only for sub-regions of the Indian subcontinent and not for the whole subcontinent. This again highlights the fact that even if monsoon rainfall is predicted to be good for the whole subcontinent using a regional climatic model that averages over the entire region, individual regions may be hit with drought/flood. Therefore, regional climatic models typically do not predict such local extreme events since they deal with averaged quantities. We extend this analogy to paleoclimatic data, where data largely represents climate of short-lived local depositional events [3,11] and does not represent the overall climate of the region. This shows that the link between regional climate and paleoclimatic data is weak.

Our observations in the previous section also support the well known fact that climate is a very complex system. However, this complexity has been studied from a different viewpoint in this paper using predictability indices. The complexity is manifest in the absence of any strong regional vs. local and seasonal vs. monthly correlations in the predictability indices.

In this paper, we have also extended the analysis carried out in our previous paper [1] for the south-west monsoon. There we had studied stations with low predictability for rainfall and analysed causes for the same. In this paper, the same analysis was performed for the north-east monsoon. We found that for most stations it is the unpredictability in pressure which leads to the unpredictability in rainfall.

We also find that the dynamics of the south-west monsoon is affected by the transition from winter to summer in the previous February-March period. This corresponds to the observation [12,3] that in years of weak glacial/ low snowfall (weak winter over the continent), the Tibetan plateau is able to warm up earlier leading to a strong monsoon. On the other hand, a delayed or weak monsoon is due to deep snow (severe winter over the continent) [13]. In the core region of the monsoon, pressure is consistent across the years (leading to higher PI_P) and it is the variability in temperature that causes the variability in rainfall [cf. Figure 4]. However, this pattern reverses during the north-east monsoon. For this monsoon, there is no winter to summer like transition for temperature prior to the monsoon. Further, since it has already been raining (during the south-west monsoon) prior to this monsoon, there is a general cooling trend in temperature which makes it predictable in the core region for this monsoon (which is now centered around south-east India). On the other hand, the pressure is observed to become unpredictable [cf. Figure 6] leading to variability in the north-eastern monsoon rainfall.

Table 1 Table 1 gives the climate predictability index (PI_C) for 31 stations spread throughout India. The values shown are the indices multiplied by a factor of 10. PI_C is listed for all months and also for 2 seasons — the South-West and North-East monsoon periods.

Station	Jan	Feb	Mar	Apr	May	Jun	Jul	Aug	Sep	Oct	Nov	Dec	SW	NE
Agra	(4,4,2)	(4,2,1)	(3,1,2)	(0,1,2)	(4,3,3)	(0,6,1)	(2,5,3)	(2,4,6)	(1,4,2)	(1,1,0)	(4,3,1)	(3,3,1)	(3,7,5)	(3,2,0)
Akola	(5,5,0)	(5,2,3)	(3,6,1)	(1,5,1)	(2,6,1)	(1,4,1)	(4,4,3)	(4,5,0)	(1,4,2)	(4,3,2)	(3,5,1)	(5,4,1)	(4,6,4)	(4,7,3)
Allahabad	(2,5,1)	(5,2,3)	(4,2,0)	(2,2,1)	(4,5,2)	(4,5,3)	(1,3,0)	(1,0,3)	(2,3,2)	(5,0,2)	(5,5,1)	(5,5,1)	(2,8,1)	(6,2,0)
Bangalore	(6,2,3)	(5,0,2)	(4,2,3)	(3,5,3)	(2,6,0)	(1,3,2)	(2,3,0)	(2,4,1)	(5,0,3)	(4,2,2)	(5,1,2)	(6,1,3)	(3,3,2)	(5,1,1)
Bikaner	(8,5,2)	(1,1,1)	(5,1,1)	(4,4,2)	(5,6,3)	(5,2,1)	(5,3,1)	(4,2,1)	(4,1,1)	(8,4,2)	(8,4,4)	(8,2,3)	(6,4,1)	(10,4,0)
Bombay	(3,5,0)	(6,6,2)	(6,6,3)	(6,0,1)	(6,6,2)	(3,4,2)	(4,4,2)	(5,5,4)	(5,4,0)	(4,4,1)	(6,6,3)	(6,6,3)	(5,7,5)	(5,7,2)
Calcutta	(6,0,0)	(8,1,2)	(8,1,3)	(5,3,1)	(8,3,2)	(6,4,2)	(6,5,2)	(6,3,4)	(8,1,0)	(6,0,1)	(8,3,3)	(10,4,3)	(9,7,2)	(9,0,4)
Cochin	(6,1,0)	(8,1,2)	(8,2,3)	(10,4,1)	(6,6,2)	(5,3,2)	(6,1,2)	(6,4,4)	(8,2,0)	(8,2,1)	(6,1,3)	(5,0,3)	(9,3,2)	(8,1,2)
Darbhanga	(5,5,3)	(6,1,4)	(4,1,0)	(1,2,1)	(4,3,2)	(3,6,1)	(1,4,0)	(5,1,2)	(5,2,1)	(4,1,2)	(6,1,2)	(8,1,2)	(5,8,1)	(7,1,1)
Darjeeling	(3,6,3)	(5,6,3)	(4,6,2)	(2,8,3)	(4,8,2)	(4,8,1)	(6,8,2)	(6,8,3)	(4,8,1)	(5,6,1)	(5,6,4)	(6,6,4)	(5,7,4)	(6,7,1)
Dibrugarh	(6,0,1)	(6,1,1)	(2,2,1)	(6,2,3)	(0,3,3)	(4,5,4)	(6,3,2)	(5,2,1)	(6,1,5)	(3,1,3)	(6,3,2)	(8,1,3)	(9,3,5)	(6,0,3)
Dwaraka	(4,6,2)	(3,6,4)	(4,4,3)	(2,2,6)	(0,5,5)	(2,0,0)	(5,1,2)	(3,2,2)	(2,1,1)	(4,3,1)	(3,5,3)	(2,3,1)	(3,2,2)	(1,2,2)
Gauhati	(6,4,1)	(6,0,4)	(2,3,2)	(5,2,5)	(2,2,2)	(1,4,2)	(5,3,3)	(4,1,1)	(4,5,3)	(2,0,0)	(3,5,2)	(8,4,2)	(5,4,1)	(4,1,0)
Hyderabad	(6,6,4)	(6,5,4)	(6,5,0)	(6,5,3)	(6,8,0)	(4,6,1)	(6,5,0)	(6,6,4)	(6,6,0)	(8,5,2)	(6,5,3)	(4,6,1)	(7,7,3)	(8,5,2)
Indore	(3,5,2)	(2,5,3)	(2,8,3)	(4,5,1)	(1,6,0)	(0,3,0)	(4,6,3)	(1,6,4)	(2,3,5)	(4,3,2)	(5,6,1)	(3,8,0)	(3,8,7)	(5,6,3)
Jagdalpur	(2,2,2)	(6,2,2)	(5,5,3)	(2,1,2)	(1,2,1)	(0,0,3)	(6,4,3)	(3,1,4)	(2,1,0)	(6,4,2)	(5,1,3)	(6,1,5)	(2,0,5)	(6,0,1)
Jaipur	(1,6,3)	(1,2,1)	(4,4,2)	(2,1,1)	(5,6,1)	(1,3,0)	(3,3,1)	(4,2,0)	(1,4,1)	(3,1,3)	(5,5,1)	(1,5,3)	(5,5,3)	(4,4,3)
Jodhpur	(2,5,5)	(3,3,3)	(3,3,1)	(1,4,3)	(5,3,3)	(3,3,3)	(1,5,4)	(4,2,1)	(1,0,1)	(1,1,1)	(1,4,2)	(0,5,3)	(2,5,0)	(0,4,1)
Kodaikanal	(4,4,2)	(4,6,0)	(5,5,3)	(6,6,3)	(5,6,3)	(5,8,1)	(8,8,3)	(8,6,1)	(6,6,2)	(6,6,1)	(5,8,3)	(4,5,2)	(9,8,1)	(6,8,3)
Kumaon	(3,4,3)	(6,3,1)	(3,5,3)	(3,5,2)	(4,4,1)	(5,4,5)	(2,5,2)	(4,3,1)	(5,5,1)	(1,3,4)	(5,6,4)	(4,3,4)	(6,6,3)	(4,4,4)
Ludhiana	(2,6,3)	(2,0,2)	(2,2,0)	(1,2,3)	(2,4,3)	(2,6,0)	(3,1,1)	(1,1,2)	(2,0,1)	(2,0,2)	(3,3,1)	(1,4,0)	(1,4,2)	(3,0,2)
Madras	(0,3,4)	(6,5,0)	(8,5,1)	(10,0,1)	(2,4,2)	(5,8,2)	(6,6,4)	(5,6,2)	(1,4,0)	(0,1,1)	(2,5,2)	(2,4,0)	(6,8,4)	(1,7,0)
Nagpur	(3,5,1)	(2,2,4)	(3,5,1)	(4,4,0)	(3,6,1)	(1,3,2)	(6,2,4)	(0,5,2)	(3,5,1)	(3,2,1)	(2,1,2)	(1,3,2)	(2,6,4)	(2,2,0)
New Delhi	(3,2,2)	(2,4,1)	(3,4,1)	(0,5,2)	(6,2,5)	(6,8,1)	(2,5,5)	(5,4,4)	(2,0,0)	(2,4,4)	(3,6,1)	(3,1,0)	(5,7,3)	(3,7,4)
Pune	(5,3,1)	(5,2,3)	(5,4,3)	(5,4,0)	(3,4,2)	(1,2,4)	(5,0,3)	(5,3,3)	(2,1,2)	(3,1,3)	(2,3,1)	(3,3,0)	(4,5,0)	(3,1,1)
Shillong	(6,4,0)	(3,1,1)	(0,2,1)	(4,3,3)	(2,1,3)	(3,1,4)	(6,3,2)	(3,2,2)	(5,3,2)	(1,0,0)	(6,2,1)	(0,1,1)	(5,2,5)	(3,0,0)
Simla	(2,5,2)	(3,1,4)	(5,5,3)	(6,4,2)	(4,2,5)	(5,2,1)	(6,1,1)	(6,0,0)	(8,2,2)	(3,4,2)	(5,1,1)	(5,1,0)	(8,2,2)	(5,2,1)
Silchar	(4,4,0)	(4,0,1)	(4,1,4)	(3,2,3)	(2,1,4)	(5,1,4)	(6,2,0)	(4,2,2)	(4,1,3)	(2,1,1)	(4,3,2)	(5,2,0)	(7,2,5)	(3,1,2)
Srinagar	(1,4,1)	(3,1,0)	(1,2,3)	(2,6,3)	(2,3,2)	(1,4,1)	(5,4,2)	(2,2,1)	(4,1,1)	(3,2,1)	(4,4,3)	(1,1,0)	(5,5,2)	(4,3,4)
Veraval	(5,5,2)	(4,6,1)	(5,6,6)	(6,3,2)	(8,5,2)	(8,2,2)	(5,2,5)	(6,3,5)	(8,1,2)	(4,3,1)	(3,4,5)	(3,2,0)	(9,4,6)	(4,4,1)
Vishakapatnam	(2,3,2)	(8,0,5)	(10,2,0)	(10,1,2)	(10,4,0)	(8,6,2)	(6,5,2)	(6,0,3)	(6,0,0)	(5,0,4)	(6,4,4)	(3,4,1)	(8,8,5)	(6,4,3)

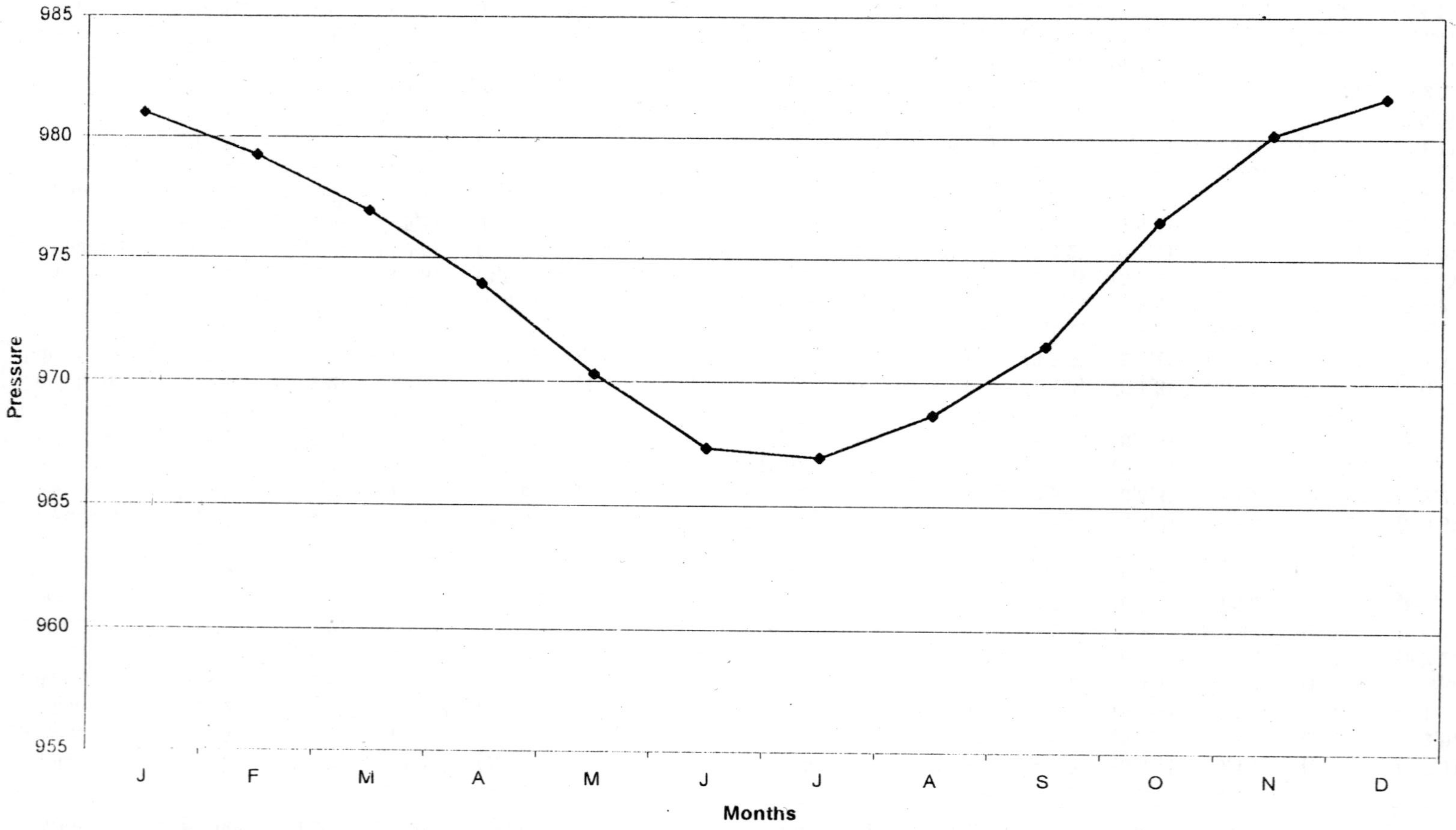

Figure 1 A graph of mean monthly pressure against the months for Nagpur.

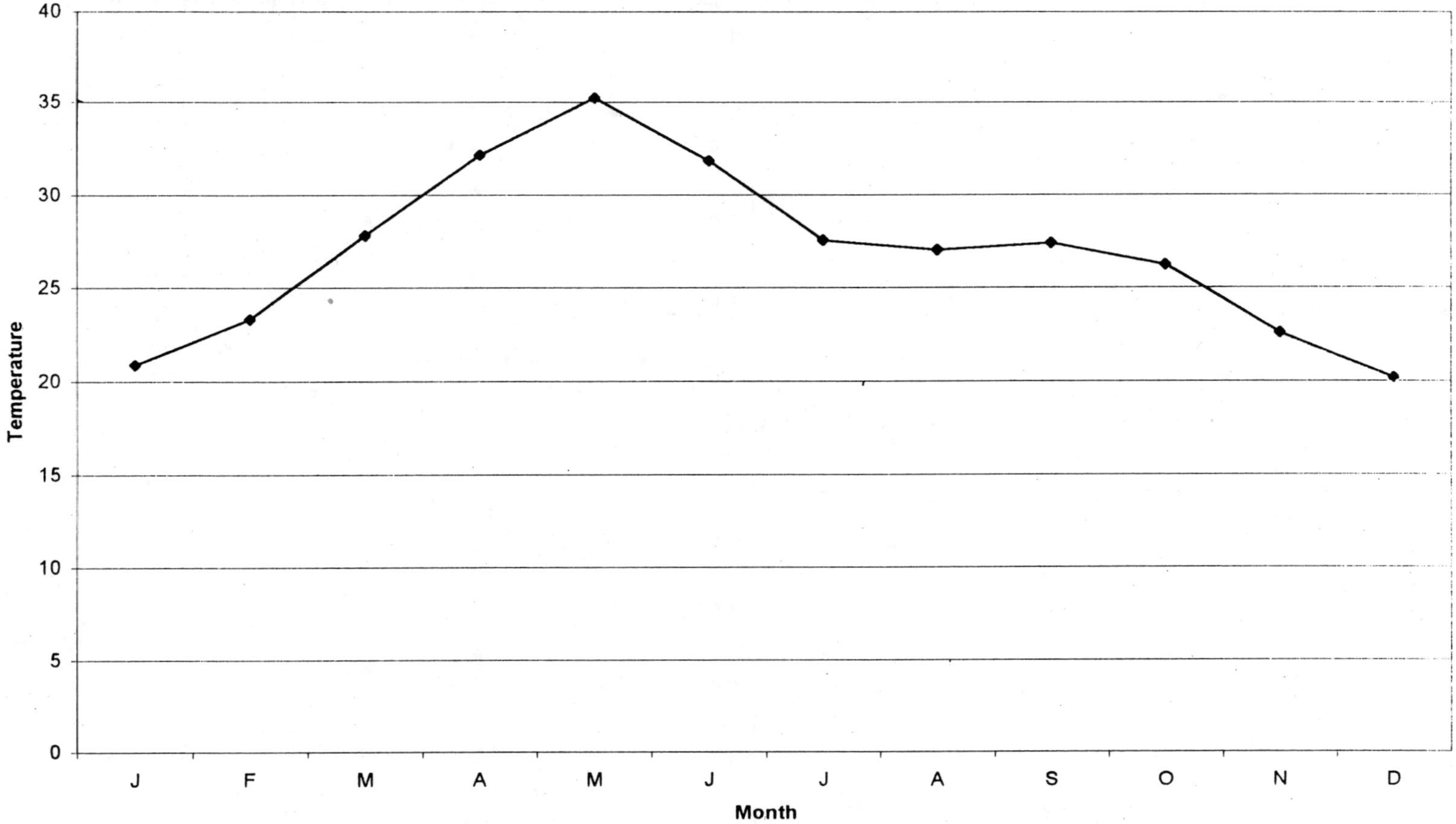

Figure 2 A graph of mean monthly temperature against the months for Nagpur.

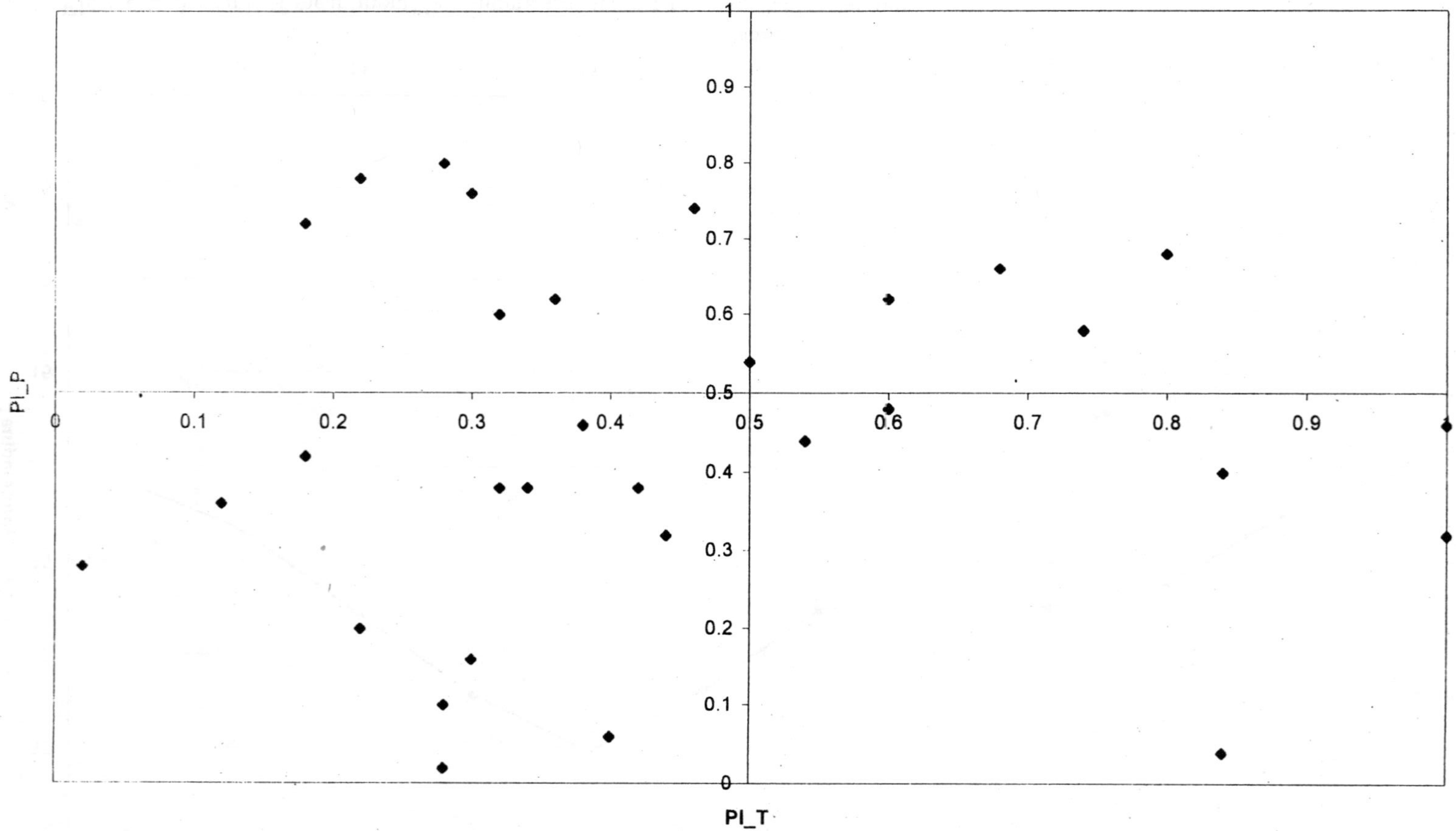

Figure 3 Scatter plot of PI_T vs. PI_P for the pre-South-West monsoon season (March – May).

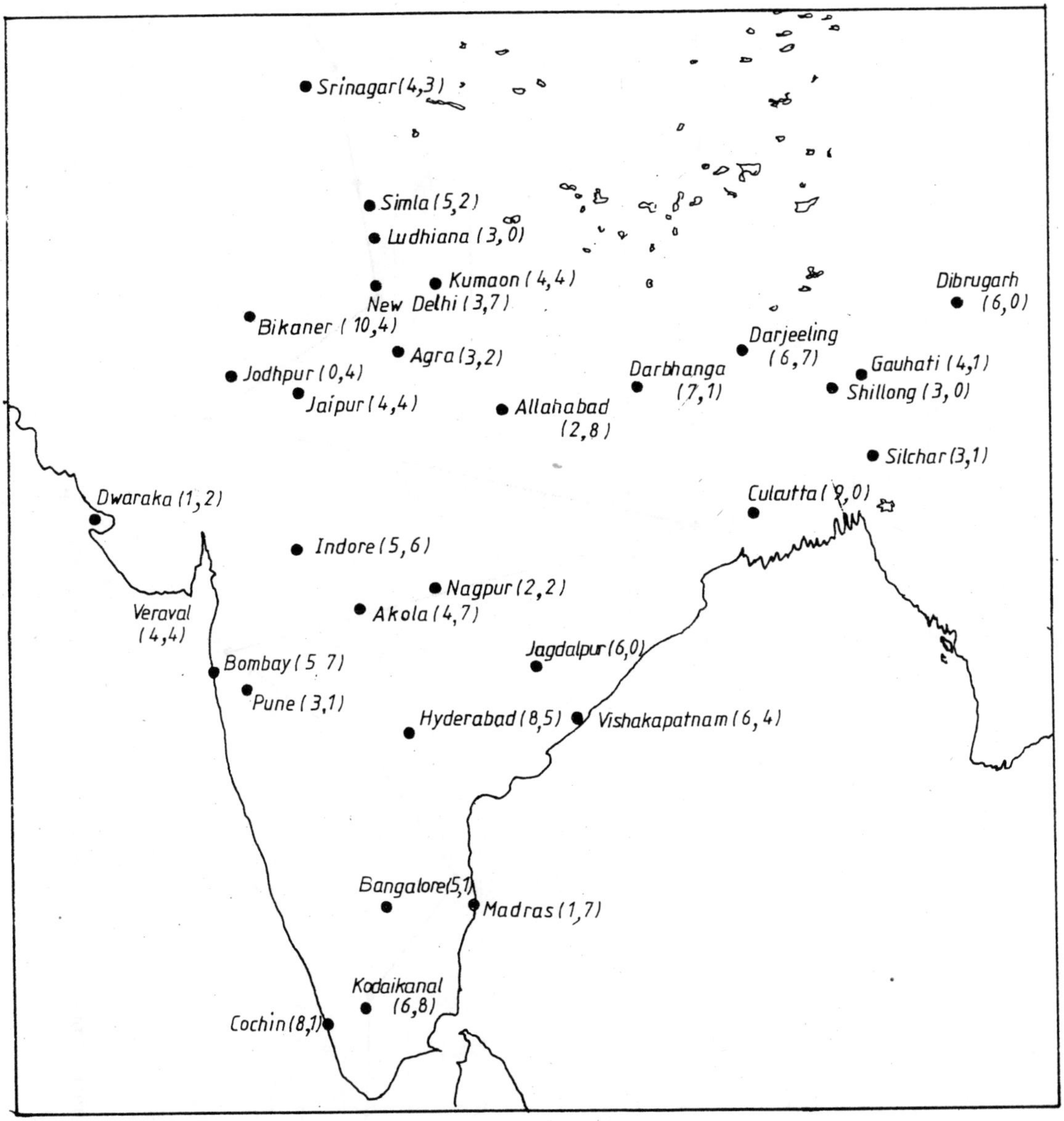

Figure 4 Regional distribution of PI_T and PI_P values during the South-West monsoon period.

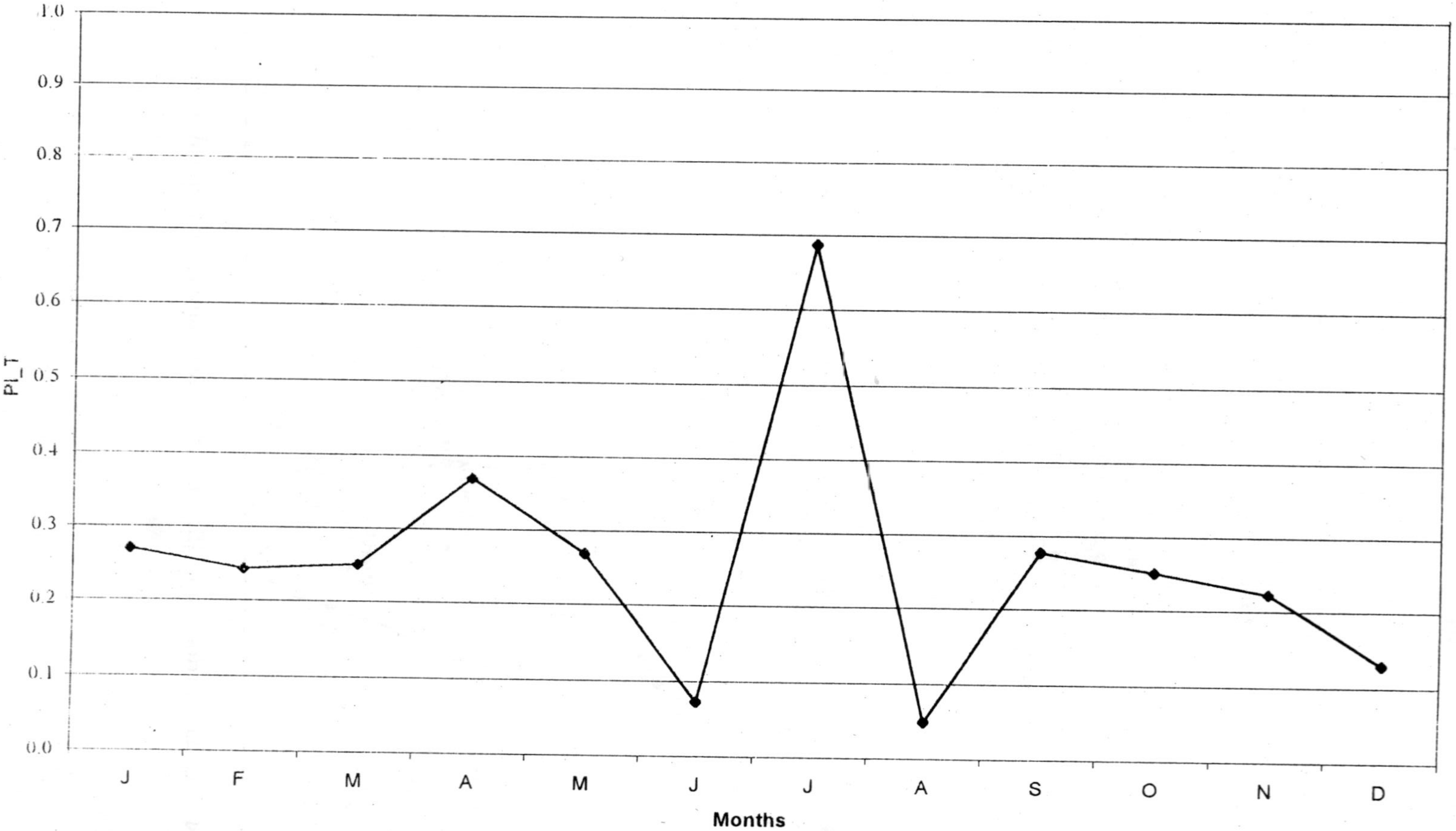

Figure 5 Temperature predictability graph for Nagpur showing the variation of the monthly PI_T values with the months.

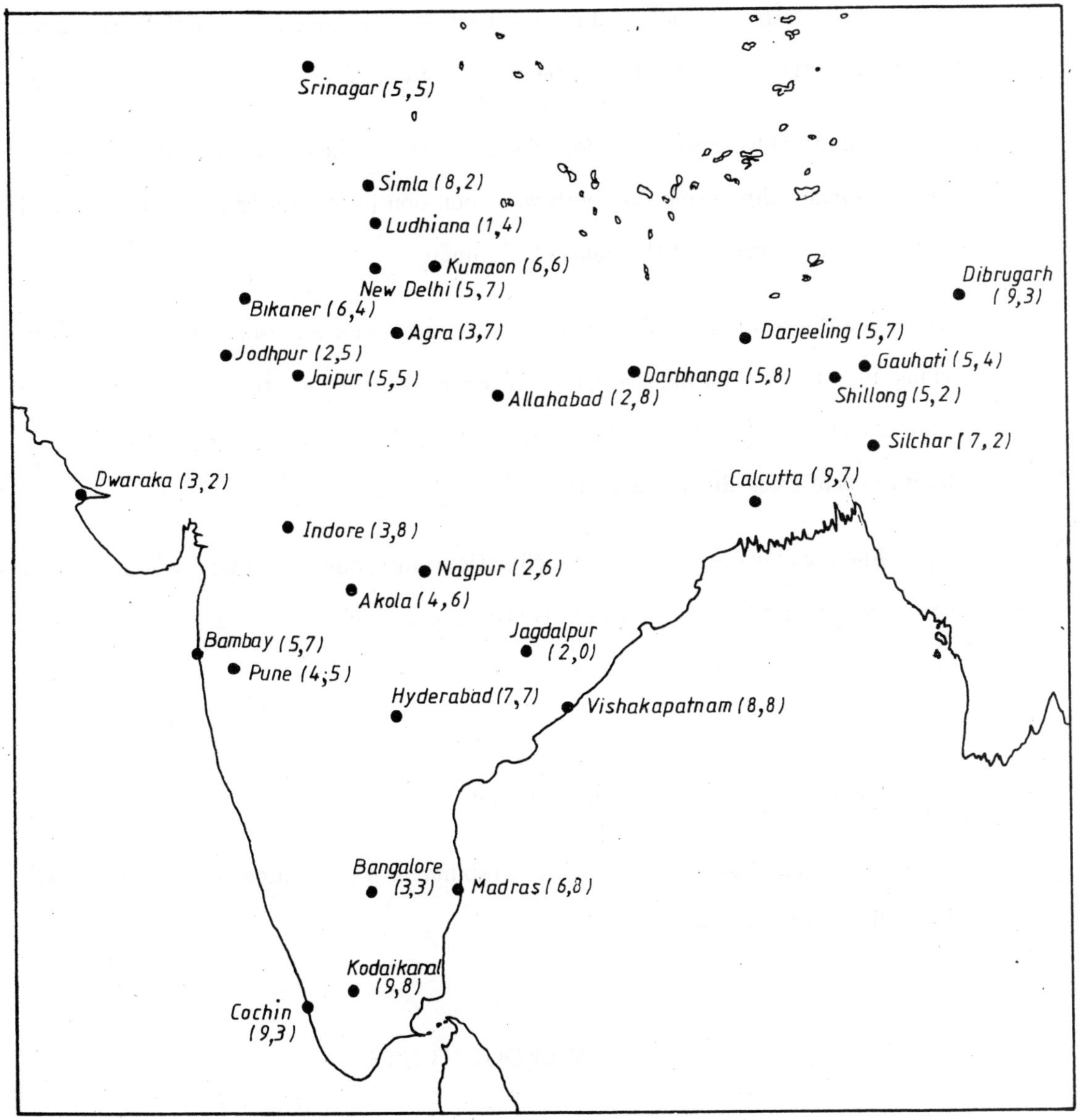

Figure 6 Regional distribution of PI_T and PI_P values during the North-East monsoon period.

VI. CONCLUSIONS

1. The mean monthly pressure for all stations has a characteristic trough invariably during the months of June and July, irrespective of the amount of rainfall the station receives and the season in which it receives the rainfall.

2. The mean monthly temperature for all stations shows the expected crest during summer. A shallow dip during the south-west monsoon (except for stations in north-east India) is a consequence of the monsoonal rainfall.

3. There is a discrepancy between predictability indices for a season and those for individual months constituting the season. There is no significant regional pattern on the subcontinental scale in the predictabilty indices. These observations point to possible limitations of gross climatic models.

4. Extending the above analysis to paleoclimatic studies, one should be cautious in inferring regional climatic patterns from continental paleoclimatic data which represents local short-lived depositional events.

5. The dynamics of south-west monsoon is apparently affected by the dynamics during the winter to summer transition period prior to the monsoon.

6. During the north-east monsoon, unpredictability in precipitation is linked to unpredictability in pressure for most stations.

ACKNOWLEDGEMENTS

We would like to thank Prof. N. V. Joshi for useful discussions. This work was supported by a research grant from DRDO under the Engineering Mathematics Programme at IISc. GR is also associated with the Jawaharlal Nehru Centre for Advanced Scientific Research, Bangalore as a Honorary Faculty.

REFERENCES

[1] Rangarajan G. and Sant D. A. A climate predictability index and its applications. *Geophys. Res. Lett.* **24** 1239-1242 (1997).

[2] Vose R. S. *et. al.* The Global Historical Climatology Network: Long-Term Monthly Temperature, Precipitation, Sea Level Pressure, and Station Pressure Data. National Oceanic and Atmospheric Administration, Asheville (1995).

[3] Overpeck J., Anderson D., Trumbore S., Prell W., The south-west Indian monsoon over the last 18000 years. *Climate Dynamics* **12** 213-225(1996).

[4] Prell W. L., Murray D. W., Clemens, S. C., Anderson D. M., *Synthesis of Results from Scientific Drilling in the Indian Ocean* Geophysical Monograph 70, American Geophysical Union, New York pp. 447-469 (1992).

[5] Hurst H. E., Black R. P., Simaika Y. M., *Long-Term Storage: An Experimental Study,* Constable, London (1965).

[6] Mandelbrot B. B., Wallis J. R. Some long-run properties of geophysical records. *Water Resources Res* **5** 321-340 (1969).

[7] Flugeman R. H. Jr. and Snow R. S. Fractal analysis of long-range paleoclimatic data: Oxygen isotope record of Pacific core V28-239. *Pure Appl Geophys* **131** 307-313 (1989).

[8] Hsui A. T., Rust K. A., Klein G. D. A fractal analysis of Quarternary, Cenozoic-Mesozoic, and Late Pennsylvanian Sea Level Changes. *J Geophys Res* **98B** 21963-21967 (1993).

[9] Turcotte D. L., *Fractals and Chaos in Geology and Geophysics,* Cambridge University Press, New York (1992).

[10] Voss R. F. In *Scaling Phenomena in Disordered Systems,* (Edited by Pynn R. and Skjeltorp A.) Plenum, New York (1985).

[11] Sirocko F., Sarnthein M., Erlenkeuser H., Lang H., Arnold M., Duplessy J. C. Century-scale events in monsoonal climate over past 24000 years. *Nature* **364** 322-324 (1993).

[12] Van Campo E., Duplessy J. C., Rossignol-Strick M. Climatic conditions deduced from a 150Kyr oxygen-isotope-pollen record from the Arabian sea. *Nature* **296** 56-59 (1982).

[13] Barnett T. P., Dumenil L., Schlese U., Roeckner E., Latif M. The effect of Eurasian snow cover on regional and global climate variations. *J Atmos Sci* **46** 661-68 (1989).

Modern Applied Mathematics
Editor: J.C. Misra

Numerical Modelling of the North Indian Ocean Circulation

Indu Jain, S.K. Dube, A.D. Rao and P.C. Sinha

Centre for Atmospheric Sciences, Indian Institute of Technology Delhi, Hauz Khas, New Delhi-110016, India

Abstract: *A three-dimensional, primitive equation, time dependent, sigma coordinate, baroclinic and free surface numerical model (Princeton Ocean Model) has been implemented for the North Indian Ocean, from 30°E to 120°E and 26°S to 25°N to study the physical, dynamical and thermodynamical processes in the region. The model uses 26 levels in vertical. The horizontal grid uses rectilinear coordinate system and an "Arakawa C" differencing scheme. The horizontal time differencing is explicit whereas the vertical differencing is implicit. The latter eliminates time constraints for the vertical coordinate and permits the use of fine vertical resolution in the surface and bottom boundary layer. After a spin-up period with mean monthly wind stress forcing derived from the 30 years (1950-79) of Comprehensive Oceanographic Atmospheric Data Sets (COADS), the model was forced with NCMRWF winds to generate a hindcast for daily May, June and July 2001.*

Princeton Ocean Model (POM) is used to simulate the currents, sea surface elevation, SST and vertical structure of temperature and salinity for SW monsoon period for the North Indian Ocean region. The model captures many of the significant general features of the region like the presence of strong westward and southwestward flowing South Equatorial Current (SEC) in the entire southern tropical Indian Ocean during southwest monsoon. The northeastward flowing Somali Current is also produced by the model and subsequently it turns eastward and southeastward at around 12°N. A weak anticyclonic gyre off Madras and large eastward transport of water off Visakhapatnam is also produced by the model. Hence, the model computed currents are reasonably in good agreement with the observed large-scale surface circulation from an analysis of ship-drift data [35]. Upwelling and downwelling zones identified by the model are in general agreement with available observations and are in consistent with the dynamics of currents in the Bay of Bengal and Arabian Sea.

A comparative study of the results with and without inclusion of fresh water discharge from one of the greatest river systems of the world, that is, Ganga-Brahmaputra-Meghana in the head of Bay during the southwest monsoon season (July) have been made. The simulation provide definitive evidence that, the combined effects of simultaneous presence of winds and discharge of fresh water during monsoon season has modifying effect on the surface circulation pattern of the Bay of Bengal.

Keywords: North Indian Ocean, Numerical model, North Indian Ocean circulation, surface current, Surface temperature and salinity.

1. Introduction

The Northern Indian Ocean is interesting due to its unique annually-reversing monsoon wind system. These monsoon winds result in an equatorial current system and currents along the Somali coast that is unique to the Indian Ocean. North Indian Ocean plays a very significant role in regulating the monsoons over the Indian subcontinent. The large scale air-sea interaction processes on monthly, seasonal and intra-seasonal time scales in the Bay of Bengal and Arabian Sea are mainly controlled by basin scale 3-dimensional circulation which generates temperature and salinity inhomogeneities in the sea and modifies the heat budget parameters at the air-sea interface.

Following the work of [6], in the last two decades there have been considerable mathematical modeling work for the Indian Ocean, e.g., [8], [14], [15], [16], and [17]. A three dimensional oceanic general circulation model [6] with seven levels in vertical direction, whereas, most of the other workers used layer models with assumption of one or more active layers over an infinitely deep motionless layer, these layer models had less vertical resolution and less prognostic equations require relatively small computational resources compared to oceanic general circulation models. A one and half layer model with two different wind forcings [16] simulated realistically the Somali Current and the associated gyre system. [15] used 23-years mean monthly wind data [5] and showed the interannual variability in Somali Current and the Indian Ocean. Simulation studies for the sub-surface circulations [14] besides the surface circulation of the Indian Ocean using a three and half layer model. The relationship between the interannual variability in model fields of central Arabian Sea with the Indian monsoon rainfall was investigated by [10].

It is now well recognized that the large scale circulation of the Bay of Bengal is greatly influenced because of the presence of large quantities of fresh water discharge from one of the greatest river systems of the world, that is, Ganga-Brahmaputra-Meghana. The fresh water discharge from these rivers is highly variable from season to season. The greatest change in the surface circulation and salinity occur during the period of maximum river drainage towards the end of southwest monsoon season. The distribution of low salinity waters at the head of Bay of Bengal varies with the direction of the currents. The counter clock-wise circulation of the southwest monsoon season transports this low-salinity water westward towards the east-coast of India as far as Visakhapatnam coast. It may therefore be expected that any variation of the water transport by these rivers will affect the circulation pattern and distribution of the physical oceanographic parameters in the Bay of Bengal. Keeping this in view, an attempt has been made towards the study of a climatological ocean circulation in the Bay of Bengal which includes the fresh water discharge in the head of Bay from the northern river systems.

In the present study, a Princeton Ocean Model (POM) has been implemented to study the physical and dynamical processes in the North Indian Ocean. Here a sigma coordinate model with high vertical resolution and relatively low resolution in the horizontal is applied. This approach differs from previous modeling efforts emphasized horizontal resolution, but were limited in the number of vertical levels.

2. Model Setting

2.1 The governing equations

The POM model, described in detail by [3], has a free surface, bottom-following vertical sigma coordinate and a turbulence sub-model for surface and bottom mixed layer dynamics [23]. It has been applied to coastal and estuarine bodies of water [2], the Gulf Stream [21], and many other oceanic regions. A description of the model code can be found in [19]. The equations which form the basis of the circulation model describe the velocity and surface elevation field, and the salinity and temperature fields. Two simplifying approximations are used [4]; first, it is assumed that the weight of the fluid identically balances the pressure (hydrostatic assumption), and second, density differences are neglected unless the differences are multiplied by gravity (Boussinesq approximation).

Consider a system of orthogonal Cartesian co-ordinates with x increasing eastward, y increasing northward, and z increasing vertically upwards. The free surface is located at $z=\eta(x,y,t)$ and the bottom is at $z=-H(x,y)$. It is a primitive equation model with a free surface, a split mode time step, and solves the following equations for the ocean velocity U, V, W, potential temperature θ, and salinity S. The continuity equation is

$$\frac{\partial U}{\partial x}+\frac{\partial V}{\partial y}+\frac{\partial W}{\partial z}=0 \tag{1}$$

The Reynolds momentum equations are

$$\frac{\partial U}{\partial t}+U\frac{\partial U}{\partial x}+V\frac{\partial U}{\partial y}+W\frac{\partial U}{\partial z}-fV=-\frac{1}{\rho_0}\frac{\partial P}{\partial x}+\frac{\partial}{\partial z}(K_M\frac{\partial U}{\partial z})+F_x \tag{2}$$

$$\frac{\partial V}{\partial t}+U\frac{\partial V}{\partial x}+V\frac{\partial V}{\partial y}+W\frac{\partial V}{\partial z}+fU=-\frac{1}{\rho_0}\frac{\partial P}{\partial y}+\frac{\partial}{\partial z}(K_M\frac{\partial V}{\partial z})+F_y \tag{3}$$

$$\rho g=-\frac{\partial P}{\partial z} \tag{4}$$

with ρ_0 the reference density, ρ the in situ density, g the gravitational acceleration, P the pressure, K_M the vertical eddy diffusivity of turbulent momentum mixing. A latitudinal variation of the Coriolis parameter, f, is introduced by use of the β plane approximation. The horizontal diffusion terms F_X, and F_y are parameterized using a Smagorinsky horizontal diffusion formulation [20].

The pressure at depth z can be obtained by integrating the vertical component of the equation of motion, (4), from z to the free surface η, and is

$$P(x,y,z,t)=P_a+g\rho_0\eta+g\int_z^0\rho(x,y,z,t)dz \tag{5}$$

Henceforth, the atmospheric pressure, P_a is assumed constant.

The conservation equations for temperature and salinity may be written as

$$\frac{\partial\theta}{\partial t}+U\frac{\partial\theta}{\partial x}+V\frac{\partial\theta}{\partial y}+W\frac{\partial\theta}{\partial z}=\frac{\partial}{\partial z}\left(K_H\frac{\partial\theta}{\partial z}\right)+F_\theta \tag{6}$$

$$\frac{\partial S}{\partial t}+U\frac{\partial S}{\partial x}+V\frac{\partial S}{\partial y}+W\frac{\partial S}{\partial z}=\frac{\partial}{\partial z}\left(K_H\frac{\partial S}{\partial z}\right)+F_S \tag{7}$$

Where θ the potential temperature and S is the salinity. The vertical eddy diffusivity for turbulent mixing of heat and salt is denoted as K_H. Using the temperature and salinity, the density is computed according to an equation of state of the form

$$\rho=\rho\,(\theta, S) \tag{8}$$

The potential density is ρ, that is, the density evaluated as a function of potential temperature and salinity but at atmospheric pressure; it provides accurate density information to calculate horizontal baroclinic gradients which enter in the pressure gradient terms and the vertical stability of the water column which enters into the turbulence closure model even in deep water when pressure effects become important.

All of the motions induced by small–scale processes not directly resolved by the model grid (subgrid scale) is parameterized in terms of horizontal mixing processes. The horizontal diffusion terms F_x, F_y, F_θ, and F_S found in (2), (3), (6), and (7) represent these unresolved processes and in analogy to molecular diffusion can be written as

$$F_x=\frac{\partial}{\partial x}\left[2A_M\frac{\partial U}{\partial x}\right]+\frac{\partial}{\partial y}\left[A_M\left(\frac{\partial U}{\partial y}+\frac{\partial V}{\partial x}\right)\right] \tag{9a}$$

$$F_y=\frac{\partial}{\partial y}\left[2A_M\frac{\partial V}{\partial y}\right]+\frac{\partial}{\partial x}\left[A_M\left(\frac{\partial U}{\partial y}+\frac{\partial V}{\partial x}\right)\right] \tag{9b}$$

and

$$F_{\theta,S}=\frac{\partial}{\partial x}A_H\frac{\partial(\theta,S)}{\partial x}+\frac{\partial}{\partial y}A_H\frac{\partial(\theta,S)}{\partial y} \tag{10}$$

One should note that F_x and F_y are invariant to coordinate rotation. While, these horizontal diffusive terms are meant to parameterize subgrid scale processes, in practice the horizontal diffusivities, A_M and A_H, are usually required to damp small-scale computational noise. The form of F_x , F_y and $F_{\theta,S}$ allows for variable A_M and A_H but thus far they have been held constant. The diffusivities are chosen so that they do not produce excessive smoothing of real features.

2.2 Turbulence closure

The governing equations contain parameterized Reynolds stress and flux terms which account for the turbulent diffusion of momentum, heat, and salt. The parameterization of turbulence in the model is based on the work of [22].

The vertical mixing coefficients, K_M and K_H, in (2), (3), (6), and (7) are obtained by using a second order turbulence closure scheme [23] which characterizes the turbulence by equations for the turbulence kinetic energy, $q^2/2$, and a turbulence macroscale, ℓ, according to,

$$\begin{aligned}\frac{\partial q^2}{\partial t} + U\frac{\partial q^2}{\partial x} + V\frac{\partial q^2}{\partial y} + W\frac{\partial q^2}{\partial z} &= \frac{\partial}{\partial z}\left(K_q\frac{\partial q^2}{\partial z}\right) \\ &+ 2K_M\left[\left(\frac{\partial U}{\partial z}\right)^2 + \left(\frac{\partial V}{\partial z}\right)^2\right] + \frac{2g}{\rho_0}K_H\frac{\partial \rho}{\partial z} - \frac{2q^3}{B_1\ell} + F_q\end{aligned} \tag{11}$$

and

$$\begin{aligned}\frac{\partial q^2\ell}{\partial t} + U\frac{\partial q^2\ell}{\partial x} + V\frac{\partial q^2\ell}{\partial y} + W\frac{\partial q^2\ell}{\partial z} &= \frac{\partial}{\partial z}\left[K_q\frac{\partial}{\partial z}\left(q^2\ell\right)\right] \\ &+ \ell E_1 K_M\left[\left(\frac{\partial U}{\partial z}\right)^2 + \left(\frac{\partial V}{\partial z}\right)^2\right] + \frac{\ell E_1 g}{\rho_o}K_H\frac{\partial \rho}{\partial z} - \frac{q^3}{B_1}\tilde{W} + F_\ell\end{aligned} \tag{12}$$

where a wall proximity function is defined as

$$\tilde{W} \equiv 1 + E_2\left(\frac{\ell}{\kappa L}\right)^2 \tag{13}$$

and where

$$(L)^{-1} \equiv (\eta - z)^{-1} + (H + z)^{-1} \qquad (14)$$

Near surfaces it may be shown that both ℓ / κ and L are equal to the distance from the surface (κ=0.4 is the von Karman constant) so that $\widetilde{W} = 1 + E_2$. Far from the surfaces where $\ell << L, W^2 \cong 1$. The length scale provided by (12) is a characteristic length of the turbulent motion at any point in space or time. The terms F_q and F_ℓ in (11) and (12) are the horizontal mixing and are parameterized analogously to temperature and salinity by using (10).

While details of the closure model are rather involved, it is possible to reduce the prescription of the mixing coefficients K_M, K_H, and K_q to the following expressions,

$$K_M \equiv \ell q s_M \qquad (15a)$$

$$K_H \equiv \ell q s_H \qquad (15b)$$

$$K_q \equiv \ell q s_q \qquad (15c)$$

The stability functions, s_M, s_H, and s_q are analytically derived, algebric relations functionally dependent upon $\partial U/\partial z$, $\partial V/\partial z$, $g\rho_o^{-1}\, \partial \rho/\partial z$, q and ℓ.

2.3 Boundary conditions

The boundary conditions at the free surface, $z=\eta(x,y)$, are

$$\rho_o K_M \left(\frac{\partial U}{\partial z}, \frac{\partial V}{\partial z} \right) = \left(\tau_{ox}, \tau_{oy} \right) \qquad (16a)$$

$$\rho_o K_H \left(\frac{\partial \theta}{\partial z}, \frac{\partial s}{\partial z} \right) = \left(\dot{H}, \dot{S} \right) \qquad (16b)$$

$$q^2 = B_1^{2/3} u_{\tau s}^2 \qquad (16c)$$

$$q^2 \ell = 0 \qquad (16d)$$

$$W = U \frac{\partial \eta}{\partial x} + V \frac{\partial \eta}{\partial y} + \frac{\partial \eta}{\partial t} \qquad (16e)$$

where (τ_{ox}, τ_{oy}) is the surface wind stress vector with the friction velocity, $u_{\tau s}$, the magnitude of the vector. The quantity $B_1^{2/3}$ is an empirical constant (6.51) arising from the turbulence closure relations. The net ocean heat flux is $\dot{H}$ and $\dot{S}$ is the surface salinity flux. On the side walls and bottom of the basin, the normal gradients of θ and S are zero so that there are no advective and diffusive heat and salt fluxes across these boundaries. At the lower boundary (b),

$$\rho_o K_M\left(\frac{\partial U}{\partial z}, \frac{\partial V}{\partial z}\right) = \left(\tau_{bx}, \tau_{by}\right) \tag{17a}$$

$$q^2 = B_1^{2/3} u_{\tau b}^2 \tag{17b}$$

$$q^2 \ell = 0 \tag{17c}$$

$$W = -U_b \frac{\partial H}{\partial x} - V_b \frac{\partial H}{\partial y} \tag{17d}$$

Where H(x,y) is the bottom topography and $u_{\tau b}$ is the friction velocity associated with the bottom frictional stress $\left(\tau_{bx}, \tau_{by}\right)$. The bottom stress is determined by matching velocities with the logarithmic law of the wall. Specifically,

$$\vec{\tau_b} = \rho_o C_D \mid \vec{V_b} \mid \vec{V_b} \tag{18}$$

with value of the drag coefficient C_D given by

$$C_D = \left[\frac{1}{\kappa} \ln(H + z_b)/z_o\right]^{-2} \tag{19a}$$

where z_b and V_b are the grid points and corresponding velocity in the grid point nearest the bottom and κ is the von Karman constant. The finer result of (18) and (19) in conjunction with the turbulent closure derived K_M is that the circulation will yield

$$\vec{V} = \left(\tau_b/\kappa u_{\tau b}\right)\ln\left(z/z_o\right) \tag{19b}$$

in the lower boundary region if enough resolution is provided. In those instances where the bottom boundary layer is not well resolved it is more appropriate to specify C_D=0.0025. The actual algorithm is to set C_D to the larger of the two values given by (19a) and 0.0025. The parameter z_o depends on the local bottom roughness; in the absence of specific information z_o=1 cm is used as suggested by [33].

Open lateral boundary conditions are problematic since one must parameterize the environment exterior to the relevant domain. Two types of open boundaries exist, inflow and outflow. Temperature and salinity are prescribed from data at an inflowing boundary, whereas at outflow boundaries,

$$\frac{\partial}{\partial t}(\theta,S)+U_n\frac{\partial}{\partial n}(\theta,S)=0 \tag{20}$$

is solved where the subscript n is the coordinate normal to the boundary. Turbulence kinetic energy and the macroscale scale quantity ($q^2\ell$) are calculated with sufficient accuracy at the boundaries by neglecting the advection in comparison with other terms in their respective equations.

2.4 Stability Constraints

The leap-frog differencing scheme is used for the time stepping introduces a tendency for the solution at even and odd time steps to split. This time splitting is removed by a weak filter, where the solution is smoothed at each time step according to

$$F_S^n = F^n + \frac{\alpha}{2}\left(F^{n+1} - 2F^n + F_S^{n-1}\right) \tag{21}$$

where α=0.05 and F_S is a smoothed solution. This technique introduces less damping than either the Euler-backward or forward stepping techniques.

The Courant-Friedrichs-Levy (CFL) computational stability condition on the vertically integrated, external mode, transport equations limits the time step as shown by [1] according to

$$\Delta t \le \frac{1}{C_t}\left(\frac{1}{\Delta x^2}+\frac{1}{\Delta y^2}\right)^{-1/2} \tag{22a}$$

where $C_t = 2(gH)^{1/2} + \overline{U}_{max}$ (22b)

$\overline{U}_{max}$ is the maximum average velocity expected. The time step criteria for the internal mode is

$$\Delta t \le \frac{1}{C_T}\left(\frac{1}{\Delta x^2}+\frac{1}{\Delta y^2}\right)^{-1/2} \tag{23}$$

where $C_T = 2C + U_{max}$, with C being the maximum internal gravity wave speed and U_{max} is the maximum advective speed.

2.5 The model grid

The model covers an analysis area lying between 26°S and 25°N and between 30°E and 120°E. The model has horizontal rectilinear grids and contains 87 x 57 grid points with grid resolution of 1° x 1° but a high vertical resolution consisting of 26 sigma levels. The σ is defined as $\sigma=(z-\eta)/(H+\eta)$, where $\eta(x,y)$ and $H(x,y)$ are the surface elevation and bottom topography, respectively.

The vertical resolution is higher in the upper layers to depict better mixed layer processes and the equatorial and near-equatorial surface current system. This approach differs from previous modeling efforts for this region, in that the emphasis is on the vertical structure of water masses and currents in the upper layers, whereas the earlier efforts using the layered approach had generally greater horizontal resolution, but were limited in the number of levels in the vertical. The horizontal time differencing is explicit, whereas the vertical differencing is implicit. The latter eliminates time constraints for the vertical coordinate and permits the use of fine vertical resolution near the surface and in shallow regions. The model has free surface and a split time step. The two-dimensional external mode that uses a short time step of 150 sec based on the CFL condition and the external wave speed. And a three-dimensional internal mode uses a long time step of 900 sec based on the CFL condition and the internal wave speed. An implicit numerical scheme in the vertical direction and a mode-splitting technique in time are adopted for computational efficiency. The horizontal finite difference scheme is staggered. At the three open boundaries (except western boundary), Sommerfeld-type radiation conditions are used.

2.6 Bathymetry and initial conditions

The model bottom topography is derived from the 5′ resolution of ETOPO5 NGDC database. A bilinear interpolation was used to reduce the topographic gradients that could otherwise cause spurious along-slope currents in a sigma coordinate model [12]. The analysis area and resulting bottom topography are shown in Fig. 1. The minimum depth of the coastal boundary is 10m; the contour interval in Figure 1 is 500m. During model spin-up, the model

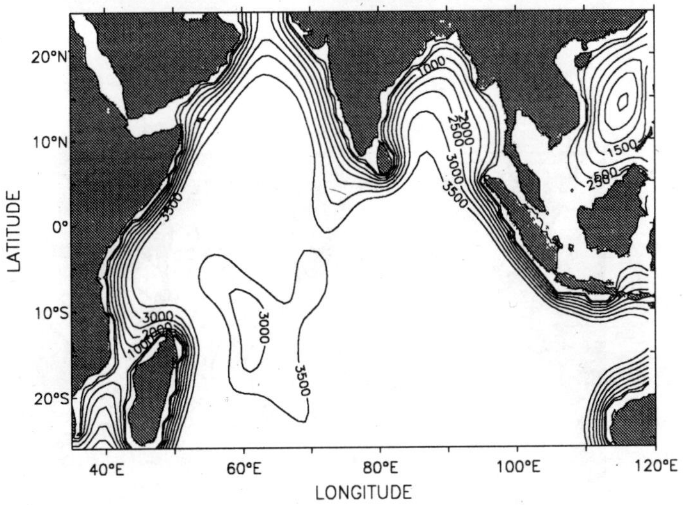

Fig. 1 Bathymetric height (m)

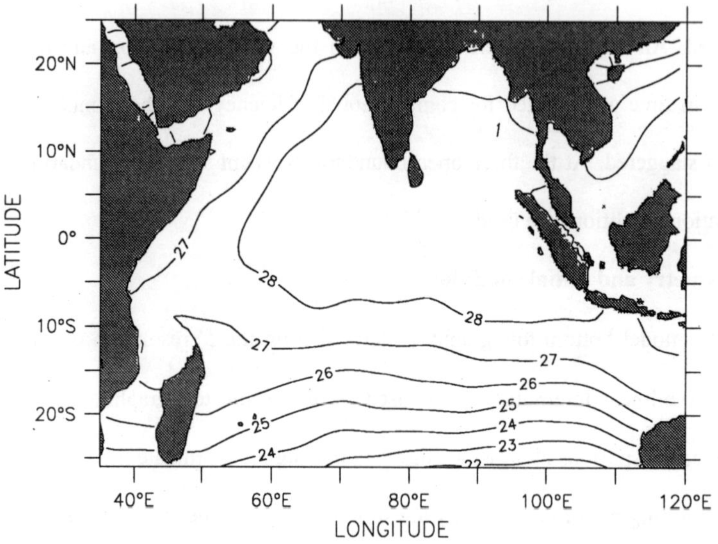

Fig. 2 (a) Surface initial temperature (^{o}C).

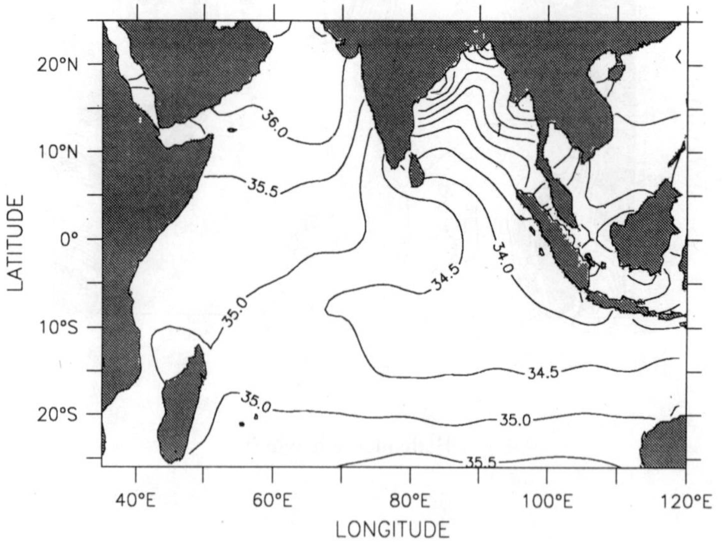

Fig.2 (b) Surface initial salinity (ppt).

(a) May

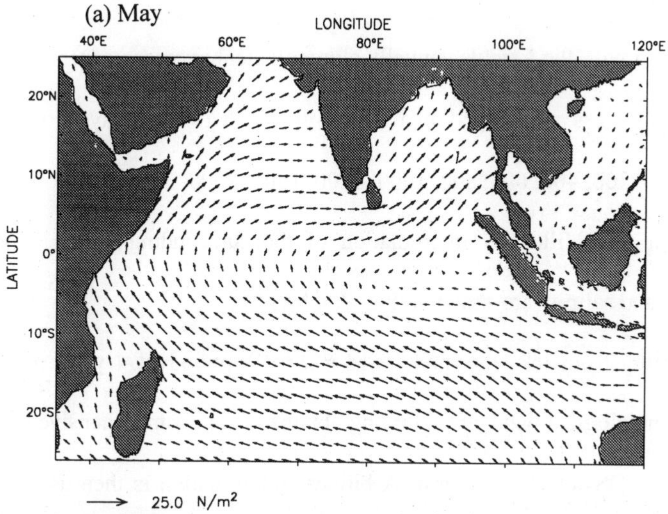

(b) June

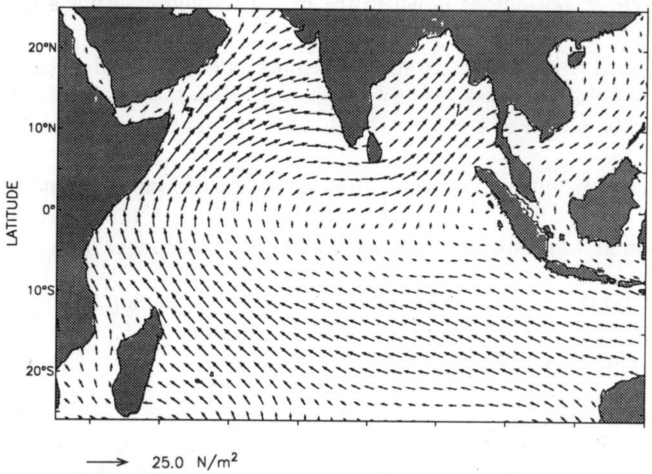

(c) July

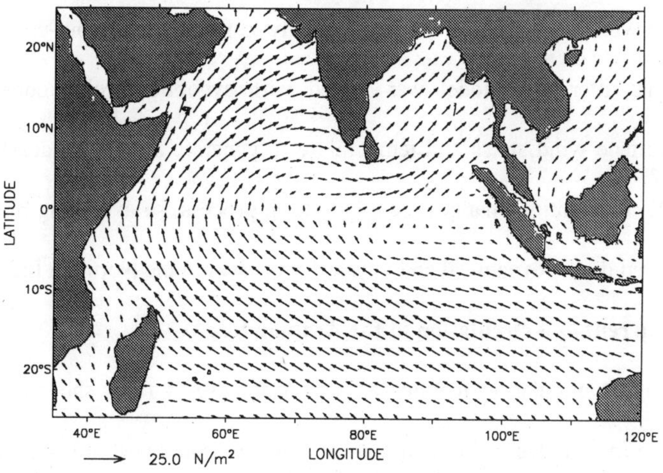

Fig. 3 NCMRWF Winds for the year 2001

was initialized with the Levitus annually climatological averages of temperature and salinity. The annual averages are mapped on a 1° x 1° grid with 19 depth layers (0, 10, 20, 30, 50, 75, 100, 150, 200, 300, 400, 600, 800, 1000, 1200, 1500, 2000, 3000, 4000). The grid points with missing data were filled by interpolation of the surrounding values. The surface initial temperature and salinity are shown in Fig. 2(a, b).

Starting from rest, the model was run for a period of 4 years and 4 months using monthly climatological wind stress forcing obtained from Comprehensive Ocean-Atmosphere Data set (COADS) on 2° x 2° grid. A bilinear interpolation is then used to bring the wind stress from their 2° resolutions down to the model resolution. After a spin-up period, the model was forced with daily sea surface temperature and wind stress analyzed by the NCMRWF for the period of 1 May – 31 July 2001, to generate a daily hindcast. Fig. 3(a-c) shows mean monthly wind-stress forcing for the months of May, June and July, respectively. The comparison of wind stress reveals that winds become stronger from pre-monsoon (May) to monsoon (July) period. During July, the wind stress distribution is similar to that of June, with slightly higher magnitudes found than that of June.

3. Results and discussion

As mentioned earlier, the model equations were integrated using mean monthly climatological wind stress, temperature and salinity as forcing for a spin-up period of 4 years and 4 months, after that model was forced with synoptic forcing consisting of daily NCMRWF surface wind stress and temperature to generate a daily hindcast for May, June and July 2001. Circulation changes are a long-term process, so averaging has been done for each month to study the dynamical and physical processes in the Northern Indian Ocean.

3.1 Somali Current and circulation in the Arabian Sea

The model circulation features as seen from Fig. 4 (a-c) show the major current structure in the Indian Ocean. In the month of June, the southwest monsoon is fully set over

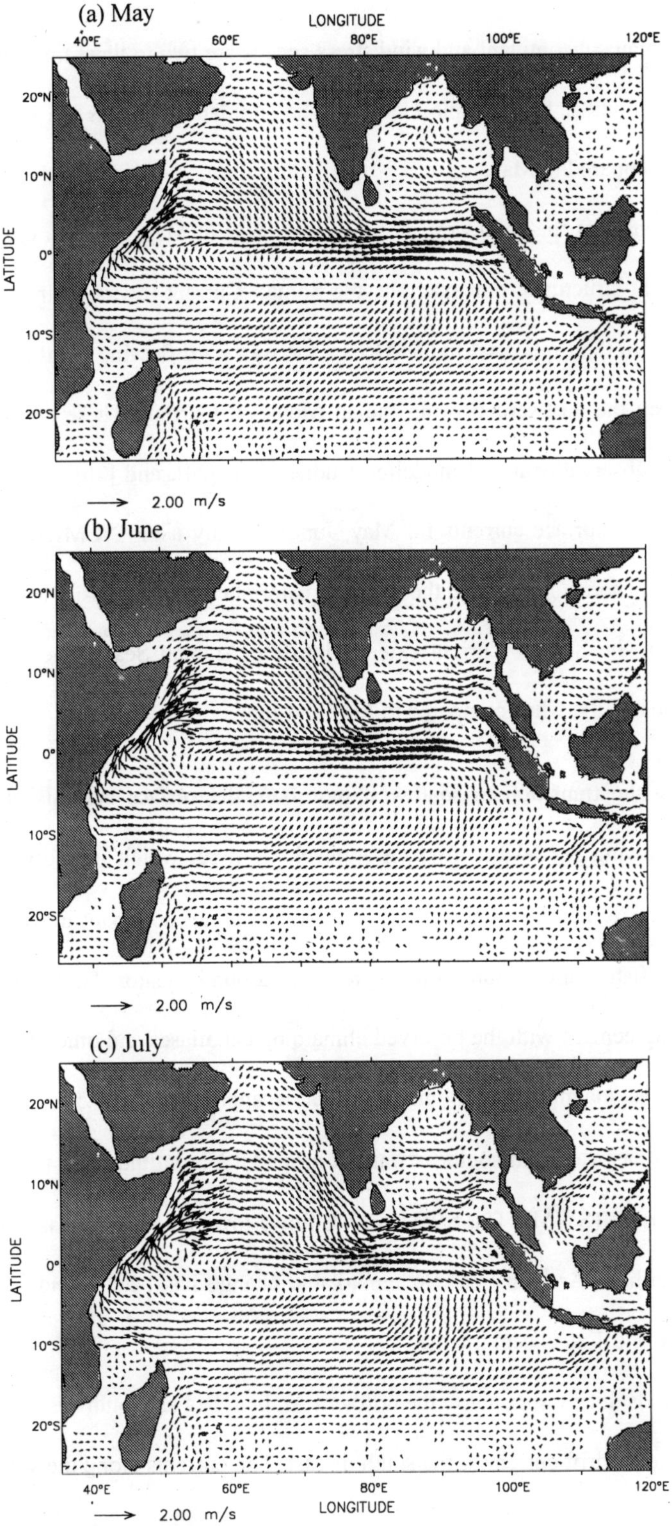

Fig. 4 Surface currents (NCMRWF) for the year 2001

the entire Indian subcontinent and wind stress speeds are two to three times higher than those of transition months, for example October. The simulated currents are intense in June and July, when compared to May.

The onset of the southwest monsoon, typically in May, result in changes in the near surface current structures. For example, the Somali Current reverses direction and flows to the northeast. In the Equatorial Indian Ocean, the westward flowing North Equatorial Current is replaced by the eastward flowing Indian Monsoon Current. These seasonal changes are seen in both observational and modeling studies [18], [24], and [30]. Fig. 4(a-c) show the model simulated surface currents for May, June and July using NCMRWF surface forcing. The Indian Monsoon Current and the South Equatorial current are easily identified. Along the African coast, the Somali Current extends into the Arabian Sea. . The simulated currents speed intense in July, when compared to May and June.

During the transition to southwest monsoon, the surface currents along the Somali coast start flowing northward. And an anticyclonic gyre, which is known as Great Whirl forms around 5°N in July whereas, this Great Whirl is not dominant in May and June [Fig. 4(a-c)]. The offshore flow south of this gyre is just around equator. These circulation features are in good agreement with the observed climatological atlases and other model studies [13], [27], [35]. The magnitude of Somali Current is approximately 1.5 m sec^{-1} .The Somali Current turns eastward and southeastward at around 12°N latitude and feeds the Southwest Monsoon Current (SWM). The SWM, which flows towards south of east, could be seen in the entire northern Arabian Sea. No remarkable change has been found in current pattern between June and July except intensity.

The winds during summer in the northern Arabian Sea are southwesterly and this would induce an Ekman drift current to the south of east. The currents along the west coast of India

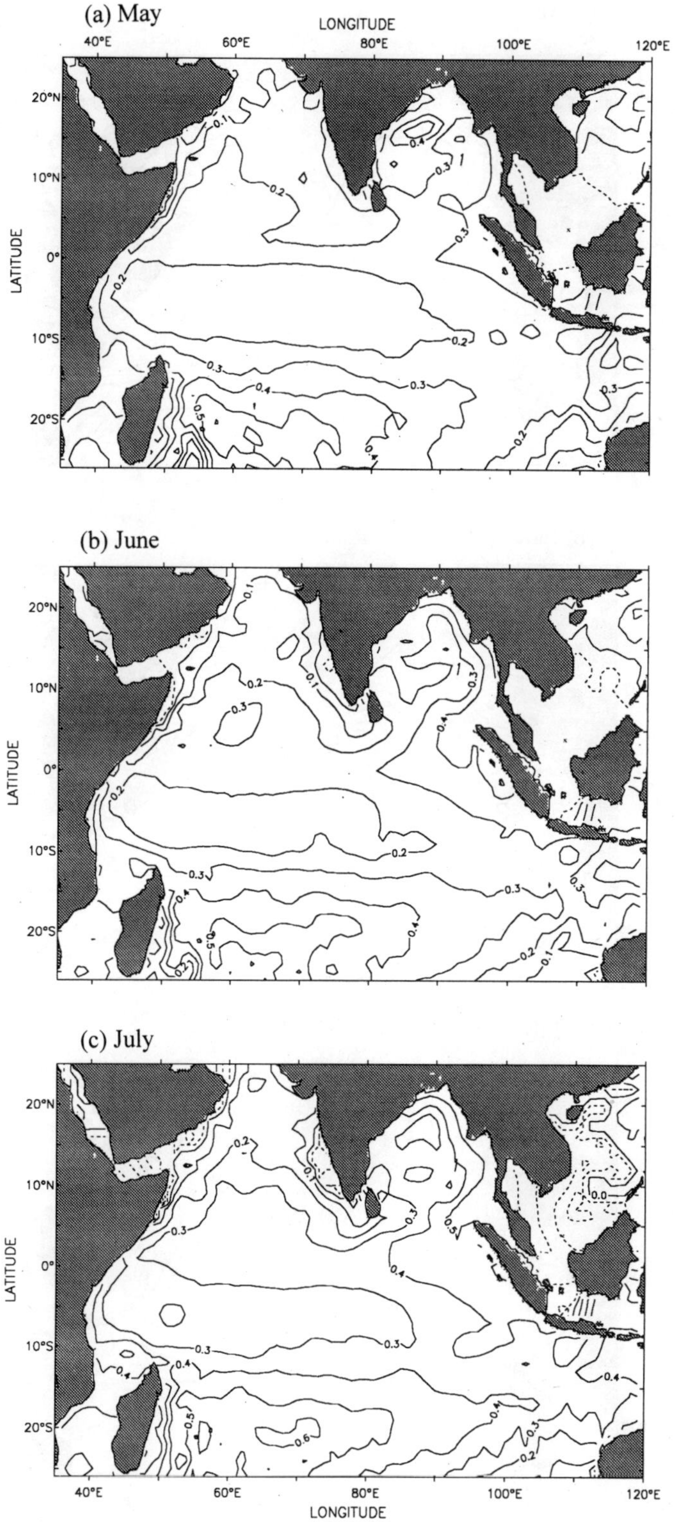

Fig. 5 Sea surface elevation (m) for the year 2001

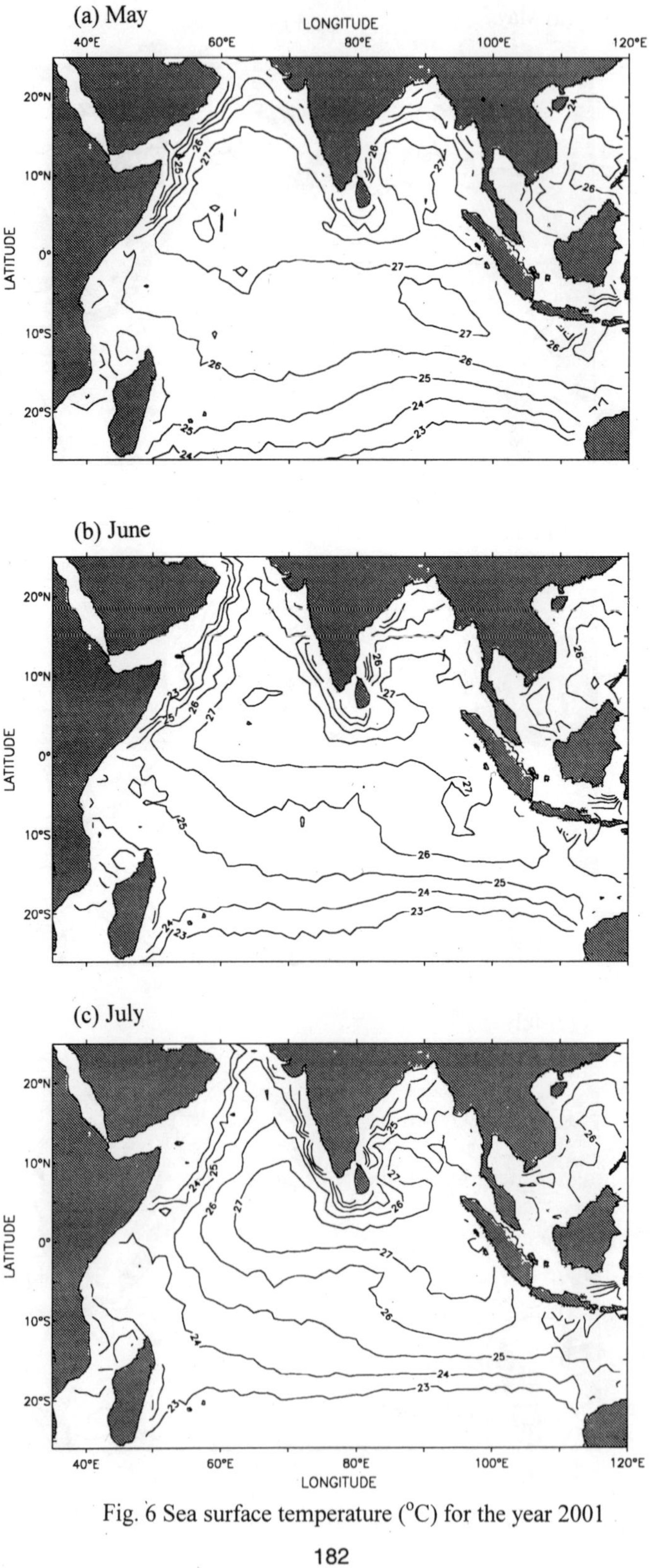

Fig. 6 Sea surface temperature (°C) for the year 2001

do not completely follow the coastline and have mostly eastward tendency, which is in general a realistic feature. In July, the southerly coastal currents along the west coast of India are very favorable to the upwelling processes and in general are favorable for lowering of sea level on the west coast. The southerly coastal current on the west coast is strong near the Indian peninsula [Fig. 4(a-c)] and therefore the lowering of the sea level [Fig. 5(a-c)] as well as the drop in temperature [Fig. 6(a-c)] is also more significant in this region, say off Cochin. This is very much consistent with the observations. Further, although during the period of study May, June and July in general favorable for upwelling and lowering of sea level and temperature drop, it may be seen that the process start in May gradually assumes a peak in July as may be seen in the Figs. 5 and 6. From May to July, winds become stronger along the African coast and coastal upwelling starts along the coast and persist during the monsoon period. In order to confirm the occurrence of upwelling along the west coast of India and east coast of Africa, the sea surface elevations for May, June and July are shown in Fig. 5(a-c). The negative elevation along the west coast of India and east coast of Africa is as a result of offshore Ekman transport. This gives an indication that the region near the coast is an upwelling zone. The computed results generally agree with the ship drift data compiled [7]

3.2 Equatorial currents

The model-produced North Equatorial Current (NEC) flows eastward in summer which is more appropriately known as Monsoon Current (MC) [Fig. 4(a-c)], the summer surface mean current pattern shows the presence of westward and southwestward flowing South Equatorial Current (SEC) in the entire tropical Indian Ocean between 5°S and 20°S latitude and is in agreement with observation. This current is mainly driven by prevailing southeast trades, which provide the major mechanical forcing for the basin-wide upper circulation in the southern tropical Indian Ocean. The SEC is the most powerful, perennial current system in the southern tropical Indian Ocean.

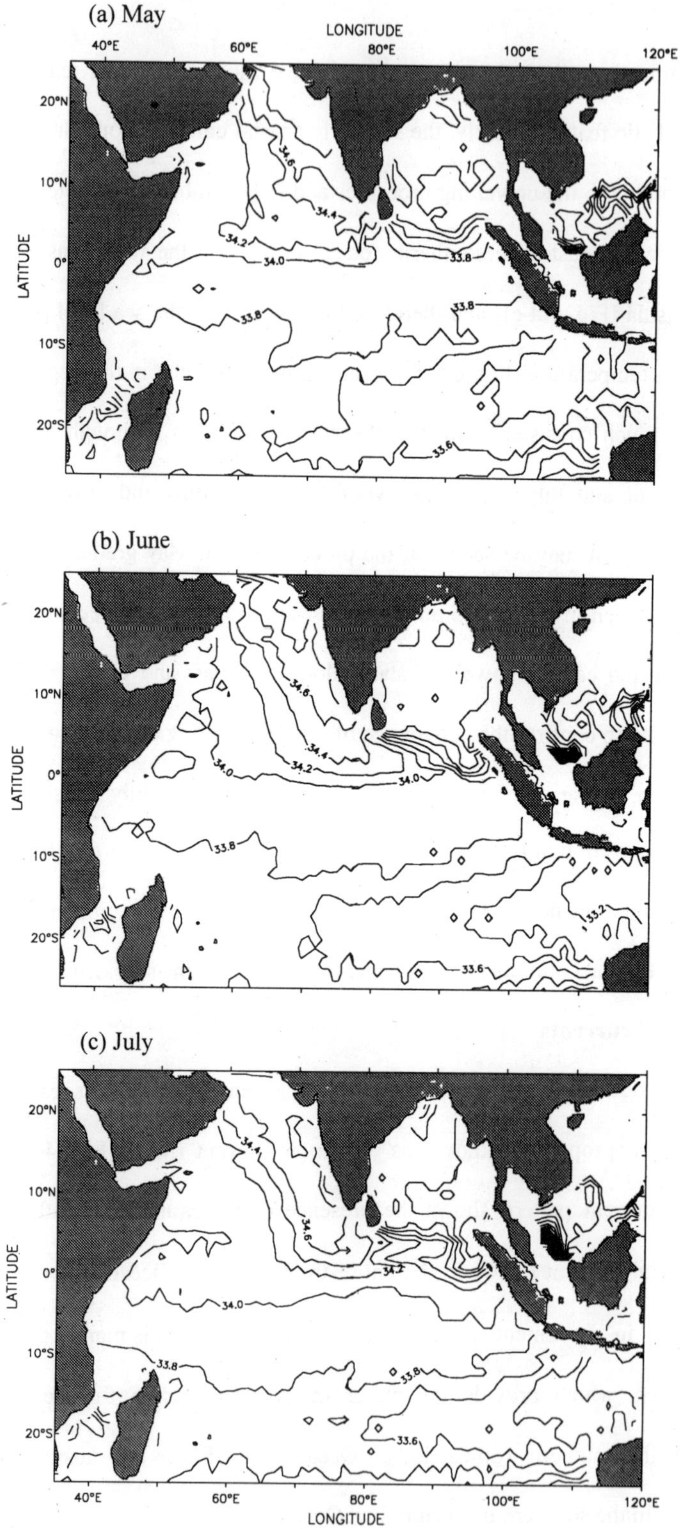

Fig. 7 Surface salinity (ppt) for the year 2001

The computed SEC flows primarily toward the south of west, although zonal component could be seen in the northern portion of SEC. The SEC, which flows south of west 8°S and 20°S splits to the east coast of Madagascar, the northward flowing branch of SEC along eastern tip of Madagascar is identified by Fig. 4, whereas other branch flowing southward. The latitude of this splitting is consistent with the observed split at 17°S [29], [31] and the model results of [34]. The southward flowing branch is called the East Madagascar Coastal Current (EMCC). The northward flowing coastal current along the east coast of Madagascar feeds into the northward flowing Somali Current (SC) via the East African Coastal Current (EACC), after traveling westward at the northern tip of Madagascar. These features can be identified by the Fig. 4(a-c). The magnitude of SEC varies between 50 cm sec^{-1} to 1 m sec^{-1}. The SEC, on reaching the East African coast, feeds initially the East African Coastal Current (EACC) and subsequently the northeastward flowing strong Somali Current (SC).

The equatorial current system provides the mechanism by which salt is exchanged between the eastern and western Indian Ocean. Fig. 7(a-c) depicts the surface salinity for the southwest monsoon season. At this time the eastward flowing Indian Monsoon Current (Fig. 4) carries saline water from the Arabian Sea into the southern Bay of Bengal. And at the same, relatively fresh waters is carried into the Arabian Sea along the African Coast by the Somali Current.

Hence the persistent equatorial currents transport salt between the eastern and western Indian Ocean. During the southwest season (Fig.7), the Equatorial Current carries saline water masses from the western to eastern Indian Ocean while the south Equatorial Current carries fresh water to the west.

3.3 Circulation in the Bay of Bengal

The model simulated surface currents in the Bay of Bengal during southwest monsoon for the month of May, June and July are shown in Fig. 4(a-c). Model computed features of the surface currents are in good agreement with the observed circulation [35]. The model reproduced the western boundary current between 81°E and 82°E, which was identified by [28] with the help of a hydrographic survey in 1977 along 13°N off Madras.

The computed western boundary current extends about 300 km off the east coast of India and it has a maximum speed of about 40 cm sec^{-1}. We find from the figure that a weak anticyclonic outdraft is simulated around 12°N off the east coast of India. Strong currents along the east coast of India having a large eastward transport of water off the Visakhapatnam coast (~16°-17°N) is in good agreement with the climatological circulation. The model also simulates a cyclonic eddy at the head Bay in July (Fig 4c), whereas it is in developing stage in June (Fig. 4b) and completely absent in May (Fig. 4a). The maximum current speed was observed along the Sri Lanka coast, where its value attained as high as 50cm/sec and cyclonic gyre at the head Bay currents were found to be as high as 30 cm/sec. The eastward moving current simulated by the model along the Sri Lanka coast is the monsoon drift current normally observed during southwest monsoon season. The simulated circulation pattern for June (Fig 4b) was analogous to that of the July, the only difference being that the intensity of the cyclonic eddy formed at the head Bay was reduced. The model computed positive sea-surface elevations in the northeastern sector of the Bay of Bengal are a result of entrainment of water towards the northeast region [Fig. 4(a-c)]. This is in good agreement with the inferences of [26]. Hence, the model simulates the major features of the circulation including eddies. The results are in good agreement with those produced by a more computationally expensive model of the Indian Ocean [25].

3.4 A comparative study with and without inclusion of fresh water discharge at head of Bay of Bengal

Two basic experiments are described in this paper. The first corresponds to the case of

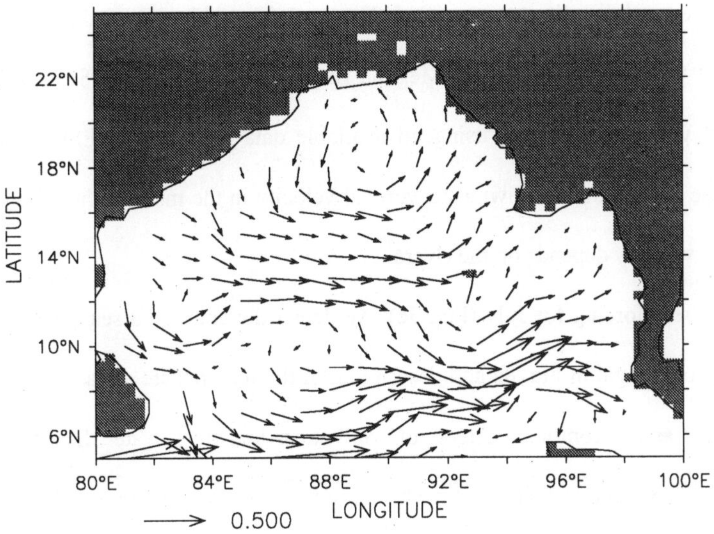

Fig. 8(a) Surface currents (m/s) without fresh water for July july

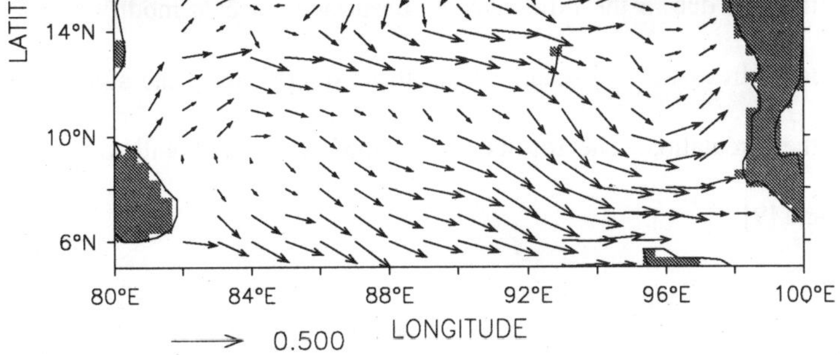

(b) Surface currents (m/s) with fresh water for July

pure wind stress forcing, the second to a combination of wind stress and forced fresh water in the head of Bay. The applied surface wind stress is determined from a bulk quadratic law in which the wind speed and direction are specified. The inward monthly averaged combined flux of fresh water out of Ganga-Brahmaputra-Meghna river system is taken as $81955 m^3/s$; this monthly averaged value is calculated using all available data from the UNESCO database. The river discharge is specified as downward vertical velocity in the model. Thus the value of downward vertical velocity depends on the rivers.

Using wind stress forcing for July (Fig. 3c), we depict the computed sea surface current vector in the Bay of Bengal in Fig. 8a and description of the features are already discussed in section 3.3. Where as the combined effects of wind-stress and fresh water discharge in the Bay is shown in Fig. 8b, shows that the currents along the coast do not completely follow the coastline and they have the tendency to move eastward. At the head of Bay, there is a clear formation of a cyclonic gyre in Fig 8a. As there is huge fresh water discharge during July from northern rivers, the presence of the counter clockwise circulation brings low-salinity waters to the coastal region up to the latitude of Visakhapatnam (Fig. 8b). From the figure, it may be noted that the circulation in the latitudes north of Visakhapatnam is completely dominated by the estuarial flow. The flow to the south of the latitude is unchanged and thereby limiting the local effect of the northern fresh water discharge. As observed (Fig. 8b) the surface circulation features notably is smooth in the latitudinal belt between 5° N - 8°N, which may be attributed due to the influence of subsurface currents modifying the surface flow. Hence the large fresh water discharge from these rivers during the summer monsoon period modifies the circulation. The results are in good agreement with the observations reported by [11] and [9].

Conclusions:

Princeton Ocean Model is used to simulate the general oceanographic features of the North Indian Ocean with synoptic forcing during southwest monsoon. On the basis of the results the following general conclusions may be made:

- The model produces the northeastward flowing Somali Current and subsequently it turns eastward and southeastward at around 10 °N [17].
- The presence of westward and southwestward flowing South Equatorial Current (SEC) in the entire southern tropical Indian Ocean [18].
- The model also produces a weak anticyclonic gyre off Madras and large eastward transport of water off Visakhapatnam. Hence, the model computed currents are reasonably in good agreement with the observed large-scale surface circulation from an analysis of ship-drift data [35].
- The currents along the west coast of the India (eastern sector of the Arabian Sea) do not completely follow the coastline and have mostly eastward tendency, which is in general a realistic feature.
- Upwelling and downwelling zones identified by the model are in good agreement with available observations and are in consistent with the dynamics of currents in the Bay of Bengal and Arabian Sea [26].
- The combined effects of simultaneous presence of winds and discharge of fresh water during monsoon season at head of Bay have modifying effect on the surface circulation pattern of the Bay of Bengal [11].
- The model is unable to capture the upwelling processes along the east coast of India due to coarse resolution in the horizontal direction.

References:

1. Blumberg, A.F. and Mellor, G.L., "A numerical calculation of the circulation in the Gulf of Mexico", Dynalysis of Princeton, Report No, **66**, 153 pp (1981a).

2. Blumberg, A.F. and Mellor, G.L., "Diagnostic and prognostic numerical circulation studies of the south Atlantic Bight", J. Geophys. Res., **88**, 4579-4592 (1983).

3. Blumberg, A.F. and Mellor, G.L., "A description of a three-dimensional coastal ocean circulation model", in Three-Dimensional Coastal Ocean Models, **4**, 208 pp., American Geophysical Union, Washington, D. C (1987).

4. Bryan, K., "A numerical method for the study of the circulation of the world ocean", J. Comput. Phys., 4, No. **3**, 347-376 (1969).

5. Cadet, D.L. and Diehl, B.C., "Interannual variability of surface fields over the Indian Ocean in recent decades", Mon. Weath. Rev., **112**, 1921-1985 (1984).

6. Cox, M.D., " A mathematical model of the Indian Ocean." Deep Sea Res., **17**, 47-75 (1970).

7. Cutler, A.N. and Swallow, J.C., "Surface currents in the Indian Ocean", Institute of Oceanographic Sciences, U.K. Report No. **187**, 8 pp, 36 charts (1984).

8. Das, P.K., Dube, S.K. and Rao, G.S., "A steady state model of the Somali Current", Proc. Indian Acad. Sci., **96**, 279-290 (1987).

9. Dube, S.K, Rao, A.D., Sinha, P.C. and Indu Jain, "Implications of climatic variations in the freshwater outflow on the wind induced circulation of the Bay of Bengal", Atmos. Environ., Vol. **29**, 2133-2138 (1995).

10. Dube, S.K., Luther, M.E. and O'Brien, J.J., " Relationships between interannual variability in the Arabian Sea and Indian monsoon rainfall", Meteor. Atm. Phys, **44**, 153-165 (1990).

11. Gopal Krishna, V.V and Sastry, J.S., "Surface circulation over the shelf off the east-coast of India during the southwest monsoon", Indian J. Mar. Sci., **14**, 62-65 (1985).

12. Haney, R.L., "On the pressure gradient force over step topography in sigma coordinate models", J.Phys.Oceanogr., **21**, 610-619 (1991).

13. Hasternath, S. and Greischar L.L., "Climate atlas of the Indian Ocean, Part-III, Upper ocean structure, The University of Wisconsin Press, 247 charts, (1989).

14. Jensen, R.G., "A numerical study of the seasonal variability of the Somali Current, Ph.D. dissertation, Florida State University, 118 pp (1990).

15. Luther, M.E., O'Brien, J.J., "Modelling the variability in Somali Current, Meso scale/Synoptic coherent structure in Geophysical turbulence, Eds. J. C. L. Nihoul and B.M. Jamart, Elsevier Science Publ., 373-386 (1989) .

16. Luther, M.E., O'Brien, J.J. and Meng, A.H., "Morphology of the Somali Current system during the southwest monsoon; Coupled ocean-atmosphere models", Ed. J. C. L. Nihoul, Elsevier Science Publ. (1985).

17. McCreary, P.J. and Kundu, P.K., "A numerical investigation of the Somali Current during southwest monsoon", J. Mar. Res., **46**, 25-58 (1988).

18. McCreary, P.J. and Kundu, P.K. and Molinari, R.L., "A numerical investigation of dynamics, thermodynamics and mixed-layer processes in the Indian Ocean", Progress in Oceanogr., **31**, 181-244 (1993).

19. Mellor, G.L., "User's guide for a three-dimensional, primitive equation, numerical ocean model", Program in Atmos. and Ocean. Sci. report, Princeton Univ., Princeton, N.J. 34 pp (1991a).

20. Mellor, G.L., and Blumberg, A.F., "Modelling vertical and horizontal diffusivities and the sigma coordinate system", Mon. Weather Rev., **113**, 1379-1383 (1985).

21. Mellor, G.L., and T. Ezer, "A gulf stream model and an altimetry assimilation scheme, J. Geophy. Res., **96**, 8779-8795 (1991).

22. Mellor, G.L. and Yamada, T., "A hierarchy of turbulence closure models for planetary boundary layers", J. Atoms. Sci., **31**, 1791-1896 (1974).

23. Mellor, G.L. and Yamada, T., "Development of a turbulence closure model for geophysical fluid problems". Rev. Geophys., **20**, 851-875 (1982).

24. Molinari, R.L., Olson, D. and Reverdin G., "Surface current distributions in the tropical Indian Ocean derived from compilations of surface buoy trajectories", J. Geophys. Res., **95**, 7217-7238 (1990).

25. Potemra, J.T., Luther, M.E., O'Brien, J.J., "The seasonal circulation of the upper ocean in the Bay of Bengal", . J. Geophys. Res., **96**, 12667-12683 (1991).

26. Rao D.P., " A comparitive study of some physical processes governing the potential productivity of the Bay of Bengal and Arabian Sea", Ph. D. thesis, Andhra University, India (1972).

27. Rao, R.R., Molinari, R.L., Festa, J.F.,"Surface meteorological and near surface oceanographics atlas of the tropical Indian Ocean", NOAA Technical Memorandum, ERL AOML-69 (1991).

28. Scherbinin, A.D., Arsen'yev V.S. and Goldin Y.A., Dokl Acad Sci, USSR, Earth Sci. Sect Engl Transl, 244 (1979).

29. Schott, F., Fieux, M., Kindle, J., Swallow, J. and Zantopp, R., "The boundary currents east and north of Madagascar, Direct measurements and model comparision, J Geophys. Res., **93**, 4963-4974 (1988).

30. Simmons, R.C., Luther, M.E., O'Brien, J.J. and Legler, D.M., "Verification of a Numerical Ocean Model of the Arabian Sea", **93**, 15437-15453 (1988).

31. Swallow, J.C., Molinari, R.L., Sruce, J.G. and Evans, R.H., "Development of near-surface flow pattern and watermass distribution in the Somali basin in response to the southwest monsoon of 1979", J. Phys. Oceanogr ., **13**, 1398-1415 (1983).

32. UNESCO, Discharge of Selected Rivers of the World (Studies in Hydrology 5). Vols. 1, 2, and 3. http://http.ncar.ucar.edu/dss/datasets/ds552.0.html.

33. Weatherly, G., and Martin, P. J., " On the structure and dynamics of the ocean bottom boundary layer", J. Phys. Oceanogr., **8**, 557-570 (1978).

34. Woodberry, K.E., Luther M.E. and O'Brien J.J., "The wind-driven seasonal circulation in the southern tropical Indian Ocean", J. Geophy. Res., **94**, C12, 17985-18002 (1989).

35. Wyrkti, K., "Oceanographic Atlas of the International Ocean Expedition", National Science Foundation, Washington DC, pp 531 (1971).

SECTION 2

WAVES AND FLUIDS

Modern Applied Mathematics
Editor: J.C. Misra

Significant Contributions of 20th Century to Boundary Conditions in Fluid Mechanics

N. Rudraiah

National Research Institute for Applied Mathematics (NRIAM), No. 492/G, 7th Cross, 7th Block (West)
Jayanagar, Bangalore-560 001, India
UGC-DSA Centre in Fluid Mechanics, Department of Mathematics, Bangalore University,
Bangalore-560 001, India

Abstract

Important boundary conditions developed during the 20th century in Fluid Mechanics are delineated. Different materials bounding the fluid are discussed. Boundary conditions in Fluid Mechanics that are going to play a challenging and an interesting role in 21st century are briefly outlined.

1. Introduction

20th century has been a golden age for both pure and applied mathematics. Many unsolved problems in both pure and applied mathematics have been tackled during that century. The problems that we study in this article are concerned with applied mathematics particularly the theoretical and experimental aspects of Fluid Mechanics. Many important problems in Fluid Mechanics have been tackled both by gaining a richer and deeper understanding of fluid flow and by exploring the interaction between the subfields like viscous and non-viscous fluids; rotational and irrotational flows, boundary layer and potential flows and so on in fluid mechanics. Now at the turn of the century and entering into the new millennium it offers time for reflection and recognition of the contributions made in fluid mechanics in general and to the boundary conditions in particular during the 20th century and their impact on one hundred years of changes in fluid mechanics through Science, Engineering, Technology, Medicine, Industrial problems and many others. The interaction between mathematics and different branches of fluid mechanics have lead both to a great insight within applied mathematics and to the broadening and deepening of fluid mechanics itself. With this objective, I have undertaken the task of briefly discussing the boundary conditions in fluid mechanics developed during the past century that have made significant contributions to the utility of fluid mechanics in improving the quality of living of an individual in a society.

Many processes and systems in Science, Engineering, and Technology involve fluids bounded either by rigid boundaries made up of different materials and of different geometries or by free surfaces. Such boundaries occur commonly in industrial, nuclear, electrical and electronic networks, biomechanics, science, engineering, geophysical and astrophysical environments. Some of these problems involve thermal energy and concentration of species like storage system, thermal insulation, solar collector, geothermal engineering, pollutant dispersal in aquifer and the storage of nuclear wastes in deep geological repositories and so on. In all these applied areas both theoretical and experimental works have been done extensively in the 20th century using suitable boundary conditions to determine the nature of fluid flow, heat and mass transfer rate in a variety of fluids and fluid saturated porous materials.

I believe that the work done in different branches of fluid mechanics using boundary conditions by my research group and myself did not happen by chance or by decree. The developments in this field are the natural results of an important and relatively recent shift in the priority of the industrial, nuclear and space research developments in India. Immediately after independence these topics have been stimulated by questions arising in problems facing the country. During the end of the 20th century these topics have been extensively diversified both in theoretical and experimental development. The main motivation for these developments is the great importance given to food and energy self-sufficiency and industrial developments. The acute shortages of energy the depletion of fossil fuels and severity of environmental pollution that have been faced in the country provide extensive scope for further research in the area of fluid mechanics. These problems involve pure fluid flow or fluid saturated porous media with the addition of the external constraints of rotation, electromagnetic field, concentration of species, heating or cooling at the boundaries bounded by rigid (permeable/impermeable) porous layer or free surfaces with and without surface tension. I, therefore, make an attempt in this article to present different boundary conditions that have been developed and used in

20th century and project some important boundary conditions that are going to play a challenging role in the 21st century.

Reviewing the existing literature and to have a critical outlook on the recent developments in boundary conditions is certainly one way of assessing the area. This however requires considerably more space and time, one which is best suited for a week long program dedicated only to fluid mechanics. An alternative to this is to focus one's attention only to one's own work. Our work covers many aspects of fluid flow, electrically conducting fluid flow and flow through and past a porous media. Not all of these can be covered in the time at my disposal. I therefore restrict only to different types of boundary conditions with particular emphasis on the limitations in the use of boundary conditions in fluid mechanics that form the back bone for solving problems arising in science and engineering using either analytical or numerical techniques. The boundary conditions will depend on the nature of materials bounding the fluid. Therefore, before discussing the boundary conditions, I briefly review different types of materials that have been used to bound a fluid.

2. Materials Bounding the Fluid

The fluid, with and without porous media, may be bounded by either rigid materials (permeable or impermeable) or by free surface (flat or variable with or without surface tension, either variable with temperature or constant). Rigid boundaries may be classified as conducting (metals), insulating (non-metals), semi-conducting, dielectrics or magnetic materials.

Since very early days, man has been using materials available for him and has learnt to shape them in the form useful to improve his lifestyle. Mixing two elements to improve the quality of individual elements came into practice, which opened up the whole field of alloy science and technology. Though the periodic table consists of several elements, only a few of them are being used in pure form. In general these elements are classified as conducting (i.e. metals), insulating (i.e. non-metals), and semiconductors. Elements of group I and II of the periodic table, transition and rare earth elements are all metallic. Some elements of the group IV are found to be semiconducting and the rest of the elements of the periodic table are insulating (i.e. non-metallic). It is known that living things are made up of only a few elements like C, H, N and O in the combined forms, and nature has synthesized more number of compounds using just these few elements. Depending on his requirements man has synthesized the compounds which are not found in nature. Organic chemists have played a significant role in synthesizing hundred and thousands of new compounds over the years. Many of these were discovered either by scientific curiosity or to find applications in pharmaceuticals, dyes, artificial fibers and so on. Recently there has been a serious effort to synthesize materials with desired physicochemical properties for desired applications. This has influenced our day to day life like housing, transportation, communication and so on. Thus a wide variety of compounds and alloys can be manufactured with a wide range of properties. Some of the properties can be tailor made by mixing the right kind of element in right proportion for many applications. Recently material science has synthesized compounds belonging to polymers, ceramics and so on materials (see Sheela Ramasesha 1999).

In this article we are interested only with rigid, electrically and thermally conducting or nonconducting materials, grain boundaries and free surfaces bounding electrically conducting or electrically nonconducting fluid. Therefore, we briefly discuss these materials before discussing the boundary conditions in fluid mechanics.

(i) Electrically conducting materials

By conducting materials, we mean materials in which either electrons or ions or both are free to move around the nuclei producing electrical conduction. If the conduction is dominated by the motion (or transport) of electrons, as in metallic conductors, we call the material as an electronic conductor. If the conduction is dominated by the motion of ions, as in liquid electrolysis, then we call that material as ion conductor or supersonic conductor. There are also materials in which both electronic and ionic conduction exists as in Silver Sulphide (Ags), then that material is called mixed conductor.

(ii) Dielectric Materials

Dielectric materials are those in which electrons are bounded to the nuclei and are called bound charges. There are two types of molecules in dielectrics. One is a polar molecule and the other is a non-polar molecule. Polar molecules are those in which the Centre of Gravity (CG) of the electron cloud does not coincide with CG of nuclei, therefore even in the absence of the external constraint of an electric field separation of charges occurs. Non-polar molecules are those in which separation of charges does not occur without the external constraint of an electric field. In general, dielectric material, which is practically an insulating type of material, is one in which separation of electrons (i.e. –ve charges) from the nuclei (i.e +ve charges) occur in the presence of an applied electric field. The separation of charges gives rise to polarization, which is defined as dipole moment per unit volume. Dipole occurs due to the separation of charges and dipole moment is defined as magnitude of the charge multiplied by separation distance. Physically, polarization accounts for the ability of the dielectrics to increase the charge storing capacity of the capacitors. The rate of polarization produces polarized current, called displacement current $\frac{\partial D}{\partial t}$ where $\vec{D} = \varepsilon \vec{E}$ is the dielectric field, $\vec{E}$ is the electric field and ε is the dielectric constant. Mathematically it is expressed as $\varepsilon = \varepsilon_0(1 + \chi_e)$ where ε_0 is the dielectric constant of free space and χ_e is the electric susceptibility. Physically ε represents the measure of the efficiency of charge storing capacity. The dielectric constant, dielectric loss factor and dielectric strength normally determine the suitability of a particular material for practical applications, particularly, for electronic applications. Another important aspect of dielectric materials, in addition to allowing propagation of electromagnetic waves, is the dielectric breakdown. In electrical industries the material that is used for insulation withstand the voltage gradient for the operating life of the system. But failures occur when an electric short circuit develops across the material and this failure is called dielectric breakdown. The breakdown strength in a dielectric depends on many factors like temperature, ambient atmosphere, material finish, composition, shape and thickness, the voltage gradient applied, the field frequency micro structure of the material and the electromagnetic wave form. Thus, by changing sample geometry, the field strength and heat dissipation patterns may control the dielectric breakdown considerably. Note that the group of dielectric materials, which exhibit spontaneous polarization, is called ferroelectrics. These materials contain permanent electric dipoles due to the presence of polar molecules. The common ferroelectric materials are barium titanate, rochelle salt, potassium dihydrogen phosphate and potassium nitrate. The other type of dielectrics where the polarization is induced due to the application of external fields like electric field, stresses and so on are called piezoelectric materials. These are used in transducers where electrical energy is converted into mechanical energy or vice versa. Most common applications of piezoelectrics are in microphones, ultrasonic generators, gas lighters, strain gauges, solar detectors, actuators and phonograph pickups. Dielectrics are rarely used as boundaries to bound a fluid. However, a vacuum bound an electrically conducting fluid, which is dielectric.

(iii) Insulating materials

The materials in which the electrical conductivity tends to zero or the electrical resistance tends to infinity are called insulating materials. In such materials the propagation of current is zero.

(iv) Semi – Conducting Materials

Semi-Conducting materials are those in which the conduction process is by means of electrons and holes where the increase in temperature decreases the resistance or increases the electrical conductivity. A hole is an electron valence with a charge of 1.6 x 10^{-9} coulombs.

In metals there are many valance electrons which may be moved through the materials under the action of an electric field. If these free electrons each have a charge of $-q_e$, then the resulting mechanical force on each electron will be $-q_e\vec{E}$ where $\vec{E}$ is the electric field. In contrast to these free electrons the electrons inside a metal have a complex motion within the crystalline lattice structure of the material. One of the consequences of the electron motion in a periodic lattice potential is the appearance of an effective mass m^* which is quite different from the rest mass of the electron. Another important result is that under the action of an applied electric field $\vec{E}$ the electrons move with a velocity $\vec{q_d}$ which is proportional to $\vec{E}$. This velocity $\vec{q_d}$ is called the drift velocity and is given by

$$\vec{q_d} = -\mu_n \vec{E}$$

where μ_n is the mobility factor and is measured in $m^2/_{vs}$. For holes the charge is q_e and the mobility is μ_p such that μ_p μ_n. In this case the electrical conductivity σ is given by

$$\sigma = -n\mu_n q_e$$

where n is the number of electrons.

(v) Grain Boundaries

Often we come across in fluid mechanics, boundaries bounding a clear fluid or a fluid saturated porous medium, having grains. A polycrystalline solid is a good example having grain boundaries, which is made up of many small sized single crystals called grains. In many solids the grains are visible only under a microscope. In some boundaries these grains offer a roughness to the boundaries and in some solids there will be voids leading to porosity of various sizes and shapes. The grain along with the porosity is called the microstructure. In some cases conduction of ions or electrons is obstructed at a grain boundary due to the lattice mismatch. This is responsible to make the electrical conductivity in a polycrystalline material to be less than that in a single crystal. These grain boundaries are encountered in many fluid mechanics problems particularly in transpiration cooling. The roughness to the boundaries due to grains gives rise to slip velocity boundary condition when it bounds the fluid. This will be discussed later.

(vi) Magnetic Materials

Like dielectric materials, magnetic materials are those in which magnetic dipole moment occurs which gives rise to magnetization $\vec{M}$, defined as

$$\vec{M} = \frac{\text{Magnetic moment}}{\text{Volume}}$$

For an element of volume $d\tau$ of n magnetic shells, the total bound current dI_b is

$$dI_b = n i_b$$

where i_b is the bound current in the coil over an area $\overrightarrow{ds_m}$ and thickness d. The magnetic moment $\overrightarrow{dm}$ is defined as

$$d\vec{m} = n \mu_0 \, i_b \, \overrightarrow{ds}_m$$

Then

$$\vec{M} = \frac{n\mu_0 i_b \overrightarrow{ds_m}}{d\tau} = \frac{\mu_0 dI_b \overrightarrow{ds_m}}{\overrightarrow{ds_m}\, d\vec{l}} = \frac{\mu_0 dI_b}{d\vec{l}}$$

$$\vec{M}.d\vec{l} = \mu_0\, dI_b$$

$$I_b = \frac{1}{\mu_0}\oint \vec{M}.d\vec{l}$$

$$\text{But } I_b = \int_S \vec{J}_m . d\vec{S},$$

where $\vec{J}_m$ is the bound current density. Then we get

$$\int_S \vec{J}_m \cdot d\vec{S} = \frac{1}{\mu_0}\oint_c \vec{M}\cdot d\vec{l} = \frac{1}{\mu_0}\int_S \nabla\times\vec{M}\, d\vec{S}$$

$$\vec{J}_m = \frac{1}{\mu_0}\nabla\times\vec{M}.$$

In day to day life we use number of devices whose function depend completely on magnetic materials. For example, video and audio applications (like tapes, reading or writing heads), AC or DC motors where magnetic fields are produced by means of an electric current which are in turn converted into mechanical energy, power transformers which delivers energy to homes and industries, telephones and telecommunication systems, loud speakers, computer data storage systems and so on. This means, magnetic materials can be designed and processed for specific applications like floppy disc drives, printers and other computer peripherals. The magnetic material is extensively used in the study of magnetic fluid.

(vii) Chiral materials

A three dimensional material is called chiral if it cannot be brought into congruence with its mirror image by translation and rotation. This material has the property of handedness and must be either left-handed or right – handed. That is, chirality is connected with handedness. Since chirality begets handedness and handedness begets optical activity (optical activity is the ability of certain materials to rotate the plane of polarization when transmitting polarized light), it is not surprising that the interaction between an electromagnetic wave and collection of randomly oriented chiral object can be such as to rotate the plane of polarization of the wave to the right or to the left depending on the handedness of the object. An object, which is not chiral, is said to be achiral. Thus all materials are either chiral or achiral. Some chiral materials occur naturally in two versions related to each other as chiral material and its mirror image. Such materials are said to be enantimorphs of each other. If a chiral material is left-handed, its enantimorphs is right-handed and vice-versa. Examples of chiral materials are the wire helix, mobius strip, the irregular tetrahedron and so on. It occurs not only in nature but also in works of arts and architecture as well as man – made articles. In nature, we find chirality on a molar scale. For example, in flowers, snails, veins, and, on a molecular scale we find it in array of sugars such as grape sugar and fruit sugar, amino acids, DNA and in certain mollusks and winding vegetation. The man made articles like screws, gloves, golfclubs, springs and so on are all chiral materials.

To obtain a better understanding of chirality and to test its future role in electrical designs, we have to examine the interaction between electromagnetic fields and chiral materials, in particular in composite materials consisting of randomly oriented chiral conductors embedded in a magnetic medium or embedded in a dielectric. Such a study is beyond scope of this article and is studied elsewhere. We are interested here only their properties and processing them for practical use. In the present era of materials processing, a vast amount of research from both theoretical and experimental point of view has been directed towards the development of particular target shapes and the synthesis of new material such as glossy and anisotropic ferrite not only for electrical and electronic instruments but also to manufacture required shape of material to bound a fluid. In addition to these another new material is electromagnetic chirality which embraces both optical activity and circular dichroism. Optical activity refers to the rotation of polarization while circular dichroism indicates the change in polarization ellipticity of optical waves by a medium.

(viii) Optical Materials

It is known that when an electromagnetic radiation impinges on a solid, it is transmitted, absorbed or scattered by the solid depending on the property of the material. A material can be transparent to a band of wavelengths of radiation while absorbing or scattering to the other wavelengths. Then, there are materials that can be used as windows for a particular range of wavelength radiation that is visible. Even in the visible region windows can be made to transmit specific colours by tuning the material band structure. For example, Sapphire is transparent over the whole visible range but when a small amount of Cr^{3+} ions is substituted for Ae^{3+}, the crystal becomes red (Called ruby). Similarly the other ions like Co^{2+} (blue –violet) , Cu^{2+} (blue-green) and Mn^{3+} (purple) can be used to produce different colours. These crystals are cut to give brilliant reflection and are used for making jewelry. A material in which the radiation is emitted with a delay after absorption is called phosphorescence and if the emission occurs soon after absorption it is called fluorescence. The materials like sulfides, oxides and tungsten can be made to fluoresce or phosphoresce by introducing impurities in controlled amounts.

The optical properties of some materials like ceramics can be changed by applying a suitable strength of electric field and is called the electro-optical effect. This effect is due to the change in the spatial distribution of electrons, which produces an anisotropic change in electronic polarizability. If an electric field is applied, the grains in the material orient themselves along a particular direction i.e., the material is polarised, and, if the electric field is removed the material comes back to its original state. There are also photochromic materials in which shining light changes the colours of the material. The energy gap between the valence and the conduction band is in the visible range of the optical spectrum. Hence when the sunlight falls on these materials, electrons in the valance band absorb the energy from the radiation and are excited to the conduction band. When these electrons relax to the valance band, they emit radiation that is visible to the nacked eye. Therefore optical materials can be used for many practical purposes.

The materials discussed in (i) to (viii) are useful in deriving the boundary conditions in fluid mechanics. Before discussing the boundary conditions, it is better to give basic equations of motion of a fluid in the general form in an electrically conducting fluid saturated porous medium.

3. Basic Equations of Motion

We consider an electrically conducting incompressible fluid saturated porous media. The basic equations for the motion of this fluid are the conservation of mass, momentum, energy and species.

The conservation of Mass

The general form of the conservation of mass valid for compressible fluid is

$$\frac{1}{\varrho}\frac{D\varrho}{Dt} + \nabla \cdot \vec{q} = 0 \tag{1}$$

For an incompressible fluid

$$\nabla \cdot \vec{q} = 0 \tag{1a}$$

$$\frac{D\varrho}{Dt} = 0 \tag{1b}$$

The conservation of momentum

$$\varrho\left[X_p\left(\frac{1}{\varepsilon}-1\right)+1\right]\frac{\partial \vec{q}}{\partial t} + \varrho\left[X_p\left(\frac{1}{\varepsilon^2}-1\right)+1\right](\vec{q}\cdot\nabla)\vec{q}$$

$$= -\nabla p + \varrho\vec{g} + \mu_m(\nabla\times\vec{H})\times\vec{H} + \mu\left[X_p(\mu'-1)+1\right]\nabla^2\vec{q} - X_p\left[\frac{\mu}{k} + \frac{\varrho C_b}{\sqrt{k}}|\vec{q}|\vec{q}\right] \tag{2}$$

where $X_p = \begin{cases} 1 & \text{for porous media} \\ 0 & \text{for fluid} \end{cases}$

Conservation of Energy

The conservation of Energy, in the absence of heat source and radiation effect, is

$$[X_p(M-1)+1]\frac{\partial T}{\partial t}+(\vec{q}\cdot\nabla)T = K[X_p(k'-1)+1]\nabla^2 T \quad (3)$$

Conservation of species

$$\frac{DC}{Dt} = D\nabla^2 C \quad (4)$$

Conservation of magnetic lines of force

$$\frac{D\vec{H}}{Dt} = (\vec{H}\cdot\nabla)\vec{q} + \sigma[X_p(\sigma'-1)+1]\nabla^2\vec{H} - \frac{\sigma}{k}\vec{H} \quad (5)$$

$$\nabla\cdot\vec{H} = 0 \quad (6)$$

Here, $\vec{q} = (u, v, w)$ is the velocity, $\vec{H} = (H_x, H_y, H_z)$ is the magnetic field, ρ is the density, p the pressure, T the temperature, C the concentration, C_b the drag coefficient, ε the permeability of the porous media, μ the viscosity of fluid, μ_{ef} the effective viscosity (i.e. Brinkman viscosity), K_{ef} the effective thermal diffusivity, K the thermal diffusivity, σ_{ef} the effective electrical conductivity, σ the electrical conductivity, D the diffusive coefficient, μ_m the magnetic permeability, $\frac{D}{Dt} = \frac{\partial}{\partial t} + (\vec{q}\cdot\nabla)T$ and

$$\mu' = \frac{\mu_{ef}}{\mu},\quad K' = \frac{K_{ef}}{K},\quad \sigma' = \frac{\sigma_{ef}}{\sigma}$$

The last term on the right-hand side of (5) is due to solid particles of the medium and obtained by applying average process as in the Brinkman equation (2) . Note that the momentum equation (2) is so general that when $X_p = 1$ it is valid for a porous medium, $X_p = 0$ valid for an electrically conducting fluid, when $(\nabla\times\vec{H})\times\vec{H} = 0$ it is valid for an ordinary fluid. From (2) we can also get pure Darcy equation, Darcy-Lapwood-Forchheimer equation and Darcy-Lapwood-Forchheimer-Brinkman equation for a conducting fluid.

4. Boundary conditions

To solve the equations (1) to (6) we need initial and boundary conditions on velocity, pressure, temperature, concentration and magnetic field. These conditions will depend on the nature of the boundaries, namely, whether they are rigid, free surface or bounded by a porous layer. These conditions are discussed one by one.

Let the fluid be contained between the planes z=0 and z=h. On these two planes, boundary conditions on velocity, temperature, concentration and magnetic field must be specified.

I Boundary conditions on velocity.

Boundary condition (here after denoted by BC) on velocity will depend on whether the boundaries are rigid or free. Regardless of the bounding surfaces, the normal component of velocity, q_n must vanish on the surfaces. Then

$$q_n = 0 \text{ for } z = 0 \text{ and } z = h \tag{7}$$

If $\vec{q} = (u, v, w)$, then the normal component is w and hence

$$w = 0 \text{ for } z = 0 \text{ and } z = h \tag{7a}$$

The remaining conditions on velocity will depend on the nature of the surfaces at z=0 and z=h and the nature of the fluid, that is whether the fluid is viscous or non-viscous.

(a) No-slip Condition at a rigid surface.

In the case of non-viscous fluid only the condition (7) is applicable. However in the case of viscous fluid, it is assumed that the fluid will adhere to the boundary so that the condition on velocity at the boundary will be that of the boundary. This is known as the no-slip condition. This condition implies that not only the normal component of velocity, q_n, but also the tangential component of velocity, q_t, will be that of the boundaries. If the rigid boundaries are stationary, then the no-slip condition is

$$q_n = 0, \; q_t = 0 \text{ at } z = 0, \; z = h \tag{8}$$

If the boundaries are moving horizontally with a uniform horizontal velocity q_u then the no-slip condition is

$$q_n = 0 \text{ and } q_t = q_u \text{ at } z = 0 \text{ and } z = h \tag{9}$$

If the boundaries are rigid and permeable and stationary, then the no-slip conditions are

$$q_n = \pm v, \; q_t = 0 \text{ at } z = 0 \text{ and } z = h \tag{10}$$

Where +v denotes uniform or fluctuating injection velocity and –v denotes suction velocity .

If the porous rigid boundaries are moving horizontally with a velocity q_0, then the no-slip condition are

$$q_n = \pm v, \; q_t = q_0 \text{ at } z = 0 \text{ and } z = h \tag{11}$$

If the boundaries are stationary i.e. $q_0=0$ then the no-slip conditions are

$$q_n = \pm v, \; q_t = 0 \text{ at } z = 0 \text{ and } z = h \tag{12}$$

In Cartesian coordinates (x, y, z) with rectangular plane at z=0 and z=h, the no-slip conditions are

$$v = \pm V, \; u = 0, \text{ and } w = 0 \tag{13}$$

If the plane is stationary (i.e v=0), we have the no-slip condition

$$u = v = w = 0 \tag{14}$$

which implies both normal and tangential components vanish at the rigid surface. Since this condition must be satisfied for all x and y on the surface, it follows from the equation of continuity

$$\frac{\partial u}{\partial x}+\frac{\partial v}{\partial y}+\frac{\partial w}{\partial z}=0$$

that $\frac{\partial w}{\partial z}=0$ on the rigid boundaries. Then the no-slip conditions in the case of rigid surfaces are

$$w=0, \quad \frac{\partial w}{\partial z}=0 \text{ at } z=0 \text{ and } z=h \tag{15}$$

The validity of no-slip condition at the rigid surface has been one of the considerable difficulties for a long time. At present it appears that no-slip condition namely the fluid immediately in contact with a stationary solid body may be taken as having no velocity relative to the solid for nearly all fluids, but the exact no-slip condition on molecular scale, remain still in doubt.

Ever since Daniel Bernoulli found that the real fluid could not slip freely over the solid body, there is doubt about the validity of the no-slip condition. Bernoulli found the large discrepancies between his theoretical results obtained using an ideal (i.e. non-viscous) fluid and those he had measured using real fluid and he attributed this discrepancy to the use of no-slip condition. The doubt that whether the no-slip condition at the solid surface is valid or not is prevailing from the 19th century. (For details see Goldstein 1952).

The following three different arguments were put forth in 19th century by various authors:

(1) The first argument states that the velocity of the fluid is the same as that of the solid and changes continuously in the fluid.

(2) The second one was concerned with the experiments on the flow of liquid through tubes conducted by Girard. He predicted a very thin layer of the fluid completely attached to the walls and assumed that outside this layer the fluid slips over it. He assumed that the thickness of the layer depends on the curvature of the wall and on the temperature. He further assumed that in the case of liquids which do not wet the wall the thickness of the layer becomes zero and the liquid slips over the surface.

(3) Using certain hypothesis Navier derived the viscous flow equations. From those hypothesis and deduced that there is a slipping at the solid boundary which is resisted by a force proportional to the relative velocity. Since the tangential stress at any point on the solid wall is the same as the stress at a neighboring internal point of the fluid, it gives rise to the boundary condition,

$$\mu \frac{\partial u}{\partial n} = \beta u \tag{16}$$

for undirectional flow along a plane wall where u is the velocity, the derivative is along the normal away from the wall and β is a constant such that $\frac{\mu}{\beta}$ is a length scale, say l. This length is zero if there is no slip. Later Poiseuille in his work on movement of the blood, found a layer of stagnant blood at the walls of the containing vessel. Hagen, earlier to Poiseuille, stated based on his experiments on the flow in capillary tubes that the velocity increases at a uniform rate from zero at the walls to a maximum in the middle. Later he sated that their exists a stagnant layer near the walls, but without slip, and found that layer has to be thinner than the thinnest writing paper. Later Helmoltz while discussing the experiment of Piotrowski on the oscillations of spheres filled with liquid, concluded, on the basis of Navier's hypothesis that though there might not be no-slip in the case of water in contact with glass, as of ether or alcohol in contact with glass or metal surface, there will be a considerable slip for water in contact with a gilt surface. Gradually, however the hypothesis finally adopted by Stokes that there is no-slip and that all points of the fluid have the same properties gained momentum. The results obtained on the basis of this hypothesis for the flow through tubes began by Stokes and carried out later by many workers gave consistent results for liquids. However, those results are not valid for kinetic theory of gases because of the existence of a slip as demonstrated by Maxwell. Maxwell had carried out some calculations using the kinetic theory of gases and arrived at a conclusion that in the absence of difference in temperature, slipping takes place and the length $l = \frac{\mu}{\beta}$ in (16) is a moderate multiple of the mean –free path L of the gas molecules. In otherworlds, in the case of gases the slip at the boundaries exist and to a first approximations no-slip exists at the boundaries in the case of liquid at atmospheric pressure. Most of the earlier works were concerned with the laminar flows. However there are numerous experiments concerning with both laminar and turbulent flows and concluded that this distinction is mainly concerned with Reynolds number only. From these experiments it follows that if slip takes place there should be another length $l = \frac{\mu}{\beta}$ defined in (16) in addition to the length h specifying the dimensions of the system. This length l may also depend on the thickness of the layer on the boundary in addition to $\frac{\mu}{\beta}$. In that case the non-dimensional quantities in the basic equations will involve the ration $\frac{l}{h}$ in addition to Reynolds number. If l varies curiously in proportion to h, the slip condition is valid and otherwise experiments reveal that $l \to 0$, that is, it is so small that its effect is negligible and hence to a good approximation no-slip condition is valid.

We note that there is a tendency among some workers to take no-slip condition as granted, and to be valid in all cases of viscous flows and we caution that it is not so straight forward. It is only an approximation and the errors involved in the results obtained (both theory and experiments) using no-slip condition for viscous liquids are negligibly small. The cases where slip arises even in the case of liquids are discussed below.

(b) The Slip condition at the boundaries bounding a Fluid

The no-slip boundary condition on the interface between a fluid and a solid discussed above is generally valid for a smooth surface binding a viscous fluid. In the following cases, it is difficult to apply the no-slip condition, and in general, slip condition is applicable:

(i) Slip condition in the case of Non-Viscous Fluid.

In the case of non-viscous (i.e. inviscid) fluid the following slip boundary conditions must satisfy on a solid boundary

(a) Kinetic Flow condition

The velocities normal to a solid boundary are zero but tangential to a boundary are not zero. In otherwords we can specify only the normal component and not the tangential component of velocity on a solid boundary bounding an inviscid fluid.

(b) In the case of free stream, u_∞, the perturbation velocities are zero or finite in the directions where the fluid is extended to infinity. For example for a body contour (see Fig 1) f = f (x, y, z) and the velocity $\vec{q} = (u, v, w)$, the condition on the surface is

$$(\vec{q} \cdot \nabla) f = 0$$

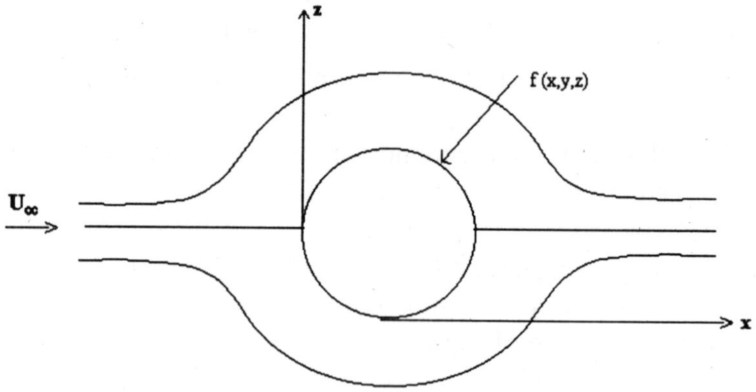

Fig. 1

That is

$$(U_\infty + u)\frac{\partial f}{\partial x} + v\frac{\partial f}{\partial y} + w\frac{\partial f}{\partial z} = 0 \quad (17)$$

where u, v, w are the perturbed velocities and it is assumed that $u \ll U_\infty$. Then (17) becomes

$$U_\infty \frac{\partial f}{\partial x} + v\frac{\partial f}{\partial y} + w\frac{\partial f}{\partial z} = 0 \quad (18)$$

For two-dimensional flow, $v = 0$, $\frac{\partial f}{\partial y} = 0$. Then (17) gives

$$w\frac{\partial f}{\partial z} = -(U_\infty + u)\frac{\partial f}{\partial x}$$

i.e.

$$\frac{w}{U_0 + u} = -\frac{\partial f / \partial x}{\partial f / \partial z} = -\left(\frac{\partial z}{\partial x}\right)_c \tag{19}$$

where c pertains to body contour and $\frac{\partial z}{\partial x}$ is the slope of the body.

For thin bodies the variation in the z-direction is negligible so that

$u(x, z) = u(x, 0)$, $w(x, z) = w(x, 0)$

and hence

$$\frac{w(x, 0)}{U_\infty + u(x, 0)} = -\left(\frac{\partial z}{\partial x}\right)_c$$

If $u << U_\infty$, we have

$$\frac{w(x, 0)}{U_\infty} = -\left(\frac{\partial z}{\partial x}\right)_c \tag{20}$$

For planar bodies, $\frac{\partial f}{\partial y} = 0$ and hence

$$\frac{w(x, y, 0)}{U_\infty} = -\left(\frac{\partial z}{\partial x}\right)_c \tag{21}$$

In (20) and (21) the elevation above the x-axis is neglected.

The same boundary conditions can be used on a solid body embedded in a densely packed porous media where the flow through porous media is governed by the usual Darcy-Lapwood equation

$$\varrho \frac{D\vec{q}}{Dt} = -\nabla p - \frac{\mu}{k}\vec{q}$$

If $k \to \infty$, we recover the Euler's equation

$$\varrho \frac{D\vec{q}}{Dt} = -\nabla p$$

governing the inviscid fluid in the absence of porous media.

(ii) Thc Slip-Boundary Condition When a Fluid is Bounded by Rough and Coated Surfaces.

When a solid boundary bounding a fluid is smooth no-slip boundary condition is applicable. For rough surface, where the scale of roughness is smaller than the scale of interest, the no-slip condition is not valid and in that case we have to use the effective –slip boundary condition applied at the mean position of the interface. This slip-boundary condition on velocity will depend on this slip-velocity proportional to the tangential stress along the surface, as in (16), the constant of proportionality is usually called slip coefficient (see Miksis and Davis 1994).

In the study of the motion of contact lines if we use a no-slip condition at a solid surface it gives rise to non-integrable rate of strain singularity at the contact line. One of the ways of removing this singularity is to use boundary velocity along the boundary i.e. slip along the boundary.

There are several models to study the effect of the roughness of the surface on the flow without contact lines. One of the efficient models is to express the roughness as a periodic modulation of the surface amplitude. In this case there are two situations, one concerned with a single fluid flowing over a surface and the other with two fluids, one fluid flowing over a rough surface and the other fluid fills the grooves have also been investigated in the literature (see Hocking 1976).

In these cases the amplitude of the surface may be large or small. It is of interest to obtain a Boundary Condition for small amplitude but arbitrary shaped roughness.

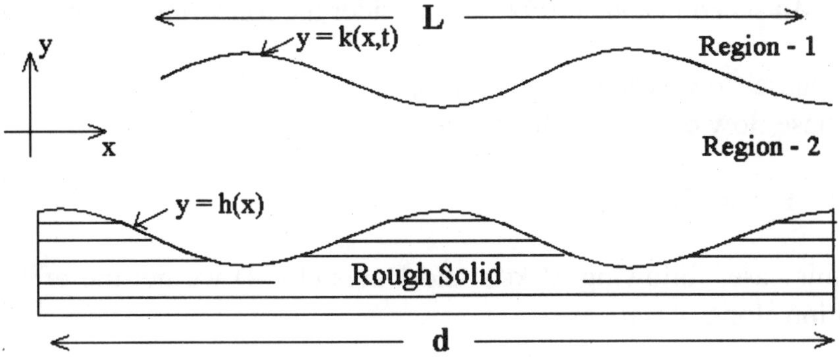

Fig. 2 Physical Configuration

To derive this slip-boundary condition consider the physical configuration shown in the figure 2. Here we consider two regions 1 and 2. Region 2 consists of a viscous incompressible fluid bounded below by a solid interface denoted by y = h(x) and above by a free surface y = k (x, t). The fluid in region 1 consists of a viscous and incompressible fluid bounded below by a free surface y = k(x, t) and extends to infinity in the upper region. The required slip-condition is derived assuming small amplitude but arbitrary shaped roughness. We assume that the flow far form the surface is prescribed and that the fluid motion is governed by the Navier-Stokes equation given by (2) with $x_p=0$, and $\mu \vec{J}\times\vec{H}=0$. Here $\vec{q}^i, \varrho^i, p^i$ and $\mu^i\ (i=1,2)$ are the velocity, density, pressure and viscosity. To derive the boundary condition following Miksis and Davis (1994) we consider two horizontal length scales in the x-direction as L and d as shown in the figure. where L is the macroscopic length scale and the roughness is characterized by a much smaller scale d. We assume

$$\varepsilon = \frac{d}{L} << 1 \qquad (22)$$

The effective boundary condition is desired in the limit of $\varepsilon \to 0$
We define the fast space variable as

$$\hat{x} = \frac{x}{\varepsilon} \tag{23}$$

and ε is the slow scale. To derive the required slip boundary condition one has to use multiple scales $\hat{x}$ and x and the method of matched asymptotic technique.

Following Miksis and Davis (1994) we consider the following two cases :

1. Single-phase flow over a rough surface
2. Two-phase flow over a rough surface

1. Single – Phase Flow

In this case, following Miksis and Davis (1994) we get the effective slip condition along $y = 0$, as

$$u(x, o, t) + \langle h \rangle \frac{\partial u(x, t)}{\partial y} = 0 \tag{24}$$

Where

$$\langle h \rangle = f(x) \tag{25}$$

is the slip coefficient. The details are omitted here and interested reader can refer to the work of Miksis and Davis (1994). Here the slip coefficient depends on the locations of the origin of the coordinate system relative to the modulation of the solid surface.

2. Two-Phase Flow over a Rough Surface

This case involves coating of fluid along the solid surface. In this case also we have to find inner and other solution using the method of matched asymptotic techniques. For this purpose it is advantages to assume that the amplitude of the surface roughness is of order δ and the thickness of region 2 is also of order δ. Our aim in this paper is only to derive the suitable slip-boundary condition and not the detailed solutions and hence the details are omitted.

In this case we have to make use of the following general boundary conditions.

On the solid boundary $y = h(x)$ the no-slip condition is valid

$$u^{(2)} = v^{(2)} = 0 \qquad (26)$$

where u and v, are respectively x and y components of velocity. As $x \to \infty$ the velocity field matches the externally imposed flow with the assumption that the external flow field is consistent and the flow fields and their derivatives are bounded in the neighborhood of $y = h(x)$. Along the free surface

$$y = \eta(x, t) \qquad (27)$$

we have the kinematic condition

$$\frac{\partial \eta}{\partial t} + u^{(i)}\frac{\partial \eta}{\partial x} - v^{i} = 0 \qquad (28)$$

where i = 1, 2, and η is the elivation from the horizontal. Further the continuity of tangential and normal stresses are imposed and they are respectively

$$\sum_{i=1}^{2}(-1)^{i}\hat{\mu}^{i}\left\{\eta_x\left(\frac{\partial u^{(i)}}{\partial x} - \frac{\partial v^{(i)}}{\partial y}\right) + \frac{1}{2}\left(\eta_x^2 - 1\right)\left(\frac{\partial u^{(i)}}{\partial y} + \frac{\partial v^{(i)}}{\partial x}\right)\right\} = 0 \qquad (29)$$

$$p^2 - p^1 + \frac{\eta_{xx}}{C_a\left(1 + \eta_x^2\right)^{\frac{3}{2}}} =$$

$$= \sum_{i=1}^{2}(-1)^{i}\hat{\mu}^{i}\left\{\eta_x^2\frac{\partial u^{(i)}}{\partial x} + \frac{\partial v^{(i)}}{\partial y} - \eta_x\left(\frac{\partial u^{(i)}}{\partial y} + \frac{\partial v^{(i)}}{\partial x}\right)\right\}\left(1 + \eta_x^2\right)^{-1}$$

where $\eta_x = \frac{\partial \eta}{\partial x}$, $\hat{\mu}^{(1)} = 1$ $\hat{\mu}^2 = \frac{\mu^{(2)}}{\mu^{(1)}} = \bar{\mu}$ (30)

p is the pressure $C_a = \frac{\mu' U}{\gamma}$ is the capillary number and γ is the surface tension. We seek the solution as a power series in δ. For example

$$\bar{u}' = \bar{u}_0^{(1)}(\hat{x}, x, \bar{y}, t) + \delta\, \bar{u}_1^{(1)}(\hat{x}, x, \bar{y}, t) + \ldots\ldots\ldots\ldots \tag{31}$$

For a leading order in δ, the no-slip conditions are

$$u_0^{(2)} = v_0^{(2)} = 0 \quad \text{along} \quad y = \bar{h}(\hat{x}, x) \tag{32}$$

Along the interface $\bar{y} = \bar{\eta}_0(\hat{x}, x, t)$ the leading order boundary conditions are

$$\frac{\partial \bar{\eta}_0}{\partial t} = 0 \tag{33}$$

$$\frac{\partial u_0^{(1)}}{\partial y} = \bar{\mu} \frac{\partial u_0^{(2)}}{\partial y} \tag{34}$$

$$\bar{p}_0^{(2)} - p_0^{(1)} + \frac{1}{C_a} \bar{\eta}_{0\hat{x}\hat{x}} = 0 \tag{35}$$

similarly we can get the next order condition. For example the next order kinematic condition along $\bar{y} = \bar{\eta}_0(\hat{x}, x, T, t)$ is

$$\frac{\partial \bar{\eta}_1}{\partial t} + \frac{\partial \bar{\eta}_0}{\partial T} + \bar{u}_0^{(i)} \bar{\eta}_{0x} - v_0^{(i)} = 0 \tag{36}$$

where T= δt is a new time scale. Similarly the other condition can be obtained. Finally we get the slip coefficient C(x, T, t) as

$$C(x, T, t) = \langle h \rangle - \left(\frac{\hat{\mu}^{(1)}}{\hat{\mu}^{(2)}} - 1 \right) D(x) - \frac{\hat{\mu}^{(1)}}{2\hat{\mu}^{(2)} A C_a} \left\langle (\eta_0 - h)^2 \frac{\partial^3 \eta}{\partial x^3} \right\rangle \tag{37}$$

where
$D(x) = <\eta - h>$ and η represents the leading order contribution to the interface height.

Note that if there were no coating film, i.e $\eta = h$ and $\mu^{(1)} = \mu^{(2)}$, then (29) gives $C(x) = <h>$. Since $\overline{C}_a$ in (35) may be small or large depending on the ratio $\frac{\varepsilon}{\delta^2}$, for large $\overline{C}_a$ i.e. $\overline{C}_a >> 1$, we can neglect the third term on the right-hand-side of (37) and obtain the slip coefficient as

$$C(x) = <h> - \left(\frac{\hat{\mu}^{(1)}}{\hat{\mu}^{(2)}} - 1 \right) D(x) \tag{38}$$

Here C is independent of time
For small $\overline{C}_a$ i.e. $\overline{C}_a << 1$, we can obtain the slip coefficient as

$$C(x, t) = <h> - \left(\frac{\mu^{(1)}}{\mu^{(2)}} - 1 \right) D(x, T) + \frac{\mu^{(1)}}{3\mu^{(2)}} \frac{\overline{A}}{A} \left[<\eta - h> - < \frac{1}{(\eta - h)^2} >< \frac{1}{(\eta - h)^3} >^{-1} \right] \tag{39}$$

where $A = \frac{\partial u_0^{(1)}(x, 0, T, t)}{\partial y}$ i.e. A is independent of $\overline{x}$ and $A \neq 0$ and $\overline{A}$ is the shear stress that the outer fluid exerts on the interface. Note that for smooth surface $<h>=0$ and for coated surface (i.e. rough surface) $<h>\neq 0$.

(iii) The Slip Boundary Condition on Velocity When the Fluid is Bounded by a Porous Layer

We often encounter the existence of a fluid adjacent to a fluid saturated porous layer in many branches of science (particularly in geophysics), biomechanics (particularly in coronary arteries, and synovial joints) and engineering environments. The flow in the region-1 above the porous layer is governed by the usual Navier-Stokes equations obtained from (2) by setting $X_p=0$ and the flow in the fluid saturated region –2 in the porous layer is governed by Darcy-Lapwood-Forchheimer-Brinkman equation obtained from (2) by setting $X_p=1$.

To postulate the slip boundary condition on velocity at the interface between Region-1 and Region-2, Beavers and Joseph (1967, here after called BJ) assumed the interface to be a "NOMINAL" surface. This surface is defined as a surface drawn in the fluid such that the external perimeters of the solid particles of the porous layer should lie on that surface. Using this assumption BJ postulated the slip velocity condition

$$\frac{du}{dy} = \alpha\left(u_B - Q\right) \qquad (40)$$

where u is the velocity in the x-direction, u_B is the slip velocity at the nominal surface and

$$Q = -\frac{k}{\mu}\frac{\partial p}{\partial x} \qquad (41)$$

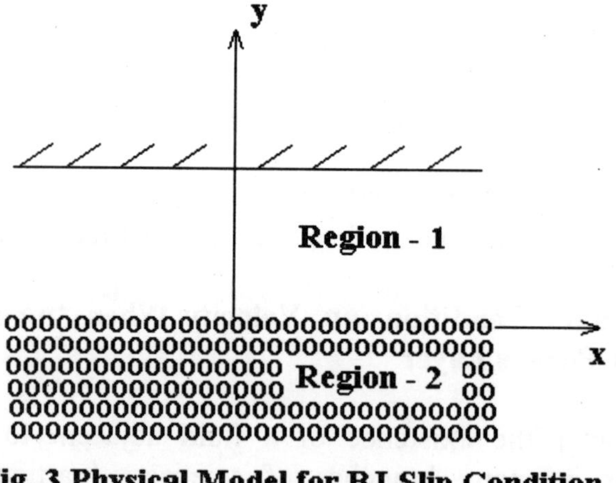

Fig. 3 Physical Model for BJ-Slip Condition

is the Darcy velocity and α is the slip coefficient which depends on the properties of the fluid and the porous layer but is independent of spatial coordinates and other quantities are defined earlier in (2). This BJ-Slip condition was verified experimentally by Beavers and Joseph (1967) and also by Beavers et al (1970), Taylor (1971), Rajasekhara (1974), Beavers et al (1974), and Ranganna(1984). A theoretical derivation of this BJ-slip condition was given by Richardrson (1971) and Rudraiah et al (1987)

(iv) Modified BJ Conditions

(a) Saffman-Modified BJ-Condition

Saffman (1971 was of the opinion that BJ condition was special to the planar geometry and in general is not correct to the order k. Saffman (1971) by performing the method of matched asymptotic analysis by matching a Darcy flow (i.e outer solution) in a porous medium with a Navier-Stokes creeping flow in the fluid near the nominal surface which is an inner solution obtained a slip velocity condition

$$\frac{du}{dy} = \frac{\alpha}{\sqrt{k}}\left(u + o(k)\right) \tag{42}$$

(b) Jones-Modified BJ condition

According to Jones(1973), BJ condition is essentially a relationship involving shear stresses rather than just velocity shear and suggested a generalised form of BJ-condition (40) as

$$\frac{\partial u}{\partial y} + \frac{\partial v}{\partial x} = \frac{\alpha}{\sqrt{k}}\left(u_B - Q\right) \tag{43}$$

where u and v are the x and y components of velocity. The condition (43) is called Jones-Modified BJ-condition.

(c) Rudraiah's (1985) Modified BJ condition

The BJ condition (40) and modified BJ-Condition (42) and (43) are valid when the porous medium is dense and its thickness is very less compared to the thickness of the fluid layer above the porous layer. Such situations commonly occur in Geophysical problem. In many practical problems particularly in Bio-Mechanics, and in some engineering problems thickness of the porous layer is either small or of the same order as that of fluid layer. To examine the effect of porous layer thickness on slip velocity at the nominal surface, particularly when the porous layer is sparsely packed i.e. porosity is large, Rudraiah (1985, 1986) using Darcy-Brinkman equation obtained a modified BJ-slip condition (known as BJR condition) as

$$\frac{du}{dy} = \left(\frac{\lambda'}{k}\right)^{\frac{1}{2}} \left[\frac{\varepsilon \lambda' Q}{\text{Sinh}\,\delta h} + (u_B - Q)\coth \delta h\right] \tag{44}$$

where h is the thickness of the porous layer, $\delta = \left(\frac{\lambda'}{k}\right)^{\frac{1}{2}}$ and λ' is a viscosity parameter. Note that (44) reduces to the BJ-condition (40) in the limit of $h \to \infty$. A comparison with (41) shows that $\lambda' = \alpha^2$ which agrees with the finding of Rudraiah et al (1987) if $\lambda' = \frac{\mu ef}{\mu}$ the viscosity ratio.

V Boundary Conditions on a Free Surface

Let the fluid be bounded by a free surface k = k (x, y, z, t). In this case the condition on velocity will depend on (a) the thickness of the layer, (b) whether the free-surface is deformable or not and (c) whether the surface has the effect of surface tension or not. These are discussed below.

(a) The thickness of the layer

If the thickness of the layer is finite say l, then the filed quantities (like velocity magnetic field, temperature and so on)will be discontinuous at the interface. Hence we cannot apply divergence and Stokes theorems to convert the surface integrals to volume integrals and the line integrals to surface integral when we use the integral form of basic equations to derive the boundary conditions. In that case one has to assume that the free surface instead of finite thickness it is of infinitesimal thickness across which the field quantities will vary instantaneously so that we can apply integrals theorems of vector analysis.

(b) Deformable surface with surface tension

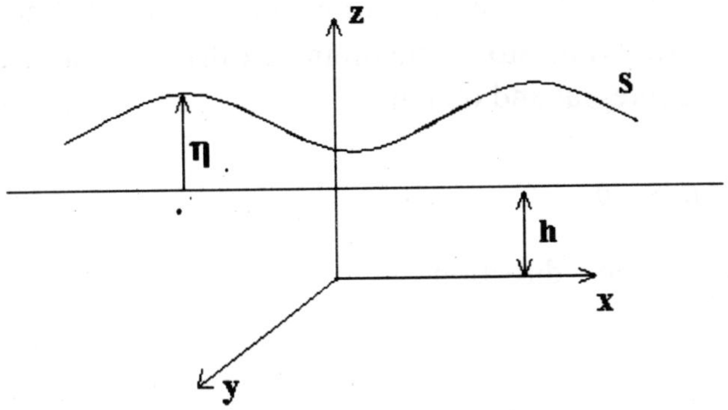

Fig 4: Physical configuaration

In this case we assume that the free surface S, be of the form

$$Z = h + \eta\ (x, y, t) \tag{45}$$

The Kinematic condition is

$$\frac{D\eta}{Dt} = 0$$

i.e.

$$\frac{\partial\eta}{\partial t} + u\frac{\partial\eta}{\partial x} + v\frac{\partial\eta}{\partial y} + w\frac{\partial\eta}{\partial z} = 0$$

Form (45) we have $\frac{\partial\eta}{\partial z} = 1$, so that

$$\frac{\partial\eta}{\partial t} + u\frac{\partial\eta}{\partial x} + v\frac{\partial\eta}{\partial y} + w = 0 \text{ on S} \tag{46}$$

The Dynamic condition.

To obtain the dynamic condition, we consider the balance of forces in a volume V within the fluid in the form

$$\vec{q}.\hat{n} = 0 \tag{47}$$

$$\int_v \vec{A}\,dV + \int_S \vec{\tau}.\hat{n}\,dS + \oint_c \gamma\hat{n}.\vec{dl} = 0 \tag{48}$$

where $\vec{A}$ is a bounded volume force consisting of body force due to gravity, electromagnetic force and the force due to rotation, $\vec{\tau}$ is the stress vector and γ is the surface tension. Applying Stokes theorem to (48). we can convert the line integral to surface integral and obtain

$$\int_v \vec{A}\,dV + \int_S \vec{\tau}.\hat{n}\,dS + \int \nabla x(\gamma\hat{k}).\hat{n}\,dS = 0 \tag{49}$$

where $\hat{k}$ is the unit vector in the z direction.

Now

$$\vec{\tau} = -pI + \nabla\vec{q} + \nabla\vec{q}^T$$

where I is the identity vector. In the absence of Volume force (i.e $\vec{A}$=0), (49) takes the form

$$\int_S \vec{\tau}.\hat{n}\,dS + \int \nabla x(\gamma\hat{k}).\hat{n}\,dS = 0 \tag{50}$$

This is true for any arbitrary surface S and hence the integrands here must be equal and we have

$$\vec{\tau} + \nabla x\,\gamma\,\hat{n} = 0$$

Then by expanding $\nabla x \gamma \hat{n}$ we get

$$\vec{\tau} = \frac{\partial\gamma}{\partial y}\hat{i} + \frac{\partial\gamma}{\partial x}\hat{j} \tag{51}$$

where $\hat{i}$ and $\hat{j}$ are the unit vectors in the x and y directions respectively.

We know that

$$\vec{\tau} = \tau_{yz}\,\hat{i} + \tau_{xz}\,\hat{j} = \mu\left(\frac{\partial u}{\partial z} + \frac{\partial w}{\partial y}\right)\hat{i} + \mu\left(\frac{\partial u}{\partial z} + \frac{\partial w}{\partial x}\right)\hat{j} \tag{52}$$

Equating (51) and (52), we get

$$\mu\left(\frac{\partial u}{\partial z}+\frac{\partial w}{\partial x}\right)=\frac{\partial \gamma}{\partial x}, \quad \mu\left(\frac{\partial V}{\partial z}+\frac{\partial w}{\partial y}\right)=\frac{\partial \gamma}{\partial y} \tag{53}$$

Since the surface tension γ is a function of temperature i.e. $\gamma = \gamma$ (T) we assume the fluid obeys Boussinesq approximation so that one can assume the variation of density of fluid ρ as

$$\rho = \rho_o [1-\alpha (T-T_o)] \tag{54}$$

where T_o is the reference temperature and ρ_o is the density of fluid for $T=T_o$. In the steady state, the density ρ_b is of the form

$$\rho_b = \rho_o[1-\alpha (T_b-T_o)] \tag{55}$$

where T_b satisfies the time independent conduction equation namely

$$\frac{d^2T_b}{dz^2} = 0 \tag{56}$$

Its solution is

$$T_b = AZ+B$$

To find the constants A and B, we assume the boundary conditions $T_b=T_L$ at $Z=0$ and $T_b=T_u$ at $Z=h$ which are the isothermal conditions called the boundary conditions of the first kind. Then the solution of (56) satisfying the above condition is

$$\frac{T_b - T_L}{T_u - T_L} = Z/h \tag{57}$$

Let $\theta = \theta(x, y, z, t)$ be the departure from the basic temperature T_b. That is

$$T = T_b + \theta$$
$$T-T_o = T_b-T_o + \theta$$

and $\rho = \rho_o[1-\alpha (T_b-T_o + \theta)]$

i.e.

$$\frac{\varrho-\varrho_0}{\varrho_0} = -\alpha\frac{T_u - T_L}{h}Z - \alpha\theta - \alpha(T_L - T_0) \tag{58}$$

Since $\gamma = \gamma(T)$

we have $-d\gamma = -\dfrac{d\gamma}{dT}\nabla T$

From (49), with proper limiting process, assuming $\rho g\hat{k}$ as the body force, we get

$$2\mu\, e_{nt} = \frac{d\gamma}{dT}\left[\nabla\left(\theta - \alpha\left(\frac{T_L - T_u}{h}\right)\eta\right)\right] \tag{59}$$

$$2\mu e_{nn} - p + \varrho_0 g h + \frac{\varrho_0 g\alpha}{2h}(T_L - T_u)\left(2\eta h - \eta^2\right) = \gamma\nabla_1 . \left\{\frac{\nabla_1\eta}{\left[1+(\nabla_1\eta)^2\right]^{1/2}}\right\} \tag{60}$$

where $\nabla_1 = \dfrac{\partial}{\partial x}\hat{i} + \dfrac{\partial}{\partial y}\hat{j}$ and e_{ij} is the rate of strain tensor in the fluid

(c) Undeformable surface with surface tension

If there is no body force (i.e. $\varrho\vec{g} = 0$) and normal stress is zero i.e $\tau_n = 2\mu\, l_{nn} - p = 0$, then (60) gives

$$\nabla_1 \cdot \left\{\frac{\nabla\eta}{\left[1+(\nabla_1\eta)^2\right]^{\frac{1}{2}}}\right\} = 0 \tag{61}$$

A solution of this is

$$\eta = 0 \tag{62}$$

This means that the free surface is undeformed i.e the free surface is flat

In the case of flat-surface with surface tension varying with temperature, the required boundary conditions can be obtained by equating the variation of surface tension with temperature across the surface to the shear stress experienced by the liquid at the free surface

Now the shear stresss given by (53) Using the fact that w vanishes for all x and y on the surface, takes the form

$$\mu\frac{\partial u}{\partial z}=\frac{\partial \gamma}{\partial x} \quad \text{and} \quad \mu\frac{\partial v}{\partial z}=\frac{\partial \gamma}{\partial y} \tag{63}$$

Now differentiating the first and second equation in (63) with respect to x and y respectively we get

$$\mu\frac{\partial}{\partial z}\left(\frac{\partial u}{\partial x}+\frac{\partial v}{\partial y}\right)=\frac{\partial^2 \gamma}{\partial x^2}+\frac{\partial^2 \gamma}{\partial y^2}$$

This using the equations continuity for incompressible fluid

$$\frac{\partial u}{\partial x}+\frac{\partial v}{\partial y}+\frac{\partial w}{\partial z}=0$$

takes the form

$$-\mu\frac{\partial^2 w}{\partial z^2}=\frac{\partial^2 \gamma}{\partial x^2}+\frac{\partial^2 \gamma}{\partial y^2} \tag{64}$$

The surface tension γ can be expressed in terms of temperature T using Pearson's law.

Pearson's Law : Pearson has formulated that γ can be expanded as the first order in powers of T at the surface in the form

$$\gamma=\gamma_0-\gamma_T\, T \tag{65}$$

where

$$-\gamma_T=\left(\frac{\partial \gamma}{\partial T}\right)_{T=T_s} \tag{66}$$

Where T_s is the surface temperature. For most liquids γ_T is positive because as the temperature rises the difference between the liquid and its vapour phase decrease.

Hence the suitable boundary conditions at the flat free surface in the presence of surface tension are

$$W = 0 \quad \text{and} \quad \mu \frac{\partial^2 w}{\partial z^2} = \sigma_T \nabla_1^2 T \tag{67}$$

Where $\nabla_1^2 = \dfrac{\partial^2}{\partial x^2} + \dfrac{\partial^2}{\partial y^2}$

Making (67) dimensionless using the scales ΔT, $\dfrac{K}{h}$, and h for temperature, velocity and distance respectively, i.e.

$$T^* = \frac{T}{\Delta T}, \quad W^* = \frac{W}{\frac{K}{h}} \left(x^*, y^*, z^*\right) = \left(\frac{x}{h}, \frac{y}{h}, \frac{z}{h}\right)$$

we get

$$\frac{\partial^2 W^*}{\partial z^{*2}} = \frac{\sigma_T \Delta Th}{\mu \kappa} \nabla_1^2 T^*$$

where k is the thermal diffusivity. Neglecting the asterisks(*) for simplicity, we get

$$\frac{\partial^2 w}{\partial z^2} = M_a \nabla_1^2 T$$

where

$$M_a = \frac{\sigma_T \Delta Th}{\mu \kappa} \tag{68}$$

is the Marangoni number.

(d) Undeformed Surface without Surface Tension

If the free surface is flat and neglecting the surface tension (i.e. $\gamma = 0$), equation (67) gives

$$w = \frac{\partial^2 w}{\partial z^2} = 0 \quad \text{at} \quad z = 0 \quad \text{and} \quad z = h \tag{69}$$

If ζ is the z-component of vOrticity, we have

$$\frac{d\varsigma}{dz} = 0 \quad \text{on the free surface} \tag{70}$$

$$\varsigma = 0 \quad \text{on the rigid surface} \tag{71}$$

The boundary conditions (67) and (68) are called stress free boundary conditions.

(e) Deformed Free Surface with Constant Surface Tension using Linear Approximation

In many situations although the free-surface deforms the surface tension γ is constant i.e. independent of temperature. If we assume η varies linearly by then the curvature term becomes

$$\gamma \nabla_1 \cdot \left[\frac{\nabla_1 \eta}{\left[1 + (\nabla_1 \eta)^2\right]^{\frac{1}{2}}} \right] = \gamma \nabla_1^2 \eta$$

and in that case no tangential i.e. shear stress act

i.e., $\tau_{xz} = \tau_{yz} = 0$ (72)

and the normal stress condition takes the form

$$2\mu e_{nn} - p - \delta\eta = \gamma \nabla_1^2 \eta \tag{73}$$

at the free surface, where δ is the body force contribution. The other condition at the free surface will remain the same.

IV Residual-Shear Slip Condition

The BJ-slip condition and the modified BJ slip conditions proposed by Saffman (1971), Jones (1973) and Rudraiah (1985, 1986) were based on the assumption of nominal surface. In many practical problems in biomechanics (particularly endothelium bounding the arteries and cartilages bounding synovial fluid in synovial joints and so on) and in industrial problems (particularly in porous lining at the interior of the bodies for effective cooling system) the assumption of nominal surface is very restrictive because there exists a variation of surface. And in that case one has to use the kinematic condition.

$$\frac{\partial \eta}{\partial t} + u\frac{\partial \eta}{\partial x} + v\frac{\partial \eta}{\partial y} + w\frac{\partial \eta}{\partial z} = 0 \tag{74}$$

where η is the elevation of the free surface from the horizontal level. Further, if we consider a composite media consisting of partly a clear fluid and partly a fluid saturated porous layer, then the shear stress in the porous layer is different from that in the clear fluid because material of the porous medium contributes to the tangential stress. In that case the assumption that no tangential stresses act on the free surface is not correct and there will be a residual stress. If $\vec{q}_1 = (u_1, v_1, w_1)$ in the clear fluid and $\vec{q}_2 = (u_2, v_2, w_2)$ are velocities in the clear fluid and the fluid saturated porous layer respectively, then

$$\left(\tau_{xz}\right)_1 - \left(\tau_{xz}\right)_2 \neq 0 \quad \text{and} \quad \left(\tau_{yz}\right)_1 - \left(\tau_{yz}\right)_2 \neq 0 \tag{75}$$

where we assume that the free surface is in the z-direction given by

$$z = h + \eta(x, y, z, t) \tag{76}$$

we know that

$$\tau_{xz} = \mu\left(\frac{\partial u}{\partial z} + \frac{\partial w}{\partial x}\right) \quad \text{and} \quad \tau_{yz} = \mu\left(\frac{\partial v}{\partial z} + \frac{\partial w}{\partial y}\right)$$

Since w vanishes for all x and y on the bounding surface it follows that

$$\tau_{(xz)_1} = \mu \frac{\partial u_1}{\partial z} \quad \text{and} \quad \tau_{(yz)_1} = \mu \frac{\partial v_1}{\partial z} \tag{77}$$

$$\tau_{(xz)_2} = \bar{\mu} \frac{\partial u_2}{\partial z} \quad \text{and} \quad \tau_{(yz)_2} = \bar{\mu} \frac{\partial v_2}{\partial z} \tag{78}$$

where $\bar{\mu}$ is the effective (or Brinkman) viscosity. Then the Residual –Shear slip conditions are

$$\mu \frac{\partial u_1}{\partial z} - \bar{\mu} \frac{\partial u_2}{\partial z} = -\frac{\beta}{\sqrt{k}} u_2 \tag{79}$$

at the free surface $z = S$

Similarly

$$\mu \frac{\partial v_1}{\partial z} - \bar{\mu} \frac{\partial v_2}{\partial z} = -\frac{\beta}{\sqrt{k}} v_2 \quad \text{at } z = s \tag{80}$$

The condition (79) and (80) are called Residual Shear Slip conditions at the free surface $z = \eta(x, y, t)$ between the clear fluid and fluid saturated porous layer. These conditions will play a significant role in the 21st century.

V. Electromagnetic Boundary Conditions

In deriving the electromagnetic boundary conditions at the interface between two media, as before, we must assume the interface be of infinitesimal thickness instead of finite thickness and that across the interface filed quantities change abruptly. The electromagnetic boundary conditions will depend on the nature of the boundaries, namely, whether they are electrically non-conducting or electrically conducting.
These cases are discussed below

Case A: The medium adjoining an electrically conducting fluid is electrically non-conducting i.e., insulating.

In this case, called the case A, no current can cross the boundary and hence

$$J_z = 0 \tag{81}$$

Further, let $\vec{H}^{ex}$ be the field in the electrically non-conducting medium. This filed must be appropriate to a vaccum and hence it may be derivable from potential. Let ψ be the magnetic potential, then

$$\vec{H}^{ex} = \nabla\psi \quad \text{where} \quad \nabla^2\psi = 0 \left(\because \nabla\cdot\vec{H}^{ex} = 0\right) \tag{82}$$

Further on the interface between the fluid and the medium, the magnetic filed must be continuous. Then the conditions on $\vec{H}$ and $\xi(= 4\pi J_z)$ are

$$\vec{H} = \vec{H}^{ex} \quad \text{and} \quad \xi = 0 \text{ on the bounding surface} \tag{83}$$

Case B. The Medium Adjoining An Electrically Conducting Fluid IS Perfect Electrical conductor.

In this case B, no magnetic field can cross the boundary and hence we must have

$$H_z = 0 \quad \text{and} \quad E_x = E_y = 0 \text{ on the surface adjoining a perfect conductor} \tag{84}$$

There electromagnetic conditions pertain to rigid boundaries. In this case there will be no-slip conditions and the conditions (81) are equivalent to

$$H_z = 0 \quad \text{and} \quad J_x = J_y = 0 \tag{85}$$

Using $\nabla\cdot\vec{J} = 0$ and the fact that $J_x = J_y = 0$ on a plane rigid boundary, it follows that $\dfrac{\partial J_z}{\partial z} = 0$. Then the electromagnetic boundary conditions are

$$H_z = 0 \quad \text{and} \quad \frac{\partial\xi}{\partial z} = 0 \text{ on the rigid boundary in the case B} \tag{86}$$

The above conditions are true for plane boundaries. However in the case of cylindrical coordinates, for example in the case of electrically conducting fluid between two coaxial cylinders, the electromagnetic boundary conditions are

$$J_r = (\nabla \times \vec{H})_r = 0 \text{ in case A} \tag{87}$$

i.e. non-conducting walls,

$$\text{and } H_r = 0 \text{ and } J_z = (\nabla \times \vec{H})_z = 0 \text{ in case B} \tag{88}$$

on conducting walls.

For axi-symmetric $\left(\text{i.e} \frac{\partial}{\partial \theta} = 0\right)$ case, condition (87) implies that

$$\frac{\partial H_\theta}{\partial z} = 0 \tag{89}$$

and the condition (88) implies that

$$\left(\frac{d}{dr} + \frac{1}{r}\right) h_\theta = 0 \text{ in the case B} \tag{90}$$

VI Boundary Conditions on Temperature, T

The boundary conditions on temperature will depend on the nature of the boundaries namely whether they are (a) thermally conducting, (b) thermally insulating (also called adiabatic). (c) maintenance of constant heat flux or (d) Radiating boundaries. These are discussed below.

(a) Thermally Conducting Boundaries

These are called isothermal boundaries where the temperature alone is specified on the boundaries.
That is

$T = T_L$ at the lower boundary (91)
$T = T_U$ at the upper boundary

Mathematically these conditions are called Dirichlet boundary conditions or the boundary condition of the first-kind.

(b) Thermally Insulating Boundaries

At the insulating boundaries, heat flux cannot exist and hence

$$\frac{dT}{dz} = 0 \text{ at the boundary} \tag{92}$$

This is called adiabatic condition. Mathematically this is called Neumann boundary condition and also called boundary condition of the second kind.

(c) Constant Heat Flux at the Boundaries

In this case

$$\frac{dT}{dz} = \frac{q}{k} \text{ at the boundaries} \tag{93}$$

where q/k is the heat flux. It is also called adiabatic boundary condition. Mathematically it is also Neumann boundary condition or the boundary condition of the second kind.

(d) Radiating Boundaries

In this case the required boundary condition is

$$\frac{dT}{dz} = \lambda T \text{ at the boundary} \tag{94}$$

where λ is the Biot number. This implies that the derivative of temperature is proportional to the temperature. It is called radiating boundary condition. Mathematically it is called boundary condition of the third kind. This boundary condition is so general that as $\lambda \to \infty$ we get isothermal condition (91) and $\lambda \to 0$ we get adiabatic condition (92).

VII Boundary Conditions on Concentration, C

The Boundary Condition on the concentration C depends on whether the boundaries an impermeable i.e. C is specified on the boundary or permeable i.e. gradient of C is specified on this boundary.
In the case of permeable boundary

$$C=C_0 \tag{95}$$

on the boundary.

This is the Dirichlet problem or Boundary condition of the first kind.

In the case of permeable boundary

$$\frac{dC_1}{dz}=\frac{dC_2}{dz}$$

on the boundary. Where C_1, C_2 are the concentrations in regions 1 and 2 respectively. This is the Neumann problem or Boundary Condition of the second kind.

There may be a transport of solute per unit area which is given by change in solute concentration C across the boundary and is represented by

$$\frac{dC}{dz}=\alpha C \tag{96}$$

on the boundary, where α is the mass transfer co-efficient. This is the radiation type of Boundary Condition or the Boundary Conditions of the third kind.

In Biomechanical problems, the solute transfer across the membrane exists and in that case, the Boundary Condition on the membrane surface is

$$\frac{dc}{dz}=h(C_i-C_0)+t_R\,v_R\,c_i \tag{97}$$

where h is diffusive solute wall permeability, t_R is the transmitance coefficient ($0< t_R<1$) characterising the molecular sizes and membrane porosity, C_i is c(x, y, z), C_0 is the concentration solute at the other side of the membrane and v_R is the filtration velocity given by

$$v_R = K\{\nabla \varrho - \sigma_s \nabla \pi_s\} \tag{98}$$

where k is the hydraulic wall permeability which may be a constant, a variable or permeability dependent, ∇p is the net permeability difference across the membrane, $\nabla \pi_s$ is the net osmotic permeability difference across the membrane and unity for no transport of solute.

In general $\pi_s = a_0 C_i + a_1 C_i^2 + a_2 C_i^3$ (99)

Where the co-efficient a_0, a_1, a_2 are in general determined from experiment.

VIII The interfacial Condition at the capillary blood vessel wall

In the study of Coronary Artery Diseases (CAD), we study the transport of cholesterol, fat substances, lipids and so on, and, in the study of cartilages in synovial joints we come across the transport of hyaluronic acid from synovial fluid to cartilages. The typical configuration of boundary of the capillary known as endothelium layer or cartilage is shown in the following figure.

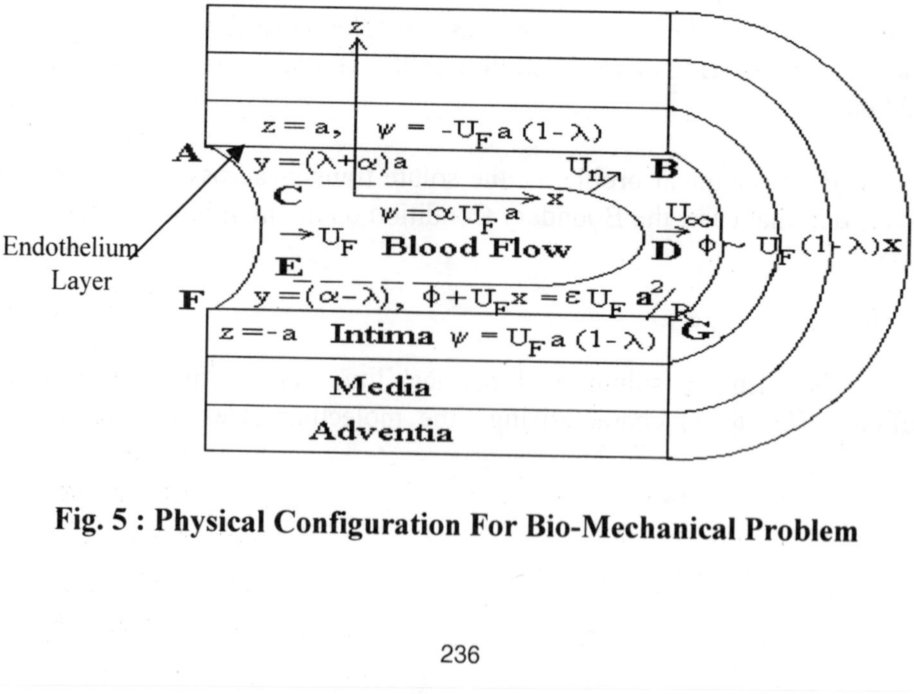

Fig. 5 : Physical Configuration For Bio-Mechanical Problem

This case can be modeled as Hele-Shaw model which exhibits fingering instability having fingering velocity u_F. In this case, a function $\varphi(x, z)$ is required which satisfies in the region between fingering and the bounding membrane wall the Laplace equation

$$\nabla^2 \varphi = 0 \tag{100}$$

and the Boundary Conditions

$$\frac{\partial \varphi}{\partial z} = 0 \quad \text{on} \quad z = \pm a \qquad -\infty < x < \infty \tag{101}$$

$$P + \frac{\gamma}{R} = 0 \qquad \text{on figering bounding CDE}$$

where ϒ – Surface tension and
R- Curvature (102)

$$\frac{\partial \varphi}{\partial n} = u_n \text{ on CDE}$$

$$\varphi \approx u_\infty x \qquad \text{as} \quad x \to \infty$$

The pressure p is related to the function ϕ by

$$P = -\frac{\mu \varphi}{k}$$

Where $k = \frac{b^2}{12}$ is the permeability and b is the gap.

Here, u_F is the fingering velocity
u_∞ is the velocity in the gap
u_n is the normal velocity on the fingering

In the fig, ψ is the stream function and λ is the width.

Conclusions

The salient features of the Boundary Conditions that we encountered in 20th century in solving many problems in fluid mechanics arising in Biomechanics and in some industrial problems have been discussed. The Residual shear slip condition that is going to play a major role in the 21th century is given. These boundary conditions consolidated here covering clear fluid, electrically conducting fluid and fluid saturated porous media may be helpful to young researchers in the field of fluid mechanics.

Acknowledgment

The work reported here is partly supported by AICTE under grant No. F.No. 8017/RDII/R&D/DEG(630)/98-99, MARCH 27, 1999 and partly by INSA Senior Scientist Programme. Their support is gratefully acknowledged

References.

1. Beavers, G. S and Joseph, D.D., J. Fluid Mech. 30, pp197, 1967

2. Beavers, G.S., Sparrow, E.M., and Magnuson, R.A., J. Basic Eng., Trans., ASME Series D, 92, pp843, 1970

3. Beavers, G.S., Sparrow, E.M., and Masha, B.A., AICHE, J., 20, pp596, 1974

4. Goldstein, S., Modern Developments In Fluid Mech., vol II, pp676, 1952

5. Hocking, L.M., J. Fluid Mech., 76, pp801, 1976

6. Jones, I.P., Camb. Phil. Soc., 73, pp231, 1973

7. Miksis, M.J., and Davis, S.H., J. Fluid Mech, 273, 125, 1994.

8. Rajasekhar, B.M., Ph.D. Thesis, Bangalore Univeristy 1971

9. Ramasesha, K. Sheela, Resonance, 1 and 2, Aug. 1999, Dec. 1999.

10. Ranganna, G., PGSAM, Bangalore University, 5, 1, 1982

11.Richardson, S. J. Fluid Mech., 49, pp327, 1971

12.Rudraiah, N., ASME J. Fluid Eng. 107, pp322, 1985

13.Rudraiah, N, Encylopedia of Fluid Mechs., Slurry Flow Tech. 5, Gulf publishers, Houston, 1986

14.Rudraiah, N., Sheela, R., Shivakumara Murthy, J.K., Arabian J. for Sci and Eng, 12, 4, pp481, 1987

15.Saffman, P., MIT studies in Apply. Maths, LNO2, 93, 1971

16.Taylor, G.I. J.Fluid Mech. 49, 319, 1971

Modern Applied Mathematics
Editor: J.C. Misra

Topics in the Geometry of Differential Equations with Applications to Water Waves

Enrique G. Reyes

Department of Mathematics, University of Oklahoma, Norman, OK 73019 USA

Abstract. We introduce conservation laws, formal, generalized and nonlocal symmetries, classical Bäcklund transformations, and pseudo-potentials, and we study them in the context of equations of pseudo-spherical type. We construct correspondences between solutions of any two equations of pseudo-spherical type, and between hierarchies of equations in this class. We present several notions of integrability, state some recent comparison results, and point out that whole classes of integrable equations possess geometrical interpretations. Finally, we consider the equations introduced by Korteweg and de Vries, Camassa and Holm, and Hunter and Saxton, describing different aspects of water waves theory, and apply our work to them.

Keywords: conservation laws, symmetries, Bäcklund transformations, equations of pseudo-spherical type, KdV equation, Camassa–Holm equation, Hunter–Saxton equation.

1. Introduction

The goal of this chapter is to introduce the reader to some recent developments in the geometric theory of differential equations. This work may be considered as a continuation of the author's contribution to the International Symposium on Challenges in Mathematical Sciences for the New Millennium, held at the Indian Institute of Technology, Kharagpur [**39**], to which the author was fortunate to assist, but it proceeds to a more leisurely pace than that paper. Besides [**39**], we would like to mention three important books which have been the author's main references for some of the topics discussed here: P. Olver's *Applications of Lie Groups to Differential Equations* [**31**], the standard reference for applications of symmetry methods to differential equations; K. Tenenblat's *Transformations of manifolds and applications to differential equations* [**46**], a recent monograph on the application of differential geometry to the study of nonlinear equations; and *Lie algebraic methods in integrable systems*, an extremely clear and inspiring introduction to the modern methods and problems of the theory of integrable systems written by an eminent Indian scientist, Prof. A. Roy Chowdhury [**41**].

A short introduction to our subject can be found in [**39**], and in the three books mentioned above. Here we would like to stress the fact that the geometric theory of differential equations began more than a hundred years ago, and that fundamental achievements of the XIXth Century research period were Sophus Lie's work on continuous groups of transformations and symmetries of systems of differential equations, and Emmy Noether's result on symmetries of the Euler–Lagrange equations and conservation laws. These seminal developments are reviewed in P. Olver's book [**31**].

The current interest in the geometry of differential equations has been strongly motivated by the discovery of "completely integrable" systems described by *partial* differential equations, like the KdV equation $u_t = u_{xxx} + 6uu_x$, and the development of the scattering/inverse scattering method (Beals, Deift, and Tomei [**4**], Beals, Deift, and Zhou [**5**], Beals and Sattinger [**6**], Faddeev and Takhtajan [**18**], Palais [**33**], Roy Chowdhury [**41**]) which allow us to deeply study these new examples [**6, 18, 33**]. Besides, a formal theory of differential equations has been developed, which provides a rigorous framework for these new endeavours (I. Krasil'shchik and A. Verbovetsky [**26**], Krasil'shchik and Vinogradov [**27**]). It seems important to present at least a few essentials of this formal theory, and we do so in Section 2. For most of this chapter, however, we follow a more intuitive approach, as in [**31**].

Today's research, represented for instance in the treatises [**4, 18, 31, 41, 46**], has certainly more avenues than the ones considered by the geometers of old. However, their work continues being crucial. We mention at this point what is probably the best known application of classical differential geometry to differential equations: Bäcklund showed in 1883 that one can construct surfaces of constant Gaussian curvature -1 starting from a given such surface. Now, there exists a correspondence between surfaces of constant Gaussian curvature -1 and (local) solutions to the sine–Gordon equation

$$u_{xt} = \sin(u);$$

in modern terminology, then, Bäcklund's construction translates into a method to obtain new solutions to the nonlinear sine–Gordon equation from old ones! Conversely, the modern theory of integrable systems has important geometrical consequences. Here we cite [**34**], in which U. Pinkall and I. Sterling show the existence of an infinite number of constant mean curvature tori by carefully analysing the sinh–Gordon equation

$$u_{xx} + u_{tt} + 2\sinh 2u = 0$$

and its "generalized symmetries", a notion to be studied in Section 3 below.

We review in this work some topics in differential equations which are motivated by the research carried out about one hundred years ago, and which have a direct impact in applications. In Sections 3 and 4 we introduce some notions which play a fundamental rôle in integrable systems: conservation laws, generalized, formal and nonlocal symmetries, and pseudo-potentials. Our main references for these topics are Olver [**31**] and Roy Chowdhury [**41**]. In Sections 5, 6, and 7 we consider a very general class of equations for which these notions can be studied from a geometrical point of view, the equations of pseudo-spherical type, and we also mention some of their generalizations. In Section 8 we briefly review classical Bäcklund transformations and construct correspondences between solutions (which satisfy a genericity condition) of any two equations of pseudo-spherical type, and between *hierarchies* of equations in this class, to stress the fact that using classical differential geometry to study differential equations is not simply an interesting point of view, but it allows one to find important new results. This part of the chapter is based on Tenenblat [**46**] and work by the author [**37, 38, 39**]. In Section 9 we present several notions of integrability, what one may take as a signal that this area of research is alive and well: even the exact notion of integrability has not been agreed upon as yet [**48**]. Some comparison results exist, however [**20, 35, 36**] and we state several of them. An interesting and intriguing fact arising from these results is that, as with the sine–Gordon equation, whole classes of integrable equations are amenable of geometrical interpretations. We certainly wonder if this is generally true for all those equations one would like to call "integrable". Finally, in Section 10 we study the equations introduced by Korteweg and de Vries [**33**], Camassa and Holm [**14**], and Hunter and Saxton [**21**], describing different aspects of water waves theory, from the point of view considered here.

To finish this introduction, we would like to point out that there exists an important area of research stemming from the impressive work of the French school led by Darboux, Vessiot, Janet, Goursat, and Cartan. The reader is referred to [**3, 11, 12, 13**] for details on these matters. We would also like to call the reader' attention to R. Palais' review article on integrable systems [**33**], and to N. Kamran's recent NSF–CBMS Washington lectures [**23**], in both of which several important aspects of the world of integrable systems which are not touched upon here, are masterly explained.

2. The background: geometry of differential equations

Our starting point is the observation that a differential equation (understood here as either a scalar equation or a system) $\Xi = 0$, in which the left hand side of the equation depends on independent variables $x_1, \ldots, x_n$, dependent variables $u^1, \ldots, u^m$, and a finite number of partial derivatives of the functions u^α with respect to the variables x_j $(1 \leq i \leq m,\ 1 \leq j \leq n)$, can be considered as determining a *locus*, contained in a suitable space in which $x_1, \ldots, x_n,\ u^1, \ldots, u^m$, and the partial derivatives of the functions u^α with respect to the variables x_j become *coordinates*, and that by restricting that locus, one can think of the equation as determining a *manifold.* Then, for instance, changes of variables in the equation correspond to changes in the coordinate system used to describe locally this manifold, and one hopes to phrase questions (and answers!) invariantly, so as to make sense in our manifold interpretation.

How can we define "invariant"? It is natural to be interested in those changes of coordinates which send solutions of an equation $\Xi = 0$ to solutions of the transformed equation $\Xi' = 0$. A very general kind of transformations with this property is defined roughly as follows: One considers a solution $u^\alpha(x_i)$ of $\Xi = 0$ as a submanifold S of the manifold determined by $\Xi = 0$, the graph of the local immersion $(x_i) \mapsto (x_i, u^\alpha(x_i))$. Then, the transformations of interest are those which change simultaneously the points $p \in S$ *and* the tangent space at $p \in S$ in such a way that, if $p \in S$ is mapped to $p' \in S'$, then the tangent space at p is mapped to the tangent space at p'. These are the *contact transformations* [**27**] which we define rigorously in (6).

On the other hand, we remark that one can also study manifolds determined by equations from a "frozen" point of view: one may be interested not in (contact) invariant results, but in a particular coordinate description of the equation at hand. For instance, it appears not to be known how to characterize an *evolution* equation $u_t = F$ in a contact invariant fashion. However, one can associate a manifold with $u_t = F$, and we do it in Section 5, after [**24, 20**].

EXAMPLE 1. Consider a first order ordinary differential equation $F(x, y, y') = 0$. We would like to think of x, y, y' as coordinates in a three dimensional space, and say that $F(x, y, y') = 0$ determines a submanifold of this space. We assume that F is a smooth function $F : \mathbf{R}^3 \to \mathbf{R}$, in which $\mathbf{R}^3$ is equipped with coordinates (x, y, p), so that the set

$$S^1 = \{(x, y, p) : F(x, y, p) = 0\}$$

is a submanifold of $\mathbf{R}^3$, provided that the Jacobian of the function F is not zero. By restricting S^1, we further assume that S^1 is connected, contractible, and that it fibers over the space of the independent variable x, so that in fact S^1 is a sub–bundle of the *trivial bundle* $\mathbf{R}^3 \mapsto \mathbf{R}$. S^1 is our basic manifold associated with the differential equation $F(x, y, y') = 0$.

It is natural to say that a solution is a *section* $x \mapsto (x, y(x))$, such that the *prolonged section* $x \mapsto (x, y(x), y'(x))$ is contained in S^1. Note that if $x \mapsto (x, y(x))$ is a local solution to $F = 0$, the differential form

$$\theta^0 = d\,y - p\,dx \tag{1}$$

vanishes identically on the prolonged section $x \mapsto (x, y(x), y'(x))$. The form θ^0 is an example of *contact form*, see Equation (5) below.

We now present some basic constructions which generalize Example 1. We follow mainly **[31, 39]**, and call the reader's attention to the more formal accounts appearing in **[26, 27]**. The reader is referred to M. Spivak **[44]** and M. do Carmo **[17]** for detailed information on the differential geometrical notions we use below.

As in Example 1, we begin by fixing a trivial fiber bundle $\pi : E \to M$, in which M is identified with the space of independent variables, and the typical fiber with the space of dependent variables. We work mainly with scalar equations in two independent variables, and therefore the reader can safely assume that E is simply $\mathbf{R}^3$ with coordinates (x, t, u) and that M is $\mathbf{R}^2$ with coordinates (x, t). Our first goal is to find a geometrical interpretation for the derivatives u_x, u_{xt}, and so forth. This was achieved in in the first half of the XXth Century by C. Ehresmann, who constructed the higher order "jet bundles" associated with the bundle E. This construction goes as follows:

We assume that the trivial bundle $E \mapsto M$ is equipped with adapted coordinates $(x_i, u^\alpha) \mapsto (x_i)$. Here and henceforth, lower case latin indices range from 1 to m, and lower case greek indices range from 1 to n. Let $s_1(x_i) = (x_i, s_1^\alpha(x_i))$ and $s_2(x_i) = (x_i, s_2^\alpha(x_i))$ be two local sections of the bundle $E \mapsto M$, defined about a point $p = (x_i)$ in M. We say that s_1 and s_2 *agree to order k at $p \in M$* if s_1, s_2 and *all* the partial derivatives of the sections s_1 and s_2 up to order k agree at p. This notion determines an equivalence relation on local sections of E. We let $j^k(s)(p)$ represent the equivalence class of the section s at p, and we call this equivalence class *the k-jet of s at p*. We then denote by $J^k(p)$ the set of all the k–jets of local sections at p, and define the *k-order jet bundle of E* as $J^kE = \cup_{p\in M} J^k(p)$.

One can prove that J^kE possesses a manifold structure and that it fibers over M: local coordinates on J^kE are $(x_i, u_0^\alpha, u_{i_1}^\alpha, u_{i_1 i_2}^\alpha, \dots, u_{i_1\dots i_k}^\alpha)$, in which

$$(2) \quad u_0^\alpha(j^k(s)(p)) = s^\alpha(p)\,, \qquad u_i^\alpha(j^k(s)(p)) = \frac{\partial s^\alpha}{\partial x_i}(p)\,, \qquad u_{i_1 i_2}^\alpha(j^k(s)(p)) = \frac{\partial^2 s^\alpha}{\partial x_{i_1}\partial x_{i_2}}(p)\,,$$

and so forth; in these coordinates, the projection maps $\pi_M^k : J^kE \to M$ and $\pi_l^k : J^kE \to J^lE$, $l \le k$ are obvious. For instance, π_M^k is simply $(x_i, u_0^\alpha, u_{i_1}^\alpha, \dots, u_{i_1\dots i_k}^\alpha) \mapsto (x_i)$.

We now define the *infinite jet bundle* $J^\infty E$ of E, as the infinite dimensional manifold equipped with canonical coordinates $(x_i, u_0^\alpha, u_{i_1}^\alpha, \dots, u_{i_1\dots i_k}^\alpha, \dots)$ obtained from the coordinates (2). Then, for instance, we can consider functions depending on an arbitrary number of derivatives of the dependent variables as functions on $J^\infty E$. Formally, $J^\infty E$ is the bundle $\pi_M^\infty : J^\infty E \to M$ obtained as the inverse limit of the tower of jet bundles $M \leftarrow E \cdots \leftarrow J^kE \leftarrow J^{k+1}E \leftarrow \cdots$ under the standard projections $\pi_l^k : J^kE \to J^lE$, $k > l$.

Any local section $s : (x_i) \mapsto (x_i, u_0^\alpha(x_i))$ of E lifts to a unique local section $j^\infty(s)$ of $J^\infty E$ called the *infinite prolongation* of s. In canonical coordinates, $j^\infty(s)$ is the section $(x_i, u_0^\alpha(x_i), u_{i_1}^\alpha(x_i), \dots, u_{i_1\dots i_k}^\alpha(x_i), \dots)$. Also, we say that a function $f : J^\infty E \to \mathbf{R}$ is *smooth*, or that f is a *differential function* **[31]** if it factors through a finite–order jet bundle, that is, if it depends on a finite number of derivatives of the dependent variables.

Next, we define vector fields on $J^\infty E$. As in standard differential geometry **[44]**, a *vector field* X on $J^\infty E$ is a derivation on the ring of smooth functions on $J^\infty E$. In local coordinates, a vector field is a formal series

$$X = \sum_{i=1}^{m} A^i \frac{\partial}{\partial x_i} + \sum_{\alpha=1}^{n} \sum_{\substack{k=0 \\ 0\le i_1<\cdots<i_k\le n}}^{\infty} B_{i_1\dots i_k}^\alpha \frac{\partial}{\partial u_{i_1\dots i_k}^\alpha}\,,$$

in which the coefficients $A^i, B^\alpha_{i_1\dots i_k}$ are differential functions. We can prolong vector fields on M to vector fields on $J^\infty E$: for a vector field X on M we define the vector field $pr^\infty X$ on $J^\infty E$ by means of

$$(pr^\infty X)_{j^\infty(s)(p)}(f) = X_p(f \circ j^\infty(s)),$$

in which f is an arbitrary differential function. This operation is fundamental, because it allows us to "split" the vector fields on $J^\infty E$ into horizontal and vertical vector fields, which in turn determine the important contact transformations mentioned above. Rigorously speaking, we say that the prolongation operation defines a connection, the *Cartan connection C on $J^\infty E$*, by stating that the horizontal lift of a vector field X on M is $pr^\infty X$. Locally, *horizontal vector fields* are linear combinations of the *total derivatives* D_{x_i}, in which $D_{x_i} = pr^\infty \partial/\partial x_i$, that is,

$$D_{x_i} = \frac{\partial}{\partial x_i} + \sum_{\alpha=1}^{n} u^\alpha_i \frac{\partial}{\partial u^\alpha_0} + \sum_{\alpha=1}^{n} \sum_{1\le i_1\le m} u^\alpha_{i_1 i} \frac{\partial}{\partial u^\alpha_{i_1}} + \sum_{\alpha=1}^{n} \sum_{1\le i_1,i_2\le m} u^\alpha_{i_1 i_2 i} \frac{\partial}{\partial u^\alpha_{i_1 i_2}} + \dots \tag{3}$$

REMARK 1. $pr^\infty[X_1, X_2] = [pr^\infty X_1, pr^\infty X_2]$ for all vector fields X_1 and X_2 on M, and therefore the Cartan connection is flat, see **[26, 27]**.

We now consider differential forms. By definition, smooth differential forms on $J^\infty E$ are the pull-backs of differential forms on $J^k E$ by the projections π^∞_k. Locally, any differential form ω on $J^\infty E$ may be written in canonical coordinates as a finite sum of terms of the form

$$\omega = A\, dx_{i_1} \wedge \dots \wedge dx_{i_p} \wedge \theta^{\alpha_1}_{I_1} \wedge \dots \wedge \theta^{\alpha_q}_{I_q}, \tag{4}$$

in which $I_1, \dots, I_q$ are multi-indices, $I_j = i_1 i_2 \dots i_{k_j}$, and A is a differential function. The one-forms θ^α_I appearing in (4) are the *basic contact one-forms*

$$\theta^\alpha_I = du^\alpha_I - \sum_{j=1}^{m} u^\alpha_{Ij} dx_j. \tag{5}$$

Note that (5) generalizes the one-form (1) considered in Example 1, and that, as in the case of (1), if we evaluate the pull-back $j^\infty(s)^*\theta_I$ we always obtain zero. We define a *contact form* as any differential form ω on $J^\infty E$ such that $j^\infty(s)^*\omega = 0$ for every local section s of E. The set $C(E)$ of all contact forms is a differential ideal in the ring of all differential forms of $J^\infty E$, and it is generated by the basic contact forms (5).

A map $\Phi : J^\infty E \to J^\infty E'$ is called a *generalized contact transformation* if it preserves contact ideals, that is, if

$$\Phi^* C(E') \subseteq C(E). \tag{6}$$

In addition, we say that Φ is a *classical contact transformation* if it covers a smooth map $\Phi_1 : J^1 E \to J^1 E'$, that is, if $\pi^\infty_1 \circ \Phi_1 = \pi^\infty_1 \circ \Phi$.

EXAMPLE 2. Consider a transformation $T : J^1 E \to J^1 E'$, in which E is the bundle $(x, t, u) \mapsto (x, t)$ and E' is the bundle $(x', t', u') \mapsto (x', t')$, of the form $x' = \phi(x, u, u_x)$, $t' = t$, $u' = \psi(x, u, u_x)$. T is a classical contact transformation if the functions ϕ and ψ satisfy

$$(\phi_x + u_x \phi_u)\psi_{u_x} = (\psi_x + u_x \psi_u)\phi_{u_x}, \qquad \phi_u \psi_{u_x} - \psi_u \phi_{u_x} \neq 0.$$

A famous example of a contact transformation of this type **[27]** is the Legendre transformation $x' = u_x$, $t' = t$, $u' = x u_x - u$.

We say that a differential form ω is *horizontal* if no forms θ_I^α appear in its local expansion (4), and finally, we define the *horizontal* exterior derivative d_H as the coboundary operator (*i.e.* $d_H \circ d_H = 0$) uniquely determined in local coordinates by the formulae

$$d_H f = \sum_{1\leq i\leq m} D_{x_i} f dx_i \ , \quad d_H(dx) = d_H(dt) = 0, \quad d_H(\theta_I^\alpha) = \sum dx_j \wedge \theta_{Ij}^\alpha \ .$$

Let us consider now a scalar partial differential equation of order $m+n$ in two independent variables,

$$\Xi\left(x,t,u,\frac{\partial u}{\partial x},\frac{\partial u}{\partial t},\ldots,\frac{\partial^{m+n}u}{\partial x^m t^n}\right)=0\ , \tag{7}$$

our main object of interest in subsequent sections. As stated at the beginning of this section, our basic remark is that this equation defines a locus in the jet bundle $J^{m+n}E$ of E. As in Example 1, we assume that we can choose an open, connected, and contractible subset S^{m+n} of this locus, that the function Ξ is smooth on a neighbourhood of S^{m+n}, and that this set fibers over the space of independent variables.

We define the *equation manifold* of $\Xi = 0$ as a sub-bundle of our infinite jet space $J^\infty E$: the *prolongations* $S^{m+n+1}, S^{m+n+2}, \ldots$ of S^{m+n} are defined by total differentiation of (7). For $k \geq m+n$,

$$\begin{aligned} S^{k+1} &= \{j^{k+1}(s)(x,t) \in J^{k+1}E : \\ &\quad j^k(s)(x,t) \in S^k \text{ and } (D_x\Xi)(j^{k+1}(s)(x,t)) = (D_t\Xi)(j^{k+1}(s)(x,t)) = 0\}. \end{aligned}$$

We assume [**26**] that each S^{k+1} is a sub-bundle of $J^{k+1}E$ which fibers over S^k. The *equation manifold* of Equation (7) is the *infinite prolongation* $\pi_M^\infty : S^\infty \to M$ of S^{m+n}, that is, the sub-bundle S^∞ of $J^\infty E$ obtained as the inverse limit of the tower $M \leftarrow S^{m+n} \leftarrow S^{m+n+1} \leftarrow \cdots$. Intuitively, the equation manifold S^∞ is the set of points in $J^\infty E$ satisfying (7) and all its (total) differential consequences.

A *local solution* of Equation (7) is a holonomic section of S^∞, namely, a section of S^∞ which is the infinite prolongation of a local section $s : (x,t) \mapsto (x,t,u(x,t))$ of E. We will use the standard phrase "$u(x,t)$ is a solution of Equation (7)" instead of "the local section $j^\infty(s)$, in which $s : (x,t) \mapsto (x,t,u(x,t))$ is a local section of E, is a local solution of Equation (7)".

A fundamental fact is that we can consider total derivatives, horizontal vector fields, and contact forms, *on* S^∞. Rigorously, what happens is this: if $\theta = j^\infty(s)(x,t)$ is a point in S^∞, the horizontal subspace C_θ of the Cartan connection of $J^\infty E$ is contained in the tangent space $T_\theta S^\infty$. Thus, we can restrict the Cartan connection to S^∞, thereby obtaining a (flat) connection on S^∞ called the *Cartan connection of* S^∞, and we can repeat our constructions for $J^\infty E$ in this new setting. For instance, for *evolution* equations $u_t = F(x,t,u,\ldots,u_{x^k})$, in which $u_{x^k} = \partial^k u/\partial x^k$, we equip S^∞ with coordinates $(x,t,u,u_x,\ldots,u_{x^k},\ldots)$ (See Example 3 below), and then the restrictions to S^∞ of the total derivatives D_x and D_t defined in (3) become

$$D_x = \frac{\partial}{\partial x} + \sum_{i=0}^{\infty} u_{x^{i+1}} \frac{\partial}{\partial u_{x^i}}\ , \qquad \text{and} \qquad D_t = \frac{\partial}{\partial t} + \sum_{i=0}^{\infty} D_x^i(F)\frac{\partial}{\partial u_{x^i}}\ . \tag{8}$$

Lastly, we observe that the operators d and d_H coincide "on solutions", that is,

$$d\left((j^\infty(s))^*\omega\right) = (j^\infty(s))^*(d_H\omega) \tag{9}$$

for every differential one-form $\omega = Xdx + Tdt$ on S^∞, and every holonomic section $j^\infty(s)$ of S^∞. In particular, (9) holds for local sections $j^\infty(s)$ of $J^\infty E$. We will use Equation (9) very often in subsequent sections.

EXAMPLE 3. Burgers' equation $u_t = u_{xx} + uu_x$ determines a locus L in J^2E,

$$L = \{(x, t, u, u_t, u_x, u_{tt}, u_{xt}, u_{xx}) : u_t = u_{xx} + uu_x\}.$$

Since $\Xi = -u_t + u_{xx} + uu_x$ is smooth on J^2E and $\Xi_{u_{xx}} \neq 0$ on J^2E, we take $S^2 = L$. Natural coordinates on the infinite prolongation S^∞ are $(x, t, u, u_x, \ldots, u_{x^k}, \ldots)$, as (total) derivatives of u with respect to t may be replaced by expressions containing only x–derivatives by using the differential consequences of $\Xi = 0$.

We now extend the theory to encompass nonlocal data. Even though this is a very natural generalization, it is very recent, while $J^\infty E$ appeared already in the early 1970's. We follow Krasil'shchik and Vinogradov **[26, 27]**. Nonlocal data appears, for example, when one considers what we will call *symmetries* of an equation $\Xi = 0$:

EXAMPLE 4. Let $\Xi = 0$ be Burgers equation $u_t = u_{xx} + uu_x$. We replace u by $u + \tau G$, in which

$$G = (2S_x - uS)e^{-\frac{1}{2}\int u dx}$$

and S is any function satisfying $S_t = S_{xx}$. We can check that, whenever $u(x,t)$ is a solution of Burgers equation, $u + \tau G$ satisfies the same equation to first order in τ. We would like to consider G as a "symmetry" of Burgers equation. But G is not a differential function, as it depends on the integral of u. We must think of G as a "nonlocal" object.

Consider then a differential equation $\Xi = 0$ with equation manifold S^∞, and let $\pi : \overline{S} \to S^\infty$ be a fiber bundle over S^∞ (fiber bundles over S^∞ are rigorously defined in **[26]**).

DEFINITION 1. The bundle π determines a covering structure (or, $\overline{S}$ is a covering of S^∞) if and only if

(a) There exists a flat connection $\overline{C}$ on the bundle $\pi^\infty_M \circ \pi : \overline{S} \to M$, and

(b) The connection $\overline{C}$ agrees with the Cartan connection C on S^∞, that is, for any vector field X on M, $\pi_*(\overline{X}) = pr^\infty(X)$, in which the vector field $\overline{X}$ on $\overline{S}$ is the unique horizontal lift of X induced by $\overline{C}$.

Points in a covering $\pi : \overline{S} \to S^\infty$ are described locally as $(x, t, u, \ldots, u_{x^m t^n}, \ldots, w^1, \ldots, w^N)$, $1 \leq N \leq \infty$, in which $(x, t, u, \ldots)$ are canonical coordinates on S^∞, and $(w^1, \ldots, w^N)$ are fiber coordinates on $\overline{S}$. Let D_x and D_t be the total derivatives operators on S^∞. The covering $\overline{S}$ is determined locally by the data $(\overline{S}, \overline{D}_x, \overline{D}_t, \pi)$, in which:

(a) $\overline{D}_x$, $\overline{D}_t$ are differential operators on $\overline{S}$ (called *total derivatives* on $\overline{S}$) of the form

$$\overline{D}_x = D_x + X_1 \quad \text{and} \quad \overline{D}_t = D_t + X_2, \tag{10}$$

where $X_i = \sum_{\beta=1}^N X^\beta_i \partial/\partial w^\beta$, and X^β_i are smooth functions on $\overline{S}$, and

(b) $\overline{D}_x$ and $\overline{D}_t$ satisfy the integrability condition

$$[\overline{D}_x, \overline{D}_t] := D_x(X_2) - D_t(X_1) + [X_1, X_2] = 0. \tag{11}$$

The horizontal distribution of $\overline{S}$ is spanned by the total derivatives $\overline{D}_x$ and $\overline{D}_t$. The fiber coordinates w^i, $1 \leq i \leq N$, are called *nonlocal variables* with respect to S^∞, and N is called the *dimension* of the covering $\pi : \overline{S} \to S^\infty$.

The action of the *horizontal exterior derivative* $\overline{d}_H$ on *horizontal* one–forms $Adx + Bdt$ is determined by

$$\overline{d}_H \overline{f} = \overline{D}_x \overline{f} dx + \overline{D}_t \overline{f} dt, \quad \overline{d}_H(dx) = \overline{d}_H(dt) = 0, \quad \overline{d}_H(w^\alpha) = X^\alpha_1 dx + X^\alpha_2 dt, \tag{12}$$

in which $\overline{f}$ is a smooth function on $\overline{S}$, and X_i^α, $i = 1, 2$, are the components of the vector fields X_i appearing in (10). Since the total derivatives $\overline{D}_x$ and $\overline{D}_t$ satisfy (11), the operator $\overline{d}_H$ satisfies $\overline{d}_H \circ \overline{d}_H = 0$.

EXAMPLE 5. Assume that the one-form $\kappa = \alpha dx + \beta dt$ satisfies $d_H \kappa = 0$ on solutions of $u_t = F$. Then, κ determines a covering $(S^\kappa, \overline{D}_x, \overline{D}_t, \pi_\kappa)$ of $u_t = F$: S^κ is locally defined by $S^\kappa = \{(x, t, u, \ldots, u_{x^m}, \ldots, w)\}$, where $(x, t, u, u_x, \ldots)$ are coordinates on the equation manifold S^∞ of $u_t = F$, and

$$(13) \qquad \overline{D}_x = D_x + \alpha \frac{\partial}{\partial w}, \qquad \overline{D}_t = D_t + \beta \frac{\partial}{\partial w}.$$

It is easy to check that $d_H \kappa = 0$ implies that (11) holds.

We are now ready to begin our excursion for the land of differential equations. As in the last part of this section, we will consider only scalar equations of two independent variables unless otherwise explicitly stated. We will be as concrete as possible, but keep in mind that, as stressed in this section, explicitness does not imply lack of theoretical understanding.

3. Symmetries and the symmetry approach to integrability

Among other many important contributions, Sophus Lie realized that one can use symmetries of differential equations to obtain special solutions of the equations of interest [**2, 31**]. We will focus on the "non-classical" symmetries which appear naturally within the framework of inverse scattering and integrable systems [**5, 26, 27, 31, 33**].

DEFINITION 2. A differential function G is a generalized symmetry of an evolution equation $u_t = F(x, t, u, \ldots, u_{x^n})$ if for any local solution $u(x, t)$, the function $u(x, t) + \tau G(u(x, t))$ satisfies $u_t = F$ to first order in τ.

Thus, G is a generalized symmetry of $u_t = F$ if and only if the equation $D_t G = F_* G$, in which for any $f(x, t, u, \ldots, u_{x^k})$, f_* is the *formal linearization* of f given by [**29, 31**]

$$(14) \qquad f_* = \sum_{i=0}^{k} \frac{\partial f}{\partial u_{x^i}} D_x^i ,$$

holds identically once all the derivatives with respect to t appearing in it have been replaced by means of $u_t = F$. Yet another way of expressing this is by saying that a second evolution equation $u_t = G$ is a generalized symmetry of $u_t = F$ if, at least formally, their flows commute, that is,

$$(15) \qquad \frac{\partial G}{\partial t} + F_* G - G_* F = 0.$$

REMARK 2. A beautiful description of generalized symmetries can be made in terms of the geometry of the equation manifold S^∞ of $u_t = F$ [**26, 27**]. It reads like this: We consider the module $D(S^\infty)$ of vector fields on S^∞, and define

$$S = \{X \in D(S^\infty) : [X, \mathrm{hor}(C)] \subset \mathrm{hor}(C)\},$$

in which $\mathrm{hor}(C)$ denotes the space of horizontal vector fields with respect to the Cartan connection of S^∞. The algebra of generalized symmetries of $u_t = F$ is the quotient Lie algebra $Sym = S/\mathrm{hor}(C)$.

We define nonlocal symmetries following [**26, 27**]:

DEFINITION 3. Let $(\overline{S}, \overline{D}_x, \overline{D}_t, \pi)$ be an N-dimensional covering of the kth order equation $u_t = F$ equipped with coordinates $(x, t, u, u_x, \ldots, w^\beta)$, $1 \leq \beta \leq N$. A nonlocal symmetry of type π of $u_t = F$ is a vector field $\overline{D}_\tau$ of the form

$$\overline{D}_\tau = \sum_{i=0}^{\infty} \overline{D}_x^i(G) \frac{\partial}{\partial u_{x^i}} + \sum_{\beta}^{N} I_\beta \frac{\partial}{\partial w^\beta}, \tag{16}$$

in which G and I_β, $1 \leq \beta \leq N$, are smooth functions on $\overline{S}$ such that the following equations hold:

$$\overline{D}_t G = \sum_{i=0}^{k} \frac{\partial F}{\partial u_{x^i}} \overline{D}_x^i(G), \qquad \overline{D}_x(I_\beta) = \overline{D}_\tau(X_1^\beta), \qquad \overline{D}_t(I_\beta) = \overline{D}_\tau(X_2^\beta). \tag{17}$$

More invariantly, a nonlocal symmetry of type π is an infinitesimal transformation of $\overline{S}$ which preserves the horizontal distribution of $\overline{S}$. Note that the first equation of (17) depends only on G and F. The vector field

$$L_\tau = \sum_{i=0}^{\infty} \overline{D}_x^i(G) \frac{\partial}{\partial u_{x^i}} \tag{18}$$

can be interpreted as a vector field on $\overline{S}$ *along* S^∞. This vector field, or simply G, is called the *shadow* of the nonlocal symmetry $\overline{D}_\tau$. In general, vector fields on $\overline{S}$ along S^∞ satisfying the first equation of (17) are called *π-shadows*. An important question is whether one can extend π-shadows to *bona fide* nonlocal symmetries, see **[26, 27]**.

EXAMPLE 6. It follows from Example 4 that the function

$$G = (2S_x - uS)\mathrm{e}^{-\frac{1}{2}\int u\,dx}, \tag{19}$$

in which $S_{xx} = S_t$, is the shadow of a nonlocal symmetry of Burgers equation $u_t = u_{xx} + uu_x$.

The existence of generalized symmetries is a straightforward test to recognize "integrable" equations (Mikhailov, Shabat and Sokolov **[29]**, Olver **[31]**, Olver and Sokolov **[32]**, Sanders and Wang **[42]**, and Krasil'shchik *et al* **[26, 27]**) which we will use below. They are also important in differential geometry: as pointed out in the Introduction, U. Pinkall and I. Sterling **[34]** showed the existence of an infinite number of constant mean curvature tori by carefully analysing the sinh–Gordon equation and its generalized symmetries. As per nonlocal symmetries, they appear naturally **[26, 27]** in the theory of equations which admit recursion operators (Definition 4 below), are related to the existence of Bäcklund transformations and linearization of differential equations, and can be used to generate "new solutions from old", as we will see in Section 10.

EXAMPLE 7. Consider again the Burgers equation $u_t = u_{xx} + uu_x$, and the (shadow of a) nonlocal symmetry

$$G = (2S_x - uS)\mathrm{e}^{-\frac{1}{2}\int u\,dx}, \tag{20}$$

in which $S_t = S_{xx}$. Solutions $u(x, t)$ which are invariant under (20) satisfy $G(u(x, t)) = 0$ **[31]**, and so the equation

$$2S_x - uS = 0 \tag{21}$$

must hold. This is the Cole–Hopf transformation between the nonlinear Burgers equation $u_t = u_{xx} + uu_x$ and the linear heat equation $S_t = S_{xx}$.

A most important issue is the construction of generalized and nonlocal symmetries. One can find generalized symmetries algorithmically [**31, 32, 20**]. Or, one can generate generalized/nonlocal symmetries by the action of a recursion operator:

DEFINITION 4. Let $u_t = F(x, u, ..., u_{x^k})$ be an autonomous evolution equation. A recursion operator is a formal pseudo–differential operator

$$\Psi = \sum_{-\infty < i \leq N} f_i D_x^i \tag{22}$$

in which the coefficients f_i are differential functions, such that the equation

$$\Psi_t = [F_*, \Psi], \tag{23}$$

holds identically on solutions of $u_t = F$.

In Equation (23), F_* is the formal linearization of F defined in (14), and Ψ_t is the formal pseudo–differential operator

$$\Psi_t = \sum_{-\infty < i \leq k} (D_t P_i) D_x^i = \sum_{-\infty < i \leq k} P_{i*} F D_x^i , \tag{24}$$

the final equality holding on solutions of $u_t = F$. The reader is referred to [**31**] for details on the calculus of formal pseudo–differential operators (22). We also remark that Definition 4 extends straightforwardly to systems of equations, see [**29, 30, 31**].

EXAMPLE 8. The Burgers equation $u_t = u_{xx} + uu_x$ admits the recursion operator [**31**]

$$\Psi = D_x + \frac{1}{2}u + \frac{1}{2}u_x D_x^{-1} ,$$

and one can check that $\Psi^n(u_x)$, $n \geq 1$, is a generalized symmetry of $u_t = u_{xx} + uu_x$.

Recursion operators were introduced by P. Olver [**31**] in 1977. Equation (23) guarantees that if G is a generalized symmetry of $u_t = F$, so is $\Psi(G)$, provided that $\Psi(G)$ is a differential function [**9, 31, 42**]. This last condition is not always satisfied: there are equations for which application of a recursion operator yields *only* shadows of nonlocal symmetries! The Bakirov system

$$u_t = u_{xxxx} + v^2, \qquad v_t = v_{xxxx} \tag{25}$$

was shown by A. Bilge in 1994 to admit a recursion operator [**10**]; however, it possesses *only one* generalized symmetry (Beukers, Sanders and Wang [**9**]).

The existence of a recursion operator (22) is a good indicator of integrability, since it implies, at least in principle, the existence of an infinite number of generalized symmetries (hence of flows commuting with the original equation). In fact, it turns out to be useful to consider a condition weaker than (23):

DEFINITION 5. Let $u_t = F$ be an nth order autonomous evolution equation. A pseudo-differential operator $\Psi = \sum_{-\infty<i}^{m} P_i D_x^i$ of order m is a formal symmetry of rank k if

$$order\,(\,\Psi_t + [\Psi, F_*]\,) \leq n + m - k \tag{26}$$

on solutions of $u_t = F$, in which Ψ_t is defined as in (24).

We call m the *order* of Ψ if $P_m \neq 0$, and we agree that the zero pseudo–differential operator has order $-\infty$. Note that if the pseudo–differential operator Ψ is simply $\Psi = c$, in which $c \neq 0$ is a constant, Ψ is a formal symmetry of infinite rank. We say that such a symmetry is *trivial*.

Definition 5 extends to the non–autonomous case [**20, 45**], to systems of equations [**29, 30, 31**], and to non–commutative differential equations [**32**].

PROPOSITION 1. *If* $u_t = Q$ *is an* m^{th} *order symmetry of an autonomous evolution equation* $u_t = F$*, then its formal linearization* Q_* *is a formal symmetry of rank* m.

Thus, Definition 5 is not vacuous. P. Olver [**31**], p.323, explains why Definition 5 is important: "Our goal is to determine explicit conditions on our evolution equation such that it will possess a formal symmetry of a prescribed rank. [...] the higher the rank of the formal symmetry, the more restrictive the conditions imposed on the evolution equation. Eventually, the existence of a formal symmetry of high enough rank will impose so many conditions on the evolution equation that it will possess a formal symmetry of infinite rank! [...] our convention that the zero pseudo–differential operator has order $-\infty$ implies that a recursion operator is the same as a formal symmetry of rank ∞."

DEFINITION 6. A system of evolution equations is formally integrable if it possesses a non–trivial formal symmetry of infinite rank.

It has been proved [**9, 42**] that for homogeneous, autonomous, polynomial, scalar equations with linear leading terms, the existence of *one* higher order symmetry guarantees formal integrability. It is also known [**29**] that an autonomous second order evolution equation is formally integrable if and only if it has a formal symmetry of rank 5, and that an autonomous third order evolution equation is formally integrable if and only if it has a formal symmetry of symmetry of rank 8. An important open problem is to investigate what rank of formal symmetry is required for an nth order evolution equation to be integrable.

We would like to present an implementation of P. Olver's remarks. We state two classification results for formally integrable equations, obtained by considering the existence of formal symmetries of sufficiently high rank. Our first result is due to Svinolupov and Sokolov, see [**20, 29**].

THEOREM 1. *Every formally integrable second order evolution equation is equivalent, under a contact transformation* $t = \chi(\bar{t})$, $x = \phi(\bar{t}, \bar{x}, \bar{u}, \bar{u}_{\bar{x}})$, $u = \psi(\bar{t}, \bar{x}, \bar{u}, \bar{u}_{\bar{x}})$, *to one of the following:*

$$u_t = u_{xx} + h(x,t)u, \tag{27}$$

$$u_t = u_{xx} + uu_x + g(x,t), \tag{28}$$

$$u_t = D_x(u_x u^{-2}), \tag{29}$$

$$u_t = D_x(u_x u^{-2} - x), \tag{30}$$

$$u_t = D_x(u_x u^{-2} + x^2 u) + xu. \tag{31}$$

Our second result is about a class of formally integrable *systems* of equations.

DEFINITION 7. A two-component system of evolution equations is called *decoupled* if one of the equations depends only on a single dependent variable.

Thus, a decoupled system can be considered as a pair of scalar equations which may be solved sequentially. For instance, the Bakirov system (25) is decoupled.

We are interested in the Boussinesq equation

$$u_{tt} = u_{xxxx} + D_x(uu_x)\ , \tag{32}$$

which can be rewritten as a system of two *coupled* evolution equations,

$$(33)\qquad \begin{cases} u_t = v_x \\ v_t = u_{xxx} + uu_x. \end{cases}$$

Equation (32) possesses a recursion operator, [**31**]. It was derived by Boussinesq in the 1870's, as a model for the uni-directional propagation of long waves in shallow water. The system (33) has an obvious scaling symmetry, $(x, t, u, v) \longmapsto (\lambda^{-1}x, \lambda^{-2}t, \lambda^2 u, \lambda v)$. The most general autonomous polynomial evolutionary system that admits such a scaling symmetry is

$$(34)\qquad \begin{cases} u_t = a_1 u_{xx} + a_2 v_x + a_3 u^2, \\ v_t = b_1 u_{xxx} + b_2 v_{xx} + b_3 uu_x + b_4 uv, \end{cases}$$

in which a_i, b_i are arbitrary constants. Can we classify all *formally integrable* systems of equations (34)? M. Foursov [**20**] answered this question in the affirmative:

THEOREM 2. *A nonlinear, non-decouplable equation of type* (34) *is formally integrable if it is equivalent (up to a scaling) to one of the following systems*

$$(35)\qquad \begin{cases} u_t = u_{xx} - v_x + \frac{1}{2}u^2 \\ v_t = u_{xxx} - v_{xx} + uu_x \end{cases}$$

$$(36)\qquad \begin{cases} u_t = u_{xx} + v_x \\ v_t = (\alpha - 1)u_{xxx} - v_{xx} + uu_x \end{cases}$$

$$(37)\qquad \begin{cases} u_t = v_x \\ v_t = \alpha u_{xxx} + uu_x \end{cases}$$

Foursov also noticed [**20**] that Equation (35) reduces to an ordinary differential equation, while (36) and (37) reduce to the Boussinesq system (33). Thus, Theorem 2 states that (33) is, in a precise sense, the *unique* integrable system among all non-decouplable systems (34)!

4. Conservation laws and pseudo-potentials

A *local conservation law* of $\Xi = 0$ is a one-form $\theta = f dx + g dt$, f, g differential functions, such that $d_H\theta = (-D_t f + D_x g)dx \wedge dt = 0$ on solutions to $\Xi = 0$ [**26, 27, 31**]; the conservation law θ is called *trivial*, if it is d_H-exact. Nonlocal conservation laws, for instance conservation laws which depend on integrations of the dependent variable u, can be also considered: a *nonlocal conservation law of type* π of $\Xi = 0$ is a one-form $\omega = \overline{f}dx + \overline{g}dt$ on the covering $\pi : \overline{S} \to S^\infty$ such that $\overline{d}_H\omega = (-\overline{D}_t\overline{f} + \overline{D}_x\overline{g})dx \wedge dt = 0$. It is called *trivial*, if it is $\overline{d}_H$-exact.

Conservation laws are used, for instance, to construct transformations between solutions of two differential equations, and are crucial for the proofs of long time existence of solutions, or for estimating the time of blow-up. We present two examples:

EXAMPLE 9. Note that if $\Xi = 0$ is of the form $u_t = F(x, t, u, \ldots, u_{x^k})$, and $\theta = f dx + g dt$ is a conservation law, then

$$\frac{d}{dt}\int f\,dx = \int f_t\,dx = \int g_x\,dx = 0\,,$$

if appropriate boundary conditions are imposed. Thus, the functional $\int f\,dx$ is *conserved* along the flow of $\Xi = 0$, and we say that f is a *conserved density* of the equation $\Xi = 0$. Consider now the equation

$$
(38) \qquad \begin{aligned} u_t - u_{txx} + 2ku_x + 3uu_x &= 2u_x u_{xx} + uu_{xxx} , \\ u(0,x) &= u_0(x) , \quad x \in \mathbf{R} , \\ u(t,x+1) &= u(t,x) , \quad t \geq 0 , x \in \mathbf{R} . \end{aligned}
$$

The functionals

$$
\int u\, dx , \quad \int u^2 + u_x^2\, dx , \quad \int \sqrt{(u - u_{xx} + k)_+}\, dx , \quad \int \sqrt{(u - u_{xx} + k)_-}\, dx ,
$$

in which $(u - u_{xx} + k)_+$ and $(u - u_{xx} + k)_-$ denote the positive and negative part of $(u - u_{xx} + k)$ respectively, are conserved. Constantin and Escher [**16**] proved that this fact implies the existence of *global solutions* to the problem (38), provided that the initial data u_0 is small and belongs to the Sovolev space H^3.

EXAMPLE 10. We consider the non-linear equation (Rosen [**40**])

$$
(39) \qquad u_t = u^2 u_{xx} ,
$$

which appears in the study of the temperature distribution in solid crystalline hydrogen. The one-form $\omega = u^{-1}dx - u_x dt$ is a conservation law of Equation (39). In fact,

$$
d\omega = (u^{-2}u_t - u_{xx})dx \wedge dt = u^{-2}(u_t - u^2 u_{xx})dx \wedge dt = 0
$$

on solutions to (39). Also, ω is not trivial, since there exists no differential function f such that $D_x f = u^{-1}$. Now, consider the change of variables [**29, 40**]

$$
(40) \qquad dx' = u^{-1}dx - u_x dt , \qquad t' = t , \qquad u' = u .
$$

The transformation (40) is well-defined *on solutions* to Equation (39) precisely because ω is a conservation law of (39). Moreover, (40) is invertible, because ω is a *non-trivial* conservation law [**29, 40**]. Extending (40) to first and second derivatives one finds

$$
u'_{x'} = uu_x , \qquad u'_{t'} = u_t + uu_x^2 , \qquad u'_{x'x'} = uu_x^2 + u^2 u_{xx},
$$

so that if $u(x,t)$ is a solution to (39), $u'(x',t')$ satisfies the heat equation $u'_{t'} = u'_{x'x'}$! The fact that (40) is invertible allows us to obtain solutions to Equation (39), see [**40**].

The precise definition of conservation laws is the following [**3, 12, 13, 26, 27**]:

DEFINITION 8. Local conservation laws of $\Xi = 0$ are the elements of the first cohomology group of the complex $(\Omega^1(S^\infty), d_H)$, in which S^∞ is the equation manifold of $\Xi = 0$, and $\Omega^1(S^\infty)$ is the space of horizontal one-forms on S^∞.

In other words, conservation laws are really equivalence classes of one-forms which are closed on solutions of the equation at hand, in which two closed one-forms are said to be equivalent, if they differ by an *exact* one-form.

We now introduce pseudo-potentials as a generalization of conservation laws:

DEFINITION 9. A real-valued function Γ is a pseudo-potential of a differential equation $\Xi(x,t,u,\ldots,u_{x^i t^j}) = 0$ if there exist smooth functions f, g depending on x, t, u, a finite number of derivatives of u, and Γ, such that the one-form $\Omega_\Gamma = d\Gamma - (f dx + g dt)$ satisfies

$$
d\Omega_\Gamma = 0 \mod \Omega_\Gamma
$$

whenever $u(x,t)$ is a solution to $\Xi = 0$.

We say that the one–form Ω_Γ is associated to the pseudo–potential Γ. Note that if the functions f, g appearing in the definition do not depend on Γ, Ω_Γ is a *bona fide* conservation law of the equation $\Xi = 0$.

Pseudo–potentials appeared for the first time in a paper by Wahlquist and Eastbrook [**47, 41**], in which they studied the integrability properties of the KdV equation $u_t = u_{xxx} + uu_x$ by means of one–forms. Pseudo–potentials can be understood geometrically in the framework of covering theory, see [**26, 27**] and references therein. We are mainly interested in *quadratic* pseudo–potentials, also called *pseudo-potentials of Riccati type*, which we define below. They are the ones which appear more often within the theory of integrable equations, and, as we will see in subsequent sections, possess a very appealing geometrical interpretation.

DEFINITION 10. We say that Γ is a quadratic pseudo–potential (or, a pseudo–potential of Riccati type) if the functions f, g appearing in the associated one–form $\Omega_\Gamma = d\Gamma - (f dx + g dt)$ are quadratic polynomials in Γ.

EXAMPLE 11. Consider the modified KdV equation $u_t = u_{xxx} + (3/2)\, u^2 u_x$. This equation admits the pseudo–potential [**46**]

$$\begin{aligned} \phi_x &= u + \eta \cos\phi \\ \phi_t &= u_{xx} + (1/2)u^3 + \eta^2 u - \eta\, u_x \sin\phi + ((1/2)\eta\, u^2 + \eta^3)\cos\phi\,. \end{aligned}$$

This pseudo–potential becomes a pseudo–potential of Riccati type under the transformation $\Gamma = \tan(\phi/2)$.

We are now ready to study our main general example of integrable equations, the class of *geometrically integrable* equations introduced by S.S. Chern and K. Tenenblat about 15 years ago. As pointed out before, the main importance this class of equations has for us is that their symmetries, conservation laws, and pseudo-potentials can be studied from the point of view of the geometry of surfaces, and that we can find new analytical results motivated by our geometrical perspective.

5. Equations of pseudo-spherical type

Connections between surfaces of constant Gaussian curvature -1 and what are now called "integrable equations" have been known since the second half of the XIXth Century. Within the modern theory of integrable systems, R. Sasaki [**43**] pointed out in 1979 that not only solutions of the sine–Gordon equation $u_{xy} = \sin u$ were related to surfaces of constant scalar curvature, but that in fact generic solutions of equations integrable by the now classical Ablowitz, Kaup, Newell and Segur (AKNS) inverse scattering scheme determine –whenever their associated linear problems are real– one–forms ϖ^α, $\alpha = 1, 2, 3$, which satisfy the structure equations of a pseudo-spherical surface.

DEFINITION 11. A two-dimensional manifold M is called a pseudo-spherical surface if there exist one-forms $\varpi^1, \varpi^2, \varpi^3$ on M that satisfy the independence condition $\varpi^1 \wedge \varpi^2 \neq 0$, and the structure equations

$$d\varpi^1 = \varpi^3 \wedge \varpi^2, \qquad d\varpi^2 = \varpi^1 \wedge \varpi^3, \qquad d\varpi^3 = \varpi^1 \wedge \varpi^2. \tag{41}$$

We recall once more that the reader may consult do Carmo [**17**] and Spivak [**44**] for information on differential forms and local geometry of surfaces. Here we point out that if the surface M is pseudo-spherical, the tensor $ds^2 = (\varpi^1)^2 + (\varpi^2)^2$ is a Riemannian metric on M, the one-form ω^3 encodes the Levi–Civita metric connection of ds^2, and Gauss "Theorema Egregium" implies that the Gaussian curvature of this metric is -1.

DEFINITION 12. A scalar differential equation in two independent variables x, t,

$$\Xi(x, t, u, u_x, \ldots, u_{x^n t^m}) = 0 \ , \tag{42}$$

is of pseudo-spherical type (or, it describes pseudo-spherical surfaces) if there exist one-forms $\omega^\alpha \not\equiv 0$,

$$\omega^\alpha = f_{\alpha 1}(x, t, u, \ldots, u_{x^r t^p})\, dx + f_{\alpha 2}(x, t, u, \ldots, u_{x^s t^q})\, dt, \qquad \alpha = 1, 2, 3, \tag{43}$$

whose coefficients $f_{\alpha\beta}$ are differential functions, such that the one-forms $\overline{\omega}^\alpha = \omega^\alpha(u(x,t))$ satisfy the structure equations (41) whenever $u = u(x,t)$ is a solution to Equation (42).

Definition 12 was introduced by Chern and Tenenblat **[15]**. We sometimes say "PSS equation" instead of "equation of pseudo-spherical type". Also, we exclude from our considerations the trivial case when the functions $f_{\alpha\beta}$, $\alpha = 1, 2, 3$, $\beta = 1, 2$, depend only on x, t.

EXAMPLE 12. Burgers equation $u_t = u_{xx} + uu_x$ is a PSS equation with

$$\omega^1 = (1/2)udx + ((1/2)u_x + (1/4)u^2)dt, \qquad \omega^2 = -\omega^3 \ = \ \eta\, dx + (\eta/2)udt,$$

in which $\eta \neq 0$ is a parameter.

We will see in Section 9 that the parameter η appearing in the last example can sometimes be interpreted as the "spectral" parameter used in the scattering/inverse scattering analysis of nonlinear equations **[4, 5, 18]**, and that PSS equations are integrability conditions of $sl(2, \mathbf{R})$-valued linear problems, which connects this geometric theory with one of the main themes of integrable systems. This motivates the following definition:

DEFINITION 13. A differential equation (or system of equations) is geometrically integrable if it describes a non-trivial one-parameter family of pseudo-spherical surfaces.

Now, we see from Definition 11 that having three one-forms ω^α satisfying (41) whenever $u(x,t)$ is a solution of $\Xi = 0$ is not enough in order to have pseudo-spherical structures: a non-degeneracy condition must also be satisfied **[37]**:

DEFINITION 14. Let $\Xi = 0$ be an equation describing pseudo-spherical surfaces with associated one-forms ω^α, $\alpha = 1, 2, 3$. A solution $u(x,t)$ of $\Xi = 0$ is I-generic if $(\omega^3 \wedge \omega^2)(u(x,t)) \neq 0$, II-generic if $(\omega^1 \wedge \omega^3)(u(x,t)) \neq 0$, and III-generic if $(\omega^1 \wedge \omega^2)(u(x,t)) \neq 0$.

EXAMPLE 13. The equation $u_t = u_{xx} + u_x$ is of pseudo-spherical type: we can set $\omega^1 = udx + u_x dt$, $\omega^2 = dx$, and $\omega^3 = udx + u_x dt$. The function $u(x,t) = x + t$ is a I- and III-generic solution, but it is not II-generic.

PROPOSITION 2. *Let $\Xi = 0$ be a PSS equation with associated one-forms ω^α. Then,*

(a) *If $u(x,t)$ is a I-generic solution, $\overline{\omega}^2$ and $\overline{\omega}^3$ determine a Lorentz metric of Gaussian curvature $K = -1$ on the domain of $u(x,t)$, with connection one-form given by $\overline{\omega}^1$.*

(b) *If $u(x,t)$ is a II-generic solution, $\overline{\omega}^1$ and $-\overline{\omega}^3$ determine a Lorentz metric of Gaussian curvature $K = -1$ on the domain of $u(x,t)$, with connection one-form given by $\overline{\omega}^2$.*

(c) *If $u(x,t)$ is a III-generic solution, $\overline{\omega}^1$ and $\overline{\omega}^2$ determine a Riemannian metric of Gaussian curvature $K = -1$ on the domain of $u(x,t)$, with connection one-form given by $\overline{\omega}^3$.*

An important idea in integrable systems is (Faddeev and Takhtajan [18]) that an equation $\Xi = 0$ is not just the integrability condition of a linear problem $\psi_x = X\psi$, $\psi_t = T\psi$, but that the "zero curvature equation" $X_t - T_x + [X,T] = 0$ is *equivalent* to $\Xi = 0$. A crucial problem –which in its full generality is still open [37]– is to express this equivalence within the context of PSS equations:

EXAMPLE 14. Use of the MAPLE Variational Calculus Package VESSIOT [1], allows one to check that Burgers equation $u_t = u_{xx} + uu_x$ describes pseudo-spherical surfaces with associated one–forms

$$\begin{aligned}\omega^1 &= ((1/2)\,u - u_x\eta - (1/4)\,{u_x}^2 u + (1/2)\,{u_x}^2\eta + u_{xx})\,dx + ((1/2)\,u_x + (1/4)\,u^2 - \\ &\quad (1/2)\,u_x\eta\,u - (1/4)\,{u_x}^3 - (1/8)\,{u_x}^2u^2 + (1/4)\,{u_x}^2\eta\,u + uu_{xx} + {u_x}^2 + u_{xxx})\,dt,\\ \omega^2 &= (\eta + (1/2)\,u_x u - u_x\eta)\,dx + 1/2\,(\eta u + {u_x}^2 + (1/2)\,u_x u^2 - u_x\eta u)\,dt,\\ \omega_{12} &= (-\eta + u_x\eta + (1/4)\,{u_x}^2u - (1/2)\,{u_x}^2\eta - u_{xx})\,dx + (-(1/2)\,\eta\,u + (1/2)\,u_x\eta\,u\\ &\quad + (1/4)\,{u_x}^3 + (1/8)\,{u_x}^2u^2 - (1/4)\,{u_x}^2\eta\,u - uu_{xx} - {u_x}^2 - u_{xxx})\,dt\,.\end{aligned}$$

Indeed, on solutions $u(x,t)$ to Burgers equation one obtains the following:

$$\begin{aligned}d\omega^1 - \omega^3\wedge\omega^2 &= \left(D_x^2\Delta + (-\eta + \eta u_x - \frac{1}{2}uu_x)D_x\Delta + (\frac{1}{2} - \frac{1}{4}u_x^2)\Delta\right)dx\wedge dt\\ d\omega^2 - \omega^1\wedge\omega^3 &= \left((-\eta + \frac{1}{2}u)D_x\Delta + \frac{1}{2}u_x\Delta\right)dx\wedge dt\\ d\omega^3 - \omega^1\wedge\omega^2 &= \left(-D_x^2\Delta + (\eta - \eta u_x + \frac{1}{2}uu_x)D_x\Delta - \frac{1}{4}u_x^2\Delta\right)dx\wedge dt,\end{aligned}$$

in which $\Delta = -u_t + u_{xx} + uu_x$. It follows that Burgers' equation is *sufficient* for the structure equations (41) to be satisfied, but it is not clear it is also necessary.

In the case of a kth order evolution equation $u_t = F(x,t,u,...,u_{x^k})$, we rule out one–forms ω^α as in Example 14 thus [24, 20]: we consider the differential ideal I_F generated by

$$du\wedge dx + F(x,t,u,...,u_{x^k})dx\wedge dt,\quad du_{x^l}\wedge dt - u_{x^{l+1}}\,dx\wedge dt,\qquad 1\le l\le k-1,$$

on a manifold J with coordinates $x,t,u,u_x,\ldots,u_{x^k}$.

DEFINITION 15. An evolution equation $u_t = F(x,t,u,...,u_{x^k})$ is *strictly pseudo-spherical* if there exist one–forms $\omega^\alpha = f_{\alpha1}\,dx + f_{\alpha2}\,dt$, $\alpha = 1,2,3$, whose coefficients $f_{\alpha\beta}$ are smooth functions on J, such that the two–forms

$$\Omega_1 = d\omega^1 - \omega^3\wedge\omega^2,\qquad \Omega_2 = d\omega^2 - \omega^1\wedge\omega^3,\qquad \Omega_3 = d\omega^3 - \omega^1\wedge\omega^2, \tag{44}$$

generate I_F.

Local solutions of $u_t = F$ correspond to integral sub-manifolds of the exterior differential system $\{I_F, dx\wedge dt\}$. It follows that if $u_t = F$ is strictly pseudo-spherical, the structure equations $\Omega_\alpha = 0$ are satisfied if and only if the two forms Ω_α are pulled–back by local solutions to $u_t = F$, *and* that the ideal I_F is algebraically equivalent to a system of differential forms satisfying the pseudo-spherical structure equations if pulled back by local solutions of $u_t = F$.

We can *classify* strictly pseudo-spherical evolution equations. Under the a priori condition $f_{21} = \eta$, in which η is a real constant (a condition dictated by inverse scattering considerations, see [5, 15] and Section 9 below) they have been characterized by Chern and Tenenblat [15] and Reyes [36]. These classification results are consequences of the fact that if $u_t = F$ is strictly pseudo-spherical, the dependence of its associated functions $f_{\alpha\beta}$ on derivatives of u

is very restricted, and moreover, these functions satisfy several identities which allow us to "reconstruct" the right hand side of the equation $u_t = F$ [**24, 38**]:

LEMMA 1. *Necessary and sufficient conditions for the* k^{th} *order equation* $u_t = F$ *to be strictly pseudo-spherical are the conjunction of* (a) *The functions* $f_{\alpha\beta}$ *satisfy the constraints*

$$f_{\alpha 1,u_{x^a}} = 0; \qquad f_{\alpha 2,u_{x^k}} = 0; \qquad f_{11,u}^2 + f_{21,u}^2 + f_{31,u}^2 \neq 0, \tag{45}$$

in which $a \geq 1$ *and* $\alpha = 1, 2, 3$; *and* (b) *The functions* F *and* $f_{\alpha\beta}$ *satisfy the identities*

$$-f_{11,u}F + \sum_{i=0}^{k-1} u_{x^{i+1}} f_{12,u_{x^i}} + f_{21}f_{32} - f_{31}f_{22} + f_{12,x} - f_{11,t} \;=\; 0, \tag{46}$$

$$-f_{21,u}F + \sum_{i=0}^{k-1} u_{x^{i+1}} f_{22,u_{x^i}} + f_{12}f_{31} - f_{11}f_{32} + f_{22,x} - f_{21,t} \;=\; 0, \qquad \textit{and} \tag{47}$$

$$-f_{31,u}F + \sum_{i=0}^{k-1} u_{x^{i+1}} f_{32,u_{x^i}} + f_{21}f_{12} - f_{11}f_{22} + f_{32,x} - f_{31,t} \;=\; 0. \tag{48}$$

Using Lemma 1, one can prove, for instance, the following classification result [**36**]:

THEOREM 3. *Let* $f_{\alpha\beta}$, $1 \leq \alpha \leq 3$, $1 \leq \beta \leq 2$ *be differential functions satisfying* (45)–(48), *and let* $f_{21} = \eta$ *be a non-zero real parameter. Suppose* $HL \neq 0$, *in which* $H = f_{11}f_{11,u} - f_{31}f_{31,u}$ *and* $L = f_{11}f_{31,u} - f_{31}f_{11,u}$, *and set* $T := f_{11,t}f_{31,u} - f_{11,u}f_{31,t}$. *The equation* $u_t = F(x,t,u,...,u_{x^k})$ *is strictly pseudo-spherical with associated one-forms* $\omega_\alpha = f_{\alpha 1}dx + f_{\alpha 2}dt$ *if, and only if,* (a) *The function* F *is given in terms of* $f_{\alpha\beta}$,

$$\begin{aligned} F \;=\; & \frac{1}{L}\sum_{i=1}^{k-1} u_{x^{i+1}}(D_x f_{22})_{u_{x^i}} + \frac{1}{HL}(f_{31}^2 - f_{11}^2)\{\sum_{i=0}^{k-2} u_{x^{i+1}}A^i + T - f_{31,u}f_{12,x} + f_{11,u}f_{32,x}\} \\ & + \frac{f_{11}}{L}(u_x f_{32,u} + f_{32,x}) - \frac{f_{31}}{L}(u_x f_{12,u} + f_{12,x}) + \frac{\eta}{H}D_x f_{22} - \frac{1}{L}(f_{31,t}f_{11} - f_{11,t}f_{31}); \end{aligned}$$

(b) *The* u_{x^j}*-derivatives of* f_{12} *and* f_{32}, $1 \leq j \leq k-1$, *are given in terms of* f_{11} *and* f_{31} *by the formulae* $f_{a2,u_{x^j}} = -(1/L)\{f_{a1}A^j - f_{a1,u}(D_x f_{22})_{u_{x^j}}\}$, *in which* $a = 1, 3$; (c) *The* u*-derivatives of* f_{12} *and* f_{32} *satisfy the equation* $f_{31,u}f_{12,u} - f_{11,u}f_{32,u} + A^0 = 0$; *and* (d) *The functions* f_{12} *and* f_{32} *satisfy the constraints*

$$f_{a2} \;=\; \frac{f_{a1}f_{22}}{\eta} + \frac{1}{H}\left(\frac{-f_{a1}}{\eta}\{\sum_{i=0}^{k-2} u_{x^{i+1}}A^i + T\} + f_{b1,u}D_x f_{22}\right) + \frac{f_{a1}}{\eta H}(f_{31,u}f_{12,x} - f_{11,u}f_{32,x}),$$

in which $a = 1, 3$; $b = 3$ *if* $a = 1$, *and* $b = 1$ *if* $a = 3$.

The functions A^j are defined recursively as follows: $A^{k-1} := 0$ and, for $0 \leq j \leq k-2$,

$$A^j = -D_x A^{j+1} + \frac{1}{L}(u_x L_u + \eta H - R)A^{j+1} - \frac{1}{L}(u_x P - \eta M - Q)(D_x f_{22})_{u_{x^{j+1}}} + f_{22,u_{x^{j+1}}}H,$$

in which $M := f_{31,u}^2 - f_{11,u}^2$; $P := f_{11,u}f_{31,uu} - f_{31,u}f_{11,uu}$; $Q := f_{11,u,x}f_{31,u} - f_{31,u,x}f_{11,u}$; and $R := f_{11,u,x}f_{31} - f_{31,u,x}f_{11}$.

REMARK 3. It follows that checking whether a given equation is of pseudo-spherical type is almost an algorithmic task. An interesting project would be to write a MAPLE computer program, say, to help with the computations, as it has been done already in the case of (generalized) symmetries [**20, 32**].

How can one generalize the notion of a PSS equation to equations with an arbitrary number of independent variables? To answer this question appears to be a remarkably difficult problem, and in fact, integrability in higher dimensions is a very intriguing and important area of research [**41**]. We present here one example of a geometrically motivated equation which is integrable by scattering/inverse scattering methods (Beals and Tenenblat, see [**46**]):

DEFINITION 16. Let $v : U \subset \mathbf{R}^n \to \mathbf{R}^n$ and h be a smooth function and a off–diagonal $(n \times n)$–matrix function on U respectively. The system of equations

$$
\begin{aligned}
& & vv^t &= 1, & & (49)\\
& & v_{i,x_j} &= v_j h_{ji} - \sum_{s=1}^{n} v_s h_{is} \delta_{ij}, & & (50)\\
& h_{ij,x_i} + h_{ji,x_j} + \sum_{s \neq i,j} h_{si} h_{sj} &&= -K v_i v_j, & i \neq j & \quad (51)\\
& & h_{ij,x_s} &= h_{is} h_{sj}, & i,j,k \text{ distinct} & \quad (52)\\
& h_{ij,x_j} + h_{ji,x_i} + \sum_{s \neq i,j} h_{is} h_{js} &&= 0, & i \neq j, & \quad (53)
\end{aligned}
$$

in which $\delta_{ij} = 1$ if $i = j$ and zero otherwise, is called the intrinsic generalized wave equation if $K = 0$, and the intrinsic generalized sine–Gordon equation if $K \neq 0$.

It is not hard to see that the system (49)–(53) does reduce to the wave and sine–Gordon equations if $n = 2$. Interesting solutions to these equations have been found by Tenenblat [**46**]. For instance, assume that $K = -1$ and that $v = (v_1, \ldots, v_n)$, $n \geq 2$, is a solution to (49)–(53) such that $v_{x_1} \neq 0$ and $v_{x_j} = 0$ for all $j \geq 2$. Tenenblat [**46**] proved that v is given by:

$$
\begin{aligned}
n = 2: \quad (v_1')^2 &= (1 - v_1^2)(1 + c - v_1^2)\\
v_2^2 &= 1 - v_1^2 \,, \quad c \in \mathbf{R} \,, \quad 1 + c > 0 \,;
\end{aligned}
$$

$$
\begin{aligned}
n = 3: \quad (v_1')^2 &= ((b^2/c^2) - v_1^2)(1 + c - b^2 - v_1^2)\\
v_2^2 &= (c^2/(1 + c^2))((b^2/c^2) - v_1^2)\\
v_3^2 &= (1/(1 + c^2))(1 + c^2 - b^2 - v_1^2) \,, \quad b, c \neq 0 \,, \quad 1 + c^2 - b^2 > 0 \,;
\end{aligned}
$$

$$
\begin{aligned}
n \geq 4: \quad v_1 &= \pm \tanh(x_1 - a)\\
v_j &= c_j \operatorname{sech}(x_1 - a) \,, j \geq 2 \,, \quad a, c_j \neq 0 \,, \textstyle\sum c_j^2 = 1 \,.
\end{aligned}
$$

The system of equations (49)–(53) possesses the following geometric interpretation:

PROPOSITION 3. *Whenever the solutions $v_i(x_1, \ldots, x_n)$ of* (49)–(53) *are real non-zero functions on an open set $U \subset \mathbf{R}^n$, the function $v : U \subset \mathbf{R}^n \to S^{n-1} \subset \mathbf{R}^n$ determines a Riemannian metric on U with constant sectional curvature K. Conversely, any such metric provides a solution of Equations* (49)–(53).

K. Tenenblat [**46**] has also determined exactly the submanifolds of $\mathbf{R}^{2n-1}$ associated with the solutions of (49)–(53) considered above.

6. Conservation laws and pseudo-potentials for PSS equations

In this section we begin to study the features of differential equations which we discussed in Sections 3 and 4, in the context of PSS equations. We start with a purely geometric proposition [**15**] which is crucial for what follows:

PROPOSITION 4. *Given a coframe* $\{\overline{\omega}^1, \overline{\omega}^2\}$ *and corresponding connection one-form* $\overline{\omega}^3$ *on a smooth surface* M*, there exists a new coframe* $\{\overline{\theta}^1, \overline{\theta}^2\}$ *on* M *and new connection one-form* $\overline{\theta}^3$ *satisfying the structure equations*

$$d\overline{\theta}^1 = 0, \quad d\overline{\theta}^2 = \overline{\theta}^2 \wedge \overline{\theta}^1, \quad \text{and} \quad \overline{\theta}^3 + \overline{\theta}^2 = 0, \tag{54}$$

if and only if the surface M *is pseudo-spherical.*

PROOF. Assume that the orthonormal frames dual to the coframes $\{\overline{\omega}^1, \overline{\omega}^2\}$ and $\{\overline{\theta}^1, \overline{\theta}^2\}$ possess the same orientation. The one-forms $\overline{\omega}^\alpha$ and $\overline{\theta}^\alpha$ are connected by means of

$$\overline{\theta}^1 = \overline{\omega}^1 \cos\rho + \overline{\omega}^2 \sin\rho, \quad \overline{\theta}^2 = -\overline{\omega}^1 \sin\rho + \overline{\omega}^2 \cos\rho, \quad \overline{\theta}^3 = \overline{\omega}^3 + d\rho. \tag{55}$$

Thus, $\overline{\theta}^1, \overline{\theta}^2, \overline{\theta}^3$ satisfying (54) exist if and only if the Pfaffian system

$$\overline{\omega}^3 + d\rho - \overline{\omega}^1 \sin\rho + \overline{\omega}^2 \cos\rho = 0 \tag{56}$$

on the space of coordinates (x, t, ρ) is completely integrable for $\rho(x,t)$, and this happens if and only if M is pseudo-spherical. □

Equations (54) and (56) determine geodesic coordinates on M. Analytically, the importance of Proposition 4 is this. If the PSS equation $\Xi = 0$ has associated one-forms $\omega^\alpha = f_{\alpha 1} dx + f_{\alpha 2} dt$, (54) and (56) imply that the Pfaffian system

$$\omega^3(u(x,t)) + d\rho - \omega^1(u(x,t)) \sin\rho + \omega^2(u(x,t)) \cos\rho = 0 \tag{57}$$

is completely integrable for $\rho(x,t)$ whenever $u(x,t)$ is a solution of $\Xi = 0$. Equations (54) and (55) then imply that for each solution $u(x,t)$ and a corresponding solution $\rho(x,t)$ of (57), the one-form $\theta^1 = \omega^1(u(x,t)) \cos\rho + \omega^2(u(x,t)) \sin\rho$ is closed. If the functions $f_{\alpha\beta}$ can be expanded as power series in a parameter η, so can $\rho(x,t)$, and therefore θ^1 determines a sequence of one-forms which are closed whenever $u(x,t)$ is a solution of $\Xi = 0$. These conservation laws may be trivial however. They may also be nonlocal, since they depend on solutions of (57). The following lemma [**36**] allows us to construct them explicitly.

LEMMA 2. *Let* $\Xi = 0$ *be a PSS equation with associated one-forms* ω^α*, and set* $\overline{\omega}^\alpha = \omega^\alpha(u(x,t))$*. Under the changes of variables* $\Gamma = \tan(\rho/2)$ *and* $\hat{\Gamma} = \cot(\rho/2)$*, the completely integrable Pfaffian system* (57) *and the one-form* $\theta^1 = \overline{\omega}^1 \cos\rho + \overline{\omega}^2 \sin\rho$ *become,*

$$-2d\Gamma = (\overline{\omega}^3 + \overline{\omega}^2) - 2\Gamma\overline{\omega}^1 + \Gamma^2(\overline{\omega}^3 - \overline{\omega}^2), \tag{58}$$

$$\Theta = \overline{\omega}^1 - \Gamma(\overline{\omega}^3 - \overline{\omega}^2) \qquad \textit{(up to an exact differential form)} \tag{59}$$

and

$$2d\hat{\Gamma} = (\overline{\omega}^3 - \overline{\omega}^2) - 2\hat{\Gamma}\overline{\omega}^1 + \hat{\Gamma}^2(\overline{\omega}^3 + \overline{\omega}^2), \tag{60}$$

$$\hat{\Theta} = -\overline{\omega}^1 + \hat{\Gamma}(\overline{\omega}^3 + \overline{\omega}^2), \qquad \textit{(up to an exact differential form)}. \tag{61}$$

Examples of conservation laws computed using (58)–(61) appear in Reyes [**35, 36**] and references therein, and we will present a new example in Section 10. An algorithm based directly on (57) was developed by Cavalcante and Tenenblat in 1988, see [**46**].

Note that Equations (58) and (60) say that Γ and $\hat{\Gamma}$ are pseudo-potentials of Riccati type for the PSS equation $\Xi = 0$!

PROPOSITION 5. *A differential equation* $\Xi = 0$ *is of pseudo-spherical type if and only if it admits a quadratic pseudo-potential.*

PROOF. As pointed out above, PSS equations admit quadratic pseudo-potentials. Assume, therefore that the equation $\Xi = 0$ admits a pseudo-potential Γ with associated one-form $\Omega_\Gamma = d\Gamma - (fdx + gdt)$ in which $f = a + b\Gamma + c\Gamma^2$ and $g = a' + b'\Gamma + c'\Gamma^2$. Then, the first order system of equations

$$\Gamma_x = a + b\Gamma + c\Gamma^2 , \qquad \Gamma_t = a' + b'\Gamma + c'\Gamma^2 ,$$

is completely integrable for Γ, and therefore the coefficients a, b, c, a', b', c' satisfy the equations

$$a_t + ba' = a'_x + b'a , \qquad b_t + 2ca' = b'_x + 2c'a , \qquad c_t + cb' = c'_x + c'b .$$

We set

$$\omega^1 = bdx + b'dt , \quad \omega^2 = (-a+c)dx + (-a'+c')dt , \quad \omega^3 = -(a+c)dx - (a'+c')dt .$$

It is straightforward to check that these one-forms satisfy the structure equations (41) whenever $u(x,t)$ is a solution of $\Xi = 0$. □

7. Generalized and nonlocal symmetries for PSS equations

Let $u_t = F$ be an kth order *strictly* pseudo-spherical equation with associated one-forms ω^α, $\alpha = 1,2,3$, and equation manifold S^∞. We let $\overline{S}$ be a covering of S^∞, and let G be a function on $\overline{S}$. We consider the one-forms $\omega^\alpha + \tau\Lambda_\alpha$ on $\overline{S}$, in which $\Lambda_\alpha = g_{\alpha1}dx + g_{\alpha2}dt$, and

$$(62) \qquad g_{\alpha1} = f_{\alpha1,u}\, G, \qquad \text{and} \qquad g_{\alpha2} = \sum_{i=0}^{k-1} f_{\alpha2,u_{x^i}} \overline{D}_x^i G, \qquad \alpha = 1,2,3.$$

The following characterization theorem for (nonlocal) symmetries of PSS equations was proven in [**39**]:

THEOREM 4. *Let $u_t = F(x,t,u,\ldots u_{x^k})$ be strictly pseudo-spherical with associated one-forms $\omega^\alpha = f_{\alpha1}dx + f_{\alpha2}dt$, $\alpha = 1,2,3$. Let G be a smooth function on a covering $\overline{S}$ of the equation manifold S^∞, and consider the deformed one-forms $\omega^\alpha + \tau\Lambda_\alpha$ defined above. They satisfy the structure equations*

$$(63) \qquad \overline{d}_H\sigma^1 = \sigma^3 \wedge \sigma^2, \quad \overline{d}_H\sigma^2 = \sigma^1 \wedge \sigma^3, \quad \overline{d}_H\sigma^3 = \sigma^1 \wedge \sigma^2,$$

up to terms of order τ^2 if and only if G is a shadow of a nonlocal symmetry of $u_t = F$.

We would like to discuss the case when a PSS equation admits an infinite number of generalized symmetries. This problem is of interest because in integrable systems we are usually concerned not with a single equation, but with a *sequence* of equations which determine pairwise commuting flows (Faddeev and Takhtajan [**18**], Roy Chowdhury [**41**]), what is called a *hierarchy* of equations. We would like to formalize this idea in the PSS equations framework [**38, 39**]:

DEFINITION 17. We consider an affine space equipped with coordinates $(x, t, \tau_1, \tau_2, \ldots)$, and we let $u_{\tau_i} = F_i$, $i \geq 0$, $\tau_0 = t$, be a countable number of evolution equations. These equations form a hierarchy of equations describing pseudo-spherical surfaces (or a hierarchy of pseudo-spherical type) if there exist differential functions $f_{\alpha\beta}$, $\alpha = 1,2,3$, $\beta = 1,2$; and $h_{\alpha i}$, $i \geq 1$, such that for all $n \geq 0$, the one-forms

$$(64) \qquad \Theta_\alpha^{(n)} = f_{\alpha1}dx + f_{\alpha2}dt + \sum_{i=1}^{n} h_{\alpha i}d\tau_i, \qquad \alpha = 1,2,3,$$

satisfy the structure equations

$$(65) \qquad d_H\Theta_1^{(n)} = \Theta_3^{(n)} \wedge \Theta_2^{(n)}, \quad d_H\Theta_2^{(n)} = \Theta_1^{(n)} \wedge \Theta_3^{(n)}, \quad d_H\Theta_3^{(n)} = \Theta_1^{(n)} \wedge \Theta_2^{(n)}.$$

In (65), $d_H\Theta_\alpha^{(n)}$ is computed by means of

(66) $$d_H g = D_x g dx + \sum_{i=0}^{n} D_{\tau_i} g d\tau_i, \quad \text{and} \quad d_H(dx) = d_H(d\tau_i) = 0, \quad i \geq 0,$$

for any differential function g; the operators D_x, D_{τ_i}, $i \geq 0$, are given by

(67) $$D_x = \frac{\partial}{\partial x} + \sum_{j=0}^{\infty} u_{x^{j+1}} \frac{\partial}{\partial u_{x^j}}, \quad \text{and} \quad D_{\tau_i} = \frac{\partial}{\partial \tau_i} + \sum_{j=0}^{\infty} (D_x^j F_i) \frac{\partial}{\partial u_{x^j}}.$$

EXAMPLE 15. We define associated functions $f_{\alpha\beta}$, $h_{\alpha i}$, $i \geq 0$, as follows: $f_{11} = 1 - u$, $f_{21} = \eta$, $f_{31} = -1 - u$, $f_{12} = -u + 1$, $f_{22} = \eta$, $f_{32} = -u - 1$, $h_{\alpha 0} = f_{\alpha 2}$, and for $i > 0$,

$$\begin{aligned} h_{1i} &= (1/2)\eta B_x^{(i)} - (1/2)B_{xx}^{(i)} - uB^{(i)} + B^{(i)}, \\ h_{2i} &= \eta B^{(i)} - B_x^{(i)}, \\ h_{3i} &= (1/2)\eta B_x^{(i)} - (1/2)B_{xx}^{(i)} - uB^{(i)} - B^{(i)}, \end{aligned}$$

in which $B^{(i)} = \sum_{j=0}^{i} B_j \eta^{2(i-j)}$, and the functions B_j, $j = 1, 2, \ldots$, are defined by the recurrence relation

(68) $$B_{0,x} = 0, \qquad B_{j+1,x} = B_{j,xxx} + 4uB_{j,x} + 2u_x B_j; \quad 0 \leq j \leq n-1.$$

This is the KdV hierarchy, see [**31, 33**]. The one–forms $\omega_i^\alpha = f_{\alpha 1}dx + h_{\alpha i}d\tau_i$ satisfy the structure equations (41) whenever $u(x,t)$ is a solution of the ith KdV equation

(69) $$u_{\tau_i} = \frac{1}{2}B_{n,xxx} + u_x B_n + 2uB_{n,x}.$$

In particular, $n = 1$ yields $B_1 = 2u$, $B^{(1)} = \eta^2 + 2u$, and (69) becomes

(70) $$u_{\tau_1} = u_{xxx} + 6uu_x.$$

PROPOSITION 6. *Let $u_{\tau_i} = F_i$, $i \geq 0$, be a hierarchy of pseudo-spherical type with associated one–forms*

(71) $$\Theta_\alpha^{(n)} = f_{\alpha 1}dx + f_{\alpha 2}dt + \sum_{i=1}^{n} h_{\alpha i}d\tau_i, \qquad \alpha = 1,2,3, \; n \geq 0.$$

The equations $u_{\tau_i} = F_i$, $i \geq 0$, describe pseudo-spherical surfaces with associated one–forms $\omega_0^\alpha = f_{\alpha 1}dx + f_{\alpha 2}dt$ if $i = 0$, and $\omega_i^\alpha = f_{\alpha 1}dx + h_{\alpha i}d\tau_i$ if $i \geq 1$.

The PSS equation $u_t = F$ will be called the *seed equation*, and the equations $u_{\tau_i} = F_i$, $i \geq 1$, the *higher* equations of the hierarchy. In analogy with the scalar case, we introduce hierarchies of *strictly* pseudo-spherical type:

Consider a countable number of evolution equations $u_{\tau_i} = F_i$, of order k_i, $i \geq 0$. For each $n \geq 0$, let $J^{(n)}$ be a manifold with coordinates $(x, t, \tau_1, \ldots, \tau_n, u, u_x, \ldots, u_{x^{M(n)}})$, where $M(n)$ is the maximum of the orders k_j, $0 \leq j \leq n$, and let $I^{(n)}$ be the differential ideal generated by the two–forms

(72) $$du \wedge dx + F_0 dx \wedge dt + \sum_{i=1}^{n} F_i dx \wedge d\tau_i,$$

(73) $$du_{x^k} \wedge dt - u_{x^{k+1}} dx \wedge dt - \sum_{j=1}^{n} D_x^k F_j d\tau_j \wedge dt, \qquad 0 \leq k \leq k_0 - 1,$$

(74) $$du_{x^k} \wedge d\tau_j - u_{x^{k+1}} dx \wedge d\tau_j - D_x^k F_0 dt \wedge d\tau_j - \sum_{l=1}^{n} D_x^k F_l \, d\tau_l \wedge d\tau_j,$$

in which for each $j = 1, \ldots, n$, $0 \leq k \leq k_j - 1$.

DEFINITION 18. The countable collection of evolution equations $u_{\tau_i} = F_i$, $i \geq 0$, is a hierarchy of strictly pseudo-spherical type if there exist smooth functions $f_{\alpha\beta}$, $\alpha = 1, 2, 3$; $\beta = 1, 2$, and $h_{\alpha i}$, $i \geq 1$, on $J^{(n)}$ such that for every $n \geq 0$, the two–forms

$$\Omega_1^{(n)} = d\Theta_1^{(n)} - \Theta_3^{(n)} \wedge \Theta_2^{(n)}, \quad \Omega_2^{(n)} = d\Theta_2^{(n)} - \Theta_1^{(n)} \wedge \Theta_3^{(n)}, \quad \Omega_3^{(n)} = d\Theta_3^{(n)} - \Theta_1^{(n)} \wedge \Theta_2^{(n)},$$

in which the one–forms $\Theta_\alpha^{(n)}$ are given by (64), generate the ideal $I^{(n)}$.

It follows that for each $i \geq 0$, the equations $u_{\tau_i} = F_i$ are strictly pseudo-spherical. Lemma 1 generalizes thus [**38**]:

PROPOSITION 7. *Necessary and sufficient conditions for the equations $u_{\tau_i} = F_i$, $i \geq 0$, to form a hierarchy of strictly pseudo-spherical type with associated functions $f_{\alpha\beta}$, $\alpha = 1, 2, 3$; $\beta = 1, 2$, and $h_{\alpha i}$, $i \geq 1$, are the conjunction of* (a) *The functions $f_{\alpha\beta}$ satisfy the constraint $f_{11,u}^2 + f_{21,u}^2 + f_{31,u}^2 \neq 0$, and for each $n \geq 0$,*

$$f_{\alpha 1, u_{x^k}} = 0,\ 1 \leq k \leq M(n);\ f_{\alpha 2, u_{x^k}} = 0,\ k_0 \leq k \leq M(n);\ h_{\alpha i, u_{x^k}} = 0,\ k_i \leq k \leq M(n);$$

and (b) *The functions F_i, $f_{\alpha\beta}$, and $h_{\alpha i}$ satisfy, for each $n \geq 0$, the identities*

$$\begin{cases} -(f_{11,t} + f_{11,u}F_0) + D_x f_{12} &= f_{31}f_{22} - f_{32}f_{21} \\ -(f_{21,t} + f_{21,u}F_0) + D_x f_{22} &= f_{11}f_{32} - f_{12}f_{31} \\ -(f_{31,t} + f_{31,u}F_0) + D_x f_{32} &= f_{11}f_{22} - f_{12}f_{21} \end{cases}$$

$$\begin{cases} -(f_{11,\tau_i} + f_{11,u}F_i) + D_x h_{1i} &= f_{31}h_{2i} - h_{3i}f_{21} \\ -(f_{21,\tau_i} + f_{21,u}F_i) + D_x h_{2i} &= f_{11}h_{3i} - h_{1i}f_{31} \\ -(f_{31,\tau_i} + f_{31,u}F_i) + D_x h_{3i} &= f_{11}h_{2i} - h_{1i}f_{21} \end{cases}$$

$$\begin{cases} -D_{\tau_i} f_{12} + D_t h_{1i} &= f_{32}h_{2i} - h_{3i}f_{22} \\ -D_{\tau_i} f_{22} + D_t h_{2i} &= f_{12}h_{3i} - h_{1i}f_{32} \\ -D_{\tau_i} f_{32} + D_t h_{3i} &= f_{12}h_{2i} - h_{1i}f_{22} \end{cases}$$

$$\begin{cases} -D_{\tau_j} h_{1i} + D_{\tau_i} h_{1j} &= h_{3i}h_{2j} - h_{3j}h_{2i} \\ -D_{\tau_j} h_{2i} + D_{\tau_i} h_{2j} &= h_{1i}h_{3j} - h_{1j}h_{3i} \\ -D_{\tau_j} h_{3i} + D_{\tau_i} h_{3j} &= h_{1i}h_{2j} - h_{1j}h_{2i}. \end{cases}$$

The importance of Proposition 7 for us is that, taken together with Theorem 4, it implies that hierarchies of strictly pseudo-spherical type generate hierarchies of pairwise commuting flows (Reyes [**38**]):

THEOREM 5. *Assume that $u_{\tau_i} = F_i$, $i \geq 0$, is a hierarchy of pseudo-spherical type with associated one–forms*

$$\Theta_\alpha^{(n)} = f_{\alpha 1}dx + f_{\alpha 2}dt + \sum_{i=1}^{n} h_{\alpha i}d\tau_i, \qquad \alpha = 1, 2, 3;\ n \geq 0. \tag{75}$$

The differential function F_j is a generalized symmetry of the equation $u_{\tau_i} = F_i$, for all $i, j \geq 0$.

8. Transformation results

We begin by defining Bäcklund transformations:

DEFINITION 19. Two equations $\Xi(x,t,u,\dots)=0$ and $\Xi'(x,t,v,\dots)=0$ are related by a Bäcklund transformation

$$F(x,t,u,u_x,v,v_x)=0\ ,\qquad G(x,t,u,u_x,v,v_x)=0\ , \tag{76}$$

if: *(i)* The system (76) is compatible for u whenever $v(x,t)$ is a solution of $\Xi'=0$, and a solution $u(x,t)$ to (76) is a solution of $\Xi=0$; *(ii)* The system (76) is compatible for v whenever $u(x,t)$ is a solution of $\Xi=0$, and a solution $v(x,t)$ to (76) is a solution of $\Xi'=0$.

Equations of pseudo-spherical type admit, in general, Bäcklund transformations. Chern and Tenenblat [**15**], and Tenenblat [**46**] constructed them exploiting the discrete symmetries of the pseudo–potential (56). The reader is referred to [**15, 46**] and [**39**] for examples. We would like to focus here on more general transformations which exist only because the equations at hand describe pseudo-spherical surfaces [**24, 38**]: an elementary result in Riemannian geometry is that any two surfaces of constant Gaussian curvature -1 are locally isometric. Since in the graphs of (generic) solutions of PSS equations can be equipped with pseudo-spherical structures, one would expect local transformations between solutions!

The space of independent variables of the equations $\Xi(x,t,u,\dots)=0$ and $\widehat{\Xi}(\widehat{x},\widehat{t},\widehat{u},\dots)=0$ will be denoted by M and $\widehat{M}$ respectively. We also denote $\omega(u(x,t))$ simply by ω, no confusion should arise.

THEOREM 6. *Let $\Xi(x,t,u,\dots)=0$ and $\widehat{\Xi}(\widehat{x},\widehat{t},\widehat{u},\dots)=0$ be two strictly pseudo-spherical evolution equations with associated one-forms $\omega^\alpha=f_{\alpha1}dx+f_{\alpha2}dt$ and $\widehat{\omega}^\alpha=\widehat{f}_{\alpha1}d\widehat{x}+\widehat{f}_{\alpha2}d\widehat{t}$ respectively, in which $\widehat{f}_{11}=\widehat{u}$. Any III-generic solution $u(x,t)$ of $\Xi=0$ determines a III-generic solution $\widehat{u}(\widehat{x},\widehat{t})$ of $\widehat{\Xi}=0$ by means of*

$$\widehat{u}\circ\Phi=\frac{1}{J}\left[(\cos\theta f_{11}+\sin\theta f_{21})\frac{\partial\psi}{\partial t}-(\cos\theta f_{12}+\sin\theta f_{22})\frac{\partial\psi}{\partial x}\right], \tag{77}$$

in which $\Phi(x,t)=(\varphi(x,t),\psi(x,t))$ is a local diffeomorphism with Jacobian J, and the functions $\theta(x,t)$, $\varphi(x,t)$, and $\psi(x,t)$ are smooth solutions of the system of equations

$$(\widehat{f}_{21}\circ\Phi)\,\varphi_x+(\widehat{f}_{22}\circ\Phi)\,\psi_x \;=\; -\sin\theta\, f_{11}+\cos\theta\, f_{21}, \tag{78}$$

$$(\widehat{f}_{21}\circ\Phi)\,\varphi_t+(\widehat{f}_{22}\circ\Phi)\,\psi_t \;=\; -\sin\theta\, f_{12}+\cos\theta\, f_{22}, \tag{79}$$

$$(\widehat{f}_{31}\circ\Phi)\,\varphi_x+(\widehat{f}_{32}\circ\Phi)\,\psi_x \;=\; f_{31}+\theta_x, \tag{80}$$

$$(\widehat{f}_{31}\circ\Phi)\,\varphi_t+(\widehat{f}_{32}\circ\Phi)\,\psi_t \;=\; f_{32}+\theta_t, \tag{81}$$

$$-(\widehat{f}_{12}\circ\Phi)\,J \;=\; (\cos\theta f_{11}+\sin\theta f_{21})\,\varphi_t \;-\; (\cos\theta\, f_{12}+\sin\theta\, f_{22})\,\varphi_x, \tag{82}$$

where $(\widehat{f}_{\alpha\beta}\circ\Phi)(x,t)$ have been evaluated by means of (77) and its derivatives.

EXAMPLE 16. We obtain the traveling wave solution of Burgers equation from solutions of the sine–Gordon equation

$$\frac{\partial^2\theta}{\partial u^2}-\frac{\partial^2\theta}{\partial v^2}=\sin\theta\cos\theta\ . \tag{83}$$

Equation (83) describes pseudo-spherical surfaces with associated functions

$$f_{11}=0,\quad f_{12}=\sin\theta,\quad f_{21}=-\cos\theta,\quad f_{22}=0,\quad f_{31}=\theta_v,\quad f_{32}=\theta_u. \tag{84}$$

We let $\theta(u,v)$ be a III–generic solution of (83), and we define $\omega(u,v)$ as a solution to

$$\theta_v+\omega_u=-\sin\omega\cos\theta,\qquad\text{and}\qquad\theta_u+\omega_v=\cos\omega\sin\theta. \tag{85}$$

(Note that (85) is a Bäcklund transformation in the sense of Definition 19). The function $\omega(u,v)$ is also a solution of Equation (83); we now introduce $\xi(u,v)$ and $\zeta(u,v)$ by means of

$$d\xi = -(\cos\omega\cos\theta\, du + \sin\omega\sin\theta\, dv), \quad \text{and} \quad d\zeta = \exp(\xi)(\cos\theta\sin\omega\, du - \sin\theta\cos\omega\, dv).$$

The functions ξ and ζ are well defined because of (85). Now, Burgers equation $\widehat{u}_{\widehat{t}} = \widehat{u}_{\widehat{x}\widehat{x}} - 2\widehat{u}\widehat{u}_{\widehat{x}}$ describes pseudo-spherical surfaces with associated functions

$$\widehat{f}_{11} = \widehat{u}, \quad \widehat{f}_{12} = -\widehat{u}^2 + \widehat{u}_{\widehat{x}}, \quad \widehat{f}_{21} = 1, \quad \widehat{f}_{22} = -\widehat{u}, \quad \widehat{f}_{31} = 1, \quad \widehat{f}_{32} = -\widehat{u}. \tag{86}$$

We define the transformation

$$\widehat{x} = \varphi(u,v) = 1 - \zeta(u,v) - \exp(\xi(u,v)), \tag{87}$$

$$\widehat{t} = \psi(u,v) = \ln|1 - \exp(\xi(u,v))| - 1 + \zeta(u,v) + \exp(\xi(u,v)). \tag{88}$$

Then, if $\tilde{\theta} = \omega(u,v) - \pi/2$, the map $\Phi : (u,v) \mapsto (\widehat{x},\widehat{t})$ given by (87) and (88) is a local diffeomorphism which satisfies the system of equations (78)–(82) (with "θ" replaced by "$\tilde{\theta}$"). Substituting into Equation (77) (again, with "θ" replaced by "$\tilde{\theta}$") we find that the pull-back of $\widehat{u}(\widehat{x},\widehat{t})$ by Φ is

$$\widehat{u}\circ\Phi = \frac{1 - \exp(\xi(u,v))}{\exp(\xi(u,v))},$$

and therefore,

$$\widehat{u}(\widehat{x},\widehat{t}) = \frac{\exp(\widehat{x}+\widehat{t})}{1 - \exp(\widehat{x}+\widehat{t})}, \tag{89}$$

the traveling wave solution of Burgers equation.

Theorem 6 appears in [**24, 38**] in a slightly different form. In particular, we remark that the assumption $\widehat{f}_{11} = \widehat{u}$ can be removed, see [**37**]. There are also transformation results for I– and II–generic solutions, see [**37, 39**].

We finish this section with a version of Theorem 6 which is valid in the case of hierarchies of pseudo-spherical type [**38**]. We first define *solutions* of hierarchies of PSS equations as follows:

DEFINITION 20. A solution of a hierarchy of pseudo-spherical type $u_{\tau_i} = F_i$, $i \geq 0$, is a sequence $\{u^{(n)}(x,t,\tau_1,\ldots,\tau_n)\}_{n\geq 0}$ of smooth functions $u^{(n)} : V^{(n)} \subset \mathbf{R}^{n+2} \to \mathbf{R}$, such that for each $n \geq 0$, (a) $u^{(n)}$ is a solution of the equations $u_{\tau_i} = F_i$, $i = 1,\ldots,n$, and (b) $u^{(n+1)}|_{V^{(n)}} = u^{(n)}$.

THEOREM 7. *Let $u_{\tau_i} = F_i(x,t,u,\ldots)$ and $\widehat{u}_{\widehat{\tau}_i} = \widehat{F}_i(\widehat{x},\widehat{t},\widehat{u},\ldots)$, $i \geq 0$, be two hierarchies of pseudo-spherical type with associated one-forms $\Theta^{(n)}_\alpha$ and $\widehat{\Theta}^{(n)}_\alpha$ respectively, $\alpha = 1,2,3$. Let $\{u^{(n)}\}$ and $\{\widehat{u}^{(n)}\}$ be solutions of $u_{\tau_i} = F_i$ and $\widehat{u}_{\widehat{\tau}_i} = \widehat{F}_i$, $i \geq 0$, and assume that $u^{(0)}(x,t)$ and $\widehat{u}^{(0)}(\widehat{x},\widehat{t})$ are III-generic. For each $n \geq 0$ there exist a local diffeomorphism $\Upsilon^{(n)} : V^{(n)} \to \widehat{V}^{(n)}$, in which $V^{(n)}$ and $\widehat{V}^{(n)}$ are open subsets of the domains of $u^{(n)}$ and $\widehat{u}^{(n)}$ respectively, and a smooth function $\mu^{(n)} : V^{(n)} \to \mathbf{R}$, such that the pull-backs of $\Theta^{(n)}_\alpha$ by $u^{(n)}$ and $\widehat{\Theta}^{(n)}_\alpha$ by $\widehat{u}^{(n)}$ satisfy*

$$\Upsilon^{(n)*}\widehat{\Theta}^{(n)}_1 = \Theta^{(n)}_1\cos\mu^{(n)} + \Theta^{(n)}_2\sin\mu^{(n)}, \tag{90}$$

$$\Upsilon^{(n)*}\widehat{\Theta}^{(n)}_2 = -\Theta^{(n)}_1\sin\mu^{(n)} + \Theta^{(n)}_2\cos\mu^{(n)} \tag{91}$$

$$\Upsilon^{(n)*}\widehat{\Theta}^{(n)}_3 = \Theta^{(n)}_3 + d\mu^{(n)}. \tag{92}$$

Moreover, the maps $\Upsilon^{(n)}$ and $\mu^{(n)}$, $n \geq 0$, can be chosen so that

$$\Upsilon^{(n+1)}|_{V^{(n)}} = \Upsilon^{(n)}, \quad \text{and} \quad \mu^{(n+1)}|_{V^{(n)}} = \mu^{(n)}, \quad n \geq 0. \tag{93}$$

9. Formal integrability, kinematic integrability, and geometry of surfaces

In this section we relate our work with a classical subject in the theory of integrable systems [**18**], the existence of linear problems associated with nonlinear equations. We start by defining kinematic integrability [**18**]:

DEFINITION 21. A differential equation $\Xi = 0$ (not necessarily of evolutionary type) is kinematically integrable if it is the integrability condition of a non–trivial one–parameter family of linear problems

$$\frac{\partial v}{\partial x} = U(\eta)v\ , \qquad \frac{\partial v}{\partial t} = V(\eta)v\ , \tag{94}$$

in which $U(\eta)$ and $V(\eta)$ are $sl(n, \mathbf{R})$–valued differential functions.

EXAMPLE 17. Consider the following version of the Boussinesq equation (33):

$$v_t = 2w_x\ , \qquad w_t = -\frac{1}{6}(v_{xxx} + 4vv_x)\ . \tag{95}$$

It is not hard to check that (95) is the integrability condition of the first order system $\psi_x = U\,\psi$, $\psi_t = V\,\psi$, with $sl(3, \mathbf{R})$–valued coefficient matrices

$$U = \begin{pmatrix} 0 & 1 & 0 \\ 0 & 0 & 1 \\ -\frac{1}{2}v_x - w + \eta & -v & 0 \end{pmatrix}, \quad V = \begin{pmatrix} \frac{2}{3}v & 0 & 1 \\ \frac{1}{6}v_x - w + \eta & -\frac{1}{3}v & 0 \\ \frac{1}{6}v_{xx} - w_x & -\frac{1}{6}v_x - w + \eta & -\frac{1}{3}v \end{pmatrix}.$$

It follows from Definition 21 that a differential equation $\Xi = 0$ is kinematically integrable if the matrix equation

$$\frac{\partial U(\eta)}{\partial t} - \frac{\partial V(\eta)}{\partial x} + [U(\eta), V(\eta)] = 0 \tag{96}$$

is identically satisfied whenever $u(x,t)$ is a solution of $\Xi = 0$, and the matrices $U(\eta)$ and $V(\eta)$ satisfy trace $U(\eta)$ = trace $V(\eta) = 0$. Equations (94) and (96) can be interpreted geometrically as a *connection*, see Faddeev and Takhtajan [**18**]. Unless explicitly indicated, in what follows we will assume that the matrices U and V appearing in Definition 21 take values in $sl(2, \mathbf{R})$.

We say that a differential equation is *strictly* kinematically integrable if it is kinematically integrable and the diagonal entries of the matrix $U(\eta)$ appearing in (94) are η and $-\eta$, as in the AKNS case. If an equation *is* integrable in the strict sense, we call η the "spectral" parameter: a strictly kinematically integrable equation is the integrability condition of a linear problem of the form

$$\left(\frac{\partial}{\partial x} - U_0\right) v = \eta\,\mathrm{diag}(1, -1)v, \quad \frac{\partial v}{\partial t} = V(\eta)v,$$

in which $U_0 = U(\eta) - \eta\,\mathrm{diag}(1, -1)$, and therefore one may hope to solve it by inverse scattering techniques [**4, 5, 18, 33**].

PROPOSITION 8. *A differential equation $\Xi = 0$ is kinematically integrable if and only if it describes a non–trivial one–parameter family of pseudo-spherical surfaces, that is, if and only if $\Xi = 0$ is geometrically integrable. Strict kinematic integrability amounts to $f_{21} = \eta$.*

PROOF. We easily see that Equations (41) hold if and only if the linear problem

$$\begin{pmatrix} dv_1 \\ dv_2 \end{pmatrix} = \frac{1}{2}\begin{pmatrix} \omega_2 & \omega_1 - \omega_3 \\ \omega_1 + \omega_3 & -\omega_2 \end{pmatrix}\begin{pmatrix} v_1 \\ v_2 \end{pmatrix} \tag{97}$$

is integrable whenever $u(x,t)$ is a solution of $\Xi = 0$, and conversely, any family of real linear problems (94) satisfying (96) whenever $u(x,t)$ is a solution of $\Xi = 0$, determines a one–parameter family of one–forms ω^α satisfying (41) whenever $u(x,t)$ is a solution of $\Xi = 0$, by setting, using an obvious notation, $\omega^1 = (U_{21}+U_{12})\,dx + (V_{21}+V_{12})\,dt$, $\omega^2 = 2U_{11}\,dx + 2V_{11}\,dt$, and $\omega^3 = (U_{21} - U_{12})\,dx + (V_{21} - V_{12})\,dt$. The extra condition $\omega^\alpha \not\equiv 0$ can be imposed by applying a transformation

$$U \mapsto PUP^{-1} + (dP/dx)P^{-1}\,, \quad V \mapsto PVP^{-1} + (dP/dx)P^{-1},$$

in which the entries of P are differential functions independent of the parameter η, to the matrices U, V appearing in (94). It is known [**18, 41**] that such a transformation will not alter the condition (96). □

This proposition suggests that we can attempt to relate formal integrability and inverse scattering by showing first that the equations at hand describe pseudo-spherical surfaces. The classification theorems for strictly PSS equations discussed in Section 5 imply that this approach is feasible. For instance, Theorem 1 allows us to find [**20**]:

THEOREM 8. *Every second order evolution equation $u_t = F(x,t,u,u_x,u_{xx})$ which is formally integrable is of pseudo-spherical type, and therefore kinematically integrable. Moreover, up to invertible contact transformations of the form*

(98) $$x' = \phi(x,t,u,u_x), \quad t' = \xi(t), \quad u' = \psi(x,t,u,u_x),$$

there exists an exhaustive list of representatives of formally integrable second order equations which are kinematically integrable in the strict sense.

What about kinematically integrable equations for which their associated linear problems are in $sl(3,\mathbf{R})$, for instance? Do they have a geometrical interpretation? We only have a partial answer to this question [**20**], which we summarize here:

THEOREM 9. *The graph of each solution $u(x,t)$, $v(x,t)$ of the Boussinesq equation* (95) *can be equipped with the structure of an affine surface. In fact, every formally integrable equation of the form* (34) *describes affine surfaces.*

We omit the proof of Theorem 9, since it would require us to make a detour into affine geometry. However, we would like to stress here that the last part of this result follows from our classification of formally integrable systems (34) considered in Section 3, and the fact that the equations appearing there can be transformed into the Boussinesq system (33). Yet another example of the usefulness of general classification theorems!

10. Integrable equations and water waves

In this final section we apply our geometric approach to three equations which appear in the study of water waves. We consider the ubiquitous Korteweg–de Vries (KdV) equation

$$u_t = u_{xxx} + 6uu_x\,,$$

describing the unidirectional propagation of waves on the free surface of shallow water under the influence of gravity, the Camassa–Holm (CH) equation [**14**]

(99) $$m = u_{xx} - u\,, \qquad m_t = -m_x\,u - 2\,m\,u_x\,,$$

also describing shallow waves but, in contradistinction with the KdV equation [**33**] admitting wave breaking solutions, and the Hunter–Saxton (HS) equation [**21**]

$$m = u_{xx}\ , \qquad m_t = -m_x\, u - 2\, m\, u_x\ , \tag{100}$$

describing weakly nonlinear unidirectional waves. Several facts about these three important equations are already known, of course. In particular, they are bihamiltonian [**31, 14, 22**] and they can be deeply investigated by scattering/inverse scattering means [**33, 7, 8**]. Moreover, it has been recently noticed that the Korteweg–de Vries, Camassa–Holm and Hunter–Saxton equations exhaust (in a precise sense) the bihamiltonian equations which can be modeled as geodesic flows on homogeneous spaces related to the Virasoro group (Khesin and Misiołek [**25**]). It is then tempting to ask whether they possess a common geometrical interpretation. This is indeed the case, as we show here. We first consider the KdV equation as a short example:

EXAMPLE 18. The KdV equation $u_t = u_{xxx} + 6uu_x$ is strictly pseudo-spherical [**43, 15**] with associated one–forms

$$\omega^1 = (1-u)\,dx + (-u_{xx} + \eta u_x - \eta^2 u - 2u^2 + \eta^2 + 2u)\,dt \tag{101}$$

$$\omega^2 = \eta\,dx + (\eta^3 + 2\eta u - 2u_x)\,dt \tag{102}$$

$$\omega^3 = (-1-u)\,dx + (-u_{xx} + \eta u_x - \eta^2 u - 2u^2 - \eta^2 - 2u)\,dt, \tag{103}$$

in which η is an arbitrary parameter. After rotating the coframe $\{\omega^1, \omega^2\}$ and changing Γ for $-\Gamma$, we can write the Pfaffian system (58) as

$$\text{(a)}\ \ \Gamma_x = -u - \eta\Gamma - \Gamma^2, \qquad \text{(b)}\ \ \Gamma_t = (\Gamma_{xx} - 3\Gamma^2\eta - 2\Gamma^3)_x .$$

Since KdV is strictly pseudo-spherical we know that if Γ solves (b), then u as given by (a) solves KdV. We thus recover the Miura transformation and the modified KdV equation.

Now, take a solution $u(x,t)$ of KdV and compute $\Gamma(x,t,\eta)$ from (a). Equation (b) is invariant under the transformation ($\Gamma \mapsto -\Gamma, \eta \mapsto -\eta$), so that $\overline{u}(x,t,\eta) = \Gamma_x(x,t,\eta) - \Gamma(x,t,\eta)\eta - \Gamma(x,t,\eta)^2$ is a one–parameter family of solutions of KdV. It follows that $\overline{u}(x,t,\eta) = u(x,t) + 2\Gamma_x(x,t,\eta)$, and therefore we also recover the classical Darboux transformation.

Note, finally, that taking derivative of $\overline{u}(x,t,\eta)$ with respect to η should yield a symmetry of the KdV equation [**29, 31**]. Moreover, we expect this symmetry to be nonlocal, since $\overline{u}(x,t,\eta)$ depends on the pseudo–potential Γ. This is indeed true, see [**28**]. We will see in detail that the same phenomenon occurs in the Camassa–Holm and Hunter–Saxton cases.

Below and henceforth, we let ϵ be a parameter, $0 \le \epsilon \le 1$, and we set $m = u_{xx} - \epsilon u$.

THEOREM 10. *The Camassa–Holm and Hunter–Saxton equations describe pseudo-spherical surfaces. Moreover, there is a one-parameter family of PSS equations to which these two equations belong.*

PROOF. We consider one–forms σ^α, $\alpha = 1, 2, 3$, given by

$$\begin{aligned}\sigma^1 &= (m - \beta + \epsilon\,\eta^{-2}(\beta - 1))\ dx \\ &\qquad + (-u_x\beta\,\eta^{-1} - \beta\,\eta^{-2} - u\,m - 1 + u\beta + u_x\,\eta^{-1} + \eta^{-2})\ dt\ ,\end{aligned} \tag{104}$$

$$\sigma^2 = \eta\,dx + (-\beta\,\eta^{-1} - \eta\,u + \eta^{-1} + u_x)\ dt\ , \tag{105}$$

$$\sigma^3 = (m+1)\ dx + \left(\epsilon\,u\,\eta^{-2}(\beta - 1) - u\,m + \eta^{-2} + \frac{u_x}{\eta} - u - \frac{\beta}{\eta^2} - \frac{u_x\beta}{\eta}\right)\ dt\ , \tag{106}$$

in which the parameters η and β are constrained by the relation

$$\eta^2 + \beta^2 - 1 = \epsilon \left[\frac{\beta - 1}{\eta}\right]^2 . \tag{107}$$

It is not hard to check that the structure equations (41) are satisfied whenever $u(x,t)$ is a solution of (99) (if $m = u_{xx} - u$) and whenever $u(x,t)$ is a solution of (100) (if $m = u_{xx}$). □

A natural way to dispense with the constraint (107) is by using a parameterization of the curve $\eta^2 + \beta^2 - 1 = \epsilon\,[(\beta - 1)/\eta]^2$. We take

$$\eta = \sqrt{\epsilon + 1 - s^2}\,, \qquad \beta = \frac{\epsilon}{s-1} - s\,. \tag{108}$$

It follows that the CH and HS equations are geometrically integrable and, by Proposition 8, *kinematically* integrable.

We turn to the quadratic pseudo-potential (60) associated with the CH and HS equations. After parameterizing the one–forms σ^α using (108), rotating the resulting forms via (55), applying the transformation $\hat{\Gamma} \mapsto \gamma + \sqrt{\epsilon + 1 - s^2}/(1 - s)$, and setting $s - 1 = 1/\lambda$, we obtain the following result:

THEOREM 11. *The CH equation* (99) *and the HS equation* (100) *admit quadratic pseudo-potentials γ determined by*

$$m = \gamma_x + \frac{1}{2\lambda}\gamma^2 - \frac{1}{2}\lambda\epsilon\,, \qquad \gamma_t = \frac{\gamma^2}{2}[1 + \frac{1}{\lambda}u] - u_x\gamma - u\,m + \epsilon[\frac{1}{2}u\lambda - \frac{1}{2}\lambda^2]\,, \tag{109}$$

in which $\lambda \neq 0$ is a parameter. Moreover, Equations (99) *and* (100) *possess the conservation law*

$$\gamma_t = \lambda\,(\,u_x - \gamma - \frac{1}{\lambda}u\gamma\,)_x\,. \tag{110}$$

In view of Example 18, it is natural to postulate the first equation of (109) as the analog of the Miura transformation for the CH and HS equations, and (110) as the corresponding "modified" equation. Equations (109) also determine simple linear problems for the CH and HS equations: the compatibility condition of the linear problem $d\psi = (X dx + T dt)\psi$, in which

$$X = \frac{1}{2}\begin{bmatrix} 0 & \epsilon\,\lambda + 2\,m \\ \lambda^{-1} & 0 \end{bmatrix}, \qquad T = \frac{1}{2}\begin{bmatrix} -u_x & -2u\,m + \epsilon\,\lambda u - \epsilon\,\lambda^2 \\ -1 - u\lambda^{-1} & u_x \end{bmatrix}, \tag{111}$$

is precisely the CH equation (99) (if $m = u_{xx} - u$) and the HS equation (100) (if $m = u_{xx}$).

We now use (109) and (110) to construct conservation laws for the CH and HS equations. Setting $\gamma = \sum_{n=1}^{\infty} \gamma_n \lambda^{n/2}$ and substituting into (109) yields the conserved densities

$$\gamma_1 \;=\; \sqrt{2}\,\sqrt{m}\,, \quad \gamma_2 \;=\; -\frac{1}{2}\ln(m)_x\,, \quad \gamma_3 = \frac{1}{2\sqrt{2}\,\sqrt{m}}[\epsilon - \frac{m_x^2}{4\,m^2} + \ln(m)_{xx}]\,, \tag{112}$$

$$\gamma_{n+1} \;=\; -\frac{1}{\gamma_1}\gamma_{n,x} - \frac{1}{2\gamma_1}\sum_{j=2}^{n}\gamma_j\,\gamma_{n+2-j}\,, \quad n \geq 3\,, \tag{113}$$

while the expansion $\gamma = \epsilon\,\lambda + \sum_{n=0}^{\infty}\gamma_n\lambda^{-n}$ implies

$$\gamma_{0,x} + \epsilon\,\gamma_0 = m\,, \quad \gamma_{n,x} + \epsilon\,\gamma_n = -(1/2)\sum_{j=0}^{n-1}\gamma_j\,\gamma_{n-1-j}\,, \; n \geq 1\,. \tag{114}$$

The conserved densities γ_n determined by (112) and (113) are all *local*. It appears that this is the first direct derivation of a sequence of local conservation laws for the CH and HS equations, see **[19]** for a different approach. On the other hand, (114) yields the local conserved densities

u, $u_x^2+u^2$, and $uu_x^2+u^3$, and a sequence of nonlocal conservation laws. The reader may notice that the first conserved densities in (112) and (114) are the ones which appeared in Example 9!

We now consider nonlocal symmetries for CH and HS constructed from γ:

THEOREM 12. *Let γ, δ and β be defined by the equations*

$$\gamma_x = m-(1/2\lambda)\gamma^2+\epsilon(1/2)\lambda\,, \qquad \gamma_t = \lambda(u_x-\gamma-(1/\lambda)u\gamma)_x\,; \tag{115}$$

$$\delta_x = \gamma\,, \qquad \delta_t = \lambda(u_x-\gamma-(1/\lambda)u\gamma)\,; \tag{116}$$

$$\beta_x = m\,e^{(1/\lambda)\delta}\,, \qquad \beta_t = e^{(1/\lambda)\delta}(-(1/2)\gamma^2+\epsilon(1/2)\lambda^2-u\,m)\,; \tag{117}$$

which are compatible on solutions of (99) *and* (100). *The systems of equations* (99), (115)–(117) *and* (100), (115)–(117), *possess the classical symmetry*

$$W=\gamma\,e^{(1/\lambda)\delta}\frac{\partial}{\partial u}+(m_x+\frac{2}{\lambda}\gamma\,m)\,e^{(1/\lambda)\delta}\frac{\partial}{\partial m}+m\,e^{(1/\lambda)\delta}\frac{\partial}{\partial\gamma}+\beta\frac{\partial}{\partial\delta}+(m\,e^{(2/\lambda)\delta}+\frac{1}{2\lambda}\beta^2)\frac{\partial}{\partial\beta}\,.$$

Thus, in particular, the evolutionary vector field $V=\gamma\,e^{(1/\lambda)\delta}\,\partial/\partial u$ is a one-parameter family of nonlocal symmetries for (99) *and* (100).

Theorem 12 can be verified using the MAPLE package VESSIOT [**1**]. We find the flow of W. For this we consider the first order system (we use independent variables ξ and η)

$$\frac{\partial x}{\partial\xi} = -e^{(1/\lambda)\,\delta(\xi,\eta)}\,, \tag{118}$$

$$\frac{\partial m}{\partial\xi} = \frac{2}{\lambda}\gamma(\xi,\eta)\,m(\xi,\eta)\,e^{(1/\lambda)\,\delta(\xi,\eta)}\,, \tag{119}$$

$$\frac{\partial\gamma}{\partial\xi} = e^{(1/\lambda)\,\delta(\xi,\eta)}\,(\frac{1}{2\lambda}\gamma(\xi,\eta)^2-\frac{1}{2}\lambda)\,, \tag{120}$$

$$\frac{\partial\delta}{\partial\xi} = \beta(\xi,\eta)-\gamma(\xi,\eta)\,e^{(1/\lambda)\,\delta(\xi,\eta)}\,, \tag{121}$$

$$\frac{\partial\beta}{\partial\xi} = \frac{1}{2\lambda}\beta(\xi,\eta)^2\,. \tag{122}$$

PROPOSITION 9. *The initial value problem* (118)–(122) *with initial conditions $\beta_0=\beta(0,\eta)$, $\gamma_0=\gamma(0,\eta)$, $\delta_0=\delta(0,\eta)$, $m_0=m(0,\eta)$, and $x_0=x(0,\eta)=\eta$, has the solution*

$$\beta(\xi,\eta) = \frac{1}{B(\eta)}\,2\,\lambda\,\beta_0\,, \tag{123}$$

$$\gamma(\xi,\eta) = \frac{1}{B(\eta)}\,[\gamma_0B(\eta)+\xi e^{(1/\lambda)\delta_0}A^+(\eta)A^-(\eta)]\,, \tag{124}$$

$$\delta(\xi,\eta) = \lambda\,\ln[\frac{4\lambda^2\,e^{(1/\lambda)\delta_0}}{(B(\eta)+\xi e^{(1/\lambda)\delta_0}A^+(\eta))(B(\eta)+\xi e^{(1/\lambda)\delta_0}A^-(\eta))}]\,, \tag{125}$$

$$m(\xi,\eta) = \frac{1}{B(\eta)^4}[B(\eta)+\xi e^{(1/\lambda)\delta_0}A^-(\eta)]^2\,[B(\eta)+\xi e^{(1/\lambda)\delta_0}A^+(\eta)]^2\,m_0\,, \tag{126}$$

$$x(\xi,\eta) = \eta+\ln[\frac{B(\eta)+\xi e^{(1/\lambda)\delta_0}A^-(\eta)}{B(\eta)+\xi e^{(1/\lambda)\delta_0}A^+(\eta)}]\,, \tag{127}$$

in which the functions B, A^+, and A^- are given by

$$B(\eta)=\xi\,\beta_0+2\,\lambda\,, \qquad A^+(\eta)=\gamma_0+\lambda\,, \qquad A^-(\eta)=\gamma_0-\lambda\,. \tag{128}$$

Now, Equation (127) determines a transformation $T : (\xi, \eta) \mapsto (x, \tau)$ in which τ is a parameter along the flow of W, given by, say,

$$\tau = \xi \ , \qquad x = h(\xi, \eta) \ . \tag{129}$$

Applying this change of variables to Equations (119)–(122), and using (118) and Equations (115), (116), and (117) for γ, δ, and β, we see that formulae (123)–(126) provide solutions for the flow equations

$$\frac{\partial m}{\partial \tau} = [\frac{\partial m}{\partial x} + \frac{2}{\lambda}\gamma m]\, e^{(1/\lambda)\delta} \ , \tag{130}$$

$$\frac{\partial \gamma}{\partial \tau} = m\, e^{(1/\lambda)\delta} \ , \tag{131}$$

$$\frac{\partial \delta}{\partial \tau} = \beta \ , \tag{132}$$

$$\frac{\partial \beta}{\partial \tau} = m\, e^{(2/\lambda)\delta} + \frac{1}{2\lambda}\beta^2 \ , \tag{133}$$

which we obtain from the formula for W in Theorem 12. It follows that finding a one–parameter family of solutions to the Camassa–Holm equation, for instance, amounts to solving *one* first order partial differential equation. More exactly, one has,

COROLLARY 1. *Let $u(x,t)$ be a solution of the CH equation. Then, the solution $u(\xi, \eta, t)$ to the initial value problem*

$$\frac{\partial u}{\partial \tau} = \gamma(x, \tau)\, e^{(1/\lambda)\delta(x,\tau)} \ , \tag{134}$$

$$u(x, t, 0) = u(x, t), \tag{135}$$

in which $\gamma(x,t,\tau)$ and $\delta(x,t,\tau)$ are determined by (124), (125), *and* (129), *is a one–parameter family of solutions to the Camassa–Holm equation* (99).

EXAMPLE 19. We begin with the Camassa–Holm equation written in the form

$$2u_\eta u_{\eta\eta} + u u_{\eta\eta\eta} = u_t - u_{\eta\eta t} + 3u_\eta u \ , \tag{136}$$

so that the "old" space variable is η. Consider an obvious solution of (136), say $u_0(\eta, t) = e^\eta$. The corresponding (pseudo)potentials γ_0, δ_0 and β_0 are given by

$$\gamma_0 = \lambda \ , \qquad \beta_0 = c \ , \qquad \delta_0 = \lambda\eta - \lambda^2 t.$$

We use these values for γ_0, δ_0 and β_0 as our initial conditions for the system (123)–(126), that is, we take

$$u_0(\eta, t) = e^\eta \ , \quad m_0 = 0 \ , \quad \gamma_0 = \lambda \ , \quad \delta_0 = \lambda\eta - \lambda^2 t \ , \quad \beta_0 = c.$$

The new space variable is then given by Equation (127), that is,

$$x(\xi, \eta) = \eta + \ln\left|\frac{\xi c + 2\lambda}{\xi c + 2\lambda + \xi e^{\eta - \lambda t} 2\lambda}\right|.$$

We now invert this equation to find a change of variables, $\eta = f(x, \tau)$, $\xi = \tau$. Taking $\beta_0 = c = 0$, we obtain,

$$\eta = x - \ln\left|1 - \tau e^{\eta - \lambda t}\right| \ , \qquad \xi = \tau.$$

Substituting into (123)–(126) again, we obtain new (pseudo)potentials, and then Corollary 1 gives us a one–parameter family of solutions of the Camassa-Holm equation (99). We find

$$u(x, t, \tau) = \lambda\tau e^{x - \lambda t} + e^x.$$

Acknowledgments

The author most warmly thanks Professor J.C. Misra for his kind invitation to contribute to this project. He is also grateful to I. Anderson, R. Beals, L. Dickey, M. Fels, N. Kamran, P. Olver, and J. Szmigielski for their interest in his work, and very many discussions on the topics discussed here.

References

[1] I.M. Anderson, "The Vessiot Handbook" (2000). Formal Geometry and Mathematical Physics Technical Report, Utah State University. Hypertext link: `http://www.math.usu.edu/~fg_mp/`

[2] I.A. Anderson, M.E. Fels, and C.G. Torre, Group invariant solutions without transversality. *Comm. Math. Phys.* 212 (2000), no. 3, 653–686.

[3] I.M. Anderson, and N. Kamran, The variational bicomplex for hyperbolic second-order scalar partial differential equations in the plane. *Duke Math. J.* 87 (1997), 265–319.

[4] R. Beals, P. Deift, C. Tomei, "Direct and inverse scattering on the line" (1988). Math. Surveys and Monographs no.28, AMS, Providence.

[5] R. Beals, P. Deift, and X. Zhou, The inverse scattering transform on the line. In "Important developments in soliton theory" (1993). A.S. Fokas and V.E. Zakharov (Eds.) Springer Series in Nonlinear Dynamics. Springer–Verlag, New York.

[6] R. Beals, and D.H. Sattinger, On the complete integrability of completely integrable systems. *Commun. Math. Phys.* 138 (1991), 409–436.

[7] R. Beals, D.H. Sattinger, and J. Szmigielski, Acoustic scattering and the extended Korteweg-de Vries hierarchy. *Advances in Math.* 140 (1998), 190–206.

[8] R. Beals, D.H. Sattinger, and J. Szmigielski, Inverse scattering solutions of the Hunter–Saxton equation. Preprint, 2001.

[9] F. Beukers, J.A. Sanders, and J.-P.Wang, One symmetry does not imply integrability. *J. Differential Equations* 146 (1998), 251–260.

[10] A.H. Bilge, A system with a recursion operator but one higher local symmetry. *Lie Groups Appl.* 1 (1994) 132–139.

[11] R.L. Bryant, S.S. Chern, R.B. Gardner, H. Goldschmidt, and P.A. Griffiths, "Exterior differential systems" (1990). MSRI Publications, Vol. 18, Springer–Verlag, New York.

[12] R.L. Bryant, and P.A. Griffiths, Characteristic cohomology of differential systems. I. General theory. *J. Amer. Math. Soc.* 8 (1995), no. 3, 507–596.

[13] R.L. Bryant, and P.A. Griffiths, Characteristic cohomology of differential systems. II. Conservation laws for a class of parabolic equations. *Duke Math. J.* 78 (1995), no. 3, 531–676.

[14] R. Camassa and D.D. Holm, An integrable shallow water equation with peaked solitons. *Phys. Rev. Lett.* 71 (1993), no. 11, 1661–1664.

[15] S.S. Chern, and K. Tenenblat, Pseudo–spherical surfaces and evolution equations. *Stud. Appl. Math.* 74 (1986), 55–83.

[16] A. Constantin and J. Escher, On the structure of a family of quasilinear equations arising in shallow water theory. *Math. Ann.* 312 (1998), 403–416.

[17] M. do Carmo, "Differential forms and applications" (1994). Translated from the 1971 Portuguese original. Universitext. Springer-Verlag, Berlin.

[18] L.D. Faddeev, and L.A. Takhtajan, "Hamiltonian Methods in the Theory of Solitons" (1987). Springer Series in Soviet Mathematics, Springer-Verlag, Berlin–New York.

[19] M. Fisher and J. Schiff, The Camassa Holm equation: conserved quantities and the initial value problem. *Phys. Lett. A* 259 (1999), no. 5, 371–376.

[20] M.V. Foursov, P.J. Olver, and E.G. Reyes, On formal integrability of evolution equations and local geometry of surfaces (2000). To appear, *Diff. Geom. Appl.*

[21] J.K. Hunter, and R. Saxton, Dynamics of director fields. *SIAM J. Appl. Math.* 51 (1991), no. 6, 1498–1521.

[22] J.K. Hunter, and Y.X. Zheng, On a completely integrable nonlinear hyperbolic variational equation. Phys. D 79 (1994), no. 2-4, 361–386.

[23] N. Kamran, The geometrical study of differential equations (2000). NSF–CBMS Lectures, Washington D.C. To appear.

[24] N. Kamran, and K. Tenenblat, On differential equations describing pseudospherical surfaces. *J. Differential Equations* 115, no. 1 (1995), 75–98.

[25] B. Khesin and G. Misiołek, Euler equations on homogeneous spaces and Virasoro orbits. Preprint, 2001.

[26] J. Krasil'shchik, and A. Verbovetsky, "Homological Methods in Equations of Mathematical Physics" (1998). Open Education, Opava. (Also Diffiety Inst. Preprint Series DIPS 7/98, and math.DG/9808130).

[27] I.S. Krasil'shchik, and A.M. Vinogradov (Eds.) "Symmetries and conservation laws for differential equations of mathematical physics" (1999). Translations of Mathematical Monographs Vol. 182, AMS, Providence.

[28] M. Leo, R.A. Leo, G. Soliani, and P. Tempesta, On the relation between Lie symmetries and prolongation structures of nonlinear field equations. *Prog. Theoretical Phys.* 105 (2001), 77–97.

[29] A.V. Mikhailov, A.B. Shabat and V.V. Sokolov, The symmetry approach to classification of integrable equations (1991). In Zakharov [**48**].

[30] Mikhailov, A.V., A.B. Shabat and R.I. Yamilov, The symmetry approach to the classification of non–linear equations. Complete lists of integrable systems. *Russian Math. Surveys* 42(4) (1987), 1–63.

[31] P.J. Olver, "Applications of Lie Groups to Differential Equations" (1993). Second Edition, Springer-Verlag, New York.

[32] P.J. Olver, and V.V. Sokolov, Integrable evolution equations on associative algebras. *Commun. Math. Phys.* 193 (1998) 245–268.

[33] R. Palais, The symmetries of solitons. *Bull. Amer. Math. Soc. (N.S.)* 34 (1997), no. 4, 339–403.

[34] U. Pinkall, and I. Sterling, On the classification of constant mean curvature tori. *Annals of Math.* 130 (1989), 407–451.

[35] E.G. Reyes, Pseudo-spherical surfaces and integrability of evolution equations. *J. Differential Equations* 147, no.1 (1998), 195–230. Erratum: *J. Differential Equations* 153, no.1 (1999), 223–224.

[36] E.G. Reyes, Some geometric aspects of integrability of differential equations in two independent variables. *Acta Applicandae Mathematicae* 64, no. 2/3 (2000), 75–109.

[37] E.G. Reyes, On generalized Bäcklund transformations for equations describing pseudo-spherical surfaces (2000). To appear, *J. Geom. Phys.*

[38] E.G. Reyes, Correspondence theorems for hierarchies of equations of pseudo-spherical type (2000). Submitted, October 2001.

[39] E.G. Reyes, On differential equations, integrability, and differential geometry" 2001. In 'Applicable Mathematics, its Perspectives and Challenges', 3–25. Proceedings of the International Symposium on Challenges in Mathematical Sciences for the New Millennium, Indian Institute of Technology, Kharagpur, India. Edited by J.C. Misra. Narosa Publishing House, New Delhi and Calcutta.

[40] G. Rosen, Non–linear heat conduction in solid H_2. *Phys. Rev.* B19 (1979), 2398–2399.

[41] A. Roy Chowdhury, "Lie algebraic methods in integrable systems" (2001). Chapman & Hall/CRC Research Notes in Mathematics 415, London, England.

[42] J.A. Sanders, J.-P. Wang, On the integrability of homogeneous scalar evolution equations. *J. Differential Equations* 147 (1998), 410–434.

[43] R. Sasaki, Soliton equations and pseudospherical surfaces. *Nucl. Phys. B* 154 (1979), 343–357.

[44] M. Spivak, "A comprehensive introduction to differential geometry. Vol. I–V" (1979). Second edition, Publish or Perish, Inc., Wilmington, Del., 1979.

[45] S.I. Svinolupov, and V.V. Sokolov, Weak nonlocalities in evolution equations. *Matematicheskie Zametki* 48(6) (1990), 91–97.

[46] K. Tenenblat, "Transformations of manifolds and applications to differential equations" (1998). Pitman Monographs and Surveys in Pure and Applied Mathematics 93, Addison Wesley Longman, England.

[47] H.D. Wahlquist and F.B. Eastbrook, Prolongation structures of nonlinear equations. em J. Math. Phys. 16 (1975), 1–7.

[48] V.E. Zakharov (Ed.), "What is integrability?" Springer Series in Nonlinear Dynamics (1991). Springer–Verlag, Berlin.

Modern Applied Mathematics
Editor: J.C. Misra

Particle-Driven Gravity Currents: Occurrence, Modelling, Open Problems and New Results

T.B. Moodie and J.P. Pascal

Pacific Institute for the Mathematical Sciences, Department of Mathematical and Statistical Sciences
University of Alberta, Edmonton, Alberta, Canada T6G 2G1

ABSTRACT. Gravity currents are flows caused by a density difference between flowing and ambient fluids. A turbidity current is a gravity current where the density difference is produced principally by solid particles suspended in the current by fluid turbulence although temperature and salinity contrasts may also play a coupled role in marine settings in producing and maintaining the density difference. These so-called gravity currents play an important role in a vast array of geophysical and engineering applications and have been the subject of extensive theoretical, numerical, and experimental investigations over the past few decades. It is our intention here to re-examine various aspects of particle-driven flows based upon a rational deployment of the standard assumptions that have been the basis of recent modelling efforts. In particular we will compare depositional patterns from turbidity currents in both deep and shallow surroundings and examine the level of velocity shear that is generated and how it depends on various parameters such as initial particle concentration and particle settling rates. We can then examine how this shear, in turn, influences the long-term patterns of deposition. We shall also show how these flows are highly baroclinic and compute the particle-generated vorticity showing its dependence on the various flow parameters. A review of related literature will be included as well as a description of some of the unresolved issues.

1. Introduction

According to Middleton [1] a gravity current consists of the flow of one fluid within another when this flow takes place because of relatively small differences in density between the fluids. The flow of rivers is not normally considered a type of gravity current because the unit weight of water is several hundred times larger than that of air. The flow of cold air beneath warm air, however, is considered to be a gravity current. A turbidity current is a gravity current in which the excess density or unit weight providing the driving buoyancy force is due principally to the presence of sediment being held in suspension by fluid turbulence. Thus whether or not a sediment-bearing gravity current is a turbidity current will depend to a great extent upon the ordering that exists between the density of the interstitial (suspending) fluid and that of the ambient fluid. For example, in the case of a bottom particle-bearing gravity current whose interstitial fluid has a density that is equal to or less than that of the ambient, the driving buoyancy forces are due solely to the presence of suspended particles and we have a turbidity current. In this case it is possible that particle settling may lead to a reduction in bulk density of the bottom flow thereby transforming the bottom current into an intrusive turbidity current at some neutrally buoyant intermediate depth [2]. Further particle settling can subsequently lead to complete buoyancy reversal, the initiation of a buoyant plume, and the creation of a surface gravity current in which the remaining particles play no role in the dynamics of the flow [3]. On the other hand, should the density of the interstitial fluid in the bottom current be greater than that of the ambient, then the turbidity and/or particle-bearing gravity current will remain buoyantly stable throughout the particle-settling phase. If this suspension is either initially or becomes sufficiently dilute through the settling of particles, then there will be a stage when these particles will play little or no role in the dynamics of the flow. The flow dynamics will be governed by the density difference that exists between interstitial and ambient fluids and we have a particle-bearing gravity current. The term turbidity current has traditionally applied only to flows of sediment suspended in water, though strictly speaking it applies also to some flows of suspended solids in air provided these solids and not temperature differences, for example, provide the driving buoyancy forces [1]. The atmospheric dust storms of North Africa known as "haboobs" are an example of a particle-bearing gravity current that is not a turbidity current because the driving buoyancy forces in this instance arise due to temperature differences [4] and not the presence of suspended particles.

These turbidity and/or particle-bearing gravity currents with their potentially extremely complex dynamics occur in a vast array of natural and human-made settings. A relatively complete catalogue of events in which they play a role is to be found in Simpson [4]. We shall describe some of these events and discuss the major difficulties inherent in their study from the point of view of modelling them either qualitatively or quantitatively. Some of the successful modelling efforts will be highlighted and open problems described.

Particle-laden submarine gravity currents (turbidity currents) are the primary mechanisms for transporting sediment from the ocean margins to the abyssal plains. Ancient accumulations of sediment formed by such flows comprise important hydrocarbon reservoirs in many locations around the globe. Accurate models are needed to predict the external geometry and internal character of such submarine gravity current deposits [5]. These models are essential if one wishes to predict the optimal location in which to drill new wells and to understand fluid flow within existing reservoir units [6]. Since these flows must have occurred and deposited in a succession of events it would be optimal that any developed models are capable of dealing with transient flow over the $O(1)$ topography that would have been laid down by a previous event. This $O(1)$ topography will, because of its contribution to streamline curvature, introduce non-hydrostatic pressures thereby ruling out the usual shallow-water models [7]. To the best of our knowledge this extension to treat $O(1)$ topography in the transient regime has not been developed to date.

All of the quantitative predictive models for these submarine particle-laden gravity currents neglect vertical stratification of sediment concentration whereas the experimental evidence for such stratification is overwhelming [8] as are the limited number of field observations [9]. Moodie *et al.* [10] developed a qualitative model that attempted to account for this vertical structure in the volume fraction of particles by introducing a "shape function" that is roughly based on the experimentally observed values of volume fraction. They found that for smaller particles the long-term areal density of deposit was relatively insensitive to the choice of shape function but that for larger particles (i.e., those with greater Stokes' settling velocities) the dependence of the deposition patterns on the shape function was more pronounced.

The model developed in [10] also presents a way to account for the initial turbulent energy of mixing present in the well-mixed suspension behind the lock in the lock release experiments that are specifically designed to simulate aspects of these submarine gravity currents [11]. This is done by means of a modified settling velocity. It was found that by accounting for initial turbulent energy in the fixed volume release the computed deposition patterns were qualitatively in close accord with the experimentally determined ones. In particular the turbulent energy model of [10] captures and explains the proximal depression in the areal density of deposit which has always been problematic for the models developed to date [6]. More detail of this turbulent energy argument will be provided later in the introduction.

When one attempts to model large-scale particulate flows such as the one produced by the 1929 Grand Banks earthquake the background rotation of the earth must be taken into account [12]. The relevant parameter here is the Rossby number $U/f\ell$, where U is the velocity scale, ℓ the length scale of the current, and f is the Coriolis parameter, and this number may be of the order of unity or less. In this case the gravitational propagation of the current in the downslope direction (corresponding to a balance between inertial and buoyancy forces) is usually arrested after some time. The final outcome is a convoluted migration path along the isobaths (corresponding to a balance between the Coriolis, inertial and buoyancy forces). These deductions concerning the flow path were obtained by Nof [12] by treating the gravity current as an isolated patch of dense and inviscid fluid situated on a parabolic bottom. The time-dependent nature of the buoyancy forces due to particle settling was totally ignored in Nof's calculations [12]. Including particle dynamics in the Grand Banks calculation would be a giant step towards an understanding of these important geological phenomena in which rotation cannot be neglected.

All of the models for transient turbidity currents that have to date been developed to explain either natural phenomena or the fixed-volume release experiments employ the assumption of dilute suspensions, that is, $\varphi << 1$ where φ is the spatially and temporally variable volume fraction of particles in suspension ([2], [5], [6], [10], [13-15]). This assumption may be valid for many situations of interest but certainly in the case of the catastrophic slump that was associated with the Grand Banks earthquake and other similar events (avalanches for example) this assumption would not be valid. For such events the current would be initiated at the highest possible volume fraction following fluidization of close-packed sediment [16].

In an interesting sequence of experiments Hallworth and Huppert [16] examined the changing properties of the particle deposits from turbidity currents as their initial volume fraction φ_0 was increased. Their experiments consisted in the release of fixed volumes of well-mixed suspensions in which the initial volume concentration of particles was gradually increased to value approaching the maximum that could be fluidized ($\varphi_0 \approx 0.45$). They found that very high-concentration flows behave quite differently from low-concentration flows and that their propagation is not accounted for by the dilute suspension models referenced above. The main new phenomenon uncovered is that suspensions with initial volume fractions larger than ≈ 0.3 propagate as rapidly moving, densely-compacted currents, but are marked by a pronounced and sudden arrest. The arrest becomes progressively closer to the release point as the initial volume fraction is increased. Turbulent mixing along the upper surface of the dense flow prior to its arrest produces a low-concentration overlying cloud which subsequently moves past the point of arrest depositing its load. The abrupt transition is reflected in the final areal density of deposit which displays an almost discountinuous step at the point of transition with mild smoothing due to deposition from the low-concentration cloud. From this set of experiments the data indicate that a critical value of the initial volume fraction of particles of $\varphi_0 \approx 0.275$ separates the different regimes of behaviour between low- and high-concentration currents. For particle-driven currents with initial volume fractions in the range $0 < \varphi_0 < 0.275$, the final run-out length of the sedimented deposit increases as φ_0 increases. Currents with initial volume fractions in the range $0.275 < \varphi_0 < 0.45$ undergo an abrupt arrest, and the final run out length of the sediment layer decreases as φ_0 increases.

The current theoretical models for time-dependent particulate flows that possibly can be used to model these lock release experiments all assume that the suspension is dilute ($\varphi << 1$), that the particles are vectically uniformly distributed, that the horizontal velocity field is depth-independent and that initial turbulence due to the mixing process that created these suspensions does not effect the subsequent flow and deposition of particles in any substantive fashion.

The use of the dilute suspension assumption for flows characterized as being of high concentration [16] is obviously going to lead to considerable error when attempting to predict the nature of any subsequent deposits. This dilute suspension assumption enables the Boussinesq approximation to be employed and when this has been coupled with the further assumption that the particles are vertically well-mixed by fluid turbulence has led some authors to arrive at a shallow-water theory for particulate flows [6]. It was, however, pointed out by Moodie *et al.* ([2], [10], [17]) that even for dilute particle-driven flows the derivation of shallow-water theory is problematic in that when particles drive the flow the horizontal velocity must of necessity acquire a vertical dependency thereby ruling out the shallow-water treatment.

The vast majority of the theoretical models, as mentioned above, rule out vertical stratification of sediment concentration. For high-concentration flows in marine environments this assumption is not a good one and particularly in the less turbulent waning stages of the flow when the deposits that form the oil reservoir units are laid down. Stratification also frequently occurs in a range of subaerial particle-laden gravity currents, such as pyroclastic flows [18], lahars [19], and snow avalanches [20]. Although there are key differences between subaerial and submarine gravity currents, the same fundamental physical principles apply and the same shortcomings are present in many of the theoretical models employed in their description.

Before we speculate on possible reasons for the behaviour of high-concentration flows as described by Hallworth and Huppert [16] we should perhaps examine the findings from some earlier related experiments by Middleton and Neal [21] which seem to have been overlooked by the above authors. In their experiments Middleton and Neal [21] examined deposits from turbidity currents produced in the laboratory by means of the classic fixed-volume lock release technique. They found that beds deposited in a horizontal channel by turbidity current surges released from a lock tend to be uniform in thickness; only in the distal part is there a rapid decrease in bed thickness. This uniform bed thickness they found is favoured by uniform grain size, high concentration, low settling velocity of grains (relative to turbidity current speed), and large lock height. They also found that for equal volumes of sediment, more highly concentrated suspensions tend not to travel as far as less concentrated suspensions and therefore, increasing concentration increases bed thickness more than it does bed length. These findings tend to agree roughly with the findings of Hallworth and Huppert [16] but because Middleton and Neal [21] did not carry out an exhaustive study of bed deposits for widely varying concentrations any more detailed comparisons of bed length versus concentration are not possible.

The experimentally obtained deposits of Hallworth and Huppert [16] do differ from those of Middleton and Neal [21] in one very distinctive way. For those measured by Hallworth and Huppert [16] and especially for those at the lower concentration levels there is a distinct proximal dip in deposit density with a local maximum occurring somewhat downstream from the point of release of the well mixed suspension. This is a characteristic that has been recorded in the majority of experimentally obtained deposits ([5], [11], [13]) and has been remarked upon by Moodie *et al.* [10] in their recent attempt to explain this persistent phenomenon. However, since few of the experimental papers actually record how the suspension behind the lock is created it is difficult to determine just what the conditions at the time of release were. If the careful experiments reported upon by Gladstone *et al.* [11] are an indication as to how these suspensions are created then one should presume that the two-phase mixture behind the lock will possess turbulent energy of mixing at the time of release. This would apparently also be the case with the experiments of Hallworth and Huppert [16] but just what conditions prevailed at the time of release in the experiments of Middleton and Neal [21] is not clear nor are these conditions specified in the earlier experiments of Middleton [22].

Moodie *et al.* [10] provided a possible explanation that could account for the proximal depression in the graphs of areal density of deposit laid down by the experimental turbidity currents. We described this mechanism briefly earlier in our introduction but because we will refer to it again in this introduction in connection with some speculations on the reasons for the geometry of deposits from high-concentration particulate flows we include more detail here.

Moodie *et al.* [10] argued that owing to the vigorous stirring that produces the well-mixed suspension behind the lock, the release volume will initially possess turbulent energy that is not present in an unstirred release volume such as those in the experimental studies of compositionally-driven flows [23]. This turbulent energy will be present even if, as in the experiments of Gladstone *et al.* [11], 'The stirrer was removed and $2s$ was allowed before current initiation ...'. Our preliminary experiments as well as our theoretical calculations suggest that decay of the turbulence happens over a much longer time scale than the $2s$ allowed for in [11]. Flows produced by unstirred releases may only acquire turbulent energy as a result of their motion after release through various instability mechanisms (Kelvin-Helmholtz, for example). The eddy motion of the imposed turbulence of mixing reduces the particle settling rates from the values that would be observed in an unstirred release volume possessing zero initial turbulent energy. The initial turbulent energy of mixing is taken into account by means of a modified ("effective") settling velocity that, over a time scale characteristic of turbulent energy decay, approaches the full Stokes settling velocity.

To determine the effective settling velocity of the particles in the suddenly released, well-mixed, fixed-volume dilute suspensions, Moodie *et al.* [10] accounted for the fact that the turbulence initially present will inhibit particle sinking by raising the mean particle potential energy at the expense of the mean turbulent kinetic energy in the suspending fluid. Because they dealt solely with dilute suspensions the back-reaction of the particles on the turbulent fluid was, at this point, ignored. This would obviously not be the case for the high-concentration suspensions of [16] and [21].

They argued that if the scale of energy injection by external stirring is much larger than that of the suspended particles, the particle size will lie well within the turbulent inertial range. They then used the transfer rate ϵ of turbulent energy density to smaller scales as a measure of the rate of energy transfer per unit volume to the suspended particles. The well-known self-similarity argument of Kolmogorov [24], which assumes that ϵ is independent of length scale within the inertial range, relates the turbulent energy spectrum $E(k)$, where k is the wavenumber magnitude, to ϵ by

$$E(k) \sim \epsilon^{2/3} k^{-5/3}. \tag{1.1}$$

It is then easily demonstrated from the incompressible Navier-Stokes equation [10] that the rate of decay of the total turbulent energy

$$E = \int E(k)dk = \frac{1}{2}\int |\boldsymbol{u}|^2 d\boldsymbol{x}, \tag{1.2}$$

is given by

$$\frac{\partial E}{\partial t} = -2\nu Z, \tag{1.3}$$

where ν is the kinematic viscosity and

$$Z = \frac{1}{2}\int |\nabla \times \boldsymbol{u}|^2 d\boldsymbol{x} \tag{1.4}$$

is the total enstrophy, or mean-squared vorticity. If it is now assumed, as in [10], that the turbulent spectrum is peaked at a characteristic wavenumber $\overline{k}$ then $Z \approx \overline{k}^2 E$ and

$$\frac{1}{E}\frac{\partial E}{\partial t} \approx -2\nu \overline{k}^2. \tag{1.5}$$

Moodie *et al.* [10] then argued that if the turbulence is strong enough, sedimentation will be inhibited until such time as the turbulent energy transfer to the particles balances the potential energy that would be lost because of particle sinking. In typical experimental situations, this balance is achieved well before the lock is opened. Moodie *et al.* [10] restricted their attention to this case. Then in a time τ the total turbulent energy transfer to the particles per unit volume, $\epsilon\tau$, must balance the buoyant potential energy differential. That is,

$$\epsilon\tau = (\rho_3 - \rho_2)g\zeta, \tag{1.6}$$

where ρ_3 is the density of the particles, ρ_2 the density of the suspending fluid, and ζ is the mean physical height difference (relative to Stokes settling) induced by the turbulent interaction. Consequently, the reduction in the settling velocity $v_t = \zeta/\tau$ induced by the turbulence is proportional to ϵ. The departure of the settling velocity from the Stokes value v_s will thus decrease exponentially at the rate δ_*, where

$$\delta_* = -\frac{1}{v_t}\frac{\partial v_t}{\partial t} = -\frac{1}{\epsilon}\frac{\partial \epsilon}{\partial t} = -\frac{3}{2}\frac{1}{E}\frac{\partial E}{\partial t} \approx 3\nu\overline{k}^2. \tag{1.7}$$

If, as assumed by Moodie *et al.* [10], sedimentation just begins to occur as the lock is opened at $t = 0$, the effective settling velocity will then decay exponentially to the Stokes value v_s according to

$$v_{\text{eff}} = v_s - v_t = v_s(1 - e^{-\delta_* t}). \tag{1.8}$$

When nondimensional variables were employed in their calculations, Moodie *et al.* [10] found for the nondimensional decay rate $\widetilde{\delta}_*$ given by

$$\widetilde{\delta}_* = \delta_*\beta H/v_s = \frac{\delta_*}{\delta}\sqrt{\frac{H}{g_1'}},$$

where $\beta = v_s/U\delta$ and $\delta = H/L$ is the aspect ratio of the flow, values in the range $0.06 - 0.14$. This finding was based on the values $\nu \approx 0.01\,\text{cm}^2/s$, $\delta = 0.1$, and β in the range 0.05 to 0.5. These values are typical of values reported in experimental work where particle size varies from 23 to $200\mu m$ and the suspending fluid is water [5]. Using this expression for the effective settling velocity, with realistic values for the decay rate, enabled Moodie *et al.* [10] to obtain patterns for particle deposition that closely resembled the lenticular deposits of the lock release experiments [11].

All of the above discussion leading to the derivation of an expression for the effective settling velocity of a particle in a turbulent field is based upon the fact that the suspension is dilute. This allows the particles to be treated as being isolated. This assumption, in turn, puts definite constraints on the range for volume fractions to which this theory strictly applies. It is known [25] that when the mean volume fraction of particles in suspension is equal to 0.01, the mean free distance between the grains is about four diameters, and the probability of collisions is small. On the other hand, at a value of 0.09 for the volume fraction, the free distance is equal to the grain diameter, and the probability becomes a certainty. The dilute suspension models referred to above are for volume fractions $\varphi_0 \approx 0.01 - 0.05$. These values are typical of those often found in nature [26] although the concentration of large natural turbidity currents is still essentially unknown [21].

Although the approach adopted by Moodie *et al.* [10] is capable of reproducing the main features of the lenticular deposits seen in both experiments and natural settings [26] and is in many ways superior to other models, it is still far from ideal in that it, like the majority of models, does not provide for a thorough understanding of the dynamical interplay between the inertial, drag, gravitational, and other forces acting on the particles and the small-scale vortical structures of the turbulent flow-field. This interplay between particles and fluid turbulence has been examined in the ideal context of heavy particles in a Burgers vortex [27]. This recent work was presumably motivated by the suggestion that much of the dynamics of turbulence can be understood by regarding it as a collection of Burgers vortices [28]. This approach may contribute somewhat to basic understanding but is limited to the dilute regime in which both the particles' effect on the fluid motion, as well as the particle-particle interaction is neglected. It thus has all of the drawbacks that the more phenomenological approaches have $(\varphi << 1)$ but with much less ability to relate to actual large scale turbidity currents.

A fundamental shortcoming of all existing models for turbidity currents is their inability to account for the behaviour observed when volume fractions of particles become high [16]. To the best of our knowledge the deposits from most dilute turbidity currents tend to be lenticular in their streamwise profile. This shape can be recovered by replacing the usual constant Stokes settling velocity by a time-dependent effective settling velocity that approaches the Stokes value from below. The time scale with which the effective settling velocity approaches the Stokes value is the scale provided by the rate of decay of the initial turbulent energy present due to mixing of the suspension [10]. The vortices associated with this initial turbulence prevent or slow the settling of particles and so contribute to the overall shape of the deposit. That it is this initial turbulence of mixing that plays the major role in creating the lenticular shape of the deposits for dilute suspensions seems clear (at least to these authors). We will speculate on its role in high-concentration flows.

If we were to suppose that the initial turbulence, generated when the heavy particles were mixed behind the lock was the major contributing mechanism inhibiting particle settling from its Stokes rate then raising initial volume fractions of particles would require a larger value for the rate of turbulent energy transfer to the particles in order to balance the buoyant potential energy differential. As the volume fraction is raised the available turbulent energy would be depleted at greater and greater rates. The depletion of this energy at shorter and shorter elapsed times in the post-release phase of the flow could account for the fact that the deposits become shorter and shorter with increasing volume fractions in the high-concentration range.

At the beginning of this introduction we stated that gravity currents involve the flow of one fluid within another when the flow is driven by buoyancy forces arising from 'small density differences' [1]. In certain applications concerned with safety (dense-gas releases, for example) the density ratio across the front can be rather high, and the usual relation for the front velocity [29] appears to give unreasonable results. On denoting the densities of the heavy and light fluids with indices 2 and 1 respectively, the thickness of the intruding layer h, and the acceleration due to gravity g, this relation is usually specified as

$$u_F = k\left(gh\,\frac{\rho_2 - \rho_1}{\rho_1}\right)^{1/2} \quad \text{or} \quad u_F = k(g'h)^{1/2} \tag{1.9}$$

using common nomenclature with u_F representing the front speed. The controversial point, and the one that Middleton avoids [1], relates to the use of ρ_1 in the denominator for flows having high density ratios across the front. For most natural flows and particularly those involving liquids the density difference is so small that it does not matter much [30]. However, there will be situations, even for compositionally-driven flows (dense-gas releases), where a density ratio of 2 to 5 would arise. In the instance of high-concentration turbidity currents the density ratios could also be in this range [16]. In that case the use of g' or the alternative g'' defined by [16]

$$g'' = g\,\frac{\rho_2 - \rho_1}{\rho_2}\,, \tag{1.10}$$

leads to quite different results. It should be noted that the use of g' here can lead to an acceleration exceeding g for high (but not unusual) density ratios; clearly an unreasonable result [31].

Given the known uncertainty in predicting heavy fluid fronts for all but low to moderate density ratios [30] we shall restrict our attention to the Boussinesq case where these ratios are close to unity. Many of our intended applications fall into this regime and the vast majority of the published literature is also for models employing the Boussinesq approximation.

We shall proceed by first surveying some of the pertinent results for compositionally-driven flows as these relate to the particle-driven flows of primary interest here. We explain the underlying assumptions in this context, introduce some of the terminology, point out possible applications, and highlight some of the shortcomings and possible ways in which they may be overcome. In particular, we examine the conditions that give rise to internal bores as these bores will have an influence on particle deposition patterns and they are known to occur naturally in the geophysical context [32]. We then critically examine some of the assumptions for particle-driven flows, develop model sets of equations for such flows in both deep and shallow surroundings, and present some new findings on vorticity generation by particulate mass in turbidity currents.

2. Fundamentals of Compositionally-Driven Flows

In this section we look at some aspects of compositional gravity currents that relate to both their compositionally- and particle-driven manifestations.

2.1. Introductory remarks

The majority of the theoretical work on compositionally-driven gravity currents from the very early calculation by von Kármán [33] up to that done by Benjamin [29] and right on through into the mid 1980s treated the gravity current as steady, existing in either an inertia-buoyancy or, at later stages in the flow, in a viscous-buoyancy balance and finally, in the very late stages of the flow, in a viscous-surface tension balance [34]. Notable exceptions to this approach were the works of Hoult [35] and Fannelop and Waldman [36], in which governing equations were established and subsequently solved rather than seeking time-independent force balances. Hoult's [35] solution was obtained in terms of a similarity variable and so the derived relationships are valid only after sufficient time has elapsed, since release of the fixed volume of fluid, for the initial geometry of that volume to be irrelevant. For particle-driven gravity currents from which particles are sedimented, a similarity solution does not exist and most authors have resorted to purely numerical solutions of approximate model equations.

Further in the realm of time-dependent homogeneous gravity flow studies, with relevance to particulate flows, is the work of Rottman and Simpson [23] on instantaneous releases of heavy bottom gravity flows. In a series of detailed experiments they studied instantaneous releases for $0 < h_0/H \leq 1$, where h_0 is the initial depth of the released heavy fluid and H the total depth of fluid in the rectangular channel. In their work they concentrated on the important issue of bore formation in the flow's transition to the self-similar phase. The key feature of their observations was that for h_0/H equal to or slightly less than unity $(0.7 \leq h_0/H \leq 1.0)$, the disturbance generated at the end wall has the appearance of an internal hydraulic drop whereas for smaller values of h_0/H ($\tilde{<} 0.7$) it is a long wave of depression. These experiments served to emphasize the importance of including the effect of the ambient fluid on the gravity current when the depth of that current is comparable to the depth of the over-lying fluid. As a possible framework for discussing these experiments they solved an initial value problem for the two-layer, shallow-water equations using the method of characteristics. Their interpretation of the experimental results in terms of the numerical simulations gives, at best, a qualitative understanding for $0 < h_0/H < 1/2$. However, for $h_0/H > 1/2$ there is little resemblance between their numerical simulations and the experimentally observed transition of the flow, through the mechanism of the internal bore, to the self-similar phase. D'Alessio *et al.* [37] employed a two-layer shallow-water model to study sudden releases for fixed volumes entering a lighter ambient fluid as a bottom gravity flow. Using MacCormack's method [38] to integrate numerically the hyperbolic system they were able to achieve very good qualitative agreement with the experimental results of Rottman and Simpson [23] for transition to self-similarity. Also, employing multiple scales arguments they were able to show analytically the dependence of internal bore formation on initial fractional depth of the release volume. The experimentally observed slowing down of the front as the internal bore overtakes it is captured by their model. Grundy and Rottman [39], in their study of the approach to self-similarity for solutions to the shallow-water equations for sudden releases of fixed volumes, employed a set of model equations that ignored inertial effects due to the presence of the lighter ambient fluid. Their subsequent numerical studies for plane flow show little resemblance to the observed flows of Rottman and Simpson [23]. Bonnecaze *et al.* [13] employing a two-step Lax-Wendroff scheme to solve the problem of sudden release of a fixed volume of homogeneous heavy fluid on a horizontal bottom achieved very good agreement with the experiments of Rottman and Simpson [23] when they took into account the inertial effects of the ambient fluid.

It should be noted here that all the theoretical and numerical work on compositionally-driven gravity flows is for a rectangular geometry. This is an obvious drawback when one wishes to employ the existing models in applications as other geometries figure prominently in many of these. For example, the cylindrical geometry of pipelines adds considerable complexity when evaluating the effects of gravity on differing densities between contiguous batch shipments of liquid hydrocarbons. The stability of these interfaces has not been treated and the induced swirl velocity that would be present in this geometry has defied attempts at analytical treatment even though the issues raised in connection with "mixing lengths" are of great economic importance to the pipeline industry [40].

The scope for reseach in gravity driven flows in geometries other than rectangular is indeed great as we are not aware of any such work in the published literature. We do not consider the simple extension to axisymmetric flows in unbounded regions [14] as constituting a true extension to the flows in rectangular geometries as these still involve but two coordinates – one for depth and one for lateral extent.

2.2. $1\frac{1}{2}$-and 2-layer models

Consider a gravity current produced by the sudden release of a heavy fluid of density ρ_2 into an ambient fluid of lesser density ρ_1 overlying a horizontal impervious bottom. The physical configuration is depicted in Figure 1, where $\eta(x,t)$ represents the displacement of the free surface from its undisturbed configuration, $\boldsymbol{u} = (u, w)$ is the fluid velocity in Cartesian coordinates with position vector $\boldsymbol{x} = (x, z)$, H is the total undisturbed depth of fluid, and $h(x,t)$ is the variable thickness of the bottom gravity current. The flow is driven by the buoyancy forces arising because of the difference between the density ρ_2 of the heavy fluid initially behind the lock and the density of the lighter ambient fluid.

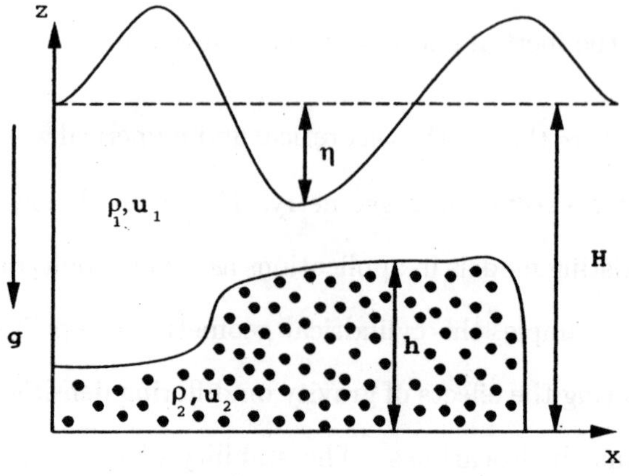

Figure 1: Geometry of the two-layer model

In our analysis of the unsteady flow of bottom gravity surges we assume, according to standard practice, that the Reynolds numbers are sufficiently large that we may neglect viscous forces in the flow dynamics [34]. Flow dynamics are then dominated by the interplay between buoyancy and inertial forces. The equations for the two layers are then written as

$$\frac{\partial u_1}{\partial x} + \frac{\partial w_1}{\partial z} = 0, \tag{2.1}$$

$$\frac{\partial u_1}{\partial t} + u_1 \frac{\partial u_1}{\partial x} + w_1 \frac{\partial u_1}{\partial z} = -\frac{1}{\rho_1}\frac{\partial p_1^*}{\partial x}, \tag{2.2}$$

$$\frac{\partial w_1}{\partial t} + u_1 \frac{\partial w_1}{\partial x} + w_1 \frac{\partial w_1}{\partial z} = -\frac{1}{\rho_1}\frac{\partial p_1^*}{\partial z}, \tag{2.3}$$

$$\frac{\partial u_2}{\partial x} + \frac{\partial w_2}{\partial z} = 0, \tag{2.4}$$

$$\frac{\partial u_2}{\partial t} + u_2 \frac{\partial u_2}{\partial x} + w_2 \frac{\partial u_2}{\partial z} = -\frac{1}{\rho_2}\frac{\partial p_2^*}{\partial x}, \tag{2.5}$$

$$\frac{\partial w_2}{\partial t} + u_2 \frac{\partial w_2}{\partial x} + w_2 \frac{\partial w_2}{\partial z} = -\frac{1}{\rho_2}\frac{\partial p_2^*}{\partial x}, \tag{2.6}$$

where the total pressure field in each layer has been written as

$$p_1 = -\rho_1 g z + p_1^*, \tag{2.7}$$

$$p_2 = \rho_1 g H - \rho_2 g z + p_2^*, \tag{2.8}$$

with p_i^* $(i = 1, 2)$ the dynamic pressure field in each layer. Here (2.1) and (2.4) are the incompressible continuity equations and the others the momentum equations. These equations must be combined with the boundary conditions for the free surface, that is,

$$p_1^*(x, H + \eta, t) = \rho_1 g(H + \eta), \tag{2.9}$$

$$w_1(x, H + \eta, t) = \frac{\partial \eta}{\partial t} + u_1(x, H + \eta, t)\frac{\partial \eta}{\partial x}, \tag{2.10}$$

the boundary conditions for the interface between the two layers,

$$p_2^*(x, h, t) = p_1^*(x, h, t) + \rho_2 g h - \rho_1 g(H + h), \tag{2.11}$$

$$w_i(x, h, t) = \frac{\partial h}{\partial t} + u_i \frac{\partial h}{\partial x} \quad (i = 1, 2), \tag{2.12}$$

and the kinematic bottom boundary condition

$$w_2(x, 0, t) = 0. \tag{2.13}$$

We now focus on all of the above equations representing the dynamics of the two-layer system with the goal of developing a set of model equations describing low-aspect ratio flows in each layer which can, by limiting processes, be made to describe both the cases of shallow and deep ambient fluids.

Although it is clear that the initial flow following release of the fixed volume of heavy fluid is a fully three-dimensional unsteady flow, soon after release, the current will have spread sufficiently that its length is very much greater than its height. The height will, at this stage, be slowly varying over the horizontal position x and in time t. To exploit this low-aspect ratio, slowly varying nature of the flow, we employ the horizontal and vertical length scales L and h_0, respectively, and introduce the small parameter $\delta \equiv h_0/L$ known as the aspect ratio [41].

We now introduce nondimensional variables according to the following scheme, wherein nondimensional variables are indicated by a tilde:

$$\left.\begin{array}{l} x = L\widetilde{x}, \quad z = h_0\widetilde{z}, \quad t = \frac{L}{U}\,\widetilde{t}, \quad h = h_0\widetilde{h}, \\ (u_1, u_2) = U(\widetilde{u}_1, \widetilde{u}_2), \qquad (w_1,\, w_2) = h_0\,\frac{U}{L}\,(\widetilde{w}_1, \widetilde{w}_2), \\ (p_1^*, p_2^*) = U^2(\rho_1\widetilde{p}_1^{\,*}, \rho_2\widetilde{p}_2^{\,*}), \qquad \eta = \frac{U^2}{g}\,\widetilde{\eta}, \\ g_1' \equiv \frac{\rho_2-\rho_1}{\rho_2}\,g, \quad U \equiv (g_1'h_0)^{1/2}, \quad H = h_0\widetilde{H}. \end{array}\right\} \tag{2.14}$$

In (2.14), g_1' is a reduced gravity and the vertical length scale h_0 is determined by the initial height of the release volume.

Employing this scheme to nondimensionalize all of our equations gives the following set

$$\frac{\partial u_1}{\partial x} + \frac{\partial w_1}{\partial z} = 0, \tag{2.15}$$

$$\frac{\partial u_1}{\partial t} + u_1\,\frac{\partial u_1}{\partial x} + w_1\,\frac{\partial u_1}{\partial z} = -\,\frac{\partial p_1^*}{\partial x}, \tag{2.16}$$

$$\delta^2\left(\frac{\partial w_1}{\partial t} + u_1\,\frac{\partial w_1}{\partial x} + w_1\,\frac{\partial w_1}{\partial z}\right) = -\,\frac{\partial p_1^*}{\partial z}, \tag{2.17}$$

$$\frac{\partial u_2}{\partial x} + \frac{\partial w_2}{\partial z} = 0, \tag{2.18}$$

$$\frac{\partial u_2}{\partial t} + u_2\,\frac{\partial u_2}{\partial x} + w_2\,\frac{\partial u_2}{\partial z} = -\,\frac{\partial p_2^*}{\partial x}, \tag{2.19}$$

$$\delta^2\left(\frac{\partial w_2}{\partial t} + u_2\,\frac{\partial w_2}{\partial x} + w_2\,\frac{\partial w_2}{\partial z}\right) = -\,\frac{\partial p_2^*}{\partial z}, \tag{2.20}$$

$$\gamma p_1^*(x, 1+\gamma\eta, t) = 1 + \gamma\eta, \tag{2.21}$$

$$w_1(x, 1+\gamma\eta, t) = \gamma\,\frac{\partial \eta}{\partial t} + \gamma u_1(x, 1+\gamma\eta, t)\,\frac{\partial \eta}{\partial x}, \tag{2.22}$$

$$\gamma p_2^*(x, h, t) = \gamma\,\frac{\rho_1}{\rho_2}\,p_1^*(x, h, t) + h - \frac{\rho_1}{\rho_2}\,(1+h), \tag{2.23}$$

$$w_i(x, h, t) = \frac{\partial h}{\partial t} + u_i\,\frac{\partial h}{\partial x} \quad (i = 1, 2), \tag{2.24}$$

$$w_2(x,0,t) = 0. \tag{2.25}$$

In the above equations and boundary conditions, tildes have been dropped from nondimensional quantities for notational convenience and $\gamma \equiv g_1'/g$ is a nondimensional parameter related to the importance of the free surface in the sense that putting it to zero filters free surface gravity waves out of the model thereby focusing attention on the internal wave processes.

From (2.17) and (2.20) we see that to $O(\delta^2)$ the pressure in both layers is hydrostatic and, as a consequence, the horizontal velocities u_1 in the upper layer and u_2 in the lower are independent of depth [41]. Using this consequence of the low-aspect nature of these flows we find, upon satisfying the dynamic and kinematic boundary conditions, that the nondimensional equations describing the dynamics of this two-layer coupled system can be written as [42]

$$\frac{\partial u_1}{\partial t} + \frac{\partial}{\partial x}\left[\frac{1}{2}u_1^2 + \eta\right] = 0, \tag{2.26}$$

$$\frac{\partial}{\partial t}(h - \gamma\eta) + \frac{\partial}{\partial x}\left[u_1(h - H - \gamma\eta)\right] = 0, \tag{2.27}$$

$$\frac{\partial u_2}{\partial t} + \frac{\partial}{\partial x}\left[\frac{1}{2}u_2^2 + h + \eta - \gamma\eta\right] = 0, \tag{2.28}$$

$$\frac{\partial h}{\partial t} + \frac{\partial}{\partial x}(hu_2) = 0. \tag{2.29}$$

The 2-layer shallow water equations (2.26) - (2.29) can now be specialized to the case of a deep upper layer ($1\frac{1}{2}$-layer) by letting $H \to \infty$ in this set to get from mass conservation that

$$\frac{\partial u_1}{\partial x} = 0 \tag{2.30}$$

or that u_1 is a function of t alone. Then, because $u_1 = 0$ at $x = 0$ we have $u_1 \equiv 0$ and

$$\frac{\partial \eta}{\partial x} = 0. \tag{2.31}$$

The equations for a deep upper layer then reduce to

$$\frac{\partial u_2}{\partial t} + \frac{\partial}{\partial x}\left[\frac{1}{2}u_2^2 + h\right] = 0, \tag{2.32}$$

$$\frac{\partial h}{\partial t} + \frac{\partial}{\partial x}(hu_2) = 0. \tag{2.33}$$

These equations are the same as those for the single layer shallow water equations but when written in dimensional variables will have the usual g replaced by g_1'. This is termed the "$1\frac{1}{2}$-layer model" approximation whereas the model specified by (2.26) - (2.29) is the "2-layer model" approximation [43].

2.3. Gravity-driven flows in shallow and deep surroundings

So that we may compare some of the salient features of both compositionally-driven and particle-driven flows we shall here examine numerically the former in some detail. We will pay particular attention to the role played by the inertia of the ambient fluid in bore generation and the limiting properties of the 2-layer model as the parameter $H \to \infty$.

We obtain numerical solutions to the initial boundary value problems for both the 2-layer and $1\frac{1}{2}$-layer models corresponding to the sudden release of a fixed volume of heavy fluid. For the 2-layer model the initial values may be expressed as

$$\left.\begin{array}{l} u_1(x,0)=0, \quad u_2(x,0)=0, \quad \eta(x,0)=0, \\ h(x,0)=\begin{cases} 1, & x \le 1, \\ 0, & x > 1, \end{cases} \end{array}\right\} \tag{2.34}$$

whereas the boundary conditions that prevent flow across the end wall at $x = 0$ are

$$u_1(0,t) = 0, \quad u_2(0,t) = 0. \tag{2.35}$$

For the $1\frac{1}{2}$-layer model the conditions are similar except that they do not involve either u_1 or η.

MacCormack's numerical procedure [38] is employed here and has been described and used on several occasions by the current authors to examine buoyancy-driven flows ([2], [10],

[37]). This method is an explicit, conservative finite difference scheme possessing second-order accuracy. Because the scheme is conservative, convergence will be to a physical weak solution of any hyperbolic system of equations. Furthermore, MacCormack's method provides sharp resolution of shocks and does not require the evaluation of the Jacobian of the flux vector. One drawback is the occurrence of nonphysical oscillations around the shock. An efficient strategy to dampen these spurious oscillations is to introduce artificial viscosity. Because adding the necessary artificial viscosity reduces the accuracy to first-order, Harten [44] proposed using an artificial viscosity term that is solution-dependent in such a way as to add significant artificial viscosity only around discontinuities. The resulting scheme then remains second-order accurate where the solution is smooth and is first-order accurate only near discontinuities.

Figure 2 displays a plot of the numerically determined gravity current depth h based on the 2-layer model for the sudden release of a fixed volume of heavy fluid when the total undisturbed fluid depth is $H = 1.2$. The time evolution of the current and the development of the internal bore are clearly displayed. This internal bore develops as a result of the vigorous reverse flow in the lighter ambient fluid. This reverse flow will be absent when we consider the sudden release problem for the $1\frac{1}{2}$-layer model because of the very large body of ambient fluid $(H \to \infty)$ which remains dynamically passive.

In Figure 3 below we have plotted the gravity current depth h versus x for the fixed time $t = 3$ and for various values of the parameter H. This plot is based on the 2-layer model for the fixed volume release specified by (2.34) and (2.35). The results displayed in Figure 3 show how the upper fluid becomes more passive as the total depth of the 2-layer model is increased while the depth of the release volume remains fixed at $h = 1.0$. By increasing H the reverse flow in the ambient fluid is diminished, through momentum conservation, and the internal bore will cease to be initiated.

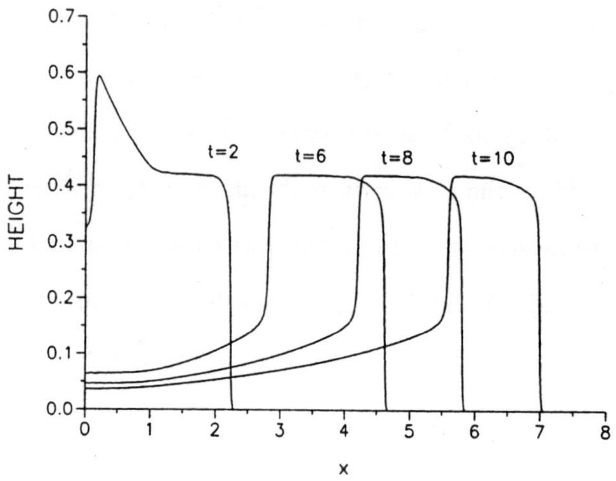

Figure 2: Plot of h versus x for $t = 2, 6, 8$ and 10 with $H = 1.2$ and $\gamma = 0.05$.

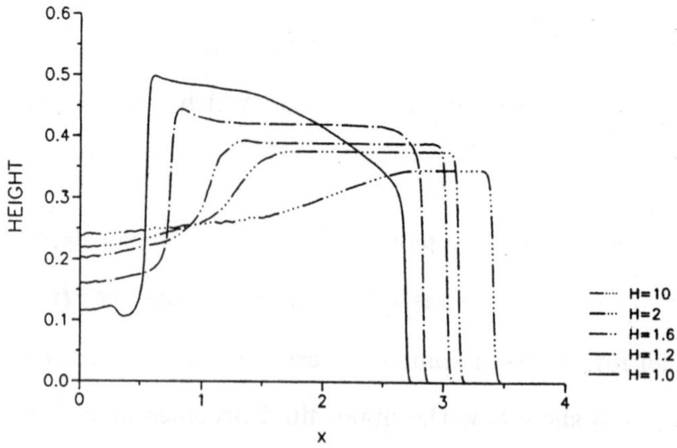

Figure 3: Plot of h versus x for $t = 3$ and $H = 1.0, 1.2, 1.6, 2.0, 10$ and $\gamma = 0.05$

Figure 4 shows the numerical solution for the gravity current produced by the sudden release of a fixed volume whose evolution is governed by the $1\frac{1}{2}$-layer model described by the pair of equations (2.32) and (2.33). We see here that no internal bore is formed. It can be shown [37] that the pair of equations describing the $1\frac{1}{2}$-layer model possess similarity solutions of the form

$$u_2(x,t) = t^{-1/3}v(\xi), \tag{2.36}$$

$$h(x,t) = t^{-2/3}f(\xi), \tag{2.37}$$

where

$$v(\xi) = \frac{2}{3}\,\xi, \tag{2.38}$$

$$f(\xi) = \frac{1}{9}\,(\xi^2 + \xi_1^2), \tag{2.39}$$

$$\xi = xt^{-2/3}, \tag{2.40}$$

and

$$\xi_1 = 3\big(1/4\big)^{1/3}. \tag{2.41}$$

The shock front location x_f is given by

$$x_f(t) = 3(1/4)^{1/3}t^{2/3} \tag{2.42}$$

beyond which u_2 and h are zero. These solutions can be interpreted as the leading terms of an asymptotic expansion for large t and used as a check on any numerical schemes.

We have demonstrated in this section that the observed internal bore associated with fixed-volume releases in the laboratory is a result of vigorous reverse flow in the ambient fluid overlying the dense underflow. In the next section we shall examine the possible role that is played by this bore in particle-driven flows and their depositional patterns as well as the role of particles in generating shear and vorticity in both 2- and $1\frac{1}{2}$-layer models of these flows.

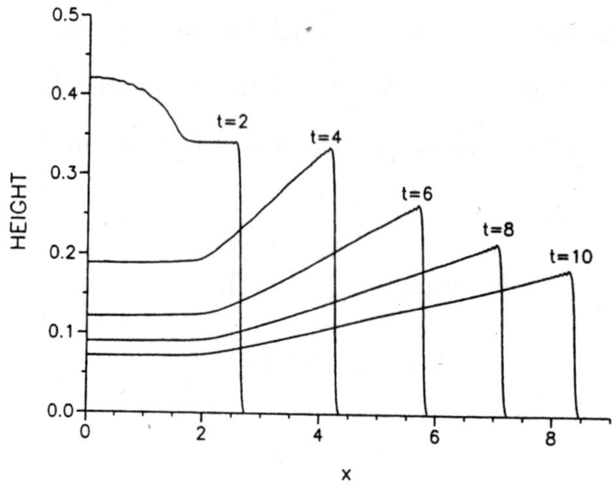

Figure 4: Plot of h versus x for the $1\frac{1}{2}$-layer model with $t = 2, 4, 6, 8, 10$ and $\gamma = 0.05$.

3. Particle-Driven Gravity Currents

We now develop models for particle-driven gravity currents in both shallow and deep surroundings. In order to do so, several simplifying assumptions will have to be made. These we will state clearly. These models apply to small scale relatively fast moving flows in rectangular geometry and so their application is principally to experiments of the nature of the fixed volume release experiments of Gladstone *et al.* [11]. Whether or not these experiments apply to naturally occurring turbidity currents is debatable. There are many problems in devising scale models of turbidity currents. In order to produce a bed of measurable thickness, relatively coarse sediment must be used. The ratio of sediment size to turbidity current size is thus much larger than in nature. It is also necessary to use relatively high concentrations of suspended sediment and the concentration of large natural turbidity currents is probably less in many cases than the concentrations used in the laboratory. Both coarse size and high concentration of sediment lead to rates of deposition that are probably much higher than in nature. However, we feel that the basic qualitative properties of these naturally occurring turbidity currents are captured by the laboratory models and hence our interest in them.

3.1. Model development

The physical configuration is as depicted in Figure 1 but with the lower fluid layer comprising a suspension of particles of density ρ_3 $(\rho_3 > \rho_2 > \rho_1)$. The flow is driven by the buoyancy forces arising because of the difference between the bulk density $\rho(\varphi)$ of the suspension and the density ρ_1 of the ambient fluid. The density of the suspension is the local volume average of the particle density and the density ρ_2 of the interstitial fluid and is given by

$$\rho(\varphi) = \rho_3\varphi + (1-\varphi)\rho_2. \tag{3.1}$$

In all our development we assume that the suspension is dilute and that the particles are vertically well mixed so that the volume fraction $\varphi = \varphi(x,t) << 1$. We assume, further, as in the compositionally-driven flows, that the Reynolds numbers are sufficiently large that we can neglect viscous forces in the flow dynamics. The flow dynamics are then dominated by the interplay between buoyancy and inertial forces. For dilute suspensions, all terms proportional to φ in the lower layer momentum equations, with the exception of the term containing the gravitational acceleration, can be discarded to obtain a Boussinesq approximation to these equations. This gives, in dimensional variables,

$$\frac{\partial u_2}{\partial t} + u_2\,\frac{\partial u_2}{\partial x} + w_2\,\frac{\partial u_2}{\partial z} = -\,\frac{1}{\rho_2}\,\frac{\partial p_2^*}{\partial x}\,, \tag{3.2}$$

$$\frac{\partial w_2}{\partial t} + u_2\,\frac{\partial w_2}{\partial x} + w_2\,\frac{\partial w_2}{\partial z} = -\,\frac{1}{\rho_2}\,\frac{\partial p_2^*}{\partial z} - \varphi g\,\frac{(\rho_3-\rho_2)}{\rho_2}\,, \tag{3.3}$$

where the total pressure field has been transformed as in (2.8).

We also require an equation describing the conservation of particles in the lower layer. We assume here, as has been assumed in most of the literature dealing with time-dependent suspension flows, that the volume fraction of particles φ is a function of x and t only with no vertical dependence. In our previous work [10], as mentioned earlier, we accounted for vertical structure in the concentration of particles by specifying a shape function for φ to provide this structure in accord with current understanding concerning the mechanics of dense underflows [45]. In such flows, turbulence will be most intense in the wall region at the bottom of the flow and relatively coarse and on the verge of collapse at the top of the flow. This shape function was then chosen to provide for an increased density of particles toward the bottom of the flow. It was found that the computed long-time areal density of deposits were fairly insensitive to the choice of shape function but very much dependent on the initial turbulent energy of mixing present in the suspension that constitutes the fixed volume release. This observation, that the areal density is relatively insensitive to the vertical distribution of particles in the flow, seems to agree with an earlier observation by Martin and Nokes [46] for crystal settling from turbulently convecting magmas that led to the vertically well-mixed assumption of previous authors ([3], [5], [13]). Martin and Nokes [46] found that even when the turbulent motions were such that the convective velocities fell to the range for the Stokes settling velocity and, hence, that particles were concentrated toward the base of the flow, the rate at which particle concentrations decayed varied little. They found that the temporal decay constant scales as the ratio of the Stokes settling velocity to the depth of the suspension and is relatively insensitive to the ratio of settling to convective velocities. This apparent insensitivity of the settling rate to the turbulence in the flow regime but strong dependence on the initial mixing may be related to the scale of the turbulence and orientation of the associated eddys [47].

In deriving this particle equation, we further assert that the particles leave the current only through the viscous sublayer at the base of the flow and with a flux $-v_{\text{eff}}\varphi$, where v_{eff} denotes the effective settling velocity of an isolated particle. The expression for v_{eff} was derived earlier and is given by (1.8).

The above assumptions concerning the suspended particles combined with conservation of particles applied to an arbitrary segment of the flow regime leads directly to the particle equation

$$\frac{\partial}{\partial t}(h\varphi) + \frac{\partial}{\partial x}\left[\varphi \int_0^{h(x,t)} u_2(x,z,t)dz\right] + \varphi v_{\text{eff}} = 0. \tag{3.4}$$

Our governing equations now consist of the usual continuity and momentum equations for the homogeneous upper layer together with (3.2), (3.3), (3.4), and

$$\frac{\partial u_2}{\partial x} + \frac{\partial w_2}{\partial z} = 0 \tag{3.5}$$

for the lower layer. The usual dynamic and kinematic boundary conditions hold at the free surface, the bottom, and across the interface between the two layers.

We now focus on all the equations with the goal of developing a set of model equations describing low-aspect ratio flows of dilute particulate suspensions in both shallow and deep surroundings. In other words we will develop 2-layer and $1\frac{1}{2}$-layer model approximations for particle-driven flows. We exploit the low-aspect ratio, slowly varying nature of these flows by employing the previously introduced parameter $0 < \delta << 1$. The dilute nature of the suspension in each case is characterized by the parameter $\varepsilon \equiv \varphi_0 << 1$ representing the initially spatially uniform volume fraction of particles in the release volume.

We now introduce nondimensional and scaled variables according to the scheme in (2.14) with the addition of

$$\varphi = \varepsilon\widetilde{\varphi}, \quad v_{\text{eff}} = \delta\widetilde{v}_{\text{eff}}, \quad g_2' = \frac{\rho_3 - \rho_2}{\rho_2}\, g, \tag{3.6}$$

where g_2' is a second reduced gravity, and the scaled quantities $\widetilde{\varphi}$ and $\widetilde{v}_{\text{eff}}$ are now $O(1)$ in both ε and δ.

Employing the scheme of (2.14) and (3.6) to render nondimensional all of the fluid equations, kinematic and dynamic boundary conditions, as well as the particle equation gives the set of equations

$$\frac{\partial u_i}{\partial x} + \frac{\partial w_i}{\partial z} = 0, \quad i = 1, 2, \tag{3.7}$$

$$\frac{\partial u_i}{\partial t} + u_i \frac{\partial u_i}{\partial x} + w_i \frac{\partial u_i}{\partial z} = - \frac{\partial p_i^*}{\partial x}, \quad i = 1, 2, \tag{3.8}$$

$$\delta^2 \Big(\frac{\partial w_1}{\partial t} + u_1 \frac{\partial w_1}{\partial x} + w_1 \frac{\partial w_1}{\partial z} \Big) = - \frac{\partial p_1^*}{\partial z}, \tag{3.9}$$

$$\delta^2 \Big(\frac{\partial w_2}{\partial t} + u_2 \frac{\partial w_2}{\partial x} + w_2 \frac{\partial w_2}{\partial z} \Big) = - \frac{\partial p_2^*}{\partial z} - \alpha \varphi, \tag{3.10}$$

$$\frac{\partial}{\partial t} (h\varphi) + \frac{\partial}{\partial x} \Big(\varphi \int_0^h u_2 dz \Big) + \varphi \frac{v_{\text{eff}}}{U} = 0, \tag{3.11}$$

$$\gamma p_1^*(x, H + \gamma \eta, t) = H + \gamma \eta, \tag{3.12}$$

$$w_1(x, H + \gamma \eta, t) = \gamma \frac{\partial \eta}{\partial t} + \gamma u_1(x, H + \gamma \eta, t) \frac{\partial \eta}{\partial x}, \tag{3.13}$$

$$\gamma p_2^*(x, h, t) = \gamma \frac{\rho_1}{\rho_2} p_1^*(x, h, t) + h - \frac{\rho_1}{\rho_2} (H + h), \tag{3.14}$$

$$w_i(x, h, t) = \frac{\partial h}{\partial t} + u_i(x, h, t) \frac{\partial h}{\partial x}, \quad i = 1, 2, \tag{3.15}$$

$$w_2(x, 0, t) = 0. \tag{3.16}$$

In the above equations the tildes have again been dropped from nondimensional and scaled quantities, $\gamma = g_1'/g$ as previously defined, and $\alpha \equiv \varepsilon g_2'/g_1'$.

Our system consisting of equations (3.7) - (3.16) involves the five nondimensional parameters $\gamma, \delta^2, \alpha, \beta = v_s/U\delta$, and H, where v_s is the Stokes settling velocity which, as we have seen, appears in the expression for v_{eff}. The parameter γ is the square of the ratio of the internal gravity wave speed to the surface gravity wave speed so that vanishingly small values of γ will filter out surface gravity waves from the model and so focus on the slower internal wave processes. The parameter δ^2 is a measure of the importance of vertical structure in the flow field for when streamlines are deflected sufficiently for the assumption $\delta^2 << 1$ to be violated, vertical structure must be introduced into the flowfield and the pressure can no longer be hydrostatic [48]. The parameter α measures, for fixed density contrasts, the concentration of particles in the release volume. The parameter β measures the ratio of the scaled Stokes settling velocity to the internal gravity wave speed. The Stokes settling velocity for an isolated spherical particle in a fluid of kinematic viscosity ν is given by

$$v_s = \frac{g_2' a^2}{18\nu},$$

where a is the diameter of the particle. Thus, for fixed density contrasts, we can think of the parameter β as a measure of particle size with decreasing values of β corresponding to decreasing particle size. The parameter H controls the depth of the ambient fluid in that the 2-layer model is for $H = O(1)$ whereas the $1\frac{1}{2}$-layer model corresponds to $H \to \infty$.

All of the previous theoretical work on transient particle-driven flows ([5], [6], [13]) makes the additional simplifying assumption that $u_2 = u_2(x,t)$.

We note here that for the flow of highly dilute suspensions $(\alpha \sim \delta^2 << 1)$ in deep surroundings $(H \to \infty)$ equations (3.7) - (3.16) give, after a lengthy but standard sequence of manipulations [2], the so-called $1\frac{1}{2}$-layer model for a monodisperse, particle-bearing flow; that is

$$\frac{\partial u_2}{\partial t} + \frac{\partial}{\partial x}\left[\frac{1}{2}u_2^2 + h\right] = 0, \tag{3.17}$$

$$\frac{\partial h}{\partial t} + \frac{\partial}{\partial x}(hu_2) = 0, \tag{3.18}$$

$$\frac{\partial \varphi}{\partial t} + u_2\frac{\partial \varphi}{\partial x} = -\frac{\varphi}{hU}v_{\text{eff}}. \tag{3.19}$$

The important thing to note here is that the dynamics of fluid and particles are decoupled so that shallow-water theory may be obtained. Equations (3.17) and (3.18) are identical to equations (2.32) and (2.33) derived previously for compositionally-driven flows in deep surroundings.

For highly dilute suspensions in shallow surroundings $(H = O(1))$ equations (3.7) - (3.16) now give the 2-layer model for particle-bearing flows consisting of equations (2.26) - (2.29) together with the particle equation (3.19) [2].

We emphasize here that in order to get shallow-water theory, whether for the $1\frac{1}{2}$- or 2-layer model, the dynamics of fluid and particles must be decoupled.

To bring in the influence of the suspended particles on the flow dynamics in a rational fashion, we increase their presence by moving into the parameter regime $\alpha = O(1)$, $0 < \delta^2 << 1$. Thus, we are allowing for a higher concentration of particles while still maintaining the low-aspect ratio nature of the flow. It is the influence of this higher concentration of particles that leads to the nonhydraulic nature of the flow rather than the curvature of streamlines as associated, for example, with flow over a sill, as studied in [48].

In this parameter regime, we neglect terms $O(\delta^2)$ in our equations and replace (3.9) and (3.10) by

$$\frac{\partial p_1^*}{\partial z} = 0, \quad \frac{\partial p_2^*}{\partial z} + \alpha\varphi = 0. \tag{3.20}$$

From the first equation in (3.20) we can easily deduce that $u_1 = u_1(x,t)$ [41] and carry out some standard manipulations [37] to get the upper-layer continuity and momentum equations in the form

$$\frac{\partial}{\partial t}(h - \gamma\eta) + \frac{\partial}{\partial x}\left[u_1(h - H - \gamma\eta)\right] = 0, \tag{3.21}$$

and

$$\frac{\partial u_1}{\partial t} + \frac{\partial}{\partial x}\left(\frac{1}{2}u_1^2 + \eta\right) = 0. \tag{3.22}$$

We now integrate in the second equation in (3.20) to get

$$p_2^* = -\alpha\varphi z + \Phi(x,t), \tag{3.23}$$

where Φ is the arbitrary function arising from the integration. The function Φ is then determined from the dynamic condition (3.14) at the interface to be

$$\Phi(x,t) = \alpha\varphi h + \frac{\rho_1}{\rho_2}\eta + h. \tag{3.24}$$

Thus the total pressure field in the lower layer, when expressed in terms of nondimensional variables, is

$$p_2(x,z,t) = \frac{\rho_1 H - \rho_2 z}{\rho_2 - \rho_1} + \alpha\varphi(h - z) + \frac{\rho_1}{\rho_2}\eta + h. \tag{3.25}$$

The above expression for p_2 shows clearly the impossibility of a depth-independent horizontal velocity field in the lower particle-bearing fluid layer. In fact, the z-dependence in the horizontal velocity field arises precisely because of the horizontal gradients in the particle concentration. This z-dependence will grow most rapidly in regions of high horizontal gradients in this concentration. Our numerical solutions of the equations of motion will bear this out.

In order to carry out the numerical computations for both the 2-layer and $1\frac{1}{2}$-layer models of particle-driven flows for the fixed-volume releases we construct the horizontal velocity field for the lower layer in the form of a power series, that is,

$$u_2(x, z, t) = \sum_{n=0}^{\infty} \theta_n(x, t) z^n, \tag{3.26}$$

where the evolution of the θ_n are to be determined from the horizontal momentum equation. Using (3.26) in the continuity equation for the lower layer, while invoking the kinematic condition (3.16) on the lower boundary we obtain the lower layer vertical velocity field as

$$w_2(x, z, t) = -\sum_{n=0}^{\infty} \frac{\partial \theta_n}{\partial x}(x, t)\, \frac{z^{n+1}}{n+1}. \tag{3.27}$$

Now substituting (3.26) and (3.27) into the horizontal momentum equation for the lower layer, combining the continuity equation with the kinematic condition at the interface and subsequently employing the expansion for u_2 in the result gives as the governing equations for the 2-layer model the pair in (3.21) and (3.22) together with

$$\frac{\partial \theta_0}{\partial t} + \frac{\partial}{\partial x}\left(\frac{1}{2}\,\theta_0^2 + \alpha\varphi h + \frac{\rho_1}{\rho_2}\,\eta + h\right) = 0, \tag{3.28}$$

$$\frac{\partial \theta_1}{\partial t} + \theta_0\,\frac{\partial \theta_1}{\partial x} - \alpha\,\frac{\partial \varphi}{\partial x} = 0, \tag{3.29}$$

$$\frac{\partial \theta_2}{\partial t} + \frac{\partial}{\partial x}\left(\frac{1}{4}\,\theta_1^2 + \theta_0\theta_2\right) - 2\theta_2\,\frac{\partial \theta_0}{\partial x} = 0, \tag{3.30}$$

$$\frac{\partial \theta_3}{\partial t} + \frac{\partial}{\partial x}(\theta_0\theta_3) - 3\theta_3\,\frac{\partial \theta_0}{\partial x} + \frac{2}{3}\,\theta_1\,\frac{\partial \theta_2}{\partial x} = 0, \tag{3.31}$$

$$\frac{\partial h}{\partial t} + \frac{\partial}{\partial x}\left[h\left(\theta_0 + \theta_1\,\frac{h}{2} + \theta_2\,\frac{h^2}{3} + \theta_3\,\frac{h^3}{4}\right)\right] = 0, \tag{3.32}$$

$$\frac{\partial \varphi}{\partial t} + \left(\theta_0 + \theta_1\,\frac{h}{2} + \theta_2\,\frac{h^2}{3} + \theta_3\,\frac{h^3}{4}\right)\frac{\partial \varphi}{\partial x} + \frac{\varphi v_{\text{eff}}}{hU} = 0. \tag{3.33}$$

Our governing equations for the 2-layer approximate model for low-aspect ratio, dilute, particle-driven flows in rectangular geometry now consist in the set of 8 coupled, nonlinear equations given by (3.21), (3.22), and (3.28) - (3.33) in the 8 unknowns $u_1, h, \eta, \varphi, \theta_0, \theta_1, \theta_2,$ and θ_3. These equations will be solved numerically subject to initial and boundary conditions corresponding to the fixed-volume, sudden-release experiments of Gladstone *et al.* [11] for well-mixed solid particle suspensions.

The governing equations for the $1\frac{1}{2}$-layer model can be obtained from those for the 2-layer one by taking $H \to \infty$ with $h = O(1)$. The set of governing equations is then reduced from those for the 2-layer model through the elimination of (3.21) and (3.22) with equation (3.28) being replaced by

$$\frac{\partial \theta_0}{\partial t} + \frac{\partial}{\partial x}\left(\frac{1}{2}\,\theta_0^2 + \alpha\varphi h + h\right) = 0; \tag{3.34}$$

whereas, equations (3.29) - (3.33) remain the same. The $1\frac{1}{2}$-layer model is now specified by a system of 6 equations in the 6 unknowns $h, \varphi, \theta_0, \theta_1, \theta_2,$ and θ_3. The dynamics of the deep upper layer is now eliminated from the model with the influence of that layer appearing only through the expression for the reduced gravity.

3.2. Numerical investigation

In this subsection we display and interpret numerical solutions to our model equations for both the 2-layer and $1\frac{1}{2}$-layer systems. These solutions will correspond to initial boundary value problems designed specifically to examine the flow and deposition from turbidity currents in both shallow and deep surroundings following the sudden release of fixed-volume, well-mixed suspensions in rectangular geometry corresponding to the lock release experiments of [11]. We shall attempt, through modelling, to include the initial turbulence of mixing and explore the effects of initial particle concentrations and particle size on various flow properties including velocity shear and vorticity. The deposition of particles and influences upon this will also be examined.

The initial values for the 2-layer approximate model problem for sudden release of the well-mixed suspension may be expressed as

$$\theta_i(x,0) = 0 \quad (i = 0,1,2,3), \quad u_1(x,0) = 0, \quad \eta(x,0) = 0, \quad x \geq 0,$$

$$h(x,0) = \begin{cases} 1, & x \leq 1, \\ 0, & x > 1, \end{cases}$$

$$\varphi(x,0) = \begin{cases} 1, & x \leq 1, \\ 0, & x > 1. \end{cases} \tag{3.35}$$

The boundary conditions that prevent flow across the end wall at $x = 0$ are

$$\theta_i(0,t) = 0 \quad (i = 0,1,2,3), \quad u_1(0,t) = 0, \quad t \geq 0 \tag{3.36}$$

and the further flux conditions to be satisfied are

$$h_x(0,t) = 0, \quad \eta_x(0,t) = 0, \quad t \geq 0. \tag{3.37}$$

These latter conditions can be expected from symmetry arguments and can be derived from the governing equations using the impermeability conditions. The initial and boundary conditions for the problem corresponding to the $1\frac{1}{2}$-layer model are as above but with the conditions on u_1 and η removed.

Again, MacCormack's numerical procedure is employed here and has already been described and used to examine buoyancy-driven flows earlier in this chapter and elsewhere ([10], [17], [37]).

In the next set of two figures we examine first the difference between gravity current structure, that is depth profiles, for particle-driven flows in shallow and deep surroundings, respectively. Then in Figure 6 we study the evolution of the volume fraction of particles in both shallow and deep surroundings. In these figures for φ and h we take the nondimensional decay rate $\widetilde{\delta}_*$ defined in the introduction to lie in the range $0.06 - 0.14$. In fact, for the calculations reported in Figures 5 and 6 we have taken $\widetilde{\delta}_* = 0.1$. A more complete argument for this choice, based on experimental parameter values is set out in Moodie *et al.* [10]. Indeed, we shall employ this value in all calculations unless there is a statement to the contrary.

Upon comparing the computed gravity current depth profiles shown in Figure 5(a) and (b) for shallow and deep surroundings respectively, the most prominent difference is the appearance of the rear bore in the case of the 2-layer model of the particle-driven flow. The mechanism leading to the creation of this bore is the same as that which was discussed earlier in this article in connection with compositionally-driven flows. One of the main effects of the bore is the thinning of the tail of the particle-driven flow. This will, in turn, lead to an acceleration in the rate at which particles are deposited from the flow since rates of deposition are inversely proportional to current depth ([10], [46]). This can be seen from the form of the particle equation displayed in (3.33). The higher rate of particle deposition from the particle-driven flow modelled by the 2-layer theory because of the proximal thinning of the current means a lower value for the bulk density and a decreased rate of advance under the influence of buoyancy forces. This can be observed in our numerical solutions presented in Figure 5(a) and (b) with the current in (b) advancing at the higher rate. The difference in rates of advance will, of course, be greatest in the earlier particle rich phase of the flow. As the particles are deposited and the flow becomes dominated by the compositionally-generated buoyancy forces $(\rho_1 \neq \rho_2)$ the rate of advance will approach each other. This is clearly evident from the numerical results shown in Figure 5.

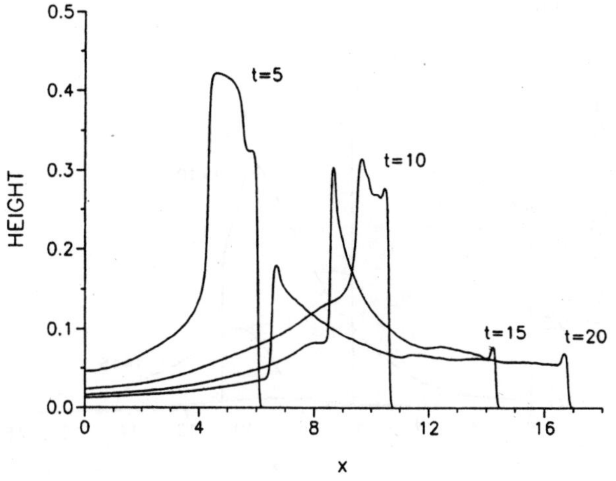

(a)

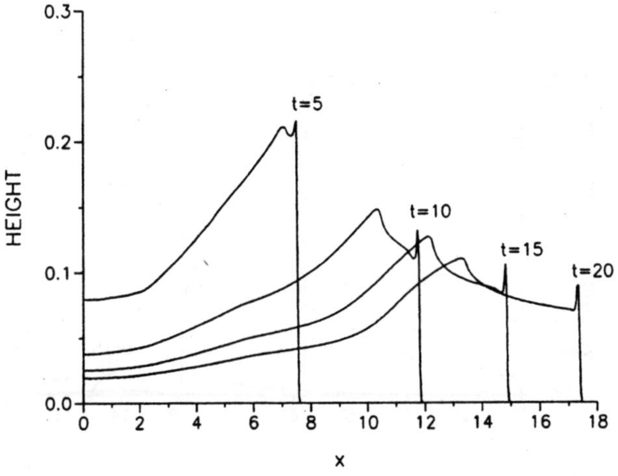

(b)

Figure 5. Graphs of gravity current depth profiles at $t = 5, 10, 15, 20$ with $\alpha = 2$, $\beta = 0.05$ for (a) the 2-layer model with $\gamma = 0.05$, $H = 1.2$ and (b) the $1\frac{1}{2}$-layer model.

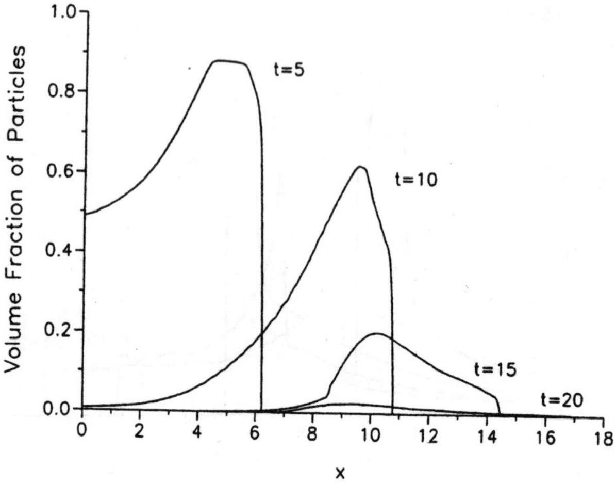

(a)

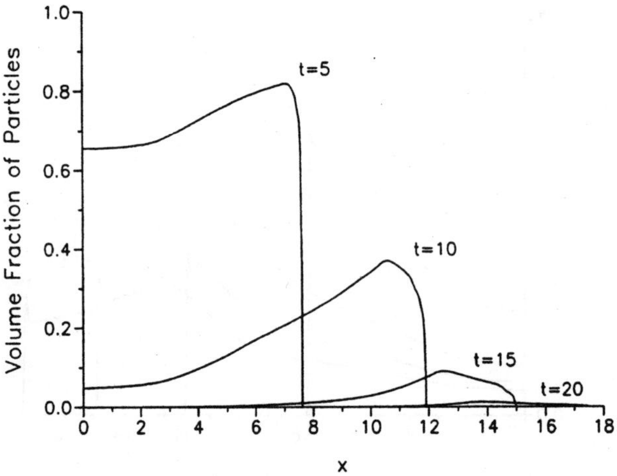

(b)

Figure 6. Volume fraction of particles for particle-driven flow at $t = 5, 10, 15, 20$ with $\alpha = 2$, $\beta = 0.05$ for (a) the 2-layer model with $\gamma = 0.05$, $H = 1.2$ and (b) the $1\frac{1}{2}$-layer model.

Figures 6(a) and (b) displays the volume fraction of particles over the flow regime that remain in suspension at various times. We observe here that in the proximal zone, where thinning due to the bore that is generated in the 2-layer model is most pronounced, the volume fraction remaining in suspension for this model is much less than that for the $1\frac{1}{2}$-layer one where no bore is formed.

In Figure 7(a) and (b) we display the computed horizontal velocity field at various stations in the flow regime for a post-release elapsed time of 20 nondimensional units for the 2-layer and $1\frac{1}{2}$-layer particle-driven flow models, respectively. The very first observation that we can make is that there is a very high level of velocity shear over much of the flow regime for both models. This high level of depth dependence in the velocity field would appear to rule out either the use of shallow water theory or depth averaging of the velocity field. When a velocity field changes by greater than 100% over the depth, the significance of a depth-averaged velocity is unclear.

The calculations based upon the two models also show that the velocity shear is greatest in the front portion of the flow and diminishes upstream as particle concentration also diminishes. This is, of course, to be expected from our theoretical work showing that shear is generated in regions of high horizontal gradients in the particle concentration. Upon comparing the results of Figure 7(a) and (b) it is clear that velocity shear in the 2-layer model spans a larger proportion of the flow regime than it does in the case of the $1\frac{1}{2}$-layer model and also that the maximum velocity variation over depth is greatest for the 2-layer model. These differences for the two models derive from the fact (see Figure 6) that particle concentration gradients over much of the flow are greater in the case of the 2-layer model. This, in turn, is a result of the end wall generated bore in the case of the 2-layer model and its ability to thin the current thereby enhancing deposition rates in the proximal zone and creating larger horizontal gradients in particle concentration.

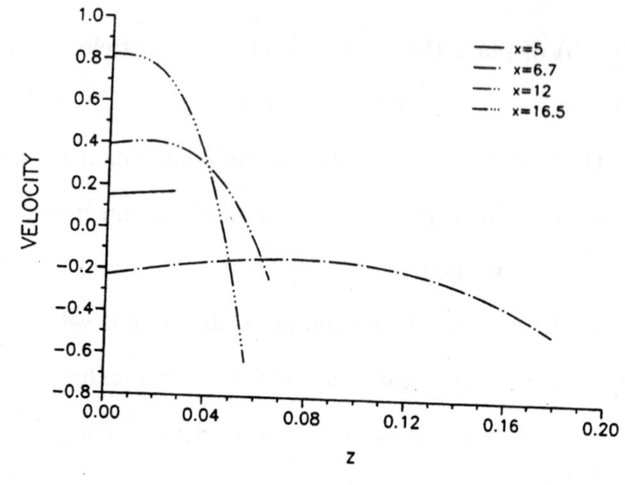

(a)

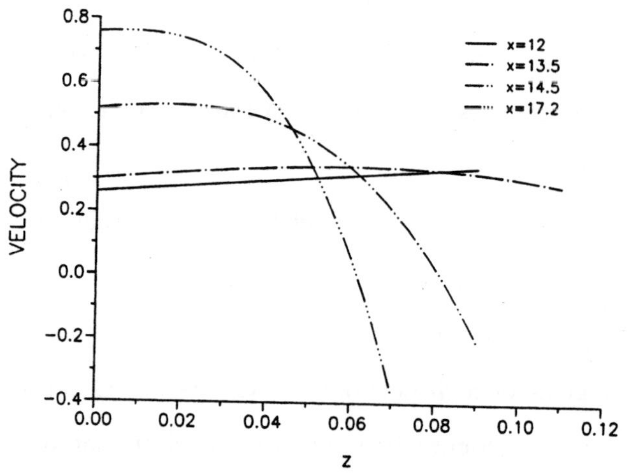

(b)

Figure 7. Variation in horizontal velocity with depth for $\alpha = 2$, $\beta = 0.05$, $t = 20$ for (a) the 2-layer model with $\gamma = 0.05$, $H = 1.2$ and (b) the $1\frac{1}{2}$-layer model.

In Figure 8 we have plotted the calculated values of the horizontal velocity field as functions of depth at $t = 10$ for three values of the parameter α governing initial volume fraction of particles. All of the calculations in Figure 8 are for the 2-layer model of particle-driven flows. The first thing that we observe is that by increasing the initial volume fraction of particles the region of velocity shear is lengthened. For example, by increasing α from 2 to 10 we increase the lateral zone of significant velocity shear from ≈ 1.0 nondimensional length units to ≈ 5.0 units, that is, by a factor of 5. We can also observe from the sequence of calculations present in Figure 8 that increasing the initial volume fraction of particles increases the rate of advance of the particle-driven bottom current.

We have carried out the same calculations for the $1\frac{1}{2}$-layer model and presented these in Figure 9. Here we see that developed velocity shear is greater for the case of the 2-layer model that is depicted in Figure 8 and we can attribute these higher shear levels to the influence of the internal bore that is created in this model due to the interaction of the reverse flow with the end wall of the rectangular tank. The numerical results from Figures 8 and 9 show that the advance of the current in the $1\frac{1}{2}$-layer model is more rapid than that for the 2-layer model. This was also found to be the case for compositionally-driven flows although the differences are greater in the particle-driven case. From these observations we infer that in the case of the 2-layer model (compositionally- or particle-driven) some of the initial potential energy arising from buoyancy forces gets depleted in the creation of the reverse flow in the thin upper fluid layer. When this loss is coupled with the increased rate of particle deposition from the 2-layer model the net effect is to increase the velocity spread of the particle-driven currents over their compositional counterparts.

There are no known experimental results against which to compare our model-based calculations for the vertical variations in the horizontal velocity field displayed in Figures 7, 8, and 9 but it is certainly reasonable to expect such variations in strongly baroclinic flows such as these particle-driven ones. In this plane flow problem there is a single component of vorticity in the y direction, ω_y, and its time rate of change following the motion is governed by the vorticity equation

$$\frac{D\omega_y}{Dt} = (\rho_3 - \rho_2)\,\frac{\partial \varphi}{\partial x}\left[\frac{\rho_2}{\rho_2 - \rho_1} + \alpha\varphi\right].$$

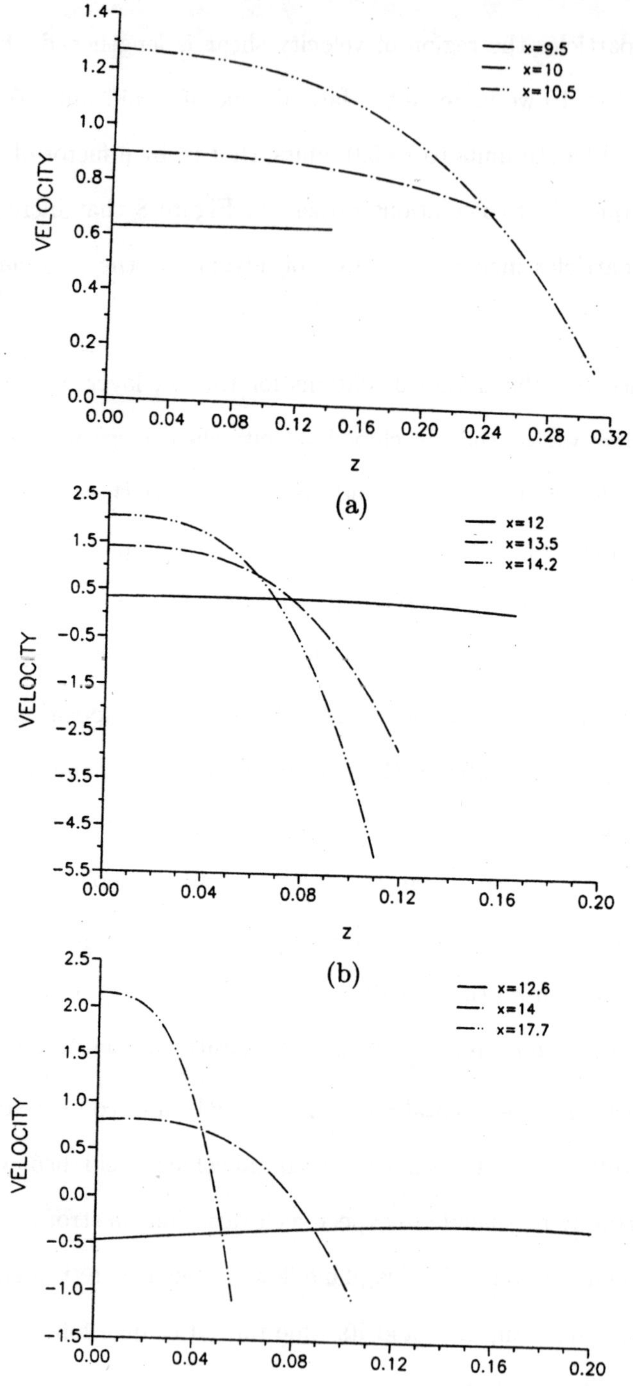

Figure 8. Variation in horizontal velocity with depth for $\beta = 0.01$, $t = 10$ for the 2-layer model with $\gamma = 0.05$, $H = 1.2$ and (a) $\alpha = 2$, (b) $\alpha = 5$, (c) $\alpha = 10$.

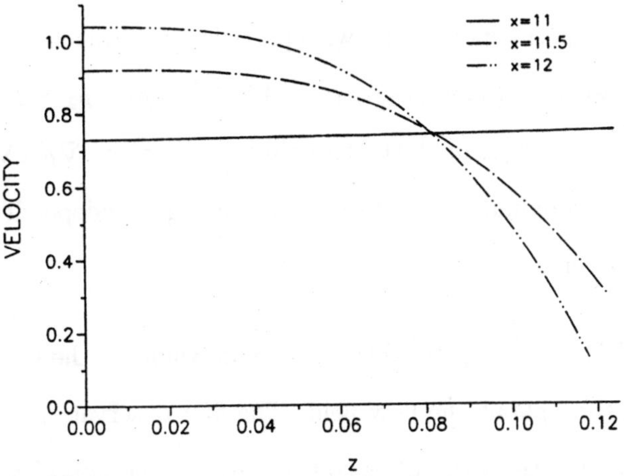

(a)

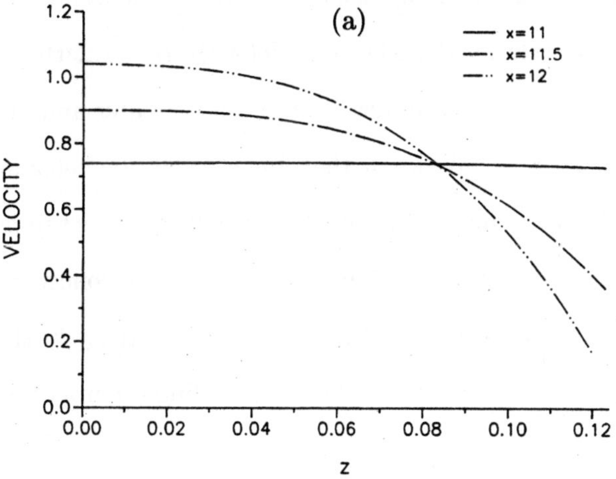

(b)

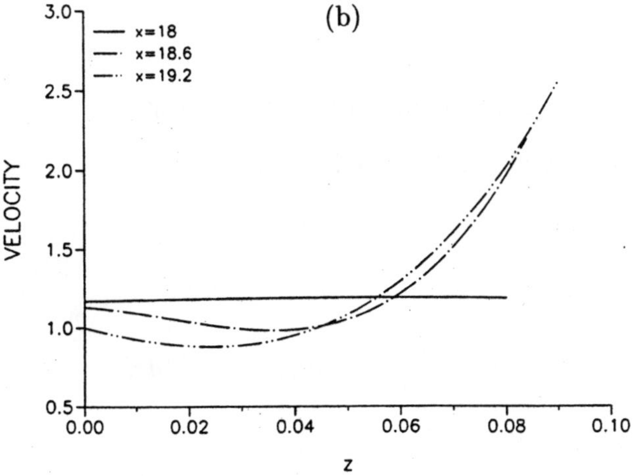

(c)

Figure 9. Variation in horizontal velocity with depth for $\beta = 0.01$, $t = 10$ for the $1\frac{1}{2}$-layer model with (a) $\alpha = 2$, (b) $\alpha = 5$, (c) $\alpha = 10$.

We note that for typical experimental values with particles of silicon carbide [11] suspended in a 1% saline solution we would have $\rho_2/(\rho_2 - \rho_1) \approx 10^2$, $\rho_3 - \rho_2 \approx 2 - 3$, and $\alpha \approx 2$. The rate of vorticity generation through the baroclinic mechanism $\nabla\rho \times \nabla p$ can therefore be quite high and so to approximate such flows as having depth-independent velocity fields will be a major source of error.

In Figures 10 and 11 we have plotted the maximum values of the vorticity over depth at the various stations throughout the flow field. The first observation that can be made from these two figures is that the vorticity is concentrated in the frontal zone of the flow in both cases. We also observe that for the 2-layer model with $H = 1.2$ the zone of significant vorticity covers a greater portion of the flow field and has greater amplitude than it does for the same model with $H = 20$. We know that for $H = 20$ the behaviour of this model approaches that of the $1\frac{1}{2}$-layer model with its dynamically quiescent upper layer, absence of an internal bore, and reduced horizontal gradients in particle concentration. It is these reduced gradients that lead to the observed lower levels of vorticity in the $H = 20$ case. These calculations again confirm the importance of including the upper layer dynamics in models for particle-driven flows.

In the final figure that follows, we plot the results of our computations to determine the areal density of deposit on the bottom after the sedimentation process has been completed, and all particles have been deposited. The nondimensional density of deposit is given by

$$m(x,t) = \int_0^t \varphi(x,\tau)\beta(1 - e^{-\delta_*\tau})d\tau,$$

wherein all variables are nondimensional, and we have adhered to our convention of omitting tildes from such quantities.

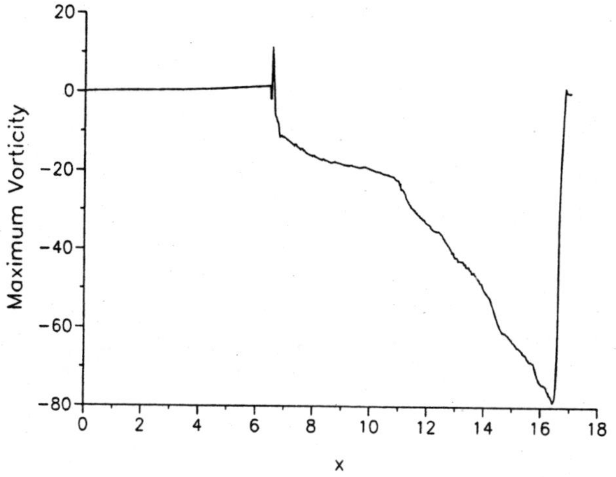

Figure 10. Downstream variation of $\max_z \omega_y$ for the 2-layer model with $\beta = 0.05$, $\alpha = 2$, $\gamma = 0.05$, $H = 1.2$ at $t = 20$.

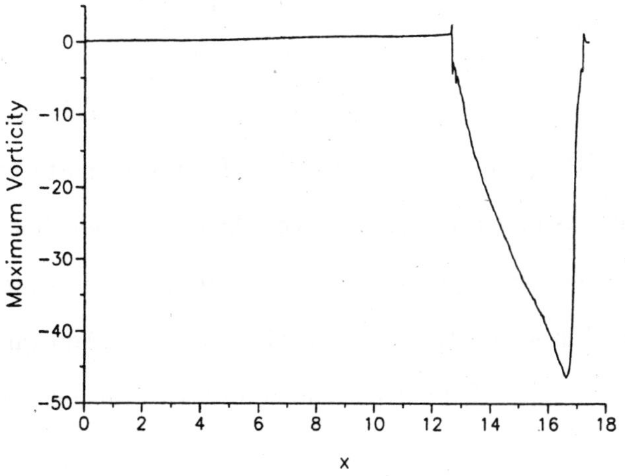

Figure 11. Downstream variation of $\max_z \omega_y$ for the 2-layer model with $\beta = 0.05$, $\alpha = 2$, $\gamma = 0.05$, $H = 20$ at $t = 20$.

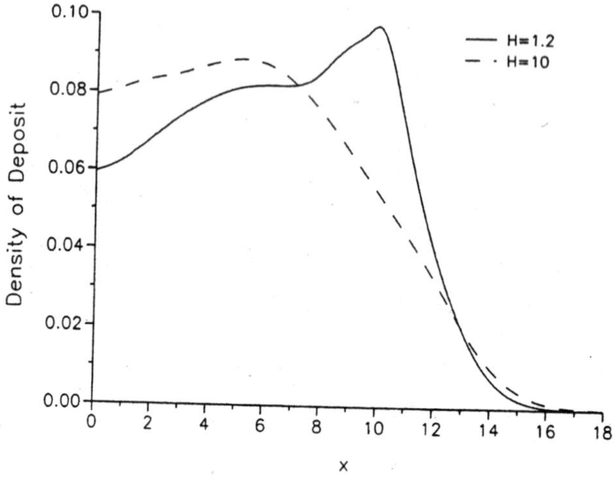

Figure 12. Nondimensional density of deposit as a function of downstream distance x for the 2-layer model with $\beta = 0.01$, $\gamma = 0.05$, $\alpha = 5$, $H = 1.2$ and 10.

In both cases we see a local maximum for the density of deposit that is downstream of the release point for the well-mixed, fixed volume suspension. This local maximum is much more pronounced in the case of the highly mobile thin upper layer of fluid wherein the bore is generated at the interface between the two layers. This property of these deposits is in accord with the computations for volume fraction of particles in suspension at various times that are displayed in Figure 6. For there it is seen that for the 2-layer model there is a large deposition of particles over a ≈ 10 unit time scale wherein these particles will be deposited in a zone that is centered at $x \approx 10$. An examination of the deposit that is depicted in Figure 12 is certainly in accord with this observation from our previous numerical experiments.

4. Concluding Remarks

In this chapter we have attempted to present a review of particle-driven gravity currents while providing a possible model for their study. The major deficiencies in our current understanding of these complex and important flows have been described. To extend our understanding will involve the addition of rotational effects to model the large scale flows that occur naturally, the inclusion of more complex geometries to deal with industrial applications, an examination of instability mechanisms and the transition to turbulence in the flows, as well as, models to deal with high-concentration flows such as those associated with the Grand Banks earthquake of 1929. We feel that all of these necessary advances will have to be made through a combination of theoretical and experimental studies.

REFERENCES

[1] G.V. Middleton, Sediment deposition from turbidity currents, *Annu. Rev. Earth Planet. Sci.* **21**: 89-114 (1993).

[2] T.B. Moodie, J.P. Pascal, and G.E. Swaters, Sediment transport and deposition from a two-layer fluid model of gravity currents on sloping bottoms, *Stud. Appl. Math.* **100**: 215-244 (1998).

[3] R.S.J. Sparks, R.T. Bonnecaze, H.E. Huppert, J.R. Lister, M.A. Hallworth, H. Mader, and J. Phillips, Sediment-laden gravity currents with reversing buoyancy, *Earth Planet Sci. Lett.* **114**: 243-257 (1993).

[4] J.E. Simpson, *Gravity Currents: In the Environment and the Laboratory*, Cambridge U.P., Cambridge, 1997 (2nd ed.).

[5] R.T. Bonnecaze, H.E. Huppert, and J.R. Lister, Patterns of sedimentation from polydispersed turbidity currents, *Proc. R. Soc. Lond.* A**452**: 2247-2261 (1996).

[6] H.E. Huppert, Quantitative modelling of granular suspension flows, *Phil. Trans. R. Soc. Lond.* **A356**, 2471-2496 (1998).

[7] T.B. Moodie, Gravity currents, *J. Comp. Appl. Math.*, to appear.

[8] M.H. Garcia, Depositional turbidity currents laden with poorly sorted sediment, *J. Hydr. Engrg.* **120**: 1240-1263 (1994).

[9] J. Zeng, D.R. Lower, D.B. Prior, W.J. Wisemen, and B.D. Bornhold, Flow properties of turbidity currents in Bute Inlet, British Columbia, *Sedimentology* **38**: 975-996 (1991).

[10] T.B. Moodie, J.P. Pascal, and J.C. Bowman, Modeling sediment deposition patterns arising from suddenly released fixed-volume turbulent suspensions, *Stud. Appl. Math.* **105**: 333-359 (2000).

[11] C. Gladstone, J.C. Phillips, and R.S.J. Sparks, Experiments on bidisperse, constant-volume gravity currents: Propagation and sediment deposition, *Sedimentology* **45**: 833-843 (1998).

[12] D. Nof, Rotational turbidity flows and the 1929 Grand Banks earthquake, *Deep Sea Res.* **43**: 1143-1163 (1996).

[13] R.T. Bonnecaze, H.E. Huppert, and J.R. Lister, Particle-driven gravity currents, *J. Fluid Mech.* **250**, 339-369 (1993).

[14] R.T. Bonnecaze, M.A. Hallworth, H.E. Huppert, and J.R. Lister, Axisymmetric particle-driven gravity currents, *J. Fluid Mech.* **294**: 93-121 (1995).

[15] R.T. Bonnecaze and J.R. Lister, Particle-driven gravity currents down planar slopes, *J. Fluid Mech.* **390**: 75-91 (1999).

[16] M.A. Hallworth and H.E. Huppert, *Abrupt transitions in high-concentration, particle-driven gravity currents, Phys. Fluids* **10***: 1083-1087 (1998).*

[17] T.B. Moodie and J.P. Pascal, Non-hydraulic effects in particle-driven gravity currents in deep surroundings, *Stud. Appl. Math.*, to appear.

[18] A. Freundt and M. Bursik, Pyroclastic flow transport mechanisms, In *From Magma to Tephra: Modelling Physical Processes of Explosive Volcanic Eruptions*, ed. A. Freundt and M. Rosi, Elsevier, Amsterdam, 173-245 (1998).

[19] K.M. Scott, Origin, behaviour and sedimentology of lahars and lahar-runout flows in the Toutle-Cavlitz river system, *U.S.G.S. Prof. Paper*, **1447** (1988).

[20] E.J. Hopfinger, Snow avalanche motion and related phenomena, *Ann. Rev. Flud Mech.* **15**, 47-76 (1983).

[21] G.V. Middleton and W.J. Neal, Experiments on the thickness of beds deposited by turbidity currents, *Jour. Sed. Petrology* **59**, 297-307 (1989).

[22] G.V. Middleton, Experiments on density and turbidity currents, III. Deposition of sediment, *Can. J. Earth Sci.* **4**, 475-505 (1967).

[23] J.W. Rottman and J.E. Simpson, Gravity currents produced by instantaneous releases of a heavy fluid in a rectangular channel, *J. Fluid Mech.* **135**, 95-110 (1983).

[24] A.N. Kolmogorov, The local structure of turbulence in an incompressible viscous fluid for very large Reynolds numbers, *C.R. Acad. Sci. U.S.S.R.* **30**, 301-306 (1941).

[25] R.A. Bagnold, Autosuspension of transported sediment; turbidity currents, *Proc. R. Soc. London* **A265**, 315-319 (1962).

[26] W.B. Dade and H.E. Huppert, Predicting the geometry of channelized deep-sea turbidites, *Geology* **22**, 645-648 (1994).

[27] B. Marcu, E. Meiburg, and P.K. Newton, Dynamics of heavy particles in a Burgers vortex, *Phys. Fluids* **7**, 399-410 (1995).

[28] J.M. Burgers, A mathematical model illustrating the theory of turbulence, *Adv. Appl. Mech.* **1**, 171-199 (1948).

[29] T.B. Benjamin, Gravity currents and related phenomena, *J. Fluid Mech.* **31**, 209-248 (1968).

[30] H.P. Gröbelbauer, T.K. Fannelop, and R.E. Britter, The propagation of intrusion fronts of high density ratios, *J. Fluid Mech.* **250**: 669-687 (1993).

[31] J.A. Fay, Some unresolved problems of LNG vapor dispersion, MIT-GRI, *LNG Safety and Research Workshop*. Gas Research Institute, Chicago (1982).

[32] F.H. Weirich, Field evidence for hydraulic jumps in subaqueous sediment gravity flows, *Nature* **332**: 626-629 (1988).

[33] T. von Kármán, The engineer grapples with nonlinear problems, *Bull. Am. Math. Soc.* **46**, 615-683 (1940).

[34] J.B. Klemp, R. Rotunno, and W.C. Skamarock, On the dynamics of gravity currents in a channel, *J. Fluid Mech.* **269**: 169-198 (1994).

[35] D.P. Hoult, Oil spreading on the sea, *Ann. Rev. Fluid Mech.* **4**, 341-368 (1972).

[36] T.K. Fannelop and G.D. Waldman, Dynamics of oil slicks, *AIAA J.* **10**, 506-510 (1972).

[37] S.J.D. D'Alessio, T.B. Moodie, J.P. Pascal and G.E. Swaters, Gravity currents produced by sudden release of a fixed volume of heavy fluid, *Stud. Appl. Math.* **96**: 359-385 (1996).

[38] R.J. LeVeque, *Numerical Methods for Conservation Laws*, Birkhäuser (1992).

[39] R.E. Grundy and J.W. Rottman, The approach to self-similarity of the solutions of the shallow-water equations representing gravity-current releases, *J. Fluid Mech.* **156**: 39-53 (1985).

[40] R.M. Baptista, F.B. de Freitas Rachid, and J.H.C. de Araujo, Estimating mixing volumes between batches in multiproduct pipelines, *Proceedings of International Pipeline Conference 2000*, Calgary (2000).

[41] J. Pedlosky, *Geophysical Fluid Dynamics*, Springer-Verlag, Berlin/New York, 1987.

[42] J.J. Stoker, *Water Waves*, Interscience, New York, 1957.

[43] P.G. Baines, *Topographic Effects in Stratified Flows*, Cambridge U.P., Cambridge, 1995.

[44] A. Harten, The artificial compression method for computation of shocks and contact discontinuities III. Self-adjusting hybrid schemes, *Math. Comp.* **32**: 363-389 (1978).

[45] G. Parker, Y. Fukushima, and H.M. Pantin, Self-accelerating turbidity currents, *J. Fluid Mech.* **171**: 145-181 (1986).

[46] D. Martin and R. Nokes, A fluid-dynamical study of crystal settling in convecting magmas, *J. Petrol.* **30**: 1471-1500 (1989).

[47] H. Tennekes and J.L. Lumley, *A First Course in Turbulence*, The MIT Press, Cambridge, MA, 1972.

[48] D.Z. Zhu and G.A. Lawrence, Nonhydrostatic effects in layered shallow water flow, *J. Fluid Mech.* **355**: 1-16 (1998).

Modern Applied Mathematics
Editor: J.C. Misra

Impulsive Free-Surface Flows and Small-Time Expansions

P.A. Tyvand[1] and T. Miloh[2]

[1]Department of Agricultural Engineering, Agricultural University of Norway, 1432 Ås, Norway
[2]Faculty of Engineering, Tel Aviv University, Ramat Aviv 69978, Israel

Introduction

When a free surface is initially at rest, it cannot start to move in an arbitrary way. Even in the limit where we assume a forced motion started impulsively with finite velocity and disregard fluid compressibility, there are strong physical restrictions on the possible fluid motion. Due to fluid inertia, the tangential velocity after the impulsive start is zero at the surface, relative to a coordinate system at rest with the undisturbed fluid. This is true for all surface particles that are not in direct contact with a boundary in forced motion tangentially to the surface. This dynamical restriction on the initial fluid motion is represented mathematically as the zero-potential condition, which follows from Bernoulli's equation integrated over the infinitesimal time interval of impulsive start.

In this review article we will cover the variety of impulsive flows due to submerged singularities or solid boundaries, impulsively put in motion or placed into the fluid. The submerged singularities can be vortices that are suddenly released in the fluid, or sources/sinks that are suddenly turned on. The rigid boundaries may be the contours of bodies that are submerged, floating or penetrating the free surface. The solid boundaries may be the seabottom, the walls of a container or a piston that are suddenly put into motion, generating a free-surface flow. If the forced motion of a surface-piercing wall is horizontal, there is a conflict at the intersection point between the wall in forced motion and the neighbouring free-surface particles. Usually this conflict results in a mathematical singularity for the surface elevation, just at the intersection between the solid wall in forced horizontal motion and the free surface in zero horizontal motion. This singularity reflects the physical necessity that there must be a region close to the wall (the inner flow regime of matched asymptotic expansions) where convective inertia cannot be neglected because of the large initial surface velocity. In the normal situation where the surface velocity is finite everywhere, the entire initial flow follows linear theory, and nonlinear effects evolve only gradually with time. Free-surface singularities arise to tell us that the assumption of linear theory being valid for the total initial flow field is physically inconsistent, even though it may be good in most of the flow field.

In an inviscid incompressible fluid, there are two different categories of impulsively started free-surface flows: (1) Flows with singularities at the free surface. (2) Regular free-surface flows, without any singularities along the free surface.

A free-surface singularity may occur at an intersection between the free surface and a region of prescribed normal velocity. This happens for example in the impulsive-wavemaker problem where a surface-piercing plate starts moving horizontally (Peregrine [1]). A singularity may also occur at the free surface next to a semi-submerged cylinder put into sway motion (Vinje [2]), and next to a flat body in vertical impact [3, p.474]. Wang and Chwang [4] considered the impulsively moving vertical cylinder, which has singularities at the waterline of the cylinder. The singularities are integrable in the sense that the impulsive force exerted by the fluid on the solid body is finite. In principle, it is possible to resolve the free-surface singularities by means of matched asymptotic expansions.

When the impulsive flow is caused by a submerged object, there is no singularity at the free surface. The present authors [5,6] have investigated analytically the case of a submerged horizontal cylinder. The same problem had earlier been studied numerically by Telste [7] and others.

According to linear theory, a surface elevation grows with a constant rate after an impulsive start with a constant normal velocity. The asymptotic series solution for the nonlinear problem is a power series in time [1,8,9], in contrast to transient flow problems where the start from rest is caused by diffusion. According to linear theory for very small times, the immediate free-surface behaviour is the piling up of an elevation growing linearly in time. Through nonlinear self-interaction at the free surface, a squared-velocity term is generated as the next order in the asymptotic time series. Further interactions will generate higher integer powers of time in the asymptotic series. In the presence of a free-surface singularity, stronger singularities are generated with each new order included in the asymptotic series.

In the present paper we attempt to unify some of the approaches and results from previous analytical papers involving submerged vortices, sources, and bodies in motion. Various authors have studied such problems numerically or experimentally, and comparisons are made with these. Our basic aim is to demonstrate how nonlinearities in the flow evolve successively after an impulsive start. We thus focus on problems where a small-time expansion can be carried out to second or higher orders. We will also discuss nonlinear impulsive free-surface flows due to piston wavemakers, flows inside containers, and flows due to bottom deformations in the sea. But we omit impact problems, where a body hits a fluid surface impulsively from the outside. This is a field of extensive recent research, but most of the analytical results are obtained by linearizing the free-surface boundary conditions.

Mathematical formulation

We consider inviscid flow in an incompressible fluid, which is initially at rest. The fluid has a free surface subject to constant atmospheric pressure. Cartesian coordinates x, y, z are introduced, where the z axis is directed upwards in the gravity field and $z = 0$ represents the undisturbed free surface. The gravitational acceleration is g, and ρ denotes the constant fluid density. The surface elevation is denoted by $\eta(x, y, t)$.

At time $t = 0$ the fluid motion starts impulsively due to motion of a body, parts of the boundary, or some singularity in the fluid being turned on. The vertical length scale for this motion is H. The velocity scale is W, and will be specified in each case as an initial velocity. Our general definition of the Froude number is:

$$F = \frac{W}{\Omega\sqrt{gH}} \qquad (1)$$

The factor Ω is a number to be chosen as 2π for two-dimensional flows and 4π for three-dimensional flows in the presence of submerged singularities (sources, sinks or vortices). In all other cases we choose $\Omega = 1$. This includes submerged bodies as well as boundaries of the fluid domain put into motion.

It is assumed that the flow is irrotational and obeys Laplace's equation:

$$\nabla^2\Phi = 0 \qquad (2)$$

$\Phi(x, y, z, t)$ is the velocity potential and the velocity field is the gradient of this potential. Lord Kelvin's circulation theorem [3, p. 273] is automatically satisfied for sources, sinks and submerged bodies without circulation, but it causes some conceptual problems in the case of vortices [10,11,12]: A material curve enclosing just one vortex or the vortex ring locally, formally violates Kelvin's theorem. This dilemma is resolved by considering a virtual body with an associated circulation moving quickly away from the free surface, leaving no signature except for its shed vortex. Then the shed vortex can be considered as suddenly appearing in the fluid, when we disregard the body that it was shed from. Such a point of view finds some support in the experiments reported by Willmarth et al. [13], where two flapping wings generated and shed a vortex pair.

From now on we work with dimensionless variables. The unit of dimensionless length is H, which is a characteristic vertical length of the initial configuration. The unit of dimensionless velocity is W, which is a characteristic initial velocity. The unit of dimensionless time is H/W, the unit of dimensionless pressure is ρW^2, and the unit of dimensionless force is $\rho W^2 H^2$.

Bernoulli's equation for the dimensionless pressure reads:

$$p = -\frac{\partial \Phi}{\partial t} - \frac{1}{2}|\nabla \Phi|^2 - \frac{z}{\Omega^2 F^2} \tag{3}$$

The kinematic free-surface condition is:

$$\frac{\partial \eta}{\partial t} + \nabla \Phi \cdot \nabla \eta = \frac{\partial \Phi}{\partial z}, \quad z = \eta(x, y, t) \tag{4}$$

The dynamic free-surface condition is:

$$\frac{\partial \Phi}{\partial t} + \frac{1}{2}|\nabla \Phi|^2 + \frac{\eta}{\Omega^2 F^2} = 0, \quad z = \eta(x, y, t) \tag{5}$$

There may also be a kinematic condition at a moving boundary, which will be specified in each case. The far-field condition is:

$$|\nabla \Phi| \to 0, \quad x^2 + y^2 + z^2 \to \infty \tag{6}$$

The initial free-surface condition for the potential is found by integrating the dynamic condition (5) over the infinitesimal time interval of the impulsive start:

$$\Phi(x, y, 0, 0) = 0 \tag{7}$$

This equipotential condition at the free surface is mathematically the most characteristic property of incompressible flows started impulsively from rest. We usually assume that the surface is initially undeformed, which is expressed as:

$$\eta(x, y, 0) = 0 \tag{8}$$

The small-time expansion

Our nonlinear initial-boundary-value problem consists of eqs. (2)-(8). An analytical solution of these equations may be found in terms of an asymptotic small-time expansion [1]:

$$(\Phi, \eta, \mathbf{R}) = (0, 0, \mathbf{R_0}) + H(t)((\Phi_0, 0, 0) + t(\Phi_1, \eta_1, \mathbf{R_1}) + t^2(\Phi_2, \eta_2, \mathbf{R_2}) + \ldots.) \tag{9}$$

H(t) is the Heaviside unit step function. In eq. (9) Φ_n and η_n are unknown functions, whereas the instantaneous position vector of a central point in the submerged object is:

$$\mathbf{R}_n = (X_n, Y_n, Z_n), \quad n = 0, 1, 2, ... \tag{10}$$

The initial position of the central point in the submerged object is at unit length below the free surface:

$$(X_0, Y_0, Z_0) = (0, 0, -1) \tag{11}$$

The velocity potential to each order is governed by Laplace's equation:

$$\nabla^2 \Phi_n = 0, \quad n = 0, 1, 2, ... \tag{12}$$

The corresponding far-field conditions are:

$$|\nabla \Phi_n| \to 0, \quad x^2 + y^2 + z^2 \to \infty, \quad n = 0, 1, 2, ... \tag{13}$$

The operator of total time differentiation at the moving free surface is:

$$\left(\frac{d}{dt}\right)_{surface} = \frac{\partial}{\partial t} + \frac{\partial \eta}{\partial t}\frac{\partial}{\partial z} \tag{14}$$

The free-surface conditions are found by successive applications of this operator. The lowest-order dynamic conditions are:

$$\Phi_0 = 0, \quad z = 0 \tag{15}$$

$$\Phi_1 = -\frac{\eta_1^2}{2}, \quad z = 0 \tag{16}$$

$$\Phi_2 = -\eta_1 \left(2\eta_2 + \frac{1}{2\Omega^2 F^2}\right), \quad z = 0 \tag{17}$$

$$\Phi_3 = -\frac{\eta_2}{3} \left(4\eta_2 + \frac{1}{\Omega^2 F^2}\right), \quad z = 0 \tag{18}$$

The last equation is a simplified version, where it is assumed that the impulsive motion starts with zero velocity and constant acceleration.

The corresponding kinematic conditions are:

$$\eta_1 = \frac{\partial \Phi_0}{\partial z}, \quad z = 0 \tag{19}$$

$$2\eta_2 = \frac{\partial \Phi_1}{\partial z}, \quad z = 0 \tag{20}$$

$$6\eta_3 = 2\frac{\partial \Phi_2}{\partial z} + \eta_1^2 \nabla^2 \eta_1 + 2\eta_1 |\nabla \eta_1|^2, \quad z = 0 \tag{21}$$

$$4\eta_4 = \frac{\partial \Phi_3}{\partial z}, \quad z = 0 \tag{22}$$

Again the last equation refers to impulsive motion with constant acceleration and zero initial velocity.

The conditions at the boundary in forced motion or at the singularity are dealt with in the next section. We may split the potential into two contributions at each order, i.e.:

$$\Phi_n = \phi_n + \psi_n, \quad n = 0, 1, 2, ... \tag{23}$$

The first term ϕ_n to each order is generated by the inhomogeneous conditions at the free surface with zero normal derivative at the body in force motion. The second term ψ_n to each order is generated by the inhomogeneous conditions at the body boundary, with the homogeneous condition $\psi_n = 0$ at the free surface.

The small-time expansion for the pressure is:

$$p = p_{-1}\delta(t) + H(t)(p_0 + p_1 t + ...) + \text{hydrostatic pressure} \tag{24}$$

The pressure impulse p_{-1} during the start from rest is given by:

$$p_{-1} = -\Phi_0 \tag{25}$$

This is to be integrated over the submerged area of a body in order to determine the force impulse. In dimensionless description the force impulse is equal to the added momentum during the impulsive start.

On geometric nonlinearity

Within our analytical small-time expansion, we have to extrapolate all variables and positions out from the initial geometric location. In linear theory this is trivial in the sense that the displacements of the moving objects will vanish in the linearisation. In nonlinear theory, the mere fact that the objects are displaced a finite distance away from their initial location counts as a nonlinear effect. Since it does not explicitly involve any dynamics, we call it geometric nonlinearity.

We have two different ways of taking geometric nonlinearity into account. For a moving singularity this can best be done by introducing an unexpanded potential ψ with its singularity located at its correct instantaneous position. This unexpanded potential satisfies the homogeneous condition ($\psi = 0$) at the undisturbed free surface, which is the exact condition at time zero. In the case of a single submerged vortex in two dimensions, such an unexpanded potential is given by [11]:

$$\psi = \frac{1}{2\pi} \arctan \frac{z - Z}{x - X} + \frac{1}{2\pi} \arctan \frac{z + Z}{x - X} \tag{26}$$

The potential $\psi(x, z, t)$ must also be expanded. This is done by inserting the expansions for $X(t)$ and $Z(t)$. The geometric nonlinearity means the finite displacement of the vortex singularity point (X, Z) away from its initial location

(0,-1). This displacement for a freely moving vortex is governed by Helmholtz's theorem. In two dimensions this means that the vortex motion follows an individual particle in its motion, and a singular vortex will follow the local velocity provided the singular vortex field is disregarded.

The geometric nonlinearity for a moving body is based on the exact boundary condition at the body surface. For a circular cylinder this exact condition at all times can be written (dot superscript denoting time derivative):

$$(\mathbf{r} - \mathbf{R}) \cdot (\nabla\Phi - \dot{\mathbf{R}}) = 0, \quad \text{along the instant cylinder wall} \tag{27}$$

The geometric nonlinearity in Φ is due to the fact that the difference between $\mathbf{R}(t)$ and $\mathbf{R}(0)$ is finite for any finite t. The geometric nonlinearity appears in the small-time expansion by succesive application of an operator of total time differentiation following the moving cylinder:

$$\left(\frac{d}{dt}\right)_{cylinder} = \frac{\partial}{\partial t} + \dot{\mathbf{R}} \cdot \nabla \tag{28}$$

For a moving singularity the geometric nonlinearity is simply the finite displacement of the singular point. For a moving body the geometric nonlinearity is less transparent, but it reveals itself to each higher order by a successive application of the convective derivative $\mathbf{R}_1 \cdot \nabla$. See Tyvand and Miloh [5] who derived the following kinematic conditions at a circular cylinder which starts impulsively from rest:

$$\frac{\partial \Phi_0}{\partial r} = \mathbf{n} \cdot \mathbf{R}_1, \quad r = R \tag{29}$$

$$\frac{\partial \Phi_1}{\partial r} = 2\mathbf{n} \cdot \mathbf{R}_2 - \mathbf{n} \cdot \nabla(\mathbf{R}_1 \cdot \nabla\Phi_0), \quad r = R \tag{30}$$

$$\begin{aligned} 2\frac{\partial \Phi_2}{\partial r} &= 6\mathbf{n} \cdot \mathbf{R}_3 - 2\mathbf{n} \cdot \nabla(\mathbf{R}_2 \cdot \nabla\Phi_0) - 2\mathbf{n} \cdot \nabla(\mathbf{R}_1 \cdot \nabla\Phi_1) \\ &\quad -\mathbf{n} \cdot \nabla(\mathbf{R}_1 \cdot \nabla(\mathbf{R}_1 \cdot \nabla\Phi_0)), \quad r = R \end{aligned} \tag{31}$$

$$\frac{\partial \Phi_3}{\partial r} = -\mathbf{n} \cdot \nabla(\mathbf{R}_2 \cdot \nabla\Phi_1), \quad r = R \tag{32}$$

This last equation is only valid for impulsive motion with constant acceleration and zero initial velocity. The dimensionless cylinder radius is R, and the length unit is the initial submergence of the cylinder centre. Note that all kinematic conditions are to be applied at the initial cylinder contour. The position vector $\mathbf{R}(\mathrm{t})$ must not be confused with the cylinder radius R.

Here $\mathbf{n}$ denotes the outward unit vector along the initial circle contour. It is equal to the radial unit vector. For the impulsive flow in a container suddenly put into steady motion, the geometric nonlinearity can be avoided by describing the flow in the inertial system that follows the container in its motion. In an unbounded fluid domain, however, it is preferable to find the solution in the system that is at rest at infinity.

In the end of this article we will develop the expansion of geometric nonlinearity when the bottom is deforming impulsively starting at time zero. This is done separately because a more general time-dependence is allowed for the forced motion.

Vortices in free motion near a free surface

Vortices in interaction with a free surface may be forced or free. A deeply submerged body may carry a net circulation which may appear as a vortex as seen from a distance. Such vortices are vortices in forced motion. In the present article we will disregard vortices in forced motion. We will only consider vortices which move freely in the inviscid fluid as governed by the Helmholtz theorem.

Physically it is not obvious how to release free vortices and put them impulsively into an inviscid fluid, because this is in conflict with Lord Kelvin's circulation theorem. This means that viscosity must play a role during the rapid vortex release as the flow is started, but the subsequent flow may still be approximately inviscid. Below we will discuss a vortex pair, a single vortex and a vortex ring. The conventional way to generate a vortex pair is by a pair of flapping wings. A vortex ring can be released from an ending pipe where the flow inside the pipe stops. A single vortex can be released as a starting vortex from an impulsively moving foil with circulation. In our theoretical model we assume that the vortex generation is so rapid that we can disregard macroscopic viscous effects, and that we can disregard all influence on the free surface from the device responsible for the vortex generation.

Novikov [14] studied vortex/free-surface interaction by linear theory. The research on nonlinear interaction between freely moving submerged vortices and a free surface started with Sarpkaya [15] and Tryggvason [16]. Many numerical and some experimental papers have been written, see the review article by Sarpkaya [17]. In most cases the impulsive release of the vortices is not very influential on their long-time interaction with the free surface. From Tyvand [11] we estimate that the impulsive type of flow due to a released vortex element occurs during a time of order $min(\sqrt{H/g}, H^2/\Gamma)$ after the start. Here H denotes the initial submergence of the vortex element, and Γ denotes the circulation around the vortex element. Most numerical simulations cover much greater time spans than this, unless there is an early breaking of the free surface.

From an analytical point of view the situation is different. Vortices released impulsively and interacting nonlinearly with the free surface can only be descibed analytically as long as the flow is of the impulsive type. We will here concentrate on analytical solutions for three basic problems where horizontal vortices are released impulsively near a free surface. In all cases, the velocity scale W is defined as the circulation around a vortex divided by its initial submergence depth.

A horizontal vortex pair

Tyvand [10] studied analytically a horizontal vortex pair (two-dimensional) approaching a free surface. Compared with earlier numerical and experimental work (Marcus and Berger [18], Telste [19], Willmarth et al. [13]) on the same problem, some clarification of the early vortex motion was found: Each of the vortices in a strong vortex pair starting from large depth tends to accelerate one another upwards as soon as they start feeling the presence of the free surface above. They will also be attracted to one another because of the early interaction with the free surface. But as the vortex pair rises further, the free-surface nonlinearity will decelerate the upward vortex motion and also try to push the vortices apart. Gravity will have the same qualitative effects of impeding the upward motion of the vortices and push them apart. While the first mutual interaction between the vortices will be an attraction, this attraction will soon be opposed by a repulsion both from free-surface nonlinearity and from gravity. Numerical simulations [18,19] show that very strong vortices ($F \ll 1$) tend to continue to move upwards as they did before they felt the presence of the free surface. So the leading-order attraction and nonlinear repulsion tend to cancel each other. For weak vortices (F of order 1 or smaller), the additional gravitational repulsion will push the vortices apart. The only way to achieve a significant attraction between the vortices, is to release a strong vortex pair at a depth smaller than their mutual horizontal distance. So far, this attraction has not been demonstrated numerically or experimentally, only analytically.

We give the first-order surface elevation due to a vortex pair being released at unit depth:

$$\eta_1(x) = \frac{2X}{\pi}\frac{X^2+1-x^2}{((x-X)^2+1)((x+X)^2+1)} \tag{33}$$

The initial positions of the two vortices are here denoted by $(x,y) = (\pm X, -1)$.

A single horizontal vortex

Tyvand [11] studied one single vortex and developed a complete third-order theory for the vortex position. The analytical paper by Tyvand included a numerical appendix on the vortex motion by R.P. Tong. The same problem had already been investigated numerically by Yu and Tryggvason [20], although they did not focus on the behaviour for small times. Tong's computations for the single vortex were carried further by Tong and Peregrine [21].

Let us give the exact third-order formulas derived by Tyvand [11] for the motion of a single vortex rotating counter-clockwise. The exact surface elevation to third order is given by:

$$
\begin{aligned}
\eta(x,t) &= \frac{x}{\pi(x^2+1)}t + \frac{x^2(x^2-3)}{4\pi^2(x^2+1)^3}t^2 \\
&+ \left(\frac{x(x^6-29x^4+35x^2+1)}{24\pi^3(x^2+1)^5} - \frac{x}{3\pi F^2(x^2+1)^2}\right)t^3 + O(t^4)
\end{aligned}
\tag{34}
$$

The derivation of the non-gravitational third-order elevation is very cumbersome, and has been performed by MATHEMATICA. The required normal derivatives for a certain class of harmonic functions defined along the x axis ($g_m(x,0) = x(x^2+1)^{-m}$, $m = 1,2,3,4$) are found in [6, appendix].

The exact third-order formulas for the position vector $(X(t), Y(t))$ of the freely moving vortex are given by:

$$X = \frac{t}{4\pi} + (1 - F^{-2})\frac{t^3}{96\pi^3} + O(t^4) \tag{35}$$

$$Y = -1 + O(t^4) \tag{36}$$

In the Froude number definition (1) the characteristic velocity W is now taken as the circulation of the vortex divided by the initial submergence depth. Eq.(35) shows an accelerating motion for supercritical vortices ($F > 1$), which will end in surface breaking. A reversal of the initial motion takes place for subcritical vortices ($F < 1$). We will now extend the small-time expansion to higher order for subcritical vortices. We include only the asymptotically leading terms as $F \to 0$. We then find:

$$X_4^{(F)} = 0 \tag{37}$$

$$Y_4^{(F)} = \frac{1}{3072\pi^4 F^2} \tag{38}$$

$$X_5^{(F)} = \frac{1}{7680\pi^5 F^4} \tag{39}$$

Superscript (F) means Froude-number dependent (i.e. gravity-dependent) quantities. In formula (39) we have neglected a term of order F^{-2}. Formula (38) shows the lowest order gravitational effect on the vertical motion, which incorporates free-surface nonlinearity and makes the vortex move slightly upwards. However, this effect will probably never dominate over the non-gravitational effects. For small Froude numbers, the vertical gravitational motion is negligible compared with the horizontal gravitational motion. So asymptotically for $F \ll 1$, the motion can be considered as purely horizontal, and it is expressed by:

$$\frac{4\pi X}{t} = 1 - \frac{1}{6}\left(\frac{t}{2\pi F}\right)^2 + \frac{1}{120}\left(\frac{t}{2\pi F}\right)^4 + O(t^5) \tag{40}$$

Assuming that we can neglect the vertical motion of a very weak vortex, it follows that its motion for large time is given by its initial motion reversed.

$$\frac{4\pi X}{t} = -1, \quad t \gg 1 \tag{41}$$

This is valid because a very weak vortex in steady motion will feel the free surface as a rigid lid. This extra information for large time allows us to improve the convergence of the approximation (40) by means of Padé approximants (Bender and Orszag [22, p. 383]). Then we find a solution that is valid for all time $t > 0$ in the asymptotic limit $F \ll 1$:

$$\frac{4\pi X}{t} = \frac{1 - \frac{1}{12}\left(\frac{t}{2\pi F}\right)^2 - \frac{1}{360}\left(\frac{t}{2\pi F}\right)^4}{1 + \frac{1}{12}\left(\frac{t}{2\pi F}\right)^2 + \frac{1}{360}\left(\frac{t}{2\pi F}\right)^4} + O(t^5) \tag{42}$$

Note that the asymptotic behaviour (41) has been incorporated twice, both to the second and to the fourth order. Figure 1 shows a comparison between eq.(42) and Tong's numerical appendix in the paper by Tyvand [11]. We choose the case $F = 0.01$, represented by the crosses in Tong's fig. 1. The agreement is excellent for both small and large times. But Tong's numerical results produces a slightly smaller final vortex velocity than our analysis. This means that a vortex of finite Froude number settles at a depth somewhat greater than its initial depth after it has reversed its motion, which is a general trend in the numerical simulations by Tong.

The later paper by Tong and Peregrine [21] displays a large-time evolution of the free surface above a freely moving vortex in the case F=0.1 (their figure 2). There is a pronounced asymmetry in the wave pattern, where the steepest and shortest waves propagate in the -x direction. There is a Doppler effect due to the vortex motion, but it is too weak to be responsible for the asymmetry. The dominating wave radiation takes part between the instant of the impulsive start and the reversal of the vortex motion, and the amount of wave energy radiated out in the +x and -x directions seem to be of comparable magnitude.

A horizontal vortex ring

Tyvand and Miloh [12] considered a horizontal vortex ring. An exact first-order theory was developed. The first-order vortex velocity and the first-order elevation were calculated exactly. The first-order elevation is:

$$\eta_1(r) = \frac{K(\lambda)}{\pi\sqrt{1+(r+R)^2}} + \left(\frac{2R(r+R)}{(1+(r+R)^2)^{3/2}} - \frac{1}{\sqrt{1+(r+R)^2}}\right)\frac{E(\lambda)}{\pi(1-\lambda^2)} \tag{43}$$

Here r denotes the radial coordinate, and R is the initial radius of the vortex ring, assuming its core is very thin. It is important to note that both r and R

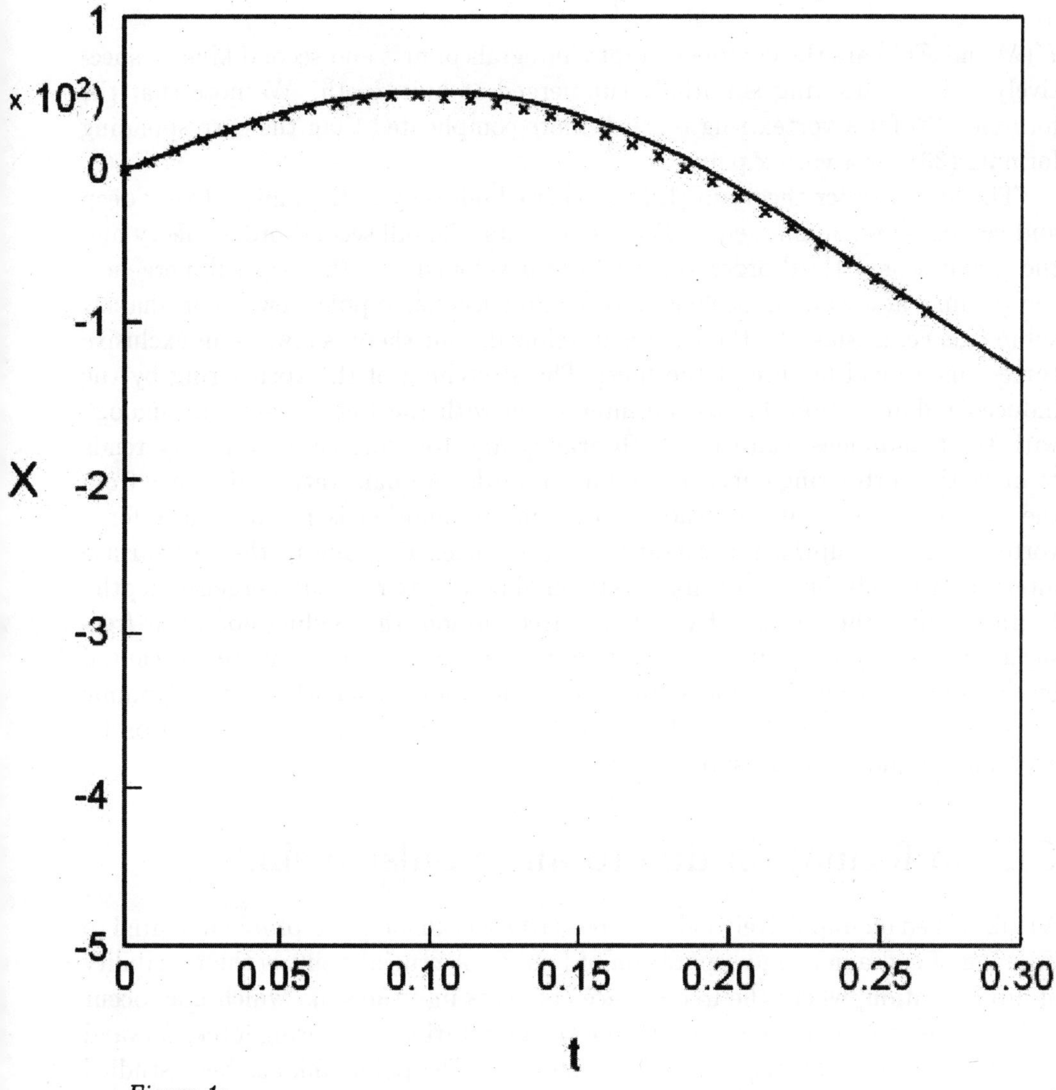

Figure 1:
The time evolution $X(t)$ for the position of a single vortex. Displayed case: F=0.01.
Solid curve: Solution in terms of Padé approximants (eq. 42). The solution is valid asymptotically as $F \ll 1$.
Crosses: Numerical results computed by Tong [11, appendix].

are dimensionless with the initial submergence depth as unit. A modified radial ccordinate λ is defined as:

$$\lambda = 2\sqrt{\frac{rR}{1+(r+R)^2}} \tag{44}$$

$K(\lambda)$ and $E(\lambda)$ are the complete elliptic integrals of first and second kind, respectively. The vortex ring is initially submerged at unit depth. We note that the formula (43) for a vortex ring is much more complicated than the corresponding formula (33) for a vortex pair.

The higher-order theory in [12] was carried out only in the limits of very deep and very shallow submergence. For these limits, the full second-order theory and the gravitational third-order theory were developed. With deep submergence, the leading flow contribution is a three-dimensional dipole flow. For shallow submergence, a quasi-2D theory was developed. But there is always an exclusive three-dimensional feature of the flow: The stretching of the vortex ring by the induced radial motion due to the interaction with the free surface. In analogy with the two-dimensional case, both gravity and free-surface nonlinearity tends to move the vortex ring outwards and downwards. A quantitative difference from the vortex pair is that the leading-order inward motion is much weaker for a vortex ring. The upward acceleration of the vortex ring due to the free-surface interaction is only important for a very shallow vortex ring submergence depth.

In all these three cases of vortices in free motion, the leading nonlinear free-surface effect tends to reduce the size of a surface mound. This reduction is caused by the convective acceleration at the free surface, or equivalently, by the dynamic pressure. This is a nonlinear effect that leads to a reduced surface elevation by avoiding a buildup of pressure.

On dip formation due to an impulsive sink

A sink turned on impulsively below a free surface is an idealized problem related to draining of containers and separation of liquids. A crucial point is the possibility of dip formation, where the free surface collapses into the sink, which may occur if the sink is strong enough to overcome the gravitational restoring force. Tyvand [23] analyzed the submerged impulsive line sink. The point sink has been studied analytically and experimentally by Miloh and Tyvand [24] as well as numerically by Xue and Yue [25]. Kim et al. [26] and Kim [27] have performed a numerical study of dip formation due to a two-dimensional sink distribution. Earlier papers by Lubin and Springer [28], Zhou and Graebel [29] present experimental and numerical investigations of transient impulsive draining of containers from a hole in the bottom.

A submerged horizontal line sink

For a line sink we have the characteristic velocity defined as:

$$W = \frac{q}{H} \tag{45}$$

Here q is the volume flux per unit length of the line sink submerged initially at a depth H below the free surface. The first-order expression for the free surface shape is: The full third-order surface elevation above a line sink was derived by Tyvand [23]:

$$\eta(x,t) = -\frac{t}{\pi(x^2+1)} + \frac{x^4+6x^2-3}{8\pi^2(x^2+1)^3}\,t^2 \\ + \frac{1-x^2}{24\pi^3(x^2+1)^2}\left(\frac{1}{F^2}-1\right)t^3 + \frac{7x^2-1}{3\pi^3(x^2+1)^5}\,t^3 + O(t^4) \tag{46}$$

We insert $x = 0$ to find the behaviour of the surface above the sink:

$$\eta(0,t) = -\frac{t}{\pi} - \frac{3}{8}\left(\frac{t}{\pi}\right)^2 + \frac{F^{-2}-9}{24}\left(\frac{t}{\pi}\right)^3 + O(t^4) \tag{47}$$

Gravity enters the problem to third order, through the Froude number. The first competition between the source drawdown and gravitational restoration thus takes place at third order. The early balance between these two effects above the sink is given by $\eta_3(0) = 0$. This determines a critical Froude number for the initial stage of dip formation:

$$F_{max} = 1/3 \tag{48}$$

The subscript *max* indicated that this value is an upper bound for the criterion of dip formation. When $F > 1/3$ the initial tendency of the free surface is to collapse towards the sink. But we cannot exclude the possibility of dip formation to occur also for lower Froude numbers, due to the nonlinear autocatalysis where an already deformed surface is accelerated into the sink. The critical Froude number $1/3$ marks the distinction between positive and negative third-order elevation at $x = 0$. This appears to be the only sharp criterion for dip formation that can be given analytically. If F is slightly below $1/3$, we cannot be certain that a dip formation can be avoided. Even though the third-order elevation is positive, it may not be sufficient to overcome the fact that both the first and the second order elevations are negative. Therefore we want to see whether the upward gravitational pull is present not only at third order, but also at fourth order. So we compute the fourth-order gravitational effect on the surface elevation:

$$\eta_4^{(F)}(x) = \frac{5x^2-1}{48\pi^4F^2(x^2+1)^4} \tag{49}$$

This fourth-order gravitational effect is a nonlinear phenomenon, since it is triggered by the second-order elevation. Evaluating it at $x = 0$ shows that the gravitational pull above the sink is always downward at fourth order:

$$\eta_4^{(F)}(0) = -\frac{1}{48\pi^4 F^2} < 0 \tag{50}$$

In third-order theory, the only thing that could prevent surface collapse by dip formation was the upward gravitational pull. This effect tends to be undermined for slightly subcritical Froude numbers, since the fourth-order gravitational effect promotes collapse instead of preventing it. We thus confirm that our analytical criterion for dip formation F>1/3 is a sufficient but not necessary condition.

Kim [27] extended the two-dimensional line sink problem [23] to a sink array distribution, solving it numerically by a truncated Fourier series in space and forward integration in time. A generalized dip formation criterion was suggested by introducing an effective Froude number which is a decreasing function of the sink width.

A bottom sink in a two-dimensional canal

Landrini and Tyvand [30] investigated the free-surface flow due to an impulsively started line sink at the bottom in the case of uniform unit depth. A full third-order analysis was carried out, leading to the critical Froude number given by:

$$F_{max} = \frac{2}{\pi^2} = 0.20264 \tag{51}$$

This is much smaller than the value 1/3 for a submerged sink, which is likely because a submerged sink produces a much more concentrated flow. Numerical simulations of the full nonlinear problem were also performed, and it predicts a continuous collapse towards the sink for all $F > 0.14$. For Froude numbers in the range $0.06 < F < 0.13$ there will not be full gravitational restoration above the sink, but surface breaking will occur, and the simulation of the inviscid model has to stop. A narrow upward jet will occur before the breaking. What will happen after the breaking in an experimental situation is an open question.

The small-time expansion of Landrini and Tyvand [30] included two other types of impulsively started bottom flux. A horizontal dipole, and a forced downwelling flow near a stagnant layer. The small-time expansions were taken to the third order. The aim was to see how the early nonlinearities evolve in time. Upwelling flows were also included in the model, but physically this is not acceptable because there will be vortex shedding, which the potential-flow model [30] did not account for.

The paper by Landrini and Tyvand [30] concentrated on the behaviour of bores generated by constant sources or sinks. This theme is omitted here because the nature of these flows is not impulsive even though they are started impulsively.

A submerged point sink

For a point sink below the free surface, the characteristic velocity W is defined as:

$$W = \frac{Q}{H^2} \tag{52}$$

Here Q is the volume flux of the point sink, and H is its initial depth of submergence. Miloh and Tyvand [24] investigated this problem by small-time expansion and found a critical value for the initial stage of dip formation:

$$F_{max} = 15^{-1/2} = 0.2582 \tag{53}$$

This Froude number is determined by the sign change of the third order elevation above the sink. The full second-order solution was derived. The third-order elevation was evaluated only at the central point above the sink. Experiments were performed, showing reasonable agreement with the theory. The theory was also extended to account for a container with a finite radius.

The problem of an impulsively started point sink submerged below a free surface in semi-infinite fluid has also been studied numerically by Xue and Yue [25]. For very small times, they confirmed our analytical criterion $F = 0.2582$ which determines zero third-order elevation above the sink. But they also found dip formation occuring for all Froude numbers greater than 0.1930, which shows that our third-order analytical criterion cannot give a physical criterion for dip formation, only an upper bound for the Froude number of dip formation. Our analytical criterion was based on a third-order analysis, but we can expect a fourth-order destabilizing effect, similar to the one found in eq.(50) for a line sink. So it is reasonable that the instability of dip formation will occur for Froude numbers significantly below the analytical value.

On bodies in forced motion

In this section we will discuss transient free-surface flows due to bodies in forced impulsive motion. The motion is assumed to be forced at all times, and we will treat in parallel the two different cases of constant velocity and constant acceleration.

We will investigate the hydrodynamic force exerted by the fluid on the body in forced motion. The unit of dimensionless force is defined by $(\rho H^2 W^2)$ in the three-dimensional case. The small-time expansion for the dimensionless hydrodynamic force $\mathbf{f}$ is:

$$\mathbf{f} = \mathbf{f_{-1}}\delta(t) + H(t)(\mathbf{f}_0 + \mathbf{f}_1 t + \mathbf{f}_2 t^2 + ...) \tag{54}$$

Here the impulsive force during the abrupt start of the motion is represented by Dirac's delta function $\delta(t)$. The hydrostatic force is omitted from this analysis, since it is not turned on at time zero.

Generally there are three different classes of bodies in forced motion: (i) The basic class considered in the present section, comprises bodies that are located in the fluid already before the motion starts. This includes any body that is submerged, floating or surface-piercing, before it is suddenly put into motion from rest. The amount of analytical work in this field is rather small and recent. (ii) The second class is "bodies" that contain the fluid; e.g. container walls or external boundaries of the fluid domain that are suddenly put into motion. This is the subject of a later section. The literature is relatively scarce. (iii) The third class comprises impact problems, where either a body penetrates the free surface from outside at time zero, or a free surface in motion impacts on a solid wall. There is a voluminous literature in this field. In the present article we choose not to discuss impact problems, where free-surface nonlinearities are usually disregarded.

Transformations linking constant velocity and constant acceleration

The literature on impulsive free-surface flows due to bodies in forced motion is divided between the cases of constant velocity and constant acceleration starting from rest. Therefore we will start with deriving transformation equations linking the two cases. This has already been done by Tyvand and Miloh [5, section 10], but unfortunately with some mistakes that are now corrected. These transformations are valid only in a coordinate system at rest with the undisturbed fluid.

In section 2 we have introduced a characteristic velocity W, which is the initial velocity for a body starting impulsively from rest. We will also study the case of a body starting from rest with a constant acceleration A. The characteristic velocity in this case is defined as:

$$W = \sqrt{AH} \tag{55}$$

Here H is the unit of dimensionless length.

Let us denote the case of constant acceleration by a star (*) as superscript. The case of constant velocity is left without superscript. The initial position is by definition that of unit submergence:

$$\mathbf{R}_0^* = \mathbf{R}_0 = (0, 0, -1) \tag{56}$$

Since we compare a constant unit velocity with a constant unit acceleration in the same direction, we have:

$$\mathbf{R}_2^* = \mathbf{R}_1/2 \tag{57}$$

Furthermore we have by definition:

$$\mathbf{R}_n^* = 0, \quad \text{n=1, n>2} \tag{58}$$

Let us give the free-surface boundary conditions for constant acceleration: The dynamic conditions are:

$$\begin{aligned} \Phi_0^* = \Phi_1^* = \Phi_2^* = 0, &\quad \text{z=0} \\ \Phi_3^* = -\eta_2^* \frac{4\eta_2^* + (\Omega F)^{-2}}{3}, &\quad \text{z=0} \end{aligned} \tag{59}$$

The corresponding kinematic conditions can be written as:

$$\begin{aligned} \eta_1^* = \eta_3^* &= 0 \\ 2\eta_2^* &= \frac{\partial \Phi_1^*}{\partial z}, \quad \text{z=0} \\ 4\eta_4^* &= \frac{\partial \Phi_3^*}{\partial z}, \quad \text{z=0} \end{aligned} \tag{60}$$

The corresponding two nontrivial conditions at the boundary of an accelerated circular cylinder are:

$$\frac{\partial \Phi_1^*}{\partial r} = 2\mathbf{n} \cdot \mathbf{R}_2^* \text{ along initial contour} \tag{61}$$

$$\frac{\partial \Phi_3^*}{\partial r} = -\mathbf{n} \cdot \nabla(\mathbf{R}_2^* \cdot \nabla \Phi_1^*) \text{ along initial contour} \tag{62}$$

A comparison with the boundary value problems for constant velocity (without superscript) gives the following transformation equations for the potential:

$$\Phi_0^* = 0 \tag{63}$$

$$\Phi_1^* = \Phi_0 \tag{64}$$

$$\Phi_2^* = 0 \tag{65}$$

$$\Phi_3^* = \frac{2}{3}\phi_1 + \frac{1}{2}\psi_1 + \frac{1}{3}\Phi_2^{(F)} \tag{66}$$

The transformation equations for the surface elevations are:

$$\eta_1^* = 0 \tag{67}$$

$$\eta_2^* = \frac{1}{2}\eta_1 \tag{68}$$

$$\eta_3^* = 0 \tag{69}$$

$$\eta_4^* = \frac{1}{3}\eta_2^{(conv)} + \frac{1}{4}\eta_2^{(geom)} + \frac{1}{4}\eta_3^{(F)} \tag{70}$$

Superscript (F) for the third-order potential denotes its Froude-number dependent (gravitational) part. The second-order elevation is separated into its two constituents:

$$\eta_2^{(conv)} = \frac{1}{2}\frac{\partial \phi_1}{\partial z}, \quad \text{z=0} \tag{71}$$

$$\eta_2^{(geom)} = \frac{1}{2}\frac{\partial \psi_1}{\partial z}, \quad z=0 \tag{72}$$

These are the elevations due to convective acceleration (superscript: *conv*) at the free surface and to geometric nonlinearity (superscript: *geom*), respectively. The corresponding zeroth-order forces will be denoted with the same superscript.

The transformation equations for the force from the fluid on the body are as follows:

$$\mathbf{f}^*_{-1} = 0 \tag{73}$$

$$\mathbf{f}^*_0 = \mathbf{f}_{-1} \tag{74}$$

$$\mathbf{f}^*_1 = 0 \tag{75}$$

$$\mathbf{f}^*_2 = \mathbf{f}_0^{(dyn)} + 2\mathbf{f}_0^{(conv)} + \frac{3}{2}\mathbf{f}_2^{(geom)} + \frac{1}{2}\mathbf{f}_1^{(F)} \tag{76}$$

Here $\mathbf{f}_{-1}$ is given by the pressure impulse (25) integrated over the surface. The forces introduced in eq. (76) result from integrating up the following pressure contributions, which are defined for the case of constant impulsive velocity:

$$p_0^{(dyn)} = -\frac{|\Phi_0|^2}{2} \tag{77}$$

$$p_0^{(conv)} = -\phi_1 \tag{78}$$

$$p_0^{(geom)} = -\psi_1 \tag{79}$$

$$p_1^{(F)} = -2\Phi_2^{(F)} \tag{80}$$

The present formulas correct the errors in the section on constant acceleration in the paper by Tyvand and Miloh [5]. A new superscript *dyn* represents the force due to the dynamic pressure from Bernoulli's equation. These formulas are adequate for bodies in translation as well as a sphere and a circular cylinder where rotational modes do not exist.

In order to generalize this analysis to include rotational modes, we need to add a new type of terms describing the mutual interaction of finite Euler angles of rotation. We must also include expressions for the torques. No general analysis of the impulsive motion of three-dimensional bodies near a free surface exists. Tyvand [31] has studied the forces and torques due to impulsive rotational modes for the relatively simple case of a torus in initially horizontal position.

It is important to note that the transformations developed here apply only to the outer solution in a matched asymptotic expansion. If there are no singularities at the free surface, the outer solution is identical to the total solution.

A surface-piercing vertical cylinder

Wang and Chwang [4] performed the first analytical study of an impulsive free-surface flow due to a three-dimensional body. They considered a vertical circular cylinder in constant acceleration from rest. The motion is along the positive x axis. We will give their basic results with our present formulation (section 2), interpreting H as the fluid depth and taking it as unit of dimensionless length. So at $z = -1$ there is a boundary condition of zero normal velocity. The dimensionless radius R is then a slenderness parameter for the cylinder.

We will now reformulate Wang and Chwang's problem for the case of constant velocity, carrying out the transformations given above. Their dimensional acceleration a_1 corresponds to our symbol A. Applying the transformation (68) to their result, the dimensionless first-order elevation around a cylinder started impulsively from rest will be:

$$\eta_1 = 2\eta_2^* = -2\cos\theta \sum_{m=1}^{\infty} \frac{K_1(\lambda_m r)}{\lambda_m K_1'(\lambda_m R)} \tag{81}$$

K_1 denotes a first-order modified Bessel function of the second kind. K_1' denotes its derivative with respect to the argument. We have introduced polar coordinates r, θ defined by:

$$\begin{aligned} r &= \sqrt{x^2+y^2} \\ \cos\theta &= x/r \\ \lambda_m &= (m-1/2)\pi \end{aligned} \tag{82}$$

The impulsive hydrodynamic force on the slender cylinder is directed along the $-x$ axis and has the magnitude:

$$f_{-1} = f_0^* = -2\pi R \sum_{m=1}^{\infty} \frac{K_1(\lambda_m R)}{\lambda_m^3 K_1'(\lambda_m R)} \tag{83}$$

The ratio between the impulsive added mass and the displaced fluid mass is thus:

$$\frac{f_{-1}}{\pi R^2} = -\frac{2}{R} \sum_{m=1}^{\infty} \frac{K_1(\lambda_m R)}{\lambda_m^3 K_1'(\lambda_m R)} \tag{84}$$

The higher-order analysis of Wang and Chwang [4] is inconsistent because there is a waterline singularity in the leading order, which needs to be resolved in terms of matched asymptotic expansions. This difficult problem is unsolved.

An initially semi-submerged horizontal cylinder

Vinje [2] investigated the difficult problem of a semi-submerged circular cylinder in impulsive motion. It is especially difficult because of the intricate geometric nonlinearities and singularities at the waterline.

Vinje [2] derived formulas for the steady-state force components. Greenhow and Li [32] had earlier derived the added momentum of an impulsively started semi-submerged cylinder. Let us summarize their formulas for impulsive sway motion in our dimensionless form:

$$\begin{aligned} f_{-1x} &= -\frac{2}{\pi} \\ f_{0y} &= -\frac{32}{15} \\ f_{1x} = -\frac{112}{315\pi} &= 0.113 \end{aligned} \tag{85}$$

The corresponding formulas for impulsive heave motion are:

$$\begin{aligned} f_{-1x} &= -\frac{\pi}{2} \\ f_{0y}^{(conv)} &= \frac{6}{5} \\ f_{0y}^{(dyn)} &= -\frac{2}{3} \end{aligned} \tag{86}$$

Vinje [2] stated only the sum (8/15) of the two last terms, but we must write them separately if we want to apply the transformations (76). The unit length is taken as the cylinder radius. The heave and sway modes of motion were treated separately by Vinje [2]. His analysis did not cover oblique motion, because the nonlinear interaction between horizontal and vertical motion was disregarded.

Geometric nonlinearity has been disregarded so far. It enters the zeroth-order force in the case of heave motion because the submergence changes with time, which is not the case for the sway mode. Taylor [33] derived an expression for the added mass of a partly submerged cylinder in impulsive motion, see also Greenhow and Ahn [34]. From Taylor's expression we derive a formula for the z-derivative of the added mass of a semi-submerged cylinder in heave motion. This gradient of the added mass gives rise to a force contribution:

$$f_{0y}^{(geom)} = \frac{2}{3} \tag{87}$$

This geometric nonlinearity force contributes to the total zeroth-order heave force, and cancels the contribution $f_{0y}^{(dyn)}$ due to dynamic pressure along the body. Thus the net force turns out to be equal to $f_{0y}^{(conv)}$ which is the leading-order effect of convective acceleration at the free surface.

However, the free-surface singularity at the leading order makes this higher-order analysis questionable. A consistent analysis must resolve the singularity by matched asymptotic expansions.

A submerged horizontal cylinder

Tyvand and Miloh [5] investigated the impulsive start of a horizontal submerged cylinder with an arbitrary radius (with initial submergence depth as unit length). The full second-order solution and the third-order gravitational contribution were found. Tyvand and Miloh [6] considered the small-cylinder limit of the same problem, where the cylinder was represented by a moving dipole. The accuracy of the small-cylinder limit was investigated in detail: It gives a good approximation for the surface elevation as long as the ratio of radius to initial submergence depth is 0.3 or less.

For a body in forced motion, the hydrodynamic force is of basic interest. Tyvand and Miloh [6] applied Fourier series in bipolar coordinates and derived an exact formula for the impulsive added-mass force per unit length of a cylinder of radius R:

$$f_{-1} = 4\pi(1 - R^2) \sum_{n=1}^{\infty} n \ \exp(-2n \ \text{arcsech}R) \tanh(n \ \text{arcsech}R) \qquad (88)$$

The initial submergence depth is unity, which requires $R < 1$. The direction of this force is opposite of the forced motion with unit velocity. Venkatesan [35] and Greenhow and Li [32] had earlier calculated this added mass by other analytical methods.

The impulsive force and the leading gravitational force for a small cylinder can be approximated by a dipole. However, the dipole limit is inconsistent as basis for predicting the zeroth-order hydrodynamic force acting on a small cylinder. There are three different types of steady-state forces on the cylinder. (i) The dynamic-pressure force due to the variation of the velocity along the cylinder contour. (ii) The force due to convective acceleration at the free surface. It is consistent with the small-cylinder dipole theory. (iii) The force contributions due to the fact that the cylinder is displaced a finite distance away from its initial position (geometric nonlinearity). There are two such forces: First the local force contribution at the cylinder due to its change of position. Secondly the force due to the finite penetration of the cylinder into the initial flow field. This is not included if the cylinder is represented by a dipole [6]. A general analysis with arbitrary cylinder size [5] is necessary to treat all force nonlinearities consistently.

From numerical analyses [7,36,37,38] and experiments [39] it is known that a large circular cylinder in rapid upward motion tends to build up a fluid layer of almost uniform radial thickness. Tyvand and Miloh [5] identified this analytically as a nonlinear phenomenon. It is primarily due to geometric nonlinearity, because the cylinder lifts the fluid more efficiently the closer it gets to the surface.

In the thesis by Moyo [40], detailed comparisons are given between a fully nonlinear boundary-integral method and our third-order analytical formulas for the surface elevation [5]. Some of Moyo's results are also published in the recent paper by Greenhow and Moyo [41] in a special journal issue devoted to violent

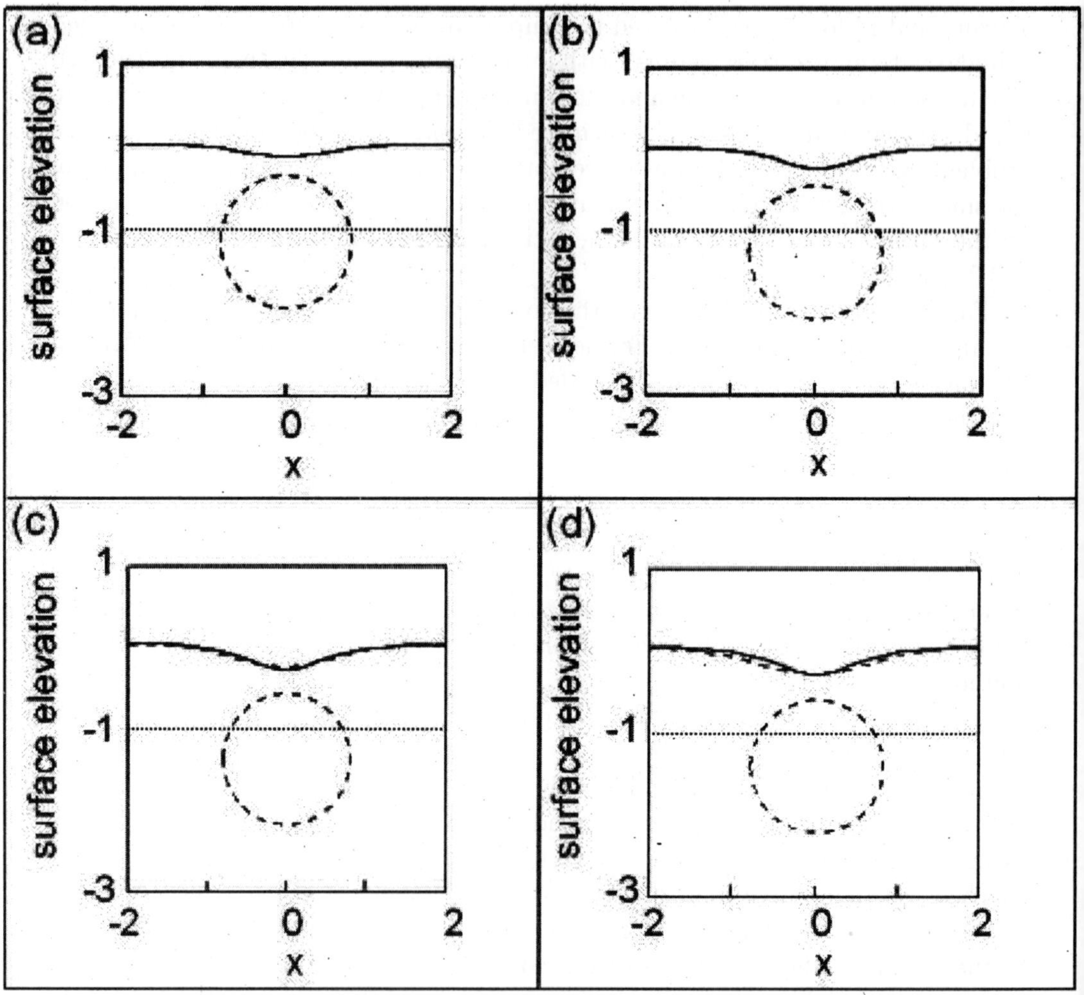

Figure 2:
Surface elevation above a horizontal circular cylinder moved impusively downwards with constant velocity, taken from Moyo [40, fig. 5.3]. Initial ratio (circle radius)/(submergence depth) of circle centre: 0.8. Froude number: $F = 0.64$. Snapshots at $t = 0.15$ (a), $t = 0.25$ (b), $t = 0.35$ (c) and $t = 0.4$ (d).
Dashed curves: Moyo's numerical computations [34]. Reproduced with permission from Dr. S. Moyo and Brunel University.
Solid curves: Third-order expansion according to Tyvand and Miloh [5].

surface motion including impact problems. The agreement is excellent for Froude numbers somewhat less than one, and for dimensionless times of order 1/2 or less. See figure 2, which shows the surface elevation above a cylinder in downward motion at Froude number 0.64. The Froude number has to be smaller than one in such a comparison, so the gravitational term dominates over the neglected non-gravitational terms in the third-order solution. On the other hand, the Froude number cannot be too small, otherwise the gravitational term blows up quickly and limits the time range of validity for our theory.

The above mentioned papers treat cylinders in forced motion. Moyo and Greenhow [42] studied numerically a neutrally buoyant cylinder moving freely after having been put into motion.

A submerged horizontal torus

Tyvand [31] performed a quasi-2D analysis of a wide torus initially submerged horizontally below a free surface, based on our analysis for a cylinder [5,6]. A thorough investigation of the accuracy of the quasi-2D theory is performed to the leading order, by comparing it with a fully three-dimensional slender-body theory. The local first-order surface elevation and the impulsive force are compared. The force prediction by quasi-2D theory is excellent when the ratio of torus radius to submergence depth is above 4. This quasi-2D theory for the torus has generally a much greater range of validity than the similar quasi-2D theory for the vortex ring [12].

The leading nonlinear interactions for the submerged torus manifest themselves in the second-order elevation and the zeroth-order (steady) forces. Tyvand [31] tabulated the zeroth-order forces and torques after a general motion has been started with all six degrees of freedom. The 6×6 interaction matrix convers all possible interactions between modes, including the self-interactions. 13 out of the 36 elements in this interaction matrix correspond to an induced heave force. Two elements correspond to inducing both a sway force and a roll torque, while two elements result in both a surge force and a pitch torque. In addition there are two matrix elements for each of the four separate modes surge, sway, roll and pitch. The last 11 matrix elements are void.

A submerged sphere

Miloh [43] has formulated a general framework for solving the impulsive motion of rigid or deformable spherical shapes below a free surface using the small-time expansion method. This is the first example of using a consistent small-time expansion for a submerged three-dimensional body. The mixed boundary-value problem consists of solving a Neumann-type oundary condition on the instantaneous spherical shape and a Dirichlet-type boundary condition on the initially undisturbed free surface. The flow field is then obtained by employing a Green function approach, Fourier-Bessel transforms and the method of multipole expansions. Analytical expressions correct to any order for the free-surface deflection can be found by inverting an infinite set of linear equations for the coefficients of the Green function expansion. The hydrodynamic force to arbitrary order can also be found from the same procedure by using the time-dependent Bernoulli equation and the Lagally theorem. A numerical example is presented for the case of a sudden spherical explosion below a free surface. The results are compared against the analytical solution based on a small-time expansion method. Good agreement between numerical and analytical results are found both for the early free-surface elevation and the pressure force exerted on the deforming sphere by the surrounding fluid.

On impulsive wavemakers and moving containers

An important class of problems is that of the sudden motion of partly filled containers with a free surface of fluid. In the limit where the width of the container is infinitely large compared with its depth, it reduces to the impulsive wavemaker problem. This is the problem where a wall meeting a free surface is suddenly put into motion.

The impulsive motion of a container or wavemaker may be with constant velocity or constant acceleration. The transformation equations (64), (68) and (74) between the two cases are still valid, as long as only the outer flow is considered according to a matched asymptotic expansion.

A piston wavemaker in surge motion

Peregrine [1] determined the flow generated by a vertical plate suddenly put into uniform horizontal motion (surge mode) in x direction. This is now considered as a fundamental problem in water wave theory: The impulsive wavemaker. With our length unit chosen as the fluid depth, the dimensionless first-order elevation is:

$$\eta_1 = 2\eta_2^* = -\frac{2}{\pi}\log\tanh\frac{\pi x}{4} \tag{89}$$

Some of Peregrine's results can be recovered from earlier work of Westergaard [44], who determined the impulsive force on the plate. In dimensionless form it reads

$$f_{-1} = f_0^* = \frac{2}{\pi^3}\sum_{n=1}^{\infty}\left(n-\frac{1}{2}\right)^{-3} = \frac{14\,\zeta(3)}{\pi^3} = 0.542755 \tag{90}$$

where ζ denotes Riemann's zeta function. Here we have also shown the results (with subscript (*)) to the leading order for a wavemaker in uniform acceleration.

The impulsive force per unit length of the wavemaker points in -x-direction, opposite to its surge motion. Since the flow is not symmetric, there will be a net pitch torque (positive in the clockwise direction) exerted on the wavemaker face about $z = 1/2$:

$$M_{-1} = 0.0535023 \tag{91}$$

Thus the resultant force (90) acts at a point $z = -0.598575$. Westergaard's theory was linear and assumed vertical dam face. Higher-order forces in the case of constant acceleration have been computed by Chwang [45]. Chwang and Housner [46] and Chwang [47] had earlier extended Westergaard's linear theory to dams with sloping face. Pohle [8,9] extended dam breaking theory by applying the Lagrangian description of motion and small-time expansion, see Stoker [48].

Peregrine's solution (89) contains a logarithmic singularity at the intersection point between the wall and the free surface. He suggested a local inner solution as the way to render a uniformly valid expansion. Eventually, King and Needham [49] took up this idea of an inner expansion near the wall. In the meantime, several papers had been written on the transient wavemaker problem. Newman [50] investigated the higher-order problem using the Lagrangian description of motion. Greenhow and Lin [39] showed experimentally that the mathematical singularity corresponds to a narrow upward wall-jet. Roberts [51] generalized the initial condition and showed how a smoother (non-impulsive) start can remove the free-surface singularity. His analysis was carried further by Yoo et al. [52] and Miles [53], to include surface tension. Frankel [54] tried to overcome the free-surface singularity by introducing fluid compressibility.

King and Needham [49] derived a complete uniformly valid solution using small-time expansion for an impulsively accelerating plate moving horizontally into a stationary fluid layer. (The problem of constant velocity will require a separate analysis.) The reason for the singularity is the neglect of nonlinear inertial terms in the outer solution. Convective inertia is taken into account in the inner solution, where it gives the dimensionless surface elevation at the intersection with the wall:

$$\eta(0,t) = -\frac{4\sigma}{\pi}t^2 \log t - \frac{2}{\pi}\sigma t^2 \left(\log(-\log t) - \log\frac{4}{\pi} + \log\frac{2s}{\pi} - 1 - \frac{\Gamma'(3/2)}{\Gamma(3/2)}\right) + o(t^2) \tag{92}$$

Here σ denotes the ratio between the acceleration of the plate and the gravitational acceleration. It is assumed to be of order unity. A simple expression valid to leading order in time was found for the surface slope at the plate:

$$\frac{\partial \eta}{\partial x} = \frac{\pi}{4 \log t} \quad \text{at x=0} \tag{93}$$

This slope is initially zero and it is negative thereafter. King and Needham [49] determined the two-term inner solution that is uniformly valid both for $t \to 0^+$

and $\chi \ll 1$:

$$\eta = -\frac{4\sigma}{\pi}t^2 \log t - \frac{2}{\pi}\sigma t^2 \left(\log(-\log t) - \log\frac{4}{\pi} + \log\frac{2\sigma}{\pi} - 1 - \frac{\Gamma'(3/2)}{\Gamma(3/2)}\right) - \frac{\pi^2\chi}{8}t^2 \tag{94}$$

Here χ is the inner horizontal variable, which is related to the physical distance x (dimensionless with depth as unit) by the relationship:

$$x = -\chi t^2 \log t \tag{95}$$

A piston wavemaker in pitch motion

We will complement Peregrine's forward moving (surging) wavemaker solution with the linear theory of a pitching wavemaker, defined by the boundary condition:

$$\frac{\partial \Phi}{\partial x} = 1 - z \quad \text{at} \quad x = 0 \quad (0 < z < 1, t \geq 0) \tag{96}$$

Note that the velocity is zero at the undisturbed free surface $z = 0$ for this pure pitch mode. This is the only way of obtaining a regular solution. Zero velocity at any other z level can be obtained by adding the surge mode with an appropriate amplitude. The pitch mode means a linear distribution of the forced normal velocity. The potential is found very easily by integrating up Peregrine's solution [1] in the z-direction:

$$\Phi_0 = -\frac{2}{\pi^3}\sum_{n=1}^{\infty}\frac{\cos(n-\frac{1}{2})\pi z}{(n-\frac{1}{2})^3}\, e^{-(n-\frac{1}{2})\pi x} \tag{97}$$

The first-order elevation is given by the formula:

$$\eta_1 = -\frac{2}{\pi^2}\sum_{n=1}^{\infty}(-1)^n \frac{e^{-(n-\frac{1}{2})\pi x}}{(n-\frac{1}{2})^2} \tag{98}$$

We note that this surface elevation has the finite value $8\pi^{-2}G = 0.742454$ at $x = 0$ (G is Catalan's constant 0.915965594..). The net impulsive force on this two-dimensional pitching wavemaker is:

$$f_{-1} = -\frac{2}{\pi^4}\sum_{n=1}^{\infty}\frac{(-1)^n}{(n-\frac{1}{2})^4} = 0.324880 \tag{99}$$

The impulsive torque about $z = -1/2$ is integrated up as:

$$M_{-1} = 0.0476432 \tag{100}$$

Thus the line of action of the resultant is located at the point $z = -0.646649$.

Axisymmetric wavemakers

The leading-order impulsive axisymmetric wavemaker problem of a vertical cylinder in water of finite (unit) depth is given as follows in cylindrical coordinates r, θ, z:

$$\nabla^2 \Phi_0 = 0, \quad -1 < z < 0, \quad r > R \tag{101}$$

$$\left.\frac{\partial \Phi_0}{\partial r}\right|_{r=R} = 1 \tag{102}$$

$$\Phi_0|_{z=0} = 0 \tag{103}$$

$$\left.\frac{\partial \Phi_0}{\partial z}\right|_{z=-1} = 0 \tag{104}$$

$$|\nabla \Phi_0| \to 0, \quad r \to \infty \tag{105}$$

The solution for this leading-order potential is given in terms of K_0, the modified Bessel function of second kind and zeroth order:

$$\Phi_0(r,z) = \sum_{n=0}^{\infty} A_n \mathrm{K}_0(k_n r) \cos k_n(z+1) \tag{106}$$

The wavenumbers k_n are defined by:

$$k_n = \left(n + \frac{1}{2}\right)\pi \tag{107}$$

Imposing eq.(102) implies that the coefficients A_n are given by:

$$A_n = \frac{8(-1)^n}{(2n+1)^2 \pi^2 \mathrm{K}_0'(k_n R)} \tag{108}$$

This implies that the first-order elevation is given by:

$$\eta_1(r) = \left.\frac{\partial \Phi}{\partial z}\right|_{z=0} = \frac{4}{\pi} \sum_{n=0}^{\infty} \frac{\mathrm{K}_0(k_n r)}{(2n+1)\mathrm{K}_1(k_n R)} \tag{109}$$

This solution is logarithmically singular at the waterline $r = R$.

The dimensionless force on an infinitesimal slice $d\theta$ of the cylinder is given by:

$$df_{-1} = \frac{16R}{\pi^3} d\theta \sum_{n=0}^{\infty} (2n+1)^{-3} \frac{\mathrm{K}_0(k_n R)}{\mathrm{K}_1(k_n R)} \tag{110}$$

In the limit $R \to \infty$ we have $df_{-1}/(R\ d\theta) = 14\ \zeta(3)/\pi^3$, in agreement with eq.(90). The force acts at a point given by:

$$z = -\left(\sum_{n=0}^{\infty} \frac{(-1)^{n+1}}{k_n^4} \frac{\mathrm{K}_0(k_n R)}{\mathrm{K}_1(k_n R)}\right) \Bigg/ \left(\sum_{n=0}^{\infty} \frac{1}{k_n^3} \frac{\mathrm{K}_0(k_n R)}{\mathrm{K}_1(k_n R)}\right) \tag{111}$$

radius R	minimum	average
.001	1999	2000
.01	199.2	200
.1	19.23	20
.2	9.230	10
.5	3.230	4
1	1.232	2
2	.2838	1
5	.0037	.4

Table 1: Minimum first-order elevation $\eta_1(0)$ for an inward axisymmetric wavemaker given as a function of the cylinder radius R and compared with the average velocity $2/R$. Eq.(116) is truncated after 1000 terms.

For large R this again gives $z = -0.598575$.

The above solution represents the outward impulsive wavemaker. We define the inward impulsive wavemaker by the following boundary value problem:

$$\nabla^2\Phi_0 = 0, \quad -1 < z < 0, \quad 0 \geq r < R \tag{112}$$

$$\left.\frac{\partial\Phi_0}{\partial r}\right|_{r=R} = -1 \tag{113}$$

$$\Phi_0|_{z=0} = 0 \tag{114}$$

$$\left.\frac{\partial\Phi_0}{\partial z}\right|_{z=-1} = 0 \tag{115}$$

The leading-order solution is simply given is given by replacing the modified Bessel function K_0 (of the second kind) in the potential (106) for the outward wavemaker with the modified Bessel function I_0 (of the first kind). This solution for the inward wavemaker is still logarithmically singular at $r = R$. The minimum value of the first-order elevation occurs at the origin $r = 0$ and is given by:

$$\eta_1(0) = \left.\frac{\partial\Phi}{\partial z}\right|_{z=0} = \frac{4}{\pi}\sum_{n=0}^{\infty}\frac{1}{(2n+1)\mathrm{I}_1(k_n R)} \tag{116}$$

The first-order solution takes care of the mass balance. It is interesting to see how this minimum value differs from the average upward flow $2/R$ across the cross-section of the cylinder. This is shown in Table 1, and illustrates how the surface flow is approximately a uniform piston flow over most of the cross-section when $R < 0.2$.

An upright cylindrical container in impulsive motion

Chwang and Wang [55] were the first to develop a nonlinear hydrodynamic analysis of impulsive free-surface flow in containers. They gave a small-time expansion for accelerated cylinders with rectangular as well as circular cross sections. However, the higher-order solutions are not valid because of the unresolved singularity in the leading order. The rectangle problem was solved independently by Roberts [56] in terms of elliptic functions, whereas Chwang and Wang [55] had applied Fourier series.

Here we will consider the problem of constant impulsive velocity, and give only the leading-order solution. Above we derived transformation equations linking the cases of constant impulsive velocity and constant impulsive acceleration. These transformations were derived for moving bodies, but they are also valid for moving containers.

The first-order elevation for a cylinder with a circular cross section is:

$$\eta_1 = 2\eta_2^* = -2\cos\theta \sum_{m=1}^{\infty} \frac{\mathrm{I}_1(\lambda_m r)}{\lambda_m \mathrm{I}_1'(\lambda_m R)} \qquad (117)$$

Here we take the same convention as for the solid cylinder, introducing $\lambda_m = (m - 1/2)\pi$. I_1 denotes the first order modified Bessel function of the first kind. I_1' denotes its derivative with respect to the argument. In eq.(117) we apply the transformation (68) to the case of constant acceleration [55]. We are basically interested in the hydrodynamic force exerted on the container walls by the fluid. It gives a virtual mass that is always smaller than the fluid mass. The ratio between the impulsive virtual mass and the total fluid mass is found to be:

$$\frac{f_{-1}}{\pi R^2} = \frac{2}{R} \sum_{m=1}^{\infty} \frac{\mathrm{I}_1(\lambda_m R)}{\lambda_m^3 \mathrm{I}_1'(\lambda_m R)} \qquad (118)$$

This relative virtual mass is shown in Table 2 as a function of the dimensionless radius R, where it is compared with formula (84) for the flow around a cylinder. The virtual mass inside a container approaches the value 1 (equal to the displaced fluid mass) as $R \to 0$ much more quickly than the added mass outside a cylinder. This is obvious because the surface outside the cylinder has much more freedom to move than that inside. On the other hand, the virtual mass tends to become the same outside and inside a cylinder when $R \ll 1$. This can be understood because both these two solutions tend to a common quasi-two-dimensional limit where the x projection of the two-dimensional wavemaker force (90) is integrated up along the contour. This gives a net force f_{-1} in the $-x$ direction given by:

$$\frac{f_{-1}}{\pi R^2} = \frac{0.542755}{R} \qquad (119)$$

Table 2 shows how the impulsive forces both outside and inside a vertical cylinder approach this formula asymptotically as $R \to \infty$. The impulsive forces for $R =$

radius R	external	internal
.001	.9988	.9995
.01	.9883	.9953
.1	.8889	.9526
.2	.7922	.9051
.5	.5800	.7630
1	.3890	.5478
2	.2299	.3002
5	.1018	.1148
10	.05259	.05591
100	.005411	.005444

Table 2: Relative added mass (vs. displaced fluid mass) for impulsive motion of a vertical circular cylinder of radius R on unit water depth. According to eq. (84) taken from Wang and Chwang [4] we give a column of results for the external flow outside a cylinder for increasing values of R. From eq. (118) taken from Chwang and Wang [55] we give a column of results for the internal flow inside a cylinder.

100 are $0.5411/R$ outside the cylinder and $0.5444/R$ inside the cylinder. The outside flow gives a smaller force because there is more free-surface area available for piling up fluid around the cylinder than the two-dimensional solution assumes. This reduces the local surface elevation and the force. Inside the cylinder, this available surface area is smaller, so the force is larger than the asymptotic limit.

In practical applications for partly filled tanks on ships, trucks or trains, values of R between 1 and 2 are common. Then the virtual mass is about 30 – 50% of the fluid mass. This means that at least half the forward momentum of the fluid inside the container will be preserved during a collision where the vehicle is suddenly brought to a halt.

A rectangular container in impulsive motion

From Chwang and Wang [55] we derive the formula for the virtual mass relative to the fluid mass in a rectangular container of dimensionless width L put into motion:

$$\frac{f_{-1}}{L} = \frac{f_0^*}{L} = -\frac{4}{L}\sum_{m=1}^{\infty}\frac{1-\cosh\lambda_m L}{\lambda_m^3 \sinh\lambda_m L} \qquad (120)$$

The fluid depth is unity. Again we apply the notation $\lambda_m = (m - 1/2)\pi$. This formula cover both the cases of constant velocity and constant acceleration.

For a rectangular cross-section (Chwang and Wang [55], Roberts [56]), it is obvious that the length of the rectangle perpendicular to the motion does not play

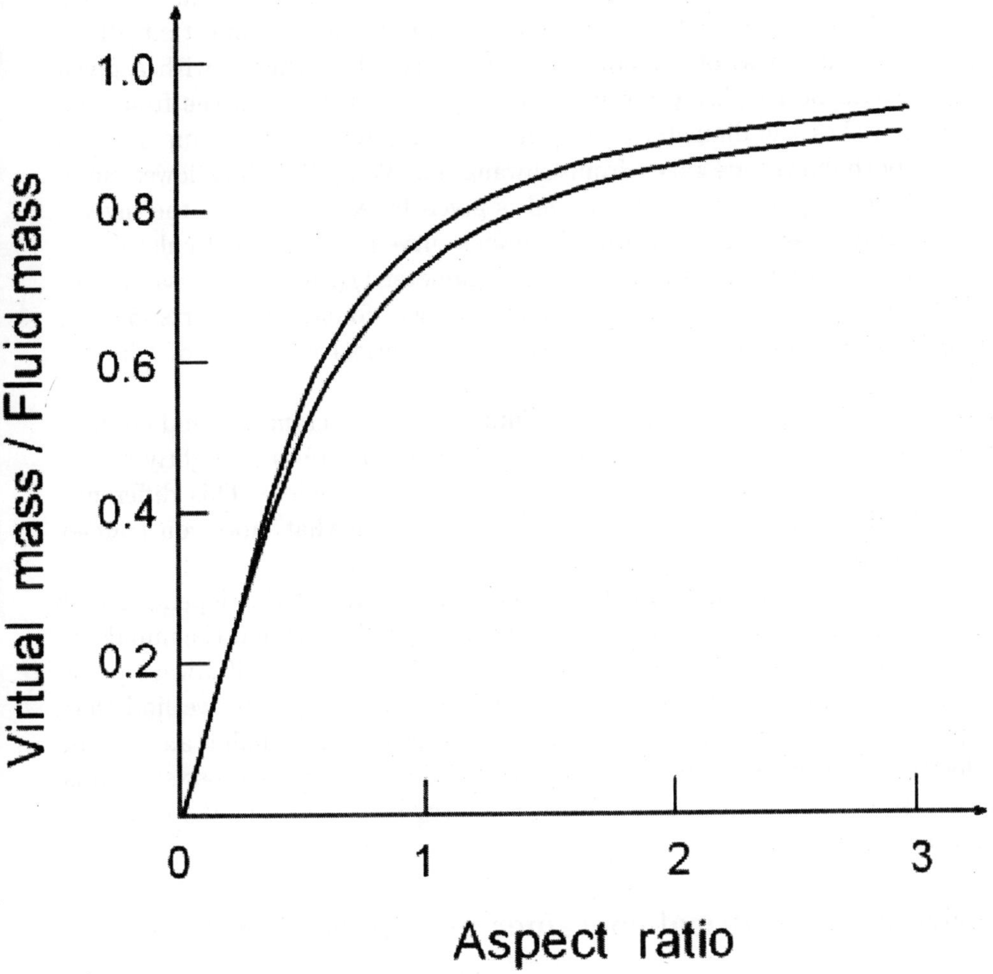

Figure 3:
Ratio of virtual mass vs. total fluid mass as a function of the aspect ratio of two different vertical cylinders. Results found from Chwang and Wang [55].
Upper curve: Circular cross-section, represented by eq.(118). Aspect ratio defined as height vs. cylinder diameter $(1/(2R))$.
Lower curve: Rectangular cross-section, represented by eq.(120). Aspect ratio defined as height vs. streamwise width $(1/L)$.

any role. By analogy it is also likely that the length of the half-axis of an elliptical cylinder which is perpendicular to the motion) is relatively unimportant. If so, the relative virtual mass of a rectangular and a circular cylinder with a given fluid level would be roughly the same, provided the gap between the front and rear walls matches the diameter of the circle. In figure 3 such a comparison is made, and both curves are taken from Chwang and Wang [55]: The lower curve represents a rectangular cross-section and is given by eq.(120), with the inverse width $1/L$ as abscissa (aspect ratio). The upper curve represents a circular cross-section and is given by (118) with the inverse diameter $1/(2R)$ as abscissa (aspect ratio). The relative added mass for the rectangle was found by Roberts [56, fig. 2], applying a different method His curve looks identical with that of Chwang and Wang [55].

The two curves shown in figure 3 for containers with rectangular and circular cross-sections are remarkably close to one another. There is a slightly larger relative virtual mass for the fluid inside the circular cylinder. This difference reflects that the fluid inside a circular boundary is somewhat more constrained than the fluid inside a rectangular boundary.

Another point of comparison between the two cylinders is the dependence of the force on the direction of the horizontal motion. It is obvious that the impulsive force from the fluid inside a circular cylinder put into motion will always point in the direction opposite of the motion. From evaluating eq.(120) we find that the impulsive force vector on a rectangular container makes a smaller angle than the velocity vector on the short side of the rectangle. The square cross-section is the only case where the impulsive force will always be opposite to the direction of impulsive motion.

A horizontal cylindrical container in impulsive motion

A half-filled circular cylinder lying horizontally is an interior-flow counterpart to the problem considered by Vinje [2]. The cylinder has unit radius and is put impulsively into unit velocity. The normal velocity along the bottom is:

$$\frac{\partial \Phi_0}{\partial r} = \cos\theta \quad \text{at r=1} \quad (\pi < \theta < 2\pi) \tag{121}$$

To derive the initial impulsive flow, we must incorporate the equipotential free-surface conditions. This is done by extending the definition of the normal derivative to be valid for the whole periphery:

$$\frac{\partial \Phi_0}{\partial r} = -\cos\theta \quad \text{at r=1} \quad (0 < \theta < \pi) \tag{122}$$

The solution of this two-dimensional problem can be written as a Fourier series:

$$\Phi_0 = -\frac{4}{\pi}\sum_{n=1}^{\infty}\frac{r^{2n}\sin 2n\theta}{4n^2-1} \tag{123}$$

The first-order surface elevation is:

$$\eta_1 = -\frac{8}{\pi}\sum_{n=1}^{\infty}\frac{nx^{2n-1}}{4n^2-1}, \quad \text{x>0} \tag{124}$$

Again there is a logarithmic singularity at the front and rear waterline points. The slope of this first-order elevation is $-8/(3\pi) = -0.8488$ at the origin, and it remains fairly constant for all $|x| < 0.4$. The flow described in a coordinate system following the container in its motion is found by adding a negative unit velocity:

$$\Phi = -\frac{4}{\pi}\sum_{n=1}^{\infty}\frac{r^{2n}\sin 2n\theta}{4n^2-1} - x \tag{125}$$

The formula for the ratio between the impulsive virtual mass and the total fluid mass inside the container is:

$$\frac{f_{-1}}{\pi/2} = \frac{32}{\pi^2}\sum_{n=1}^{\infty}\frac{n}{(4n^2-1)^2} = \frac{4}{\pi^2} = 0.4053 \tag{126}$$

Only about 40 % of the fluid momentum is changed instantaneously due to the motion of the container. Eventually all the fluid will get the same velocity as the container. Thus the remaining 60 % of the momentum transfer will not take place impulsively, but gradually during the course of the forced cylinder motion.

Figure 4 shows the instantaneous streamline pattern after the container has been put into motion. Figure 4 (a) gives the streamline pattern in the coordinate system at rest with the undisturbed fluid. Here we see that the horizontal velocity is zero at the free surface. There is a stagnation point at the middle of the free surface, where the only two oblique streamlines meet each other at a right angle. Figure 4 (b) gives the streamline pattern in the coordinate system following the container in its motion. Here the streamlines make oblique angles with the free surface.

A triangular container in impulsive motion

A particularly simple case is a two-dimensional container shaped as an isosceles triangle with a right angle in the bottom. The sides of the triangle are defined as $z = \pm x - 1$. Thus the length of the two equal sides is chosen as $\sqrt{2}$, and the lowest point of the triangle is $(x, z) = (0, -1)$ defining the depth as unity.

This wedge container is filled with fluid so that the undisturbed free surface is located in $z = 0$. This container is impulsively put into unit velocity in the $+x$ direction. The induced impulsive velocity potential is:

$$\Phi_0 = -xz \tag{127}$$

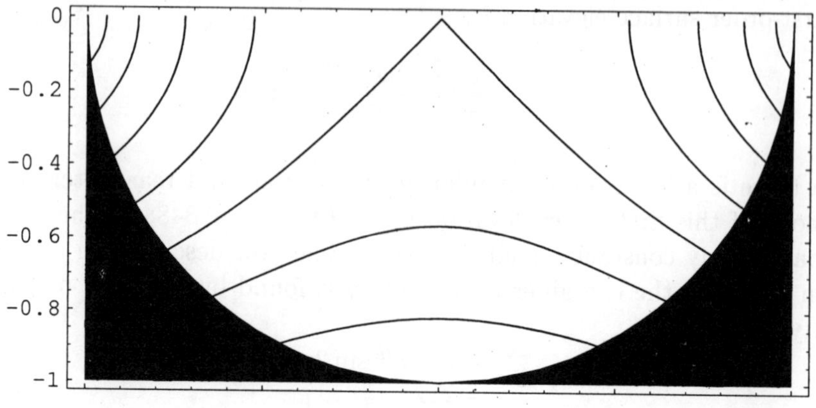

(a)

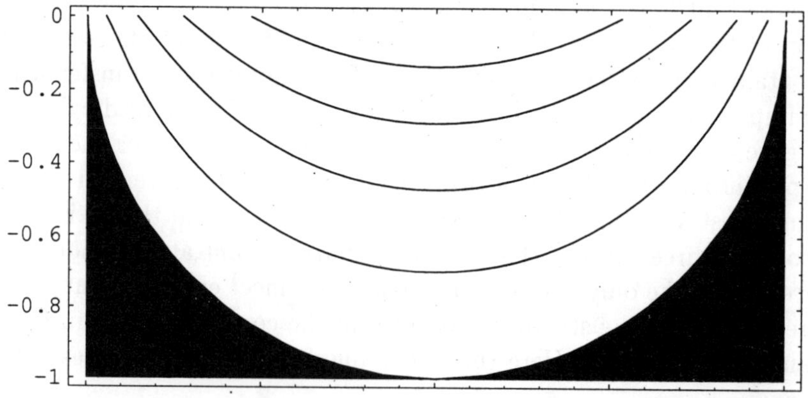

(b)

Figure 4:
Streamline patterns for the first-order free-surface flow in a half-filled horizontal circular cylinder impulsively put into horizontal motion.
a) Instantaneous streamline pattern in the system at rest with the undisturbed free surface.
b) Streamline pattern in the coordinate system following the container in its motion.

The first-order surface elevation is:

$$\eta_1 = -x \tag{128}$$

We note that there is no free-surface singularity to the first order. To leading order in the small-time expansion, the free surface is sloping at a uniform angle $-\alpha(t) = \arctan(t)$. This first-order slope of 1 can be compared with the previous example of the half-filled circular container, where there was a first-order slope of 0.8488 around $x = 0$.

When we describe the induced flow from a reference system following with the container, the total velocity potential will be:

$$\Phi = -xz - x \tag{129}$$

The associated streamfunction is $(x^2 - z^2)/2 - z$. The dimensionless virtual mass of the fluid put into impulsive motion inside the container is found to be:

$$f_{-1} = \frac{1}{3} \tag{130}$$

This is the same as the ratio between the virtual mass and the fluid mass inside the container, since the latter one is unity. This value for the relative virtual mass is somewhat smaller than the result for the semicircle, which seems reasonable.

Figure 5 shows the instantaneous streamline pattern after the container has been put into motion. Figure 5 (a) gives the streamline pattern in the coordinate system at rest with the undisturbed fluid, where we see that the horizontal velocity is zero at the free surface. Figure 5 (b) gives the streamline pattern in the coordinate system following the container in its motion.

It is interesting to compare the streamline patterns in figure 5 with those of the half-filled horizontal cylinder shown in figure 4. The most important difference is that the flow is regular at the waterline for the triangular container (figure 4), while it is singular at the waterline for the cylinder. Other pecularities for the triangular container shown in figure 5 (a) are that exactly one half of the fluid volume contain streamlines ending at the free surface, and that the two streamlines meeting in the stagnation points are straight lines with slope angles of $\pi/4$.

Gravitational release of fluid bodies

In this section we will consider gravitational release, where a finite fluid body is kept at rest for negative times. At time zero, the flow starts because the fluid body is released or unwrapped to flow with a free surface under gravity. At the same time, there may be inner of outer solid boundaries which are put into motion. In general there is a combination of forced flow and gravitational flow.

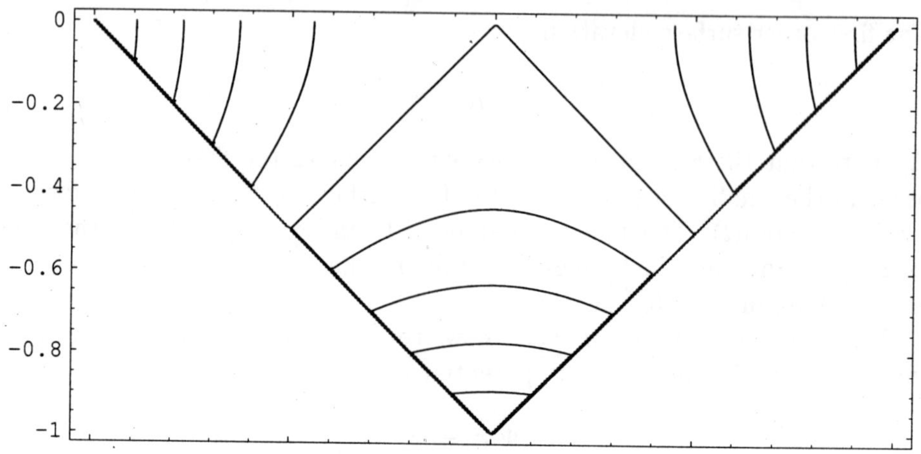

(a)

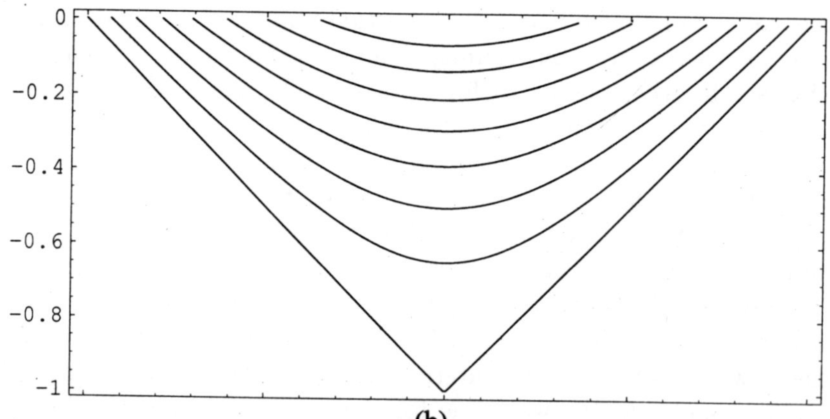

(b)

Figure 5:
Streamline patterns for the first-order free-surface flow in a cylindrical container with an initially triangular cross section. The triangle is isosceles, composed of two walls of slope angles $\pm\pi/4$ plus the undisturbed free surface of the fluid. The container is impulsively put into horizontal motion.
a) Instantaneous streamline pattern in the system at rest with the undisturbed free surface.
b) Streamline pattern in the coordinate system following the container in its motion.

The Froude number of the moving device will measure the relative importance of forced flow compared with gravitational flow.

In ship hydrodynamics this type of problem arises in connection with the behaviour of water from spilling waves on ship decks. This is called the green water problem.

Another relevant type of problem in marine hydrodynamics is the water exit for large velocities where a fluid film surrounds the body after it has lost contact with the sea below.

Dam-breaking problems

Dam breaking is the class of problems where a finite packet of fluid is released from rest at time zero, and the subsequent flow is purely gravitational. Mathematically it is equivalent to an impulsive start where a boundary starts moving with constant acceleration. Pohle [9] and Penney and Thornhill [57] proposed analytical solutions for the initial dam-breaking flow of rectangular, semi-cylindrical and hemispherical fluid packets. There is a logarithmic singularity at the fluid front, where the free surface meets the bottom, and this has not yet been resolved by asymptotic expansions. We therefore consider only the leading-order outer solutions.

Spreading of a semi-cylindrical fluid packet along a plane

We will first consider in detail the special case a gravitational release of a two-dimensional fluid packet: an initial semicircular shape with radius H. The time scale of this flow problem is gravitational, as surface tension is neglected. So we introduce dimensionless equations with units of length, time and velocity given as:

$$H, \ \sqrt{H/g}, \ \sqrt{gH} \tag{131}$$

The boundary shape will be described in polar coordinates. The surface shape is described by $r = R(\theta, t)$. The function $R(\theta, t)$ is expanded like η in the perturbation scheme for an initially flat surface. The dimensionless boundary conditions in polar coordinates are:

$$\frac{\partial R}{\partial t} + R^{-2}\frac{\partial R}{\partial \theta}\frac{\partial \Phi}{\partial \theta} = \frac{\partial \Phi}{\partial r}, \quad r = R(\theta.t) \tag{132}$$

$$\frac{\partial \Phi}{\partial t} + \frac{1}{2}|\nabla\Phi|^2 + R\sin\theta = 0, \quad r = R(\theta, t) \tag{133}$$

In addition we have the condition of an impermeable bottom:

$$\frac{\partial \Phi}{\partial \theta} = 0, \quad \theta = 0, \pi \tag{134}$$

The conditions to each order in time can be derived by successively applying the operator:

$$\left(\frac{\partial}{\partial t}\right)_{surface} = \frac{\partial}{\partial t} + \frac{\partial R}{\partial t}\frac{\partial}{\partial r} \tag{135}$$

When the motion starts from rest, there is no zeroth-order potential, so that $\Phi_0 = 0$ and $R_1 = 0$. The leading nontrivial potential is of first order and generates a second-order surface deflection:

$$\Phi_1 = -\sin\theta, \quad r = 1, \quad 0 < \theta < \pi \tag{136}$$

$$2R_2 = \frac{\partial \Phi_1}{\partial r}, \quad r = 1 \tag{137}$$

By the method of images, the bottom condition (134) can be incorporated by extending the condition (136) to the whole circle and write it as a Fourier series:

$$\Phi_1 = -\frac{2}{\pi} + \frac{4}{\pi}\sum_{n=1}^{\infty}\frac{\cos 2n\theta}{4n^2-1}, \quad r = 1, \quad 0 < \theta < 2\pi \tag{138}$$

The outer solution for the boundary shape is then:

$$R(\theta, t) = 1 + \frac{4}{\pi}t^2\sum_{n=1}^{\infty}\frac{n\cos 2n\theta}{4n^2-1} + O(t^4) \tag{139}$$

This infinite series can be summed up [57]:

$$R(\theta, t) = 1 - \frac{t^2}{\pi}\left(1 + \cos\theta\ \log(\tan\frac{\theta}{2})\right) + O(t^4) \tag{140}$$

The surface shape (139) has a logarithmic singularity at the bottom $y = 0$ ($\theta = 0, \pi$). The dry bottom along $y = 0$ ($r > 1$) thus appears to be immediately covered by a fluid film once the flow has started. This unphysical result for our outer expansion is related to the neglect of convective inertia at small times, and could possibly be overcome by introducing an inner expansion along the lines of King and Needham [49] in their analysis of the wavemaker problem.

The formula for the surface shape (139) has physical significance as it converges everywhere except at the bottom, and it involves a finite fluid transport. The second-order elevation is zero at the angle $\theta = 33.5^o$. The downwelling fluid volume from the upper part of the packet ($33.5^o < \theta < 146.5^o$) is given by $0.422\ t^2$. It is checked numerically to be equal to the total fluid volume flowing outwards along the bottom on both sides. The total dimensionless fluid volume is $\pi/2$. At the instant $t = 0.5$ this gives a fraction 6.7% of the total fluid mass flowed away from the initial semicircular domain. This fraction increases rapidly with time, but the asymptotic solution will break down as t approaches 1. The peak of the fluid packet has the dimensionless position:

$$R(\frac{\pi}{2},t) = 1 - \frac{t^2}{\pi} + O(t^4) \tag{141}$$

This shows that a fluid particle on the top will accelerate downwards with an initial acceleration that is 63.7% of the gravitational acceleration.

Spreading of a hemispherical fluid packet along a plane

Following Penney and Thornhill [57], we give the leading order outer solution for an initially hemispherical fluid packet released in a gravity field to flow over a horizontal rigid plane. The units of dimensionless variables are still given by eq. (131), where the dimensional initial radius of the fluid packet is H. We introduce spherical coordinates r, θ, ϕ where $\theta = 0$ represents the vertical direction. The outer solution for the boundary shape $r = R(\theta, t)$ is given by

$$R(\theta,t) = 1 - \frac{t^2}{2}\sum_{n=1}^{\infty}(-1)^n \frac{(4n+1)(2n)!}{2^{2n}(2n-1)(n+1)!(n-1)!} P_{2n}(\cos\theta) \tag{142}$$

where P_{2n} is the Legendre polynomial of order $2n$. Penney and Thornhill [57] have compared this solution for spherical geometry with the previous solution for cylindrical geometry. By summing up the series (142) for $\theta = 0$ they found that the initial acceleration of the peak of an hemispherical fluid packet is 82.8% of the gravitational acceleration. This is significantly greater than the value 63.7% for the semi-circular cylinder. This can be understood from the conservation of energy, since there is less kinetic energy generated horizontally when a hemispherical peak collapses by falling a given distance than when a semi-cylindrical peak falls the same distance. The smaller the horizontal motion relative to the vertical motion, the closer the vertical acceleration will be to a free fall under gravity.

These solutions for the half-cylindrical and hemispherical fluid packets are impulsive in the sense that gravity may be assumed to be turned on impulsively. Alternatively there could be no gravity and the bottom plane could start an impulsive vertical motion with constant velocity. The solution for the first-order deformation $R_1(\theta)$ would then be twice the present second-order solutions for $R_2(\theta)$ (eqs.140 and 142). This is a similar relationship as the transformation (68) between the elevations η_1 and η_2^*.

2D fluid bodies with an internal solid body put into motion

Tyvand and Landrini [58] investigated the problem of a rigid circular cylinder inside a cylindrical fluid packet. The packet is released at time $t = 0$ to flow freely due to gravity, and at the same time its concentric internal body is put in forced motion. The notation and equations are the same as in the previous problem, and the initial radius of the fluid packet is chosen as unity. The initial dimensionless radius of the inner solid cylinder is denoted by ϵ. A complete third-order analysis was developed for the surface shape of the fluid body. The result to second order is:

$$R(\theta,t) = 1 + H(t)\left(t\frac{2\epsilon^2}{1+\epsilon^2}\cos(\theta-\alpha)\right.$$
$$\left.+t^2\left(\frac{2\epsilon^2}{(1+\epsilon^2)^2(1+\epsilon^4)}\cos 2(\theta-\alpha) - \frac{\epsilon^4}{(1+\epsilon^2)^2} - \frac{1}{2F^2}\frac{1-\epsilon^2}{1+\epsilon^2}\sin\theta\right) + O(t^3)\right) \tag{143}$$

The Froude number F is here defined with the initial radius of the fluid cylinder as length scale. The inner cylinder has a direction of motion which makes an angle α with the horizontal direction. We note that the second-order surface deformation depends on gravity. The omitted third-order deformation includes nonlinear interaction between forced flow and gravitational flow. These early gravitational effects occur because the initial free surfacc is not horizontal. At an initially horizontal free surface, the leading gravitational deformation occurs at third order in a small-time expansion.

The exact three-term asymptotic small-time solution for the force from the fluid on the solid cylinder was found to he:

$$\mathbf{f} = -\delta(t)\ \pi\epsilon^2\frac{1-\epsilon^2}{1+\epsilon^2}\mathbf{n} - H(t)\frac{\pi}{F^2}\frac{\epsilon^2(1-\epsilon^2)}{1+\epsilon^2}\mathbf{k}$$
$$+H(t)\left(t\ 4\pi\frac{\epsilon^4(1-3\epsilon^2+\epsilon^4+\epsilon^8)}{(1+\epsilon^2)^4(1+\epsilon^4)}\mathbf{n} + O(t^2)\right) \tag{144}$$

Here we let $\mathbf{n}$ denote the unit vector in the direction of motion of the solid inner cylinder. The vertical unit vector is $\mathbf{k}$. We note that the zeroth-order force is purely gravitational, whereas the first-order force is independent of gravity. So there are no interaction terms between gravitational flow and forced flow to this order. Numerical simulations of the fully nonlinear problem were performed, and gave good agreement with the analytical formulas for small time. For larger times, the time history of the force has a complicated time dependence.

Bøhler, Landrini and Tyvand [59] investigate the similar problem where the initial position of the solid cylinder is eccentric compared with the fluid packet. In this case the zeroth-order force will no longer be purely gravitational.

Impulsive bottom deformations

As a model for tsunami generation due to earthquakes, the early free-surface flow due to impulsive bottom deformations has been studied [60]. The class of bottom deflections being studied is of the separable type:

$$\xi(x, y, t) = Z(x, y)T(t) \tag{145}$$

Physically we then consider a bottom deformation of a certain shape which grows in time as prescribed by the time function $T(t)$. We assume that $T'(t)$ tends to zero sufficiently fast so that the gravitational flow is only weak during the generation $(0 < t < t_0)$. We assume no motion of the bottom for $t > t_0$. Since there is no lasting imposed flow, we choose gravitational units for the dimensionless quantities. Then the Froude number will be unity by definition, and the dynamic free-surface condition (5) reduces to:

$$\frac{\partial \Phi}{\partial t} + \frac{1}{2}|\nabla\Phi|^2 + \eta = 0, \quad z = \eta(x, y, t) \tag{146}$$

The kinematic surface condition is given by (4) as before.

The kinematic condition at the bottom is:

$$\frac{\partial \xi}{\partial t} + \nabla\Phi \cdot \nabla\xi = \frac{\partial \Phi}{\partial z} \quad \text{at} \quad z = -1 + \xi(x, y, t) \tag{147}$$

We now attempt to generate an asymptotic series approximating the surface elevation $\eta(x, y, t)$. If we tentatively assume each term to be separable in their time and space-dependencies, the general asymptotic series with n terms will look as follows :

$$\eta(x, y, t) = H(t)\sum_{m=1}^{n} \eta_m(x, y)T_m(t) \tag{148}$$

Here $H(t)$ denotes the Heaviside unit step function, which is defined equal to zero for $t \leq 0$ and equal to one for $t > 0$. The asymptotic expansion is useful only if $t_0 < 1$.

Each term is governed by two equations that need to be determined subject to the given bottom deflection $\xi(x, y, t) = Z(x, y)T(t)$ and to the boundary conditions listed above. The kinematic free-surface condition (4) will now be linearized. This is compatible with full second-order theory, and includes the third-order gravitational flow. This linearized kinematic condition should make the velocity potential separable to each order:

$$\Phi(x, y, z, t) = H(t)\sum_{m=1}^{n} \Phi_{m-1}(x, y, z)T'_m(t) \tag{149}$$

Here the time dependencies are the derivatives of the time dependencies of the corresponding surface elevations so that each $\Phi_m(x, y, z)$ meet the requirement:

$$\eta_m(x, y) = \left.\frac{\partial \Phi_{m-1}}{\partial z}\right|_{z=0} \quad \text{m=1, 2,...} \tag{150}$$

Problem to each order

With the restrictions we have put on $\xi(x, y, t)$, the asymptotic series (148) may be useful up to $t = t_0$. Below the boundary value problems for the space dependent parts of the velocity potentials are listed along with the corresponding surface elevations. Laplace's equation is valid to each order. Also listed are the differential equations the time dependent parts must satisfy. There are three possible agents generating the flow to each order: The forced bottom motion, nonlinearity at the free surface, and gravity. The two former agents are active in second order, and their time dependencies are in general different. Therefore the second order problem must be split into two sub-problems, so that the total second-order elevation is written $\eta_{2B}(x, y)T_{2B}(t) + \eta_{2S}(x, y)T_{2S}(t)$, where the first contribution is due to the bottom nonlinearity and the second contribution to the free-surface nonlinearity. The associated first-order potential is $\Phi_{1B}(x, y, z)T'_{2B}(t) + \Phi_{1B}(x, y, z)T'_{2B}(t)$

The first-order potential is governed by:

$$\Phi_0|_{z=0} = 0$$

$$\left.\frac{\partial \Phi_0}{\partial z}\right|_{z=-1} = Z(x, y)$$

The second-order bottom contribution is governed by:

$$\Phi_{1B}|_{z=0} = 0 \qquad (151)$$

$$\left.\frac{\partial \Phi_{1B}}{\partial z}\right|_{z=-1} = \frac{\partial}{\partial x}\left[Z(x, y)\frac{\partial \Phi_0}{\partial x}\right]_{z=-1} + \frac{\partial}{\partial y}\left[Z(x, y)\frac{\partial \Phi_0}{\partial y}\right]_{z=-1} \qquad (152)$$

The second-order surface contribution is governed by:

$$\Phi_{1S}|_{z=0} = -\frac{1}{2}\eta_1^2 \qquad (153)$$

$$\left.\frac{\partial \Phi_{1S}}{\partial z}\right|_{z=-1} = 0 \qquad (154)$$

In third order we take only the gravitational effect into account. It is governed by:

$$\Phi_2|_{z=0} = -\eta_1 \qquad (155)$$

$$\left.\frac{\partial \Phi_2}{\partial z}\right|_{z=-1} = 0 \qquad (156)$$

The differential equations governing the time factors to each order are as follows:

$$T_1(t) = T(t),\ T_{2B}(t) = \frac{1}{2}T(t)^2,\ T''_{2S}(t) = (T'(t))^2,\ T''_3(t) = T(t) \qquad (157)$$

Initial flows for different geometries in 2D

Tyvand and Storhaug [61] derived various Green functions for initial flows with the equipotential condition at the initial free surface. A variety of geometries were investigated, including rectangular boxes, sloping beaches with or without an abrupt change in slope, bottom ridges and curved basins. Here we will give three of the most elementary solutions.

The solution for uniform unit depth with concentrated normal velocity of unit flux into the fluid at the bottom point $(x, z) = (\xi, -1)$ leads to:

$$\left.\frac{\partial \Phi}{\partial z}\right|_{z=0} = \frac{1}{2}\text{sech}\frac{\pi(x-\xi)}{2} \tag{158}$$

This is the normal velocity at the free surface, which is the most important part of the solution, and it corresponds to the first-order surface elevation from a steady unit source.

If the uniform fluid layer is interrupted by a vertical wall at $x = 0$, it is straighforward to find the solution due to a bottom source by taking the solution (158) and add a mirror image with respect to $x = 0$. But if the source is on the wall, a separate analysis is needed, and it leads to the free-surface velocity:

$$\left.\frac{\partial \Phi}{\partial z}\right|_{z=0} = \frac{\sin(\pi\xi/2)}{\sin^2(\pi\xi/2) + \sinh^2(\pi x/2)}\cosh(\pi x/2) \tag{159}$$

The source with unit flux is now located in the point $(0, -\xi)$. If this Green function is integrated with respect to ξ over the depth, it will produce Peregrine's formula (89) for a piston wavemaker.

We finally consider a beach with uniform slope α, with a unit source located at the bottom with distance ξ from the waterline. Its normal velocity at the free surface is:

$$\left.\frac{\partial \Phi}{\partial z}\right|_{z=0} = \frac{1}{\alpha\xi}\left(1 + \left(\frac{x}{\xi}\right)^{\frac{\pi}{\alpha}}\right)^{-1}\left(\frac{x}{\xi}\right)^{\frac{\pi}{2\alpha}-1} = \frac{1}{2\alpha x}\text{sech}\frac{\pi\log(x/\xi)}{2\alpha} \tag{160}$$

This is the initial vertical velocity of the free surface. It is important to note its behaviour at the waterline $x = 0$. When $0 < \alpha < \pi/2$, the waterline velocity is zero. When $\alpha = \pi/2$ there is a vertical wall where waterline velocity is finite. When $\pi/2 < \alpha < \pi$ there is an outward slope (overhanging cliff) and the waterline velocity is singular. This would require the application of matched asymptotic expansions to resolve the singularity. Physically this means that the flow near the slope is a narrow jet, where the convective inertia at the surface cannot be neglected the way it is done in the outer flow field.

Third-order theory for initially uniform depth in 2D

Tyvand, Miloh and Haugen [60] have developed a third-order theory for an impulsive bottom displacement $Z(x)T(t)$. The first-order solution in terms of Green functions follows from eq. (158):

$$\eta_1(x) = \frac{1}{2}\int_{-\infty}^{\infty} Z(x')\,\mathrm{sech}\left(\frac{\pi(x-x')}{2}\right) dx' \tag{161}$$

The Green function solution for the third-order gravitational flow is:

$$\eta_3(x) = \frac{1}{2}\int_{-\infty}^{\infty} Z(x')\left[\frac{\pi(x-x')}{2}\tanh\left(\frac{\pi(x-x')}{2}\right) - 1\right]\mathrm{sech}\left(\frac{\pi(x-x')}{2}\right) dx' \tag{162}$$

The non-gravitational third-order flow was neglected in [60]. The second-order solution cannot be found in terms of Green functions, but only as Fourier transform. Therefore the full Fourier transform solution has been found:

$$\eta_1(x) = \frac{1}{2\pi}\int_{-\infty}^{\infty} \mathrm{sech}\,(k)\mathcal{Z}(ik)e^{ikx}dk \tag{163}$$

$$\eta_{2B}(x) = -\eta_{2S}(x) = \frac{1}{4\pi^2}\int_{-\infty}^{\infty}\int_{-\infty}^{\infty} \frac{(k+K)\tanh k}{\cosh(k+K)}\mathcal{Z}(ik)\mathcal{Z}(iK)e^{i(k+K)x}dK\,dk \tag{164}$$

$$\eta_3(x,y) = -\frac{1}{2\pi}\int_{-\infty}^{\infty} \frac{k\tanh k}{\cosh k}\mathcal{Z}(ik)e^{ikx}dk \tag{165}$$

It is remarkable that the second-order terms due to bottom effects and surface effects are of equal magnitude. This is valid only for two-dimensional flow. In the equations above Z is the position dependent part of the bottom deflection and $\mathcal{Z}$ represents the corresponding Fourier transform:

$$\mathcal{Z}(ik) = \int_{-\infty}^{\infty} Z(x)e^{-ikx}dx \tag{166}$$

Impulsive flows due to bottom deformations in 3D

Miloh, Tyvand and Zilman [62] have derived some Green functions due to impulsive bottom displacements in three dimensions. Rectangular boxes, sloping beaches, cylindrical and spherical bottom geometries have been investigated. Here we will only show the simplest result for a uniform unit depth: The source point is taken as (x_0, y_0, z_0) where $z_0 = -1$. The surface velocity is:

$$\frac{\partial\Phi}{\partial z}(x,y;x_0,y_0,-1) = \frac{1}{\pi}\sum_{n=0}^{\infty}(-1)^n\frac{2n+1}{((x-x_0)^2+(y-y_0)^2+(2n+1)^2)^{3/2}} \tag{167}$$

By the Fourier-Bessel transform, the surface velocity can be written alternatively as an integral:

$$\frac{\partial \Phi}{\partial z}(x, y; x_0, y_0, -1) = \frac{1}{2\pi} \int_0^\infty \lambda \ \mathrm{sech}(\lambda) \ \mathrm{J}_0(\lambda\sqrt{(x-x_0)^2 + (y-y_0)^2}) \ d\lambda \quad (168)$$

Tyvand, Miloh and Haugen [60] derived a third-order small-time expansion for this case of initially horizontal bottom, expressing each order as Green function integrals and/or Fourier integrals. The special class of axisymmetric bottom deflections was investigated in detail, and some numerical evaluations of the integrals were shown.

Concluding remarks

In this review article a broad variety of free-surface flows starting from rest has been covered. It has been attempted to follow the early evolution of nonlinear effects analytically by small-time expansions. But in the cases where the leading-order solution has a singularity at the free surface, the higher-order analysis is not reliable when the inner expansion of the leading-order solution has been disregarded. Only King and Needham [49] have succeeded in deriving an inner solution, for a wavemaker in constantly accelerating surge motion. But their inner solution is taken only to the leading order of the matched asymptotic expansion. So far, nobody has developed a consistent higher-order outer solution when the leading order outer solution has a free-surface singularity. To develop matched asympototic expansions is a common mathematical challenge for the two-dimensional wavemaker in surge motion, the axisymmetric wavemaker, the surface-piercing vertical cylinder in forward motion, the cylindrical container put into motion, the dam-breaking problem, and the impulsive source flow at a slope with angle above $\pi/2$.

Comparisons have been made between the various singular flows. The induced net forces and torques have been integrated up. These flows with logarithmic singularities are integrable in the sense that they give finite forces on the walls and bodies.

We have investigated three impulsive wavemaker flows which are logarithmically singular at the waterline: The surge mode of the piston wavemaker, and the similar modes for the outward and inward axisymmetric wavemaker. In addition we have also identified a non-singular wavemaker: The pitch mode of the piston wavemaker.

We have mentioned three cases of impulsive container motions which are logarithmically singular at the free-surface waterline: The vertical circular cylinder, the rectangular box, and the half-filled horizontal circular cylinder. In addition we have identified a non-singular impulsive container motion: The two-dimensional flow in an isosceles triangle with two slope angles of $\pi/4$ and a right angle at the bottom. This case is exceptionally simple in that it has a uniform surface slope, growing linearly in time according to the leading-order solution.

We have seen that there are often singularity problems where a solid boundary meets the free surface. The analytical problem is simpler when there is no intersection between a moving object and the free surface, because there is no free-surface singularity. These are the problems of submerged bodies, submerged vortices in free motion, sources or sinks located in the fluid or at the bottom, and impulsive bottom deformations. We have covered the various analytical solutions of such problems, and mentioned the existing comparisons with numerical solutions.

In this article we have chosen to disregard all time-periodic free-surface problems with a forced oscillation being given. We note that the leading order impulsive flow corresponds to the high-frequency limit of oscillating bodies, see Newman [63]. Oscillatory problems may have an impulsive start, but this will influence only the early transients of the flow. The initial-value problems of the Cauchy-Poisson type are also disregarded in the present article, even though they can be considered as impulsive in the sense that gravity is suddenly turned on at time zero.

Another related field that has been disregarded in the present article is the motion and deformation of bubbles and cavities near a free surface. The initiation of such flows may be impulsive. But the great amount of work published in this field would break the limits of the present article. To our knowledge no small-time expansions have been published for bubble and cavity flows interacting with a free surface.

References

[1] D.H. Peregrine, Flow due to a vertical plate moving in a channel. Unpublished note (1972).

[2] T. Vinje, On small-time expansion of nonlinear free surface problems. *J. Engng. Math.* **28** (1994), 173-190.

[3] G.K. Batchelor, *An Introduction to Fluid Dynamics.* Cambridge Univ. Press (1967).

[4] K.-H. Wang and A.T. Chwang, Nonlinear free-surface flow around an impulsively moving cylinder. *J. Ship Res.* **33** (1989), 194-202.

[5] P.A. Tyvand and T. Miloh, Free-surface flow due to impulsive motion of a submerged circular cylinder. *J. Fluid Mech.* **286** (1995), 67-101.

[6] P.A. Tyvand and T. Miloh, Free-surface flow generated by a small submerged circular cylinder starting from rest. *J. Fluid Mech.* **286** (1995), 103-116.

[7] J.G. Telste, Inviscid flow about a cylinder rising to a free surface. *J. Fluid Mech.* **182** (1987), 149-168.

[8] F.V. Pohle, The Lagrangian equations of hydrodynamics: solutions which are analytic functions of the time. Ph.D. thesis, New York Univ. (1950).

[9] F.V. Pohle, Motions of water due to breaking of a dam, and related problems. U.S. National Bureau of Standards, Gravity Waves, N.B.S. Circular 521 (1952).

[10] P.A. Tyvand, On the interaction between a strong vortex pair and a free surface. *Phys. Fluids* **A2** (1990), 1624-1634.

[11] P.A. Tyvand, Motion of a vortex near a free surface. *J. Fluid Mech.* **225** (1991), 673-686 (appendix by R.P. Tong).

[12] P.A. Tyvand and T. Miloh, Axisymmetric interaction between a vortex ring and a free surface. *Phys. Fluids* **6** (1994), 224-238.

[13] W.W. Willmarth, G. Tryggvason, A. Hirsa and D. Yu, Vortex pair generation and interaction with a free surface. *Phys. Fluids* **A1** (1989) 170-172.

[14] Y.A. Novikov, Generation of surface waves by discrete vortices. *Izv., Atmos. Ocean Phys.* **17** (1981), 709-714.

[15] T. Sarpkaya, Trailing-vortex wakes on the free surface. In: Proc. 16th Symp. on Naval Hydrodyn. National Academy Press, Washington (1986), pp. 38-50.

[16] G. Tryggvason, Deformation of a free surface as a result of vortical flows. *Phys. Fluids* **31** (1988), 955-957.

[17] T. Sarpkaya, Vorticity, free surface, and surfactants. *Ann. Rev. Fluid Mech.* **28** (1996), 83-128.

[18] D.L. Marcus and S.A. Berger, The interaction between a counter-rotating vortex pair in vertical ascent and a free surface. *Phys. Fluids* **A1** (1989), 1988-2000.

[19] J.G. Telste, Potential flow about two counter-rotating vortices approaching a free surface. *J. Fluid Mech.* **201** (1989),259-278.

[20] D. Yu and G. Tryggvason, The free-surface signature of unsteady, two-dimensional vortex flows. *J. Fluid Mech.* **218** (1990), 547-572.

[21] R.P. Tong and D.H. Peregrine, The generation of surface waves by a free vortex. In: *Proc. Nonlinear Water Waves Workshop*, Bristol (1991), pp. 164-171.

[22] C.M. Bender and S.A. Orszag, *Advanced Mathematical Methods for Scientists and Engineers.* McGraw-Hill, New York (1978).

[23] P.A. Tyvand, Unsteady free-surface flow due to a line source. *Phys. Fluids* **A4** (1992), 671-676.

[24] T. Miloh and P.A. Tyvand, Non-linear transient free-surface flow and dip formation due to a point sink. *Phys. Fluids* **A5** (1993), 1368-1375.

[25] M. Xue and D.K.P. Yue, Nonlinear free-surface flow due to an impulsively-started submerged point sink. *J. Fluid Mech.* **364** (1998), 325-347.

[26] M.J. Kim, H.T. Moon, S.K. Choi, Y.B. Lee, H.Y. Nam and M. Cho, Time-dependent free-surface flows due to a sink: A quantitative analysis of the drain-size effect. *J. Korean Phys. Soc.* **31** (1997), 235-240.

[27] M.-J. Kim, Numerical study of dip formation using sink array model. *Int. J. Modern Phys. C* **8** (1997), 1249-1262.

[28] B.T. Lubin and G.S. Springer, The formation of a dip on the surface of a liquid draining from a tank. *J. Fluid Mech.* **29** (1967) 385-390.

[29] Q.-N. Zhou and W.P. Graebel, Axisymmetric draining of a cylindrical tank with a free surface. *J. Fluid Mech.* **221** (1990), 511-532.

[30] M. Landrini and P.A. Tyvand, Generation of water waves and bores by impulsive bottom flux. *J. Engng Math.* **39**, 131-170.

[31] P.A. Tyvand, Impulsive motion of a wide torus submerged below a free surface. *J. Ship Res.* **41** (1997), 195-209.

[32] M. Greenhow, M. and Y. Li, Added masses for circular cylinders near or penetrating fluid boundaries - review, extension and application to water-entry, -exit and slamming. *Ocean Eng.* **14** (1987), 325-348.

[33] J.L. Taylor, Some hydrodynamic inertia coefficients. *Phil. Mag.* **9** (1930), 161-183.

[34] M. Greenhow and S.I. Ahn, Added mass and damping of circular-cylinder sections. *Ocean Engng.* **15** (1988), 495-504.

[35] S.K. Venkatesan, Added mass of two cylinders. *J. Ship. Res.* **29** (1985), 234-240.

[36] M. Greenhow, Water-entry and -exit of a horizontal circular cylinder. *Appl. Ocean Res.* **10** (1988), 191-198.

[37] M. Greenhow, A complex variable method for the floating-body boundary-value problem. *J. Comp. Appl. Math.* **46** (1993), 115-128.

[38] A.G. Terent'ev, Nonstationary motion of bodies in a fluid. *Proc. Steklov Inst. Math.* **186** (1991), 211-221.

[39] M. Greenhow and W.-M. Lin, Nonlinear free surface effects: experiments and theory. Report 83-19. MIT, Dept. of Ocean Engineering (1983).

[40] S. Moyo, Hydrodynamic interaction of horizontal circular cylinders with a free-surface. Ph. D. Thesis, Dept. of Math. and Statist., Brunel Univ. (1996).

[41] M. Greenhow and S. Moyo, Water entry and exit of horizontal circular cylinders. *Phil. Trans. R. Soc. London* **A355** (1997), 551-563.

[42] S. Moyo and M. Greenhow, Free motion of a cylinder moving below and through a free surface. *Applied Ocean Res.* **22** (2000), 31-44.

[43] T. Miloh, A note on impulsive sphere motion beneath a free-surface. *J. Engng. Math.***41** (2001), 1-11.

[44] H.M. Westergaard, Water pressures on dams during earthquakes. *Trans. ASCE* **98** (1933), 418-433.

[45] A.T. Chwang, Nonlinear hydrodynamic pressure on an accelerating plate. *Phys. Fluids* **26** (1983), 383-387.

[46] A.T. Chwang and G.W. Housner, Hydrodynamic pressures on sloping dams during earthquakes. Part 1. Momentum method. *J. Fluid Mech.* **87** (1978), 335-341.

[47] A.T. Chwang, Hydrodynamic pressures on sloping dams during earthquakes. Part 2. Exact theory. *J. Fluid Mech.* **87** (1978), 343-348.

[48] J.J. Stoker, *Water Waves.* J. Wiley & Sons, New York, pp. 513-522 (1957).

[49] A.C. King and D.J. Needham, The initial development of a jet caused by fluid, body and free-surface interaction. Part 1. A uniformly accelerating plate. *J. Fluid Mech.* **268** (1994), 89-101.

[50] J.N. Newman, The impulsive motion of a wavemaker. Unpublished note (1981).

[51] A.J. Roberts, Transient free-surface flows generated by a moving vertical plate. *Q. J. Mech. Appl. Math.* **40** (1987), 128-158.

[52] S.W. Joo, W.W. Schultz, and A.F. Messiter, A.F. An analysis of the initial-value wavemaker problem. *J. Fluid Mech.* **214** (1990), 161-183.

[53] J.W. Miles, On the initial-value problem for a wavemaker. *J. Fluid Mech.* **229** (1991), 589-601.

[54] I. Frankel, Compressible flow induced by the transient motion of a wave-maker. *Z. Angew. Math. Phys.* **41** (1990), 628-655.

[55] A.T. Chwang and K.-H. Wang, Nonlinear impulsive force on an accelerating container. *J. Fluids Engng.* **106** (1984), 233-240.

[56] A.J. Roberts, Initial flow of liquid in an accelerating tank. *J. Eng. Mech.* **114** (1988), 175-180.

[57] W.G. Penney and C.K. Thornhill, The dispersion, under gravity, of a column of fluid supported on a rigid horizontal plane. *Phil. Trans. R. Soc. London* **A244** (1952), 285-311.

[58] P.A. Tyvand and M. Landrini, Free-surface flow of a fluid body with an inner circular cylinder in impulsive motion. *J. Engng. Math.* **40** (2001), 109-140.

[59] E.M. Bøhler, M. Landrini and P.A. Tyvand, Free-surface flow of a fluid body with an eccentric cylinder in impulsive motion. Manuscript in preparation (2001).

[60] P.A. Tyvand, T. Miloh and K.B. Haugen, Impulsive tsunami generation by rapid bottom deflections at initially uniform depth. To be printed in: *Underwater Ground Failures on Tsunami Generation, Modeling, Risk and Mitigation.* Proc. NATO Advanced Research Workshop, Istanbul (2001).

[61] P.A. Tyvand and A.R.F. Storhaug, Green functions for impulsive free-surface flows due to bottom deflections in two-dimensional topographies, *Phys. Fluids* **12** (2000), 2819-2833.

[62] T. Miloh, P.A. Tyvand and G. Zilman, Green functions for initial free-surface flows due to 3D impulsive bottom deflections. Submitted to *J. Engng. Math.* **43** (2002), 57-74.

[63] J.N. Newman, *Marine Hydrodynamics*, MIT Press, Cambridge, Mass. (1977).

Modern Applied Mathematics
Editor: J.C. Misra

Internal Wave Propagation

Bruce R. Sutherland

Department of Mathematical and Statistical Sciences, University of Alberta

Abstract

The current development of linear, weakly nonlinear and fully nonlinear theories of the propagation of internal waves in continuously stratified fluid is reviewed and new analyses are presented that demonstrate the usefulness of parcel arguments in developing an intuition for small-amplitude internal wave dynamics. Directions for future research are suggested.

KEYWORDS: internal waves, ray theory, instability

1 Introduction

Internal waves are characterised by periodic undulations within a fluid whose density decreases with height. The governing dynamics are essentially the same as those of surface waves: both move due buoyancy restoring forces. However, whereas surface waves are confined to the interface between water and air, internal waves can propagate both horizontally and vertically.

Internal waves occur throughout the atmosphere and ocean on scales as small as meters and as large as thousands of kilometers. Under suitable atmospheric conditions, clouds help visualise the waves. Where the waves push air upwards near the wave crests the air expands in the lower pressure and it cools due to thermodynamic effects. If the air itself is initially saturated with water vapour, clouds form as the vapour condenses into water droplets in the cooling air. Conversely, the droplets evaporate near the wave troughs where the air moves downwards into higher pressure and so heats up under compression. A typical image of internal waves visualised by clouds is shown in Figure 1.

The equation describing the motion of internal waves in the atmosphere and ocean are given by a remarkably simple set of partial differential equations. The Navier-Stokes equations express conservation of momentum:

$$\rho_{\mathrm{T}} \frac{D\mathbf{u}}{Dt} = -\nabla p_{\mathrm{T}} + \rho_{\mathrm{T}}\mathbf{g} + \mathbf{F}_{\mathrm{rot}} + \mu\nabla^2\mathbf{u}. \tag{1}$$

Here ρ_{T} is the density of the fluid, $\mathbf{u} = (u, v, w)$ denote the components of west-to-east (u), south-to-north (v) and vertical (w) motion corresponding to directions $\hat{\mathbf{x}}$, $\hat{\mathbf{y}}$ and $\hat{\mathbf{z}}$, respectively. The pressure is denoted by p_{T}, $\mathbf{g} = -g\hat{\mathbf{z}}$ is the acceleration due to gravity, μ is the molecular viscosity, and $\mathbf{F}_{\mathrm{rot}}$ denotes external forces that occur due to the Earth's rotation, namely Coriolis forces.

We neglect the effect of vertically propagating sound waves which is done by assuming the fluid is incompressible. Thus the continuity equation, which

Figure 1: Enhanced photograph showing internal waves visualised by clouds as they propagate over floating icepacks in the Arctic ocean. The picture was taken at 9pm GMT July 23, 1999 from an airplane window while flying south of Baffin Island in northern Canada. The wave crests propagate downward and to the right in the picture.

expresses conservation of mass, is written

$$\frac{D\rho_{\mathrm{T}}}{Dt} = 0, \tag{2}$$

and consequently we have

$$\nabla \cdot \mathbf{u} = 0. \tag{3}$$

In these equations the material derivative,

$$\frac{D}{Dt} \equiv \frac{\partial}{\partial_t} + \mathbf{u} \cdot \nabla, \tag{4}$$

introduces nonlinearity through the "advective" terms, denoted by the second term on the right-hand-side of (4).

The five coupled partial differential equations in five unknowns, $(\mathbf{u}, \rho, p)$ describe a wide range of phenomena, including waves and turbulence occurring on length scales as small as millimeters and as large as thousands of kilometers.

In this study, we will focus on internal wave phenomena on scales ranging from tens of meters to planetary scales. A straightforward scaling analysis reveals that viscosity and diffusion (for example of salt water or heat) are negligibly important compared with the advective and pressure gradient terms in the equations.

We further simplify the equations by subdividing the pressure and density fields into background fields ($\overline{p}(z)$ and $\overline{\rho}(z)$, respectively) and perturbation fields ($p(\mathbf{x}, t)$ and $\rho(\mathbf{x}, t)$, respectively), and we assume that the background fields are in "hydrostatic balance". That is, we assume the fields are related by

$$\frac{d\overline{p}}{dz} = -\overline{\rho}g. \tag{5}$$

This is simply the statement that, in a stationary fluid, the pressure at a given vertical level is set by the weight of fluid per unit area lying above that level.

Thus, substituting $\mu = 0$ and $p_{\rm T} = \overline{p} + p$ and $\rho_{\rm T} = \overline{\rho} + \rho$ into (1) gives

$$\rho_{\rm T}\frac{D\mathbf{u}}{Dt} = -\nabla p - \rho g\hat{\mathbf{z}} + \mathbf{F}_{\rm rot}, \tag{6}$$

and the continuity equation becomes

$$\frac{D\rho}{Dt} + w\frac{d\overline{\rho}}{dz} = 0. \tag{7}$$

Further approximations may be employed resulting in different classes of wave motion:

- If one assumes the flow is "hydrostatic" (meaning the perturbation as well as background pressure and density fields are in balance) then one sets $\partial p/\partial z = -\rho g$ in (6). Hydrostatic internal waves propagate with frequencies much smaller than the natural frequency of oscillation of a stratified fluid, the "buoyancy frequency". Waves with frequencies close to the buoyancy frequency are said to be "non-hydrostatic".

- If one assumes the flow is "Boussinesq" (meaning that density variations are important only as they affect the buoyancy term) then one sets $\rho_{\mathrm{T}} \to \rho_0$, a constant, in the terms before the material derivatives in (6). Boussinesq internal waves occur in fluids whose density decreases by only a small fraction of the total density with height.

- If one assumes the flow is "anelastic", one instead sets $\rho_{\mathrm{T}} \to \overline{\rho}$ in (6). This approximation accounts for the increase in amplitude of waves with altitude in the atmosphere as the density of air decreases by a significant fraction from the troposphere to the middle and upper atmosphere.

- If one assumes the waves are small amplitude, then the nonlinear terms in the material derivative may be neglected. For example, if there is no mean background flow, then $D/Dt \to \partial/\partial t$.

We will consider the consequences of these different approximations in the following sections.

2 Linear Theory

The linear theory for internal waves can be approached either through "fluid parcel" consideration, in which the local forces acting upon an infinitesimal patch of fluid is considered, or through partial differential equations whose solutions express the cumulative effects of conservation principles. Both approaches are discussed below.

2.1 Parcel Arguments

The motion of small amplitude internal waves can be understood in part from a "parcel" argument, as illustrated in Figure 2. Consider first a fluid parcel (a infinitesimal patch of fluid) that is vertically displaced upward by a small amount, δz, in a surrounding stratified fluid whose density, $\overline{\rho}(z)$, decreases linearly with height. The parcel, of density ρ_0, finds itself surrounded by less dense fluid and therefore it experiences a restoring force due to buoyancy. Newton's laws reveal the consequent motion:

$$\rho_0 \frac{d^2 \delta z}{dt^2} = -(\delta \rho) g. \qquad (8)$$

Here g is the acceleration due to gravity and $\delta\rho$ is the density difference between the fluid parcel and the surrounding fluid at its displaced position

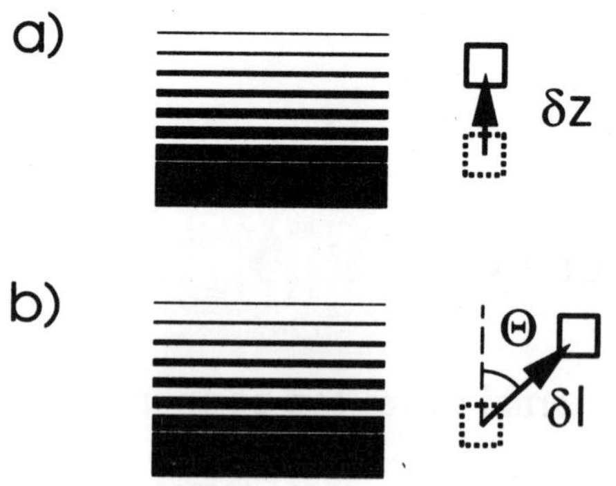

Figure 2: Schematic illustrating forces acting upon a fluid parcel that is displaced a) vertically and b) at an angle to the vertical in a stratified fluid.

The density difference can be written in terms of δz. Because $|\delta z|$ is infinitessimally small $\delta\rho = -(d\bar{\rho}/dz)\delta z$. Thus (8) becomes

$$\frac{d^2\delta z}{dt^2} + N^2\delta z = 0, \tag{9}$$

in which, by definition

$$N^2 \equiv -\frac{g}{\rho_0}\frac{d\bar{\rho}}{dz}. \tag{10}$$

Because (9) is a spring equation, it is apparent that a vertically displaced fluid parcel will oscillate in a linear stratified fluid at a frequency N, which is proportional to the square root of the background density gradient. Consequently, the value N is called the buoyancy frequency (though it is also sometimes called the Brunt-Väisälä frequency after the scientists who first examined these dynamics [75, 9]). Typical values of N in the atmosphere and ocean are around $10^{-2}\,\mathrm{s}^{-1}$, so vertical oscillations occur with a period on the order of tens of minutes.

Now consider the case in which a fluid parcel is displaced a small distance δl along a line at an angle Θ to the vertical, as shown in Fig. 2b. Furthermore, assume that the fluid motion is constrained to move along this line. Although this may seem to be an unphysical assumption, in § 2.3 a fluid parcel displaced by an internal wave is shown to undergo exactly this diagonal linear motion.

Newton's laws for the force along the line give

$$\rho_0\frac{d^2\delta l}{dt^2} = -(\delta\rho)g\cos\Theta. \tag{11}$$

The $\cos\Theta$ term enters by resolving the downward gravitational force into the along-line direction. As before, $\delta\rho$ can be written in terms of δl by the relation $\delta\rho = -(d\bar{\rho}/dz)\delta l \cos\Theta$. Thus (11) becomes

$$\frac{d^2\delta l}{dt^2} + N^2 \cos^2\Theta \delta l = 0. \tag{12}$$

This new spring equation shows that fluid motion along a line at an angle to the vertical occurs with frequency

$$\omega = N \cos\Theta, \tag{13}$$

which, indeed, is the dispersion relation for internal waves when it is recognised that $\cos\Theta$ can be written in terms of the wavenumber vector $\mathbf{k} = (k, l, m)$ and its horizontal component $\mathbf{k_H} = (k, l)$ by

$$\cos\Theta = |\mathbf{k_H}|/|\mathbf{k}| \equiv \sqrt{\frac{k^2 + l^2}{k^2 + l^2 + m^2}}. \tag{14}$$

The advantage of the derivation of (13) is that it provides physical insight into the dynamics of internal waves. Specifically, the reason the frequency of shallow waves, with $\Theta \sim 90^o$, is small is because only a small component of its motion goes toward resisting gravitational forces.

2.2 The dispersion relation

An immediate consequence of the dispersion relation is that internal waves cannot propagate with frequency faster than N. The phase vector for internal waves (whose direction is perpendicular to lines of constant phase and whose magnitude is the speed of crests moving in this direction) is

$$\mathbf{c_p} \equiv \frac{\omega}{|\mathbf{k}|}\frac{\mathbf{k}}{|\mathbf{k}|} = \frac{N}{|\mathbf{k}|}\cos\Theta(\cos\phi\cos\Theta, \sin\phi\cos\Theta, \sin\Theta). \tag{15}$$

In the last of these expressions, ϕ is defined implicitly by $\tan\phi = l/k$, and represents the angle between the horizontal component of the wavenumber vector and the k-axis. Note the components of this vector is different from the phase speeds one would measure if the waves were observed at a fixed location. For example, in a vertical time series showing the passage of waves past a fixed horizontal location, the crests would move vertically at a fixed horizontal position with speed ω/m.

Another useful quantity derived from the dispersion relation is the group velocity, $\mathbf{c_g}$. This denotes the velocity at which energy is transported by the waves. Explicitly

$$\mathbf{c_g} \equiv \nabla_k\omega \equiv \left(\frac{\partial\omega}{\partial k}, \frac{\partial\omega}{\partial l}, \frac{\partial\omega}{\partial m}\right) = \frac{N}{|\mathbf{k}|}\sin\Theta(\cos\phi\sin\Theta, \sin\phi\sin\Theta, -\cos\Theta). \tag{16}$$

Thus the magnitude $|\mathbf{c_g}| = (N/|\mathbf{k}|)\sin\Theta$, which may be compared with $|\mathbf{c_p}| = (N/|\mathbf{k}|)\cos\Theta$.

An initially surprising consequence of these equations is that the group and phase velocities are orthogonal to each other and that if the wave transports energy upwards, then the phase lines move downwards, and *vice versa*.

This is seen quite explicitly in experiments in which a cylinder oscillates

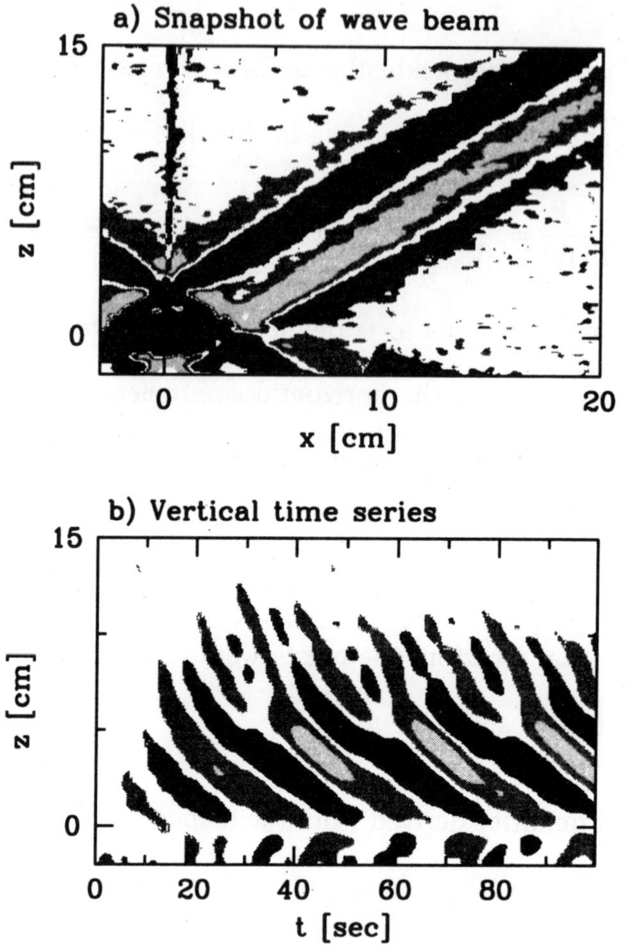

Figure 3: a) One of the four branches of waves generated when a cylinder oscillates with frequency $\omega = 0.54\,\mathrm{s}^{-1}$ in a tank filled with uniformly stratified salt water solution whose salinity decreases with height as characterised by the buoyancy frequency $N = 1.0\,\mathrm{s}^{-1}$. The beam propagates at an angle $\sim 57^o$ to the vertical consistent with the predicted angle $\Theta = \cos^{-1}(\omega/N)$. b) Time series of the wave beams showing the evolution of the waves along a vertical slice 12.5 cm to the right of the oscillating cylinder.

vertically up and down at a fixed frequency, ω, in a stratified fluid. Provided $\omega < N$, the waves propagate away from the cylinder along four beams which form a "St. Andrews Cross" pattern with the cylinder at the center of the cross [48]. One branch of the waves emanating from the upper cylinder are revealed in the "synthetic schlieren" image shown in Fig. 3a [69]. The wavebeam emanates upward and to the right of the cylinder, transporting energy in a direction parallel to the wave crests and in the direction of $\mathbf{c_g}$. In contrast, surface waves transport energy in a direction perpendicular to the wave crests. The time series in Fig. 3b shows that, whereas the group velocity is directed upward, the crests of the waves move downward over time.

Another interesting consequence of the dispersion relation is that the vertical component of the group velocity is zero for waves with $\Theta = 0$ and $\Theta = 90^o$. The maximum vertical group velocity occurs at an intermediate angle $\Theta = \tan^{-1}(1/\sqrt{2}) \simeq 35^o$

2.3 Partial Differential Equations

Much information can be gleaned from the dispersion relation of waves. However, one must solve the equations of motion as given by a coupled set of partial differential equations in order to predict the inter-relationship between components of the velocity field and perturbation density field, and to predict energy and momentum transport by the waves.

The linearised equations of motion provide insight into the behaviour of small amplitude internal waves. For simplicity in this discussion, we neglect the effects of viscosity and we assume that the vertical variation of the background density is sufficiently long compared with the scale of the waves that we can adopt the "Boussinesq approximation": density variations are important only as they influence buoyancy forces. The Boussinesq approximation is suitable for the study of ocean, in which the density changes by less than 10% from the surface to ocean floor. In the atmosphere, the approximation is useful only over scales less than approximately 10 km [63].

If we further assume that there is no mean background flow and that the fluid is uniformly stratified (the buoyancy frequency is constant), the motion of small-amplitude internal waves is given by differential equations describing the conservation of mass assuming the fluid is incompressible, as given by (7 and (3), and the linearised form of the equations describing the conservation of momentum:

$$\begin{aligned}\rho_0\frac{\partial u}{\partial t} &= -\frac{\partial p}{\partial x},\\ \rho_0\frac{\partial v}{\partial t} &= -\frac{\partial p}{\partial y},\\ \rho_0\frac{\partial w}{\partial t} &= -\frac{\partial p}{\partial x} - g\rho;\end{aligned} \tag{17}$$

In these equations ρ_0 is a characteristic density of the fluid.

These equations are conveniently rewritten in matrix form in which the notation $\partial_t \equiv \partial/\partial t$, *etc.*, is adopted:

$$
\begin{pmatrix}
\partial_t & 0 & 0 & 0 & \frac{1}{\rho_0}\partial_x \\
0 & \partial_t & 0 & 0 & \frac{1}{\rho_0}\partial_y \\
0 & 0 & \partial_t & \frac{g}{\rho_0} & \frac{1}{\rho_0}\partial_z \\
0 & 0 & -\frac{\rho_0}{g}N^2 & \partial_t & 0 \\
\partial_x & \partial_y & \partial_z & 0 & 0
\end{pmatrix}
\begin{pmatrix} u \\ v \\ w \\ \rho \\ p \end{pmatrix} = 0. \tag{18}
$$

For nontrivial solutions, the determinant of the matrix in (18) must be zero. That is

$$
\left[\partial_{ttt}\nabla^2 + N^2{\nabla_H}^2\partial_t\right] b = 0 \tag{19}
$$

in which b is one of the five basic state fields (u, v, w, ρ, p), and $\nabla_H \equiv (\partial_x, \partial_y)$ is the horizontal component of the gradient operator.

Fourier transforming gives the dispersion relation for internal waves in uniformly stratified, incompressible fluid. We assume periodic solutions of the form

$$
b(x, y, z, t) = A_b \exp[\imath(kx + ly + mz - \omega t)], \tag{20}
$$

in which it is understood that the actual wave structure is given by the real part of b. Substituting (20) in (19) gives $-\imath\omega^3|\mathbf{k}|^2 + N^2\imath\omega|\mathbf{k_H}|^2 = 0$. Taking only non-stationary solutions, we find

$$
\omega^2 = N^2\frac{|\mathbf{k_H}|^2}{|\mathbf{k}|^2},
$$

which is exactly the dispersion relation given by (13) and (14).

Effectively, the dispersion relation is an eigenvalue of the differential matrix problem (18). The corresponding eigenvectors give the inter-relationship between the basic state fields. For wave theory in general, these are known as the "polarisation relations".

Defining the vertical displacement field ξ implicitly through the relation $w = \partial\xi/\partial t$, the amplitude of the eigenfunctions, A_b, can each written in terms of the the amplitude of the vertical displacement field A_ξ. These are listed in Table 1.

Immediately apparent from the table is that the horizontal and vertical velocity fields are in phase with each other. Thus fluid parcels undergo linear (rather than circular) oscillatory motion due to waves. This is consistent with the assumption made above in deriving the dispersion relation from a parcel argument.

The vertical transport of horizontal momentum per unit mass in the x-direction is the cross-correlation, $\langle uw \rangle$, between the u and w fields. It is found by taking the average over one period of oscillation of the product between the real parts of the two fields. This can be done either by integrating in time ($\frac{1}{T}\int_0^T \mathrm{Re}(u)\mathrm{Re}(w)\,dt$), in space ($\frac{1}{\lambda_x}\int_0^{\lambda_x} \mathrm{Re}(u)\mathrm{Re}(w)\,dx$), or in phase, by setting $\phi = \mathbf{k}\cdot\mathbf{x} - \omega t$:

$$\begin{aligned}
\langle uw \rangle &\equiv \frac{1}{2\pi}\int_0^{2\pi} \mathrm{Re}\left(A_u e^{\imath\phi}\right)\mathrm{Re}\left(A_w e^{\imath\phi}\right)\,d\phi \\
&\equiv \frac{1}{2\pi}\int_0^{2\pi} \left[\frac{1}{2}\left(A_u e^{\imath\phi} + A_u{}^\star e^{-\imath\phi}\right)\right]\left[\frac{1}{2}\left(A_w e^{\imath\phi} + A_w{}^\star e^{-\imath\phi}\right)\right] \\
&\equiv \frac{1}{2}\mathrm{Re}\left(A_u A_w{}^\star\right) \\
&= \frac{1}{4}N^2 \sin 2\Theta |A_\xi|^2. \qquad (21)
\end{aligned}$$

Table 1: Polarisation relation and physically relevant correlations between basic state fields for small-amplitude Boussinesq internal waves in the (x, z) plane with no background rotation. As well as the basic state fields, the table lists values of the streamfunction (ψ), vorticity (ζ), the change in the squared buoyancy frequency due to stretching of isopycnals by waves (ΔN^2) and its time derivative (${N^2}_t$). Each field b is characterised by the magnitude and phase of its complex amplitude A_b.

Defining equation	Relation to vertical displacement	
Vertical displacement: ξ	complex amplitude: A_ξ	
$\rho = -\frac{d\overline{\rho}}{dz}\xi$	$A_\rho = (-d\overline{\rho}/dz)A_\xi$	$= \frac{\rho_0}{g}N^2 A_\xi$
$\mathbf{u} = \nabla \times (\psi\hat{\mathbf{y}})$	$A_\psi = -\frac{\omega}{k}A_\xi$	$= -\frac{N}{k}\cos\Theta A_\xi$
$w = \frac{\partial\xi}{\partial t}$	$A_w = -\imath\omega A_\xi$	$= -\imath N \cos\Theta A_\xi$
$u = -\frac{\partial\psi}{\partial z}$	$A_u = \imath\frac{m\omega}{k}A_\xi$	$= \imath N \sin\Theta A_\xi$
$\zeta = -\nabla^2\psi$	$A_\zeta = -\frac{\omega}{k}\lvert\mathbf{k}\rvert^2 A_\xi$	$= -Nk\sec\Theta A_\xi$
$\Delta N^2 = -\frac{g}{\rho_0}\frac{\partial\rho}{\partial z}$	$A_{\Delta N^2} = -\imath m N^2 A_\xi$	$= -\imath N^2 k \tan\Theta A_\xi$
${N^2}_t = \frac{\partial \Delta N^2}{\partial t}$	$A_{{N^2}_t} = m\omega N^2 A_\xi$	$= -N^3 k \sin\Theta A_\xi$

Momentum transport $\quad \rho_0 \langle uw \rangle = \frac{1}{4}\rho_0 N^2 \sin 2\Theta \lvert A_\xi\rvert^2$

Wave-induced mean flow $\quad -\langle \zeta\xi \rangle = \frac{1}{2}Nk\sec\Theta\lvert A_\xi\rvert^2$

Here, the star denotes the complex conjugate. Thus waves with $\Theta = 0$ or $\Theta = 90^o$ transport no vertical momentum. For waves with the same value of the maximum vertical displacement, A_ξ, those with $\Theta = 45^o$ transport the greatest momentum.

Another consequence of the polarisation relations is that the density field is $\pi/2$ out of phase with the velocity fields. Thus internal waves transport momentum but not mass. This is shown explicitly by following the same mathematical procedure as that used in deriving (21): the vertical transport of mass is given by

$$\langle \rho w \rangle \equiv \frac{1}{2} \mathrm{Re}\,(A_\rho A_w{}^\star) = 0. \tag{22}$$

Another useful quantity, particularly for the study of finite-amplitude waves, is the wave-induced mean flow. For surface waves, this is also known as "Stokes drift" and denotes the steady unidirectional flow of water in the direction of wave propagation. This current, which is typically much smaller than the phase speed of the waves, is proportional to the square of the wave amplitude. In the recent study of Hamiltonian fluid dynamics, the wave-induced mean flow is known more generally as the "pseudomomentum" [2, 60].

For internal waves the wave-induced mean flow, for example in the x-direction, is given explicitly in terms of the cross-correlation between the vertical displacement field (ξ) and the y-component ($\hat{\mathbf{y}} = \hat{\mathbf{z}} \times \hat{\mathbf{x}}$) of the vorticity field, $\zeta = (\nabla \times \mathbf{u}) \cdot \hat{\mathbf{y}}$ [59, 64]

$$-\langle \zeta \xi \rangle \equiv \frac{1}{2} \mathrm{Re}\,(A_\zeta A_\xi{}^\star) = \frac{1}{2} N k \sec \Theta |A_\xi|^2 \tag{23}$$

2.4 Ray theory

The effect upon wave propagation of non-uniform background flow $\overline{U}(z)$ and non-uniform stratification $N^2(z)$ can be described in an analytic theory provided the background varies vertically on scales much larger than the vertical extent of the waves. The mathematics are simplified in this limit by employing the "WKB" (or sometimes the "WKBJ", or sometimes the "Liouville-Green") approximation [34]. (The first to study this problem were Liouville [37] and Green [21], but later their results were independently rederived.)

Without derivation, the salient result is given here for waves propagating in the (x, z) plane. The path of the waves, as described by the function $z(x)$, is given by solving the differential equation

$$\frac{dz}{dx} = \frac{c_{gz}}{c_{gx} + \overline{U}} \tag{24}$$

[CHECK] In which the components of the group velocity are given by (16).

A key to anticipating the form of solutions of (24) is to realise that, due to the horizontal spatial invariance of the equations of motion and their boundary conditions, the horizontal wavenumber, k, does not change following the motion of the waves, a consequence of Fourier series analysis. Likewise, temporal invariance implies that the absolute frequency, ω, (the frequency observed at a fixed point in space, otherwise called the "intrinsic frequency") of the waves do not change, although the relative frequency (the frequency observed moving with the background flow, otherwise called the "extrinsic frequency" or, more commonly, the "Doppler-shifted frequency") may change as the wavepacket propagates in shear flow.

In two particularly well studied cases one assumes that N^2 is constant and U varies linearly with height as $U(z) = sz$ (a uniform shear flow). Suppose that an internal wave packet is situated initially at the origin and moves upward and to the right. In the case with $s > 0$, one finds that the wavepacket asymptotically approaches a vertical level z_c where the background flow speed, $\overline{U}(z_c)$, equals the wavepacket's constant horizontal phase speed, ω/k. A straightforward calculation reveals

$$z_c = \frac{N}{sk} \cos \Theta. \tag{25}$$

This height is known as the "critical level". A more detailed analysis reveals that as waves approach the critical level, their amplitude increases until the waves overturn and break. Thus a background flow, such as the jet stream in the atmosphere, can act to shield the upper level flow from waves. Near the critical level breaking waves deposit momentum to the mean flow and thus locally accelerate the fluid, which further shields the upper level from waves. A dramatic example of these dynamics are manifest in laboratory experiment that models the quasi-biennial oscillation in the equatorial stratosphere [58].

Present-day models of the general circulation of the Earth's atmosphere employ heuristic schemes based, in part, upon the above remarks in order to include the effects of drag upon the atmosphere by internal waves that originate from mountainous terrain and which break near critical levels in the middle and upper atmosphere [36, 54, 43].

Though successful in enabling the models to predict more realistic wind speeds, recent work suggests this heuristic approach is too simplistic. For example, Shutts [61, 62] has shown that waves can pass through a critical level if the flow speed veers with height (that is, if we relax the constraint that the flow moves only in the x-direction and instead, for example, allow the flow to blow increasingly in the y-direction at increasing vertical levels).

Ray theory has garnered renewed interest recently in the study of small-scale internal waves that are refracted by a spectrum of large-scale, slow frequency internal waves, which effectively impose a time-varying background mean flow [19, 7, 8, 6, 10]. A significant consequence of this work is to show that critical levels may not pose a barrier to transport by waves if the shear is time dependent. Continuing research may help to explain, for example, the observed spectrum of internal waves in the middle atmosphere [25]).

In a second well-studied case, one considers the circumstance in which a uniformly stratified flow has uniform but negative shear ($s < 0$). In this circumstance, waves with frequency ω and horizontal wavenumber k that move upward from $z = 0$ are Doppler-shifted to increasing frequencies $\Omega = \omega - k\overline{U} = \omega - ksz$. Eventually, the waves reach a level z_r at which the Doppler-shifted frequency equals the buoyancy frequency. Explicitly

$$z_r = \frac{N}{k(-s)}(1 - \cos\Theta) \tag{26}$$

Similarly, wave reflection may occur in a fluid with uniform flow ($\overline{U} = 0$) but whose stratification weakens with height. For example, if $N^2(z) = {N_0}^2(1 - z/H)$, then

$$z_r = H(1 - \cos\Theta) \qquad (27)$$

In either circumstance, one finds that the waves reflect from the level z_r and thereafter transport energy and momentum back to where they originated. When one examines the structure of the wave near the reflection level itself, one finds that the slope dz/dx becomes infinitely large, in violation with the conditions required by the WKB-approximation. At the point of reflection there are three distinct wave groups: the incident and reflected wave below z_r and an "evanescent" wave whose amplitude decreases exponentially with height above z_r. This structure is known as a "caustic" [34].

Ray theory has proven successful in reproducing the paths followed by experimentally generated internal wave beams in uniformly stratified, uniform shear flows [29] and in stationary, non-uniformly stratified fluids [70]. The latter experiments also examined the consequences of violating the WKB-approximation, showing that internal waves reflect from a level lower than z_r if N^2 decreases rapidly with height compared with the extent of the internal wave beam.

Ray theory has also proven successful in predicting the path of small-amplitude internal wave groups in uniform shear as measured from fully nonlinear numerical simulations [66]. However, as discussed in section 4, the theory breaks down for waves with amplitudes as small as 1 percent of the horizontal wavelength.

2.5 Anelastic effects

To this point we have considered small-amplitude wave propagation as described within the limits of the Boussinesq approximation. If we consider the propagation of waves in the atmosphere (the density of which decreases in height by approximately 1/2 every 6 km), it is necessary to relax the constraint that density variations are significant only in the buoyancy term of the equations of motion. Instead we allow $\rho_{\mathrm{T}} \rightarrow \overline{\rho}$ in (6), This is the "anelastic approximation".

Thermodynamics play an important role for atmospheric waves in general. As air moves up and down the fluid is heated and cooled, respectively, and so the density of fluid parcels vary. A quantity that characterises these effects is the potential temperature:

$$\theta = T\left(\frac{p}{p_0}\right)^{-2/7} \tag{28}$$

The derivation of (28) can be found in any standard textbook on dynamic meteorology (*e.g.* see Gill [20] §3.7). Here T and p are the temperature and pressure of a fluid parcel at some vertical level and p_0 is a reference value of pressure, typically taken to be $p_0 = 1000$ millibars, the pressure at sea level. Physically, θ represents the temperature a fluid parcel would have if brought adiabatically to sea level. This is a particularly useful quantity because it is conserved following the adiabatic motion of a fluid parcel. Note that while the temperature in the lower atmosphere decreases with height, the potential temperature generally increases with height.

In the anelastic approximation, the squared buoyancy frequency is defined in term of the background potential temperature, $\overline{\theta}(z)$, by

$$N^2 = \frac{g}{\overline{\theta}} \frac{d\overline{\theta}}{dz}. \tag{29}$$

There are two main differences between this definition and that given by (10): $\overline{\theta}$ rather than a constant reference value, θ_0, appears in the denominator of the fraction before the derivative, and there is no negative sign on the right-hand side of the definition. The latter is consistent with the observation that in a stable (non-convective) atmosphere, potential temperature increases with height. Thus N^2 as given by (29) is generally positive.

Here we study the simplest circumstance of an isothermal atmosphere (whose temperature T_c is independent of height) with no mean flow. In this case, thermodynamics and hydrostatic balance dictates that the background pressure and density are proportional to $\exp(-z/H_s)$ where $H_s = RT_c/g$ is called the "scale-height" ($\simeq 8.4$ km in the atmosphere using the value of the ideal gas constant R). One finds that the corresponding background potential temperature is $\overline{\theta} = \theta_0 \exp[\frac{2}{7}(z/H)]$ and $N^2 = (2/7)g/H_s$.

The partial differential equations describing the motion of small-amplitude internal waves is given, in matrix form, by

$$\begin{pmatrix} \partial_t & 0 & 0 & 0 & \frac{1}{\overline{\rho}}\partial_x \\ 0 & \partial_t & 0 & 0 & \frac{1}{\overline{\rho}}\partial_y \\ 0 & 0 & \partial_t & \frac{g}{\overline{\rho}} & \frac{1}{\overline{\rho}}\partial_z \\ 0 & 0 & -\overline{\rho}/gN^2 & \partial_t & 0 \\ \partial_x & \partial_y & \partial_z & 0 & 0 \end{pmatrix} \cdot \begin{pmatrix} u \\ v \\ w \\ \rho \\ p \end{pmatrix} = 0. \tag{30}$$

As waves propagate upward they transport momentum, the magnitude of which, according to (21), is proportional to the squared amplitude of the waves. Because the density of the atmosphere decreases with height, the waves must increase in amplitude with height in order to conserve momentum. Hence we assume solutions of the form

$$f(x, y, z, t) = f_0 \exp(\imath(\mathbf{k} \cdot \mathbf{x} - \omega t)] \exp(z/2H), \tag{31}$$

and substitute into (30).

The resulting matrix eigenvalue problem gives the dispersion relation for anelastic waves:

$$\omega^2 = N^2 \frac{|\mathbf{k_H}|^2}{|\mathbf{k}|^2 + \Gamma^2}, \tag{32}$$

in which $\Gamma = (3/14)H_s{}^{-1}$. In the limit $H_s \to \infty$, (32) is identical to the dispersion relation for Boussinesq waves, (13).

The significant difference in the dynamics is in the wave structure, as given by (31). In reality, the waves do not grow in amplitude indefinitely, but ultimately they overturn and break. As well as critical-level absorption, discussed in §2.4, this is the second well-studied mechanism for wave breaking and deposition of momentum to the mean flow in the atmosphere.

2.6 Background rotation effects

For internal waves that evolve on time scales comparable to the rotational period of the Earth, it is also important to include the effect of Coriolis forces. For simplicity, we return here to the Boussinesq approximation and write the resulting equations of motion in matrix form by

$$\begin{pmatrix} \partial_t & -f & 0 & 0 & \frac{1}{\rho_0}\partial_x \\ f & \partial_t & 0 & 0 & \frac{1}{\rho_0}\partial_y \\ 0 & 0 & \partial_t & \frac{g}{\rho_0} & \frac{1}{\rho_0}\partial_z \\ 0 & 0 & -\frac{\rho_0}{g}N^2 & \partial_t & 0 \\ \partial_x & \partial_y & \partial_z & 0 & 0 \end{pmatrix} \begin{pmatrix} u \\ v \\ w \\ \rho \\ p \end{pmatrix} = 0 \tag{33}$$

Here the constant $f = 2\Omega \sin \phi_0$ is the Coriolis parameter, which depends upon the rotational frequency of the Earth ($\Omega = 2\pi/(1\,\text{day})$) and the local latitude ϕ_0. (At the North Pole, $\phi_0 = 90^o$ and so $f = 2\Omega$.) The parameter f is sometimes called the "inertial frequency" because gravity has negligible affect upon fluid parcels moving near this frequency and thus they move only under their own inertia. In the atmosphere and oceans at mid-latitudes, $f \simeq 10^{-4}\,\text{s}^{-1}$, two orders of magnitude smaller than typical values of N.

Assuming plane wave solutions of the form (20) the matrix eigenvalue problem gives the dispersion relation

$$\omega^2 = N^2 \cos^2 \Theta + f^2 \sin^2 \Theta \tag{34}$$

where Θ is again defined by (14).

Obviously, in the case $f = 0$ we recover (13). Equation (34) shows that f is a low-frequency cut-off: internal waves cannot propagate with frequencies lower than f. Internal waves with frequency close to f are called "inertia-gravity waves".

The phase lines of inertia-gravity waves waves are almost horizontal ($\Theta \sim 90^o$). A study of the corresponding polarisation relations reveal that the u and v fields are 90° out of phase implying that fluid parcels prescribe near-horizontal elliptical paths during the passage of the waves.

Measurements of internal waves in the atmosphere have revealed that most of their energy spectrum is contained near inertial frequencies. For this reason, and also because only recently large-scale atmospheric simulations have been able to resolve the spatial structure of inertia-gravity waves, there have been numerous theoretical and numerical studies have been performed to examine their dynamics [15, 33]. However, because the waves are necessarily long period and exist typically over long scales, simplified models which assume stationary background flows are of questionable realistic value.

3 Weakly Nonlinear Theory

In the preceding section it is assumed throughout that the waves are of small amplitude. This is a useful approximation because the analytic solutions of the resulting linearised equations can be found for a wide variety of circumstances.

Weakly nonlinear theory describes the dynamics of waves with amplitudes that are non-negligible but still sufficiently small that analytic solutions can be found through perturbation theory. Typically, weakly nonlinear dynamics involve wave-wave interactions, for which the superposition principle breaks down, and interactions between waves and the wave-induced mean flow.

Although a great deal of work has examined the weakly nonlinear dynamics of water waves [56, 24, 3] relatively little research has gone into studying the corresponding internal wave dynamics.

3.1 Modulational instability

Weakly nonlinear dynamics act to modify the dispersion relation for waves:

$$\omega_{\text{wnl}} = \omega(\mathbf{k}) + \omega_2(\mathbf{k})A^2, \tag{35}$$

in which the first term on the right-hand side of the equations is the dispersion relation determined from linear theory (*e.g.* equation (13)). Considering for now the motion in the x-direction, a periodic wave train is said to be "modulationally stable" (remaining periodic for all time) if [76, 77]

$$\omega_2 \frac{\partial^2 \omega}{\partial k^2} > 0 \tag{36}$$

Indeed, for deep water waves it has been shown that the waves are always modulationally unstable, even for water waves of infinitessimally small amplitude. However, because the growth rate of the instability is proportional to amplitude, in realistic circumstances it is unlikely to manifest itself during the life-cycle of very small amplitude waves. The structure of this instability was described by Benjamin and Feir [3] who showed that the energy in the spectrum repeatedly transfers to different "sideband" wavenumbers and then transfers back to the original wavenumber following what is known more generally as the Fermi-Pasta-Ulam (FPU) recurrence phenomena [16].

The modulational stability of Boussinesq, non-rotating internal waves has been examined analytically by Grimshaw [23] and both analytically and numerically by Sutherland [67]. The latter has shown that the criterion (36) predicts that horizontally periodic, vertically compact internal waves are modulationally unstable to the growth of vertical sideband wavenumbers if their frequency is higher than that of the waves with the fastest vertical group velocity, that is, if

$$\Theta < \Theta_c = \tan^{-1}(2^{-1/2}) \simeq 35^o. \tag{37}$$

Indeed fully nonlinear numerical simulations show the emergence of FPU recurrence for these waves and show that waves with $\Theta > \Theta_c$ remain modulationally stable. Horizontally compact internal waves are predicted to be modulationally unstable to the growth of horizontal sideband wavenumbers if $\Theta > \Theta_c = \sin^{-1}(2^{-1/3}) \simeq 53^o$.

3.2 Wave-wave interactions

Modulational instability does not imply wave breaking, only non-steadiness of a periodic wavetrain. However, other weakly nonlinear dynamics result in overturning and breaking waves.

In general, three waves of wavenumber vector $\mathbf{k}_i$ and frequencies ω_i, for $i = 1, 2, 3$, may interact in what is called a "resonant triad" if $\mathbf{k}_3 = \mathbf{k}_1 \pm \mathbf{k}_2$

and $\omega_3 = \omega_1 \pm \omega_2$ (*e.g.*, see Phillips [57]). This amounts to requiring $\cos\Theta_3 = \cos\Theta_1 \pm \cos\Theta_2$, in which Θ_i is given by (14). In practise three possible resonant couplings are of interest: under "elastic scattering", energy in a high frequency wave is transported to another with nearly equal and opposite wavenumber vector; under "induced diffusion", energy in a low frequency wave is transported to a high frequency wave; under "parametric subharmonic instability", energy is extracted from a low wavenumber mode and deposited into to high frequency waves with nearly equal and opposite wavenumbers.

Observations of internal waves in the ocean suggest that elastic scattering and induced diffusion both play an important role in governing the energy spectrum of waves in the ocean interior. Both mechanisms act on time scales comparable to one wave period [50].

For internal waves that are periodic everywhere in space, internal waves are found to be parametrically unstable even at infinitessimally small amplitudes [28, 38]. The instability grows through resonant coupling between the initial waves and their superharmonics (whose horizontal wavenumbers are an integer multiple of the initial waves). Through this coupling, the superharmonic waves extract energy from the initial waves and grow in amplitude until ultimately they break.

Though mathematically interesting, the study is not relevant to realistic circumstances for two reasons. First, typical instability growth rates are so small that parametric instabilities would not likely develop significantly between the time that the waves are generated and when they break. Second, parametric instability is the dominant mode of instability only in the case of plane-periodic waves. Localised internal wavepacket dynamics have been found to be governed dominantly due to interactions between the waves and the wave-induced mean flow [67].

3.3 Breaking instabilities

Internal waves become unstable and break if their amplitude is sufficiently large. Linear theory predicts overturning occurs if the amplitude is so large that $\partial \rho_{\mathrm{T}}/\partial t \equiv d\bar{\rho}/dz + \partial \rho/\partial z > 1$ somewhere in the flow field. Using the polarisation relations in Table 1, the "overturning condition" can be derived. This states that waves propagating at angle Θ to the vertical are unstable if the amplitude A_ξ of the vertical displacement field exceeds A_{OT} where

$$A_{\mathrm{OT}} = \frac{1}{2\pi} \cot \Theta. \tag{38}$$

Due to interactions between waves and the wave-induced mean flow, simulations show that internal waves in fact are unstable at much smaller amplitudes unless Θ is close to 90°. The criterion for instability is called the "self-acceleration" condition, which states that the waves become unstable if the wave-induced mean flow exceeds the horizontal group velocity. Using

(16) and (23), it is apparent that the waves are unstable if

$$A_\xi > A_{\mathrm{SA}} \equiv \frac{\lambda_x}{2\pi\sqrt{2}} \sin 2\Theta, \tag{39}$$

in which $\lambda_x = 2\pi/k$ is the horizontal wavelength. Thus waves with $\Theta \simeq 0$ ($\omega \simeq N$) are unstable to overturning and breaking even at infinitessimally small amplitudes. Even for the most stable waves, with $\Theta = 45^o$, the waves are unstable if their amplitude is larger than 11% of the horizontal wavelength. Numerical simulations show that the onset of instability is rapid, occurring within 5 buoyancy periods, $2\pi/N$.

3.4 Critical Level absorption

Weakly nonlinear models for internal waves in stratified shear flow have been examined by a number of authors [40, 41, 12, 31], there emphasis being on examining the flows whose basic states are near the marginal stability boundary; that is, for stratified shear flows with the minimum value of the gradient Richardson number $\mathrm{Ri} \equiv N^2/(\overline{U}')^2 \simeq 1/4$ [46, 26]. In a analytical-numerical study, Lamb and Pierrehumbert [31] showed that two-dimensional internal waves originating from topography underlying a critical layer could transmit through the layer, in a manner reminiscent of coherent light waves been transmitted from a resonant cavity in a laser.

3.5 Wave reflection

The interactions between waves and the wave-induced mean flow can act to allow waves to transmit well above a reflection level, z_r that would occur in linear theory when a wavepacket moves upward in a background flow with uniform but negative shear, $s < 0$. We hypothesize that this occurs because the wave-induced mean flow creates a local mean shear that counteracts the background shear. A simple calculation reveals that the positive shear associated with the wave-induced mean flow is equal to $|s|$ when

$$A_{\mathrm{WR}} = \sqrt{2\sigma_z k_x \frac{s}{N} \cos^3 \Theta \frac{N}{k}}, \tag{40}$$

in which σ_z is the vertical width of the wavepacket [66].

3.6 Laboratory Experiments

At present few experiments have been performed to examine the weakly non-linear dynamics of internal waves. Resonant interactions between short and long waves were examined analytically and experimentally by Koop and Redekopp [30]. and between equal and opposite propagating wave beams by Teoh *et al* [73]. The latter observed the generation of an evanescent disturbance in the overlapping region which consequently grew and ultimately broke.

Weakly nonlinear effects are also apparent where the upward and downward propagating wavebeams near the cylinder overlap [69, 72]. In particular, waves with double the frequency of oscillation of the cylinder have been observed in tandem with the primary wavebeams. Though this could be due to wave-wave interactions, tracking the origin of the secondary beams to the cylinder itself implies that the beams are created due to interactions between the waves and the viscous boundary layer surrounding the cylinder.

4 Nonlinear Theory

Unlike linear and weakly nonlinear theories, the dynamics of fully nonlinear finite-amplitude internal waves, with very few and trivial exceptions, can be examined only through analysis of numerical simulations in which the Navier-Stokes equations are iteratively solved in time.

4.1 Wave modulation

The propagation of two-dimensional finite-amplitude, Boussinesq internal waves in stationary, uniformly stratified fluid has been examined by Sutherland [67]. This study both confirmed the validity of the modulational stability criterion (37) as well as the self-acceleration condition (39). The latter was shown to be successful in predicting the stability to breaking of horizontally periodic, vertically compact wavepackets. Horizontally and vertically compact wavepackets were found to be stable at moderately larger amplitudes than predicted however, particularly for waves with small Θ.

4.2 Critical level absorption

At a critical level, z_c, ray theory predicts the waves approaching from below should break and be absorbed by the mean flow. However, as finite-amplitude waves approach a critical level, one finds that the wavepackets may be partially transmitted across z_c or partial reflect below z_c.

Recently, computer speeds and memory have increased enough to run high-resolution numerical simulations of internal waves incident upon a critical level [5, 78, 79, 14]. The simulations show that the breaking occurs due to convective overturning modified by shear. The instability is inherently three-dimensional. The incident wave energy is ultimately partitioned into a three main proportions associated with 1) acceleration of the mean flow, 2) irreversible mixing of the local stratification, and 3) a reflected wave packet. In their study of three-dimensional waves, Winters and d'Asaro [79] showed that the proportion of the wavepacket that deposits energy to the mean flow is sensitive to the wave amplitude and the relative strength of the shear at the critical level.

4.3 Wave reflection

The nonlinear dynamics of finite-amplitude internal waves incident upon a reflecting level has been studied for the case of two-dimensional, Boussinesq waves. Sutherland [64] examined the dynamics of waves in a stationary background fluid whose buoyancy frequency decreased from one constant value to another constant value over a short vertical distance. A related study examined the reflection of waves in a uniformly stratified fluid in a background flow that decreased from one constant value to another over a short distance [65]. In both cases it was found that a greater proportion of the incident wavepacket was transmitted across the reflecting level if the waves were of large amplitude. Even if no reflecting level existed, a larger proportion of finite-amplitude waves were found to reflect from the varying region of the background wind or stratification. Both nonlinear phenomena occurred because the wave-induced mean flow Doppler-shifted the frequency of the incident waves.

The dynamics of finite-amplitude Boussinesq internal waves in a uniform shear flow were examined by Sutherland [66]. Surprisingly this study revealed that horizontally and vertically compact wavepackets may propagate well beyond a reflection level at an approximately constant vertical speed. The results were consistent with the transmission condition (40).

5 Conclusions

Though much is understood about the dynamics of small-amplitude waves, a great deal is unknown about the propagation of finite-amplitude internal waves. Some specific outstanding problems are listed here.

- Conditions for modulational instability and breaking due to self-acceleration have not yet been derived for anelastic waves or inertia-gravity waves.

- The transmission condition has been derived and tested for waves in a uniformly stratified shear flow. No such test has yet been performed for waves in a uniform flow with decreasing stratification with height.

- Though some fully nonlinear simulations of two-dimensional non-hydro static waves have been performed, their three-dimensional counterparts remain to be examined. In particular, it would be interesting to study finite-amplitude effects as wavepackets approach a critical level in a veering wind.

- Large-eddy simulations (LES) model the interactions of waves near a critical layer without resolving the breaking in detail [54, 43, 49]. The emphasis of these studies is to evaluate whether heuristic parameterisations of wave breaking can indeed account for discrepancies between observed flow speeds and those predicted by models that ignore the effect of internal waves. Though such parameterisations have been successful in improving model predictions, the results of high-resolution simulations imply that the heuristic methods employed are not always trustworthy. Future research should attempt to bring LES and high resolution approaches together in a way that allows for speedy but reliable parameterisations of breaking waves.

- Few laboratory experiments have been performed to examine finite amplitude effects, in part because probes give amplitude measurements at a point and classical schlieren visualisation methods [47] reveal phase but not amplitude information. Using the recently developed "synthetic schlieren" method, one can measure amplitudes of a two-dimensional wave field everywhere in space and time [13]. Further experimental progress will be made with the ongoing extension of this technique to measure three-dimensional wave fields [17].

This review mentions only briefly here other significant and active research areas in internal wave theory, including their interaction with rigid boundaries ([74, 4]) and turbulence ([32, 27, 53]), and their generation by flow over topography ([55]), convection ([1]), shear instability ([35, 42, 44, 18, 68, 71]) and geostrophic adjustment ([39]).

The focus here has been on internal waves in a continuously stratified fluid, the assumption being that the vertical extent of the stratified fluid is significantly longer than the vertical extent of the wavepackets. Nonetheless, in many natural circumstances internal waves are manifest at the interface between two fluids of different (but comparable) density. With the realization that such large-amplitude waves occur in the atmosphere (*e.g.* at atmospheric inversions in the form of Australia's Morning Glory [11, 45]) and ocean (*e.g.* at the thermocline and generated by tidal flow over the continental shelf ([51, 52]), the mathematics of solitary waves have been brought to bear upon these problems [22].

References

[1] M. J. Alexander. A simulated spectrum of convectively generated gravity waves: Propagation from the tropopause to the mesopause and effects on the middle atmosphere. *J. Geophys. Res.* 101: 1571–1588 (1996).

[2] D. G. Andrews and M. E. McIntyre. An exact theory of nonlinear waves on a Lagrangian-mean flow. *J. Fluid Mech.* 89: 609–646 (1978).

[3] T. B. Benjamin and J. E. Feir. The disintegration of wavetrains on deep water. *J. Fluid Mech.* 27: 417–430 (1967).

[4] P. Bouruet-Aubertot, J. Sommeria, and C. Staquet. Instabilities and breaking of standing internal gravity waves. *J. Fluid Mech.* 285: 265–301 (1995).

[5] R. J. Breeding. A nonlinear model of the break-up of internal gravity waves due to their exponential growth with height. *J. Geophys. Res.* 77: 2681–2692 (1972).

[6] D. Broutman, C. Macaskill, M. E. McIntyre, and J. W. Rottman. On Doppler-spreading models of internal waves. *Geophys. Res. Lett.* 24: 2813–2816 (1997).

[7] D. Broutman and W. R. Young. On the interaction of small-scale oceanic internal waves with near-inertial waves. *J. Fluid Mech.* 166: 341–358 (1986).

[8] D. L. Bruhwiler and T. J. Kaper. Wavenumber transport: Scattering of small-scale internal waves by large-scale wavepackets. *J. Fluid Mech.* 289: 379–405 (1995).

[9] D. Brunt. The period of simple vertical oscillations in the atmosphere. *Quart. J. Roy. Meteorol. Soc.* 53: 30–32 (1927).

[10] G. Buckley, D. Broutman, J. W. Rottman, and S. Eckermann. On the importance of weak steady shear in the refraction of short internal waves. *Geophys. Res. Lett.* 26: 2877–2880 (1999).

[11] D. R. Christie, K. J. Muirhead, and A. L. Hales. Intrusive density flows in the lower troposphere: a source of atmospheric solitons. *J. Geophys. Res.* 84: 4959–4970 (1979).

[12] S. M. Churilov and I. G. Shukhman. Nonlinear stability of a stratified shear flow: a viscous critical layer. *J. Fluid Mech.* 180: 1–20 (1987).

[13] S. B. Dalziel, G. O. Hughes, and B. R. Sutherland. Whole field density measurements. *Experiments in Fluids* 28: 322–335 (2000).

[14] A. Dörnbrack, T. Gerz, and U. Schumann. Turbulent breaking of overturning gravity waves below a critical level. *Appl. Sci. Res.* 54: 163–176 (1995).

[15] T. J. Dunkerton. Shear instability of internal inertia-gravity waves. *J. Atmos. Sci.* 54: 1628–1641 (1997).

[16] E. Fermi, J. Pasta, and S. Ulam. Studies of nonlinear problems I, Los Alamos Report LA 1940, 1955. In A. C. Newell, editor, *reproduced in Nonlinear Wave Motion* Amer. Math. Soc., Providence, RI (1974).

[17] M. R. Flynn, K. Onu, and B. R. Sutherland. Internal wave generation by a vertically oscillating sphere. *J. Fluid Mech.* page submitted (2001).

[18] D. C. Fritts. Shear excitation of atmospheric gravity waves. *J. Atmos. Sci.* 39: 1936–1952 (1982).

[19] D. C. Fritts. The transient critical-level interaction in a Boussinesq fluid. *J. Geophys. Res.* 87: 7997–8016 (1982b).

[20] A. E. Gill. *Atmosphere-Ocean dynamics.* Academic Press San Diego (1982).

[21] G. Green. On the motion of waves in a variable canal of small depth and width. *Trans. Cambridge Philos. Soc.* [See Mathematical Papers, pp. 223–230. Macmillan, London, 1871] (1838).

[22] R. H. J. Grimshaw. Nonlinear internal gravity waves in a slowly varying medium. *J. Fluid Mech.* 54: 193–207 (1972).

[23] R. H. J. Grimshaw. The modulation and stability of an internal gravity wave. Technical Report 32 Department of Mathematical Sciences, University of Melbourne (1975b).

[24] K. Hasselmann. On the non-linear energy transfer in a gravity wave spectrum. Part 1. *J. Fluid Mech.* 12: 481–500 (1962).

[25] C. O. Hines. The saturation of gravity waves in the middle atmosphere. Part II: Development of the Doppler-spread theory. *J. Atmos. Sci.* 48: 1360–1379 (1991).

[26] L. N. Howard. Note on a paper by John W. Miles. *J. Fluid Mech.* 10: 509–512 (1961).

[27] A. V. Ivanov, L. A. Ostrovsky, I. A. Soustova, and L.Sh. Tsimring. Interaction of internal waves and turbulence in the ocean. *Dyn. Atmos. Ocean* 7: 221–232 (1983).

[28] J. Klostermeyer. Two- and three-dimensional parametric instabilities in finite amplitude internal gravity waves. *Geophys. Astrophys. Fluid Dyn.* 64: 1–25 (1991).

[29] C. G. Koop. A preliminary investigation of the interaction of internal gravity waves with a steady shearing motion. *J. Fluid Mech.* 113: 347–386 (1981).

[30] C. G. Koop and L. G. Redekopp. The interaction of long and short internal gravity waves: Theory and experiment. *J. Fluid Mech.* 111: 367–409 (1981).

[31] K. Lamb and R. T. Pierrehumbert. Steady-state nonlinear internal gravity-wave critical layers satisfying an upper radiation condition. *J. Fluid Mech.* 238: 371–404 (1992).

[32] P. H. LeBlond. On the damping of internal gravity waves in a continuously stratified ocean. *J. Fluid Mech.* 25: 121–142 (1966).

[33] M.-Pascale Lelong and Timothy J. Dunkerton. Intertia-gravity wave-breaking in three dimensions. Part I: Convectively stable waves. *J. Atmos. Sci.* 55: 2473–2488 (1998).

[34] M. J. Lighthill. *Waves in Fluids.* Cambridge University Press Cambridge, England (1978).

[35] R. S. Lindzen. Stability of a Helmholtz velocity profile in a continuously stratified infinite Boussinesq fluid – applications to clear air turbulence. *J. Atmos. Sci.* 31: 1507–1514 (1974).

[36] R. S. Lindzen. Turbulence and stress owing to gravity wave and tidal breakdown. *J. Geophys. Res.* 86: 9707–9714 (1981).

[37] J. Liouville. Sur le développement des fonctions ou parties de fonctions en séries .... *J. Math. Pure Appl.* 2: 16–35 (1837).

[38] P. N. Lombard and J. J. Riley. On the breakdown into turbulence of propagating internal waves. *Dyn. Atmos. Ocean* 23: 345–355 (1996).

[39] Z. Luo and D. C. Fritts. Gravity-wave excitation by geostrophic adjustment of the jet stream. Part II: Three-dimensional forcing. *J. Atmos. Sci.* 50: 104–115 (1993).

[40] S. A. Maslowe. The generation of clear air turbulence by nonlinear waves. *Stud. App. Maths* 51: 1–16 (1972).

[41] S. A. Maslowe. Critical layers in shear flows. *Ann. Rev. Fluid Mech.* 18: 405–432 (1986).

[42] G. Mastrantonio, F. Einaudi, D. Fua, and D. P. Lalas. Generation of gravity waves by jet streams in the atmosphere. *J. Atmos. Sci.* 33: 1730–1738 (1976).

[43] N. A. McFarlane. The effect of orographically excited gravity wave drag on the general circulation of the lower stratosphere and troposphere. *J. Atmos. Sci.* 44: 1775–1800 (1987).

[44] M. E. McIntyre and M. A. Weissman. On radiating instabilities and resonant overreflection. *J. Atmos. Sci.* 35: 1190–1196 (1978).

[45] A. Menhofer, R. K. Smith, M. J. Reeder, and D. R. Christie. "Morning-Glory" disturbances and the environment in which they propagate. *J. Atmos. Sci.* 54: 1712–1725 (1997).

[46] J. W. Miles. On the stability of heterogeneous shear flows. *J. Fluid Mech.* 10: 496–508 (1961).

[47] D. E. Mowbray. The use of schlieren and shadowgraph techniques in the study of flow patterns in density stratified liquids. *J. Fluid Mech.* 27: 595–608 (1967).

[48] D. E. Mowbray and B. S. H. Rarity. A theoretical and experimental investigation of the phase configuration of internal waves of small amplitude in a density stratified liquid. *J. Fluid Mech.* 28: 1–16 (1967a).

[49] J. Muench and E. Kunze. Internal wave interactions with equatorial deep jets. Part II. acceleration of the jets. *J. Phys. Oceanogr.* 30: 2099–2110 (2000).

[50] W. Munk. Internal wave and small-scale processes. In B. A. Warren and C. Wunsch, editors, *Evolution of Physical Oceanography* 264–291, MIT Press, Cambridge, MA (1981).

[51] A. R. Osborne and T. L. Burch. Internal solitons in the Andaman Sea. *Science* 208: 451–460 (1980).

[52] L. A. Ostrovsky and Yu. A. Stepanyants. Do internal solitons exist in the ocean? *Rev. Geophys.* 27: 293–310 (1989).

[53] L. A. Ostrovsky and D. V. Zaborskikh. Damping of internal gravity waves by small-scale turbulence. *J. Phys. Oceanogr.* 26: 388–397 (1996).

[54] T. N. Palmer, G. J. Shutts, and R. Swinbank. Alleviation of a systematic westerly bias in general circulation and numerical weather prediction

models through an orographic gravity drag parametrization. *Quart. J. Roy. Meteor. Soc.* 112: 1001–1039 (1986).

[55] W. R. Peltier and T. L. Clark. The evolution and stability of finite-amplitude mountain waves. Part II: Surface wave drag and severe downslope windstorms. *J. Atmos. Sci.* 36: 1498–1529 (1979).

[56] O. M. Phillips. On the dynamics of unsteady gravity waves of finite amplitude. Part 1. The elementary interactions. *J. Fluid Mech.* 9: 193–217 (1960).

[57] O. M. Phillips. *The Dynamics of the Upper Ocean.* Cambridge University Press Cambridge 2nd edition (1966).

[58] R. A. Plumb and A. D. McEwan. The instability of a forced standing wave in a viscous, stratified fluid: A laboratory analogue of the quasi-biennial oscillation. *J. Atmos. Sci.* 35: 1827–1839 (1978).

[59] J. F. Scinocca and T. G. Shepherd. Nonlinear wave-activity conservation laws and Hamiltonian structure for the two-dimensional anelastic equations. *J. Atmos. Sci.* 49: 5–27 (1992).

[60] T. G. Shepherd. Symmetries, conservation laws, and Hamiltonian structure in geophysical fluid dynamics. *Adv. Geophys.* 32: 287–338 (1990).

[61] G. J. Shutts. Gravity-wave drag parametrization over complex terrain:

the effect of critical level absorption in directional wind shear. *Quart. J. Roy. Meteorol. Soc.* 121: 1005–1021 (1995).

[62] G. J. Shutts. Stationary gravity-wave structure in flows with directional wind shear. *Quart. J. Roy. Meteorol. Soc.* 124: 1421–1442 (1998).

[63] E. A. Spiegel and G. Veronis. On the Boussinesq approximation for a compressible fluid. *Astrophys. J.* 131: 442–447 (1960).

[64] B. R. Sutherland. Internal gravity wave radiation into weakly stratified fluid. *Phys. Fluids* 8: 430–441 (1996).

[65] B. R. Sutherland. Propagation and reflection of large amplitude internal gravity waves. *Phys. Fluids* 11: 1081–1090 (1999).

[66] B. R. Sutherland. Internal wave reflection in uniform shear. *Q.J.R.M.S.* 126: 3255–3287 (2000).

[67] B. R. Sutherland. Finite-amplitude internal wavepacket dispersion and breaking. *J. Fluid Mech.* 429: 343–380 (2001).

[68] B. R. Sutherland, C. P. Caulfield, and W. R. Peltier. Internal wave generation and hydrodynamic instability. *J. Atmos. Sci.* 51: 3261–3280 (1994).

[69] B. R. Sutherland, S. B. Dalziel, G. O. Hughes, and P. F. Linden. Visualisation and measurement of internal waves by "synthetic schlieren". Part 1: Vertically oscillating cylinder. *J. Fluid Mech.* 390: 93–126 (1999).

[70] B. R. Sutherland, G. O. Hughes, S. B. Dalziel, and P. F. Linden. Internal waves revisited. *Dyn. Atmos. Ocean* 31: 209–232 (2000).

[71] B. R. Sutherland and P. F. Linden. Internal wave generation by flow over a thin barrier. *J. Fluid Mech.* 377: 223–252 (1998).

[72] B. R. Sutherland and P. F. Linden. Internal wave excitation by a vertically oscillating elliptical cylinder. *Phys. Fluids* page submitted (2001).

[73] S. G. Teoh, G. N. Ivey, and J. Imberger. Laboratory study of the interaction between two internal wave rays. *J. Fluid Mech.* 336: 91–122 (1997).

[74] S. A. Thorpe. The generation of alongslope currents by breaking internal waves. *J. Phys. Oceanogr.* 29: 29–38 (1999).

[75] V. Väisälä. " Uber die wirkung der windschwankungen auf die pilotbeobachtungen. *Soc. Sci. Fenn. Commentat. Phys.-Math.* 2: 19–37 (1925).

[76] G. B. Whitham. A general approach to linear and nonlinear dispersive waves using a Lagrangian. *J. Fluid. Mech.* 22: 273–283 (1965).

[77] G. B. Whitham. *Linear and Nonlinear Waves.* John Wiley and Sons, Inc. New York (1974).

[78] K. B. Winters and E. A. D'Asaro. Two-dimensional instability of finite

amplitude internal gravity wave packets near a critical level. *J. Geophys. Res.* 94: 12709–12719 (1989).

[79] K. B. Winters and E. A. D'Asaro. Three-dimensional wave instability near a critical level. *J. Fluid Mech.* 272: 255–284 (1994).

SECTION 3

MATHEMATICAL PHYSICS

Modern Applied Mathematics
Editor: J.C. Misra

Isogroups and Exact Solutions of Higher Order Boltzmann Equation and its Integrability

O.P. Bhutani* and Lipika Roy Chowdhury[1]

Department of Mathematics, Indian Institute of Technology, Hauz Khas, New Delhi-110016, India

Abstract

Herein the physically interesting and mathematically significant higher order non linear Boltzman equation.

$$\left(\frac{\partial}{\partial t}+1\right)\left(-\frac{\partial}{\partial \xi}\right)^{p} \psi(\xi, t)=\frac{\Gamma(2p)}{\Gamma(p)} \psi^{2}(\xi, t)$$

where $p \in I^{+}$ has been analysed via the isovector approach and examined for integrability through Painleve analysis. After constructing the components of the isovector field, the invariant groups of transformations are determined through orbital equations which are utilized to derive certain exact solutions corresponding to different values of the arbitrary constants, involved in the evaluation of isovector components. Further, performing Painleve PDE test on the equation it is shown that due to the complex values of the resonances, the equation is not completely integrable. Nevertheless, closed form solutions are obtained.

1. INTRODUCTION

The Boltzmann equation describes the time evolution of the one particle distribution function in a dilute gas of particles, interacting through binary collision only (Ernst,)[1]. Considering only the spatially uniform distributions, the rate of change of velocity distribution function f(v,t) is determined by the collision term. The collision term is expressed in terms of the differential scattering cross-section or equally the transition

* Author for correspondece: Tel: 011-6591472
Email: adhessbhutani@hotmail.com
[1] Present Address: Blk, 136# 101-140, Yishun Ring Road, Singapore – 760 136.

rate. These quantities are determined by the microscopic interactions between the particles. An explicit form of the cross section or transition rate defines a model, described by a model-Boltzmann equation, hereafter called the Boltzmann equation without introducing any ambiguity. While explicit expressions of the kernel in Boltzmann equation are known for stochastic models, it is not so for deterministic models, so that the usual method of applying Fourier transform fails in such cases. However, for a special case, generalized Laplace transform (Laplace transform of Fourier representation) is applicable and the equation is written as

$$(\partial_t + 1)\phi(y) = [B(p,p)y^{2p-1}]^{-1}\int_0^y dy'[y'(y-y')]^{p-1}\ \phi(y')\phi(y-y') \tag{1.1}$$

where $\phi(y)$ replaces $\phi(y,t)$, which is the generalized Fourier transform of the energy distribution function, due to the occurrence of $k^2 = 2y$ in Bobylev's derivation only in a particular way and k is a variable introduced by the Fourier transform as

$$\phi(k,t) = \int dv\ \exp\{-ik, v\}\ f(v,t)$$

The number p is equal to half the translation degree of freedom per particle or the number of internal states, depending upon the model.

For $p \in I^+$ Eq (1.1)becomes

$$(\partial_t + 1)\,(-\partial_\xi)^p\,\psi(\xi,t) = \frac{\Gamma(2p)}{\Gamma(p)}\psi^2(\xi,t) \tag{1.2}$$

where

$$\psi(y,t) = y^{p-1}\phi(y,t)/\Gamma(p).$$

The report (Ernst)[1] states that there are no closed form solutions of this equation, though expansions in terms of Laguerre polynomials are available for the cases p = 1,2.

Vijayakumar (1989)[2] found a scale invariant transformation for pde (1.2) and obtained exact solution by solving the reduced ode in the following form:

$$\psi = \frac{A\xi^{-p}e^{-t}}{1 + Ae^{-t}} \tag{1.3}$$

where A is an arbitrary constant

We take up the case p = 2 in the next few sections and deduce via isovector approach the solution (1.3) for p = 2 from one of the numerous transformed odes. Transformations have been generalized to order p, whenever possible.

2. Isovector Approach and Similarity Reductions

The equation under consideration here is

$$(\partial_t + 1)\ (-\partial_\xi)^2 \psi = 6\psi^2 \tag{2.1}$$

Writing $\qquad \psi_\xi = y$ and $\psi_{\xi\xi} = y_\xi = z$ $\qquad$ (2.2)

this system is equivalent to the following system of exterior forms (for details see Bhutani et al.)[3,4].

$$\begin{aligned}
\alpha_1 &= d\psi \wedge dt - y d\xi \wedge dt \\
\alpha_2 &= dy \wedge dt - z d\xi \wedge dt \\
\alpha_3 &= dz \wedge d\xi + (6\psi^2 - z)\ d\xi \wedge dt
\end{aligned} \tag{2.3}$$

To check whether the ideal $I\{\alpha_1, \alpha_2, \alpha_3\}$ is a closed one or not, we need to calculate exterior derivatives of α_i's. It is easily seen that

$$\begin{aligned}
d\alpha_1 &= -dy \wedge d\xi \wedge dt = \frac{1}{z}\alpha_2 \wedge dy \\
d\alpha_2 &= -\, dz \wedge d\xi \wedge dt = -\alpha_3 \wedge dt \\
d\alpha_3 &= 12\psi\ d\psi \wedge d\xi \wedge dt - dz \wedge d\xi \wedge dt = -12\psi\ \ \alpha_1 \wedge d\xi - \alpha_3 \wedge dt
\end{aligned} \tag{2.4}$$

From Eqs (2.4) we conclude that $I\{\alpha_1, \alpha_2, \alpha_3\}$ is a closed ideal,. Next, a vectorfield in E^5 space is assumed in the following form

$$V = V^\xi \partial_\xi + V^t \partial_t + V^\psi \partial_\psi + V^y \partial_y + V^z \partial_z \tag{2.5}$$

The transport property for the Lie derivatives provides

$$\wp_V(\alpha_i) = \lambda_j \alpha_j \qquad i, j = 1,2,3 \tag{2.6}$$

where Lie derivative is calculated using

$$\wp_V(\alpha_i) = d\ (V \rfloor \alpha_i) + V \rfloor d\alpha_i \tag{2.7}$$

From eqs (2.6) and (2.7) we get the following system of pdes in V^8's

$$V^t = V^t(t) \tag{i}$$

$$V_z^{\psi} = yV_z^{\xi} \quad \text{(ii)}$$

$$V^y = zV_y^{\psi} + V_{\xi}^{\psi} - y\left(V_{\xi}^{\xi} - V_{\psi}^{\psi}\right) - y^2V_{\psi}^{\xi} - yzV_y^{\xi} \quad \text{(iii)}$$

$$V_z^y = z\,V_z^{\xi} \quad \text{(iv)}$$

$$V^z = V_{\xi}^y + yV_{\psi}^y + z(V_y^y - V_{\xi}^{\xi}) - yzV_{\psi}^{\xi} - z^2V_y^{\xi} \quad \text{(v)}$$

$$V_y^{\xi} = 0 = V_{\psi}^{\xi} \quad \text{(vi)}$$

$$V_y^z = 0 = V_{\psi}^z \quad \text{(vii)}$$

$$V_t^{\xi} + (6\psi^2 - z)V_z^{\xi} = 0 \quad \text{(viii)}$$

$$V^z + V_t^z + (6\psi^2 - z)V_z^z = 12\psi V^{\psi} + (6\psi^2 - z)V_t^t \quad \text{(ix)}$$

(2.8)

The system (2.8) comprising of Eqs (i)-(ix) is solved to get the isovector components as follows

(vi) in (viii) $\Rightarrow V^{\xi} = V^{\xi}(\xi)$

From (ii), (x), (iv) and (iii), we get

$$V^{\psi} = V^{\psi}(\xi, t, \psi), \qquad V^y = V^y(\xi, t, \psi, y), \quad V^z = V^z(\xi, t, z)$$

Using these in the expressions available for V^{ψ}, V^y and V^z we get

$$V^{\psi} = \frac{\psi}{2}\left(V_{\xi}^{\xi} + S^1(t)\right) + S^2(\xi, t)$$

$$V^z = \frac{z}{2}\left(-3V_{\xi}^{\xi} + S^1(t)\right) + S_{\xi\xi}^2$$

$$V^y = \frac{\psi}{2}V_{\xi\xi}^{\xi} + \frac{y}{2}\left(-V_{\xi}^{\xi} + S^1\right) + S_{\xi}^2 \quad (2.9)$$

Using Eqs (2.9) in Eq (ix) of system (2.8) we get

$$V^{\xi} = \frac{b}{4}\xi + c$$

$$V^t = ae^t + b + d$$

$$V^{\psi} = -\frac{1}{2}(b + 2ae^t)\psi$$

$$V^y = -\frac{1}{4}(3b + 4ae^t)y$$

$$V^z = -(b + ae^t)z \qquad (2.10)$$

The orbital equations for this vector field are solved to get similarity transformations. We consider different cases depending on the values of the arbitrary constants in Eqs (2.10)

Case 1: a, b, c, d ≠ 0

We have

$$\frac{dt^*}{ds} = ae^{t^*} + b + d; \quad t^*(0) = t$$

$$\frac{d\xi^*}{ds} = \frac{b}{4}\xi^* + c; \qquad \xi^*(0) = \xi$$

$$\frac{d\psi^*}{ds} = -\frac{1}{2}\left(b + 2ae^{t^*}\right)\psi^*; \qquad \psi^*(0) = \psi$$

which yield the new independent variable

$$x = \left(\frac{a}{b+d} + e^{-t}\right)\left(\xi + \frac{4c}{b}\right)^{\frac{4(b+d)}{b}} \qquad (2.11a)$$

and the new dependent variable

$$v(x) = \psi e^t\left(\xi + \frac{4c}{b}\right)^{-\frac{2(b+2d)}{b}} \qquad (2.11b)$$

The reduced ode in this case is

$$8(b+d)^2 x^2 v''' + 10(b+d)(3b+4d)xv'' + (3b+4d)(5b+8d)v' + 3b^2v^2 = 0 \qquad (2.12)$$

Case 2: a = 0, b ≠ -d

In this case, solutions of orbital equations are

$$t^* - t = (b+d)s$$

$$\xi^* + \frac{4c}{b} = \left(\xi + \frac{4c}{b}\right)e^{\frac{b}{4}s}$$

$$\psi^* = \psi e^{-\frac{b}{2}s}$$

Eliminating the parameter s, we get the similarity transformation

$$x = \left(\xi + \frac{4c}{b}\right)^{\frac{4(b+d)}{b}} e^{-t}$$

$$v(x) = \psi\left(\xi + \frac{4c}{b}\right)^2 \qquad (2.13)$$

which reduce the pde (2.1) to

$$bx^3v'''' - 2(3b + 2d)\, x^2v''' + 2(9b + 8d)xv' - 24(b + d)\,(v - v^2) = 0$$

Putting x = e^z and denoting the differntiation w.r.t., z by dot(.), the above ode in further reduced to

$$b\ddot{\ddot{v}} - (9b + 8d)\dddot{v} + 2(13b + 10d)\dot{v} - 24(b + d)\,(v - v^2) = 0 \qquad (2.14)$$

Using standard substitution the order of ode (2.14) can be reduced by one.

Case 3 **b = 0; a, c, d ≠ 0**

$$V = (ae^t + d)\,\partial_t + c\partial_\xi - ae^t\psi\partial_\psi - ae^t y\partial_{y'} - ae^t z\partial_z$$

Solving the orbital equations

$$\frac{dt^*}{ds} = ae^{t^*} + d;$$

$$\frac{d\xi^*}{ds} = c;$$

$$\frac{d\psi^*}{ds} = -ae^{t^*}\psi^*$$

with the initial condition

$$t^*(0) = t, \quad \xi^*(0) = \xi, \quad \psi^*(0) = \psi$$

we get the similarity transformation

$$x = \left(a + de^{-t}\right)e^{\frac{d}{c}\xi} \qquad (2.15)$$

$$X(x) = \psi e^t\left(a + de^{-t}\right)$$

which transforms pde (2.1) to

$$x^3X''' + 2x^2X'' - \frac{6c^2}{d^3}X^2 = 0 \tag{2.16}$$

Substitution

$$X = e^z$$

transforms Eq (2.16)to

$$\dddot{X} - \ddot{X} - \frac{6c^2}{d^3}X^2 = 0. \tag{2.17}$$

Another similarity transformation defined by

$$x = a\left(a + de^{-t}\right)e^{\frac{d}{c}\xi}$$

$$Y(x) = \psi e^{t - \frac{d}{c}\xi}$$

transforms Eq (2.1) to

$$x^2Y''' + 5xY'' + 4Y' + \frac{6c^2}{d^3}Y^2 = 0 \tag{2.18}$$

Further, writing (2.15a) in the following form

$$y = \text{In}x = \frac{d}{c}\xi + \text{In}(a + de^{-t}) \tag{2.19}$$

and applying the transformation

$$y = \frac{d}{c}\xi + \text{In}(a + de^{-t})$$

$$v(y) = \psi e^t (a + de^{-t})$$

to the p-th order Boltzmann Eq (1.2), we get

$$v^{(p+1)} - v^{(p)} - \frac{\Gamma(2p)}{\Gamma(p)} \frac{c^p}{d^{p+1}} v^2 = 0 \tag{2.20}$$

Putting p = 1, we get the reduction for the Krook-Wu model:

$$v'' - v' - \frac{c}{d^2}v^2 = 0 \tag{2.21}$$

Putting v′ = F(v) this reduces to Abel's equation of the second kind (Murphy (1960)[6].

Case 4: **b= 0 = d; a, c ≠ 0**

For the case under consideration, the transformation is

$$x = a\xi + ce^{-t} \qquad (2.22)$$

$$X(x) = \psi e^{t}$$

which reduces the pde (2.1) to

$$a^2 cX''' + 6X^2 = 0 \qquad (2.23)$$

and the pde (1.2) to the following ode:

$$a^p cX^{(p+1)} + \frac{\Gamma(2p)}{\Gamma(p)} X^2 = 0 \qquad (2.24)$$

From Eq (2.24), the reduction for Krook-Wu model can be deuced quite easily. Infact, its integration yields a first order ode in quadrature form:

$$(X')^2 + \frac{2}{3ac} X^3 = C_{00} \qquad (2.25)$$

Case 5: **a = 0 = b**

The similarity variable in this case is of travelling wave type:

$$x = d\xi - ct \qquad (2.26)$$

$$\psi(x) = \psi(\xi, t)$$

Hence, we get the ode as

$$\psi'''(x) - \frac{1}{c}\psi'' + \frac{6}{cd^2}\psi^2 = 0 \qquad (2.27)$$

Generalizing it to the p-th order equation

$$\psi^{(p+1)} - \frac{1}{c}\psi^{(p)} + \frac{\Gamma(2p)}{\Gamma(p)} \frac{1}{cd^p}\psi^2 = 0 \qquad (2.28)$$

From Eq (2.28) we again get Abel's equation of second kind for the case p =1, after one integration.

Case 6: **d = -b**

The transformation is

$$x = \text{In}\left(\xi + \frac{4c}{b}\right) + \frac{b}{4a} e^{-t} \qquad (2.29)$$

$$X(x) = \psi e^{t}\left(\xi + \frac{4c}{b}\right)^{2}$$

The transformed equation is

$$b(X''' - 5X'' + 6X') + 24aX^{2} = 0 \tag{2.30}$$

Table - 1 lists these similarity reductions.

Sr. No.	Constants	Similarity Transformations	Reduced odes
1	a, b, c, d $\neq 0$	$x = \left(\frac{a}{b+d} + c^{-t}\right)\left(\xi + \frac{4c}{b}\right)^{\frac{4(b+d)}{b}}$ $v(x) = \psi e^{t}\left(\xi + \frac{4c}{b}\right)^{\frac{-2(b+2d)}{b}}$	$8(b+d)^2x^2v'''+10(b+d)(3b+4d)xv''$ $+(3b+4d)(5b+8d)v' + 3b^2v^2 = 0$
2	a = 0 b $\neq$ -d	$x = \left(\xi + \frac{4c}{b}\right)^{\frac{4(b+d)}{b}} e^{-t}$ $v(x) = \psi\left(\xi + \frac{4c}{b}\right)^{2}$	$bx^3v'''-2(3b+2d)x^2v''+2(9b+8d)xv'$ $-24(b+d)(v-v^2) = 0$
3	b = 0 a, d, c $\neq$ 0	$x = (a+de^{-t})\, e^{\frac{d}{c}\xi}$ $v(x) = \psi\, e^{t}(a+de^{-t})$	$X^3v'''+2x^2v'' - \frac{6c^2}{d^3}v^2 = 0$
4	b=0=d a, c $\neq$ 0	$x = a\,\xi + ce^{-t}$ $v(x) = \psi\, e^{t}$	$A^2\, cv''' + 6v^2 = 0$
5	a= 0 = b	$x = d\,\xi - ct$ $\psi(x) = \psi(\xi, t)$	$\psi''' - \frac{1}{c}\psi'' + \frac{6}{cd^2}\psi^2 = 0$
6	d = -b	$x = \ln\left(\xi + \frac{4c}{b}\right) + \frac{b}{4a}e^{-t}$ $v(x) = \psi e^{t}\left(\xi + \frac{4c}{b}\right)^{2}$	$B(v'''-5v''+6v') + 24av^2 = 0$

Table 1: Similarity Reductions of the Third order Boltzmann Equation

3. Similarity Solutions of the Boltzmann Equation

Some of the odes obtained in the last section have been solved to get exact solutions of Eqs (1.2) and (2.1) in the special cases p = 1, 2. We list them below;

Eq (2.12) has a solution

$$V = x^{-1}$$

which implies

$$\psi(\xi, t) = \left(\frac{b+d}{a}\right)\left(\xi + \frac{4c}{b}\right)^{-2} e^{-t}\left\{1 + \left(\frac{b+d}{a}\right)e^{-t}\right\}^{-1} \tag{3.1}$$

Putting p = 2 in solution (1.3), we get a similar expression with

$$A = (b + d)/a$$

$$C = 0$$

implying, this transformation can be directly deduced by scale invariance.

Moving on to Case 4, a solution of Eq (2.23) is

$$X = 10a^2c\,(x + c_0)^{-3}$$

which yields a solution of pde (2.1) as

$$\psi = 10a^2c\left(a\xi + ce^{-t} + c_0\right)^{-3} e^{-t} \tag{3.2}$$

Similarly, a solution of Eq (2.24) can be expressed as

$$X = 2\,(-)^p\,(2p + 1)\,ca^p\,(x + c_0)^{-(p+1)}, \tag{3.3}$$

from which solutions for the case p = 1, 2 are easily deduced.

Further, Eq (2.25) is in quadrature form. For the case $C_{00} = 0$, we get the power type solutions, which is a special case of Eq. (3.3). For $C_{00} \neq 0$, solutions are in terms of Jacobi's elliptic functions and can not be obtained using scale invariance.

In Sections 1 - 3 we have used the isovector approach and given some exact solutions. In Section 4 we use the weak Painleve property for the pde (2.1).

4. Painleve Analysis of Eq (1.2)

As explained in Bhutani et.al.[5] we seek solutions of Eq (2.1) in the form

$$\psi = \phi^m \sum_0^\infty u_j \phi^j \tag{4.1}$$

To calculate m, we put

$$\psi = u_0 \phi^m \tag{4.2}$$

into the equation and balance the highest order derivative term and most nonlinear term, this yields

$$m = -3$$

$$u_0 = 10\, \phi_{,\xi}^2\, \phi_{,t} \tag{4.3}$$

To calculate the resonances we put

$$\psi = u_0 \phi^{-3} + p\, \phi^{r-3} \tag{4.4}$$

in the equation and collect coefficients of ϕ^{r-6} to get

$$(r+1)\ (r^2 - 13r + 60) = 0$$

which give r = -1, and two complex values. Thus, following Steeb and Euler[7], we remark that Eq (2.1) does not pass the Painleve test. We can infer that, as there are no real resonances, there is no need for u_j s to be arbitrary for any values of j. Thus, we seek solution in the form

$$\psi = \sum_{j=0}^{\infty} u_j(\xi, t)\, \phi^{j-3}(\xi, t) \tag{4.5}$$

which on substitution in Eq (2.1) yields

$$\begin{aligned} &u_{j-3,\xi\xi t} + (j-5)u_{j-2,\xi\xi}\phi_{,t} + 2(j-5)\, u_{j-2,\xi t}\phi_{,\xi} + 2\,(j-4)\,(j-5)\, u_{j-1,\xi}\phi_{,\xi}\phi_{,t} \\ &+ 2(j-5)\, u_{j-2,\xi}\phi_{,\xi t} + (j-5)\, u_{j-2,t}\, \phi_{,\xi\xi} + (j-4)\,(j-5)\, u_{j-1,t}\, \phi_{,\xi}^2 \\ &+ (j-4)\,(j-5)\, u_{j-1}\, \phi_{,\xi\xi}\, \phi_{,t} + (j-5)\, u_{j-2}\, \phi_{,\xi\xi t} + (j-3)\,(j-4)\,(j-5)\, u_j\, \phi_{,\xi}^2\, \phi_{,t} \\ &+ 2(j-4)\,(j-5)\, \phi_{,\xi}\, \phi_{,\xi t}\, u_{j-1} + u_{j-3,\xi\xi} + 2(j-5)\, u_{j-2,\xi}\, \phi_{,\xi} \\ &+ (j-5)\, u_{j-2}\, \phi_{,\xi\xi} + (j-4)\,(j-5)\, u_{j-1}\, \phi_{,\xi}^2 - 6 \sum_{k=0}^{j} u_{j-k}\, u_k = 0 \end{aligned} \tag{4.6}$$

For j = 0, we get

$$u_0 = -\,10\, \phi_{,\xi}^2\, \phi_{,t}$$

For j = 1,

$$u_1 = \frac{5}{4}\left[5\,\phi_{,\xi\xi}\,\phi_{,t} + 6\phi_{,\xi}\,\phi_{,\xi t} + \phi_{,\xi}^{2} + \phi_{,\xi}^{2}\,\frac{\phi_{,tt}}{\phi_{,t}}\right] \tag{4.7}$$

For j =2,

$$\begin{aligned} 2(2u_0 + \phi_{,\xi}^{2}\,\phi_{,t})\,u_2 &= -2u_1^2 + 4\,\phi_{,\xi}\,\phi_{,t}\,u_{1,\xi} + 2u_{1,t}\,\phi_{,\xi}^{2} + 2\,\phi_{,\xi\xi}\phi_{,t}\,u_1 \\ &+ 4\,\phi_{,\xi}\,\phi_{,\xi t}\,u_1 + 2\,\phi_{,\xi}^{2}\,u_1 - u_{0,\xi\xi}\phi_{,t} - 2u_{0,\xi}\,\phi_{,\xi t} \\ &- u_{0,t}\,\phi_{,\xi\xi} - u_0\,(\phi_{,\xi\xi} + \phi_{,\xi\xi t}) - 2\,\phi_{,\xi}\,(u_{0,\xi} + u_{0,\xi t}) \end{aligned} \tag{4.8}$$

and for j = 3

$$\begin{aligned} 6u_0\,u_3 &= -6u_1\,u_2 + 2\,\phi_{,\xi}\,\phi_{,t}\,u_{2,\xi} + \phi_{\xi}^{2}\,\dot{u}_{2,t} + u_2\,(\phi_{,\xi\xi}\phi_{,t} + 2\phi_{,\xi}\phi_{,\xi t} + \phi_{,\xi}^{2}) \\ &- u_{1,\xi\xi}\phi_{,t} - 2u_{1,\xi t}\,\phi_{,\xi} - 2\,\phi_{,\xi t}\,u_{1,\xi} - u_{1,t}\,\phi_{,\xi\xi} \\ &- u_1\,\phi_{,\xi\xi t} - u_1\,\phi_{,\xi\xi} - 2\,u_{1,\xi}\,\phi_{,\xi} - u_1\,(\phi_{,\xi\xi t} + \phi_{,\xi\xi}) \end{aligned} \tag{4.9}$$

etc.

Taking

$$\phi = a\;\xi + \beta(t) \tag{4.10}$$

we get

$$u_0 = -10a^2\,\beta'(t)$$

$$u_1 = \frac{5}{4}\,\frac{a^2}{\beta'}\,(\beta'' + \beta')$$

$$u_2 = -\,\frac{5}{8.38}\,\frac{a^2}{\beta'^3}\,(4\beta'''\,\beta' - 9\beta''^2 - 6\beta'\,\beta'' - \beta'^2) \tag{4.11}$$

etc.

We consider different cases now to specify β:

Case 1

Assuming $u_1 = c_0$ (a constant), Eq (4.11) yields

$$\frac{\beta''}{\beta'} + 1 = \frac{4c_0}{5a^2}$$

which on integration yields

$$\beta = \frac{5a^2 b_0}{(4c_0 - 5a^2)} e^{\left(\frac{4c_0}{5a^2} - 1\right)t} + a_0 \tag{4.12}$$

We require that u_2 vanishes identically, hence using (4.11) we get

$$c_0 = 0 \quad \text{or} \quad c_0 = a^2$$

Subcase 1 $\quad c_0 = 0$

Here we get

$$u_0 = -10\, a^2\, b_0\, e^{-t}$$

and

$$u_1 = 0 = u_2 = u_3 = \ldots$$

so that

$$\phi = a\,\xi - b_0\, e^{-t} + a_0$$

and hence

$$\psi = -\,10a^2\, b_0\, e^{-t}\,(a\,\xi - b_0\, e^{-t} + a_0)^{-3} \tag{4.13}$$

Subcase 2 $\quad c_0 = a^2$

In this case

$$\beta = -5b_0\, e^{-\frac{1}{5}t} + a_0$$

$$u_0 = -10\, a^2\, b_0\, e^{-\frac{1}{5}t}$$

$$u_1 = a^2$$

$$u_2 = 0 = u_3 = \ldots$$

so that

$$\psi = a^2 \left(a\xi - 5b_0\, e^{-\frac{1}{5}t} + a_0 \right)^{-3} \left(-15\, b_0 e^{-\frac{1}{5}t} + a\xi + a_0 \right) \qquad (4.14)$$

Case 2 Assuming $\beta'' = -\beta'$. We get

$$u_0 = -10a^2 \beta' \phi^{-3}$$

$$u_1 = 0 = u_2 = \ldots$$

which implies

$$\psi = \frac{10\, a^2\, A\, e^{-t}}{(a\,\xi + Ae^{-t} + B)^3} \qquad (4.15)$$

5. Concluding Remarks

Following Steeb and Euler we can conclude that due to the complex values of the resonances, the Boltzmann Equation (1.2) for the case p = 2, is not completely integrable. But, as shown in the last section, closed form solutions can still be obtained. Further, the comparison with isovector approach shows that one of the solutions obtained there is similar to solutions (4.13) and (4.15). However, equation (4.14) represents a new solution. Similar results were obtained by Moussa[9] for the case p = 1.

Thus, the present study has enabled us to arrive at new exact solutions to the third order Boltzmann equation, one of which can be generalized to p-th order equation. Some of the transformations which work for p-th order equation, also give reductions of the Krook-Wu model by putting p = 1. Some reductions in this case, result in Abel's equation of the second kind.

Acknowledgement

OPB and LRC gratefully acknowledge the respective financial support received through INSA 'Senior Scientist' Schemes and CSIR (India) Research Fellowship Scheme.

References

1. Ernst, M.H. Phys. Reports 78 (1), 1 (1981).

2. Vijayakumar, K. Ph.D. Thesis, I.I.T. Delhi (1989).

3. Bhutani, O.P. & Vijayakumar, K. Int. J. Engg. Sci., 20, 375 (1990).

4. Bhutani, O.P.; & Roy–Chowdhury, Lipika Computers and Mathematics with Applications 40, 1349 (2000).

5. Bhutani, O.P.; Moussa M.H. M & Vjayakumar, K. Int. J. Engg. Sci., 33, 331 (1995).

6. Murphy, G.M. "Ordinary Differential Equations and their Solutions", D. Van Nostrand Canada (1960).

7. Steeb, W.H. & Euler N "Nonlinear Evolution Equations and Painleve' Test" World Scientific (1988).

8. Moussa, M.H.M. Ph.D. Thesis, I.I.T. Delhi (1991).

Modern Applied Mathematics
Editor: J.C. Misra

Recent Studies in Molecular and Particle Modelling

Donald Greenspan

Mathematics Department, University of Texas at Arlington

ABSTRACT

In this paper we study in detail new computer models for primary vortex flow, for turbulent flow, for soliton collision, and for the motion of a top on a smooth surface. Other applications, like microdrop collision, stress of a slotted copper plate, contact angle of adhesion, biological self reorganization, the bounce of an elastic ball, and elastic snap-through are mentioned and referenced appropriately.

Key words: molecule, turbulence, soliton, conservation laws

1 Physical, Mathematical and Computational Preliminaries

1.1 Introduction.

In an invited lecture at the national American Physical Society meeting in 1977, physics Nobelist Richard Feynman [3] stated unequivocally that quantum mechanical principles placed few important restrictions on how a stable atom can be held in place, and, indeed, that a stable atom on the whole has a comparatively definite position fixed by its comparatively massive nucleus. In this spirit, N-body problems are studied in this paper using molecules and classical molecular potentials and are also studied using collections of molecules, called particles, with related potentials derived through conservation of mass and energy. The basic mathematical problem for all our models is the N-Body problem, which, for completeness, is stated now in generality as follows.

In cgs units and for $i = 1, 2, ..., N$, let P_i of mass m_i be at $\vec{r}_i = (x_i, y_i, z_i)$, have velocity $\vec{v}_i = (v_{i,x}, v_{i,y}, v_{i,z})$, and have acceleration $\vec{a}_i = (a_{i,x}, a_{i,y}, a_{i,z})$ at time $t \geq 0$. Let the positive distance between P_i and P_j, $i \neq j$, be $r_{ij} = r_{ji} \neq 0$. Let the force on P_i due to P_j be $\vec{F}_{ij} = \vec{F}_{ij}(r_{ij})$, so that the force depends only on the *distance* between P_i and P_j. Also, assume that the force $\vec{F}_{ji}$ on P_j due to P_i satisfies $\vec{F}_{ji} = -\vec{F}_{ij}$. . Then, given the initial positions and velocities of all the P_i, $i = 1, 2, 3, ..., N$, the general N-body problem is to determine the motion of the system if each P_i interacts with all the other P_j's in the system.

1.2 Classical Molecular Forces

Classical molecular forces behave, in general, as follows [4]:

(a) when two close molecules are pulled apart, they attract;
(b) when pushed together, they repel; and
(c) repulsion is of a greater order of magnitude than is attraction.

There are a variety of classical molecular potentials for the interactions of molecules and from these classical molecular force formulas can be derived [13]. The potential which has received the most attention is the Lennard-Jones potential, that is,

$$\phi(r_{ij}) = 4\epsilon\left[\frac{\sigma^{12}}{r_{ij}^{12}} - \frac{\sigma^6}{r_{ij}^6}\right] \text{ erg,} \tag{1.1}$$

in which r_{ij} is measured in angstroms. In (1.1), the term with exponents 12 is the repulsion term and the term with exponents 6 is the attraction term. As r_{ij} goes to zero, the resulting motion is volatile.

For example, an approximate Lennard-Jones potential for water vapor molecules is

$$\phi(r_{ij}) = (1.9646)10^{-13}\left[\frac{2.725^{12}}{r_{ij}^{12}} - \frac{2.725^6}{r_{ij}^6}\right] \text{erg} \quad \left(\frac{\text{grcm}^2}{\text{sec}^2}\right) \tag{1.2}$$

in which, with respect to (1.1), $4\epsilon = (1.9646)10^{-13}$ and $\sigma = 2.725$. The force $\vec{F}_{ij}$ exerted on P_i by P_j is then

$$\vec{F}_{ij} = (1.9646)10^{-5}\left[\frac{12(2.725^{12})}{r_{ij}^{13}} - \frac{6(2.725^6)}{r_{ij}^7}\right]\frac{\vec{r}_{ji}}{r_{ij}} \text{ dynes} \quad \left(\frac{\text{grcm}}{\text{sec}^2}\right). \tag{1.3}$$

Note that $F_{ij} = \| \vec{F}_{ij} \| = 0$ implies that $r_{ij} = 3.06$ Å, which is the equilibrium distance, with repulsion prevailing if $r_{ij} < 3.06$ and attraction prevailing if $r_{ij} > 3.06$. Note also that in a regular triangle with edge $r_{ij} = 3.06$ Å, the altitude is 2.65 Å.

1.3 Equations of Motion for Water Vapor Molecules.

In molecular mechanics we simulate the motion of a system of molecules using classical molecular potentials and Newtonian mechanics. Thus the motion for a single water vapor molecule P_i acted on by a single water vapor molecule P_j, $i \neq j$, is then

$$m_i\vec{a}_i = (1.9646)10^{-5}\left[\frac{12(2.725^{12})}{r_{ij}^{13}} - \frac{6(2.725^6)}{r_{ij}^7}\right]\frac{\vec{r}_{ji}}{r_{ij}}\,. \tag{1.4}$$

Since the mass of a water molecule is $(30.103)10^{-24}$ gr, the latter equation is equivalent to

(1.5) $$\vec{a}_i = (160.33)10^{19}\left[\frac{818.90}{r_{ij}^{13}} - \frac{1}{r_{ij}^{7}}\right]\frac{\vec{r}_{ji}}{r_{ij}} \qquad \left(\frac{\text{cm}}{\text{sec}^2}\right).$$

Recasting the latter equation in Å/(ps^2) yields

(1.6) $$\vec{a}_i = (160330)\left[\frac{818.90}{r_{ij}^{13}} - \frac{1}{r_{ij}^{7}}\right]\frac{\vec{r}_{ji}}{r_{ij}} \qquad \left(\frac{\text{Å}}{\text{ps}^2}\right).$$

On the molecular level, however, the effective force on P_i is local, that is it is determined only by close molecules. But, because of our interest in turbulence, in which the forces are so great that attraction is of lesser importance, we will take *local* to mean the solution of the equation

(1.7) $$\frac{dF_{ij}}{dr_{ij}} = 0.$$

The solution to this equation yields $r_{ij} = 3.39$ Å. Thus, for $r_{ij} \geq D = 3.39$ Å, we choose $\vec{F}_{ij} = \vec{0}$. Note that it is more common to choose $D = 3\sigma$.

From (1.6), then, the dynamical equation for water molecule P_i will be

(1.8) $$\frac{d^2\vec{r}_i}{dt^2} = (160330.)\sum_{\substack{j\\ j\neq i}}\left[\frac{818.90}{r_{ij}^{13}} - \frac{1}{r_{ij}^{7}}\right]\frac{\vec{r}_{ji}}{r_{ij}}\ ;\ r_{ij} < D.$$

The equations of motion for a system of water molecules are then

(1.9) $$\frac{d^2\vec{r}_i}{dt^2} = (160330)\sum_{\substack{j\\ j\neq i}}\left[\frac{818.90}{r_{ij}^{13}} - \frac{1}{r_{ij}^{7}}\right]\frac{\vec{r}_{ji}}{r_{ij}}\ ;\ i = 1, 2, 3, ..., N;\ r_{ij} < D.$$

Note that on the molecular level gravity can be neglected since 980 cm/sec^2 = $(980)10^{-16}$ Å/ps^2. Note also that in simulating the motion of a system of N molecules, nonlinear equations (1.9) forewarn us that we will have to solve *large* systems of nonlinear, second order, ordinary differential equations (2N in two dimensions and 3N in three dimensions), and that these will have to be solved numerically. The method used is the leap frog method [7], and the recursion formulas are given explicitly as follows: Choose a positive time step Δt and let $t_k = k\Delta t,\ k = 0, 1, 2, \ldots$. For $i = 1, 2, 3, ..., N$, let P_i have mass m_i and at t_k let it be at

$$\vec{r}_{i,k} = (x_{i,k}, y_{i,k}, z_{i,k})\ ;$$

have velocity

$$\vec{v}_{i,k} = (v_{i,k,x}, v_{i,k,y}, v_{i,k,z}),$$

and have acceleration

$$\vec{a}_{i,k} = (a_{i,k,x}, a_{i,k,y}, a_{i,k,z}).$$

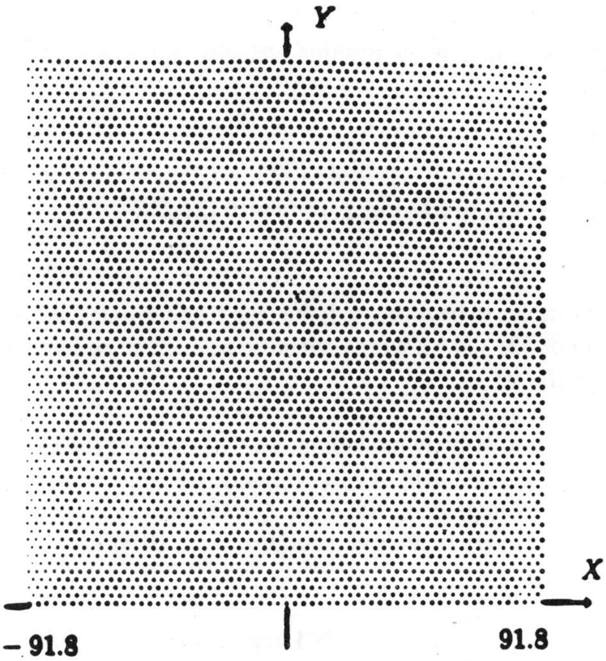

Figure 1 The vapor array.

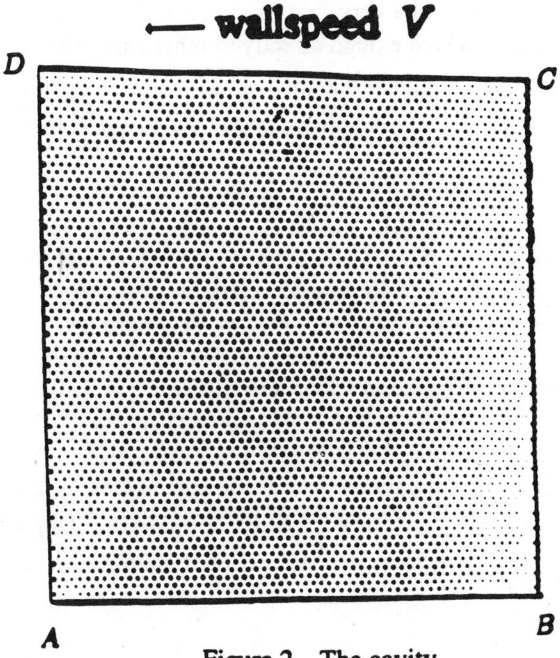

Figure 2 The cavity.

The leap frog formulas, which relate position, velocity and acceleration are

$$(1.13) \qquad \vec{v}_{i,\frac{1}{2}} = \vec{v}_{i,0} + \tfrac{1}{2}(\Delta t)\vec{a}_{i,0}, \qquad \text{(Starter)}$$

$$(1.14) \qquad \vec{v}_{i,k+\frac{1}{2}} = \vec{v}_{i,k-\frac{1}{2}} + (\Delta t)\vec{a}_{i,k}, \quad k = 1, 2, 3, ...$$

$$(1.15) \qquad \vec{r}_{i,k+1} = \vec{r}_{i,k} + (\Delta t)\vec{v}_{i,k+\frac{1}{2}}, \quad k = 0, 1, 2, 3... \; .$$

In all the examples to be discussed, the leap frog formulas will be implemented on a Digital Alpha 275 or 533 (64 bit word) scientific pc and can be reproduced readily by anyone with a scientific pc.

In our approach to N-body modelling, we will consider the following three cases:

(A) N small (say, $1 \le N \le 200$)
(B) N large (say, $200 < N \le 10000$)
(C) N very large (say, $10000 < N < 10^{20}$) .

2. N Large

2.1 A Cavity Problem

We begin with N large. The very first example will be discussed in great detail, the ideas and methodology of which extend directly to other examples.

Let us first construct a regular triangular grid of edge length 3.06 Å, and with altitude 2.65 Å. We determine 4235 grid points in the XY plane as follows:

$$x(1) = -91.8, \; y(1) = 0,$$
$$x(i) = x(i-1) + 3.06, \; y(i) = 0, \; i = 2, 61$$
$$x(62) = -90.27, \; y(62) = 2.65$$
$$x(i) = x(i-1) + 3.06, \; y(i) = 2.65, \; i = 62, 121$$
$$x(i) = x(i-121), \; y(i) = y(i-121) + 5.30, \; i = 122, 4235.$$

At each point $(x(i), y(i))$ we set a water vapor molecule P_i, $i = 1, 4235$. This array is shown in Figure 1.

To complete the initial data, recall that in two dimensions the rms velocity v for a water vapor molecule at $150^\circ C$ is $v = 6.23$ Å/ps . Each molecule is now assigned this speed in either the X or Y direction, determined at random, with its ($\pm$) sign also determined at random

Now, in two dimensions consider the square of side 183.6 Å, shown in Figure 2. The interior is called the cavity or the basin of the square. This basin encloses the molecular fluid shown in Figure 1. The four sides are called the walls. The top wall , alone, CD, is allowed to move . It moves in the X direction at a constant speed V, called the wallspeed. Also it is allowed an extended length so that the molecular fluid is always completely enclosed by four walls. Then the cavity problem is to describe the gross motion of the fluid for various choices of V , which will be given in Å /ps.

For laboratory studies of the cavity problem see the treadmill apparatus of Freitas, Street, Findikakis and Koseff [5].

2.2 Computational Considerations

In all of our examples, the following computational considerations are implemented.

For time step Δt (ps), and $t_k = k\Delta t, k = 0, 1, 2, \ldots,$ two problems must be considered relative to the computations. The first problem is to prescribe a protocol when, computationally, a molecule has crossed a wall into the exterior of the cavity. For each of the lower three walls, we will proceed as follows (no slip condition). The position will be reflected back symmetrically, relative to the wall, into the interior of the basin, the velocity component tangent to the wall will be set to zero and the velocity component perpendicular to the wall will be multiplied by -1. If the molecule has crossed the moving wall, then its position will be reflected back symmetrically, its Y component of velocity will be multiplied by -1, and its X component of velocity will be increased by the wallspeed V.

The second problem derives from the fact that an instantaneous velocity field for molecular motion is Brownian. In order to better interpret gross fluid motion , we will introduce average velocities as follows. For J a positive integer, let particle P_i be at $(x(i,k), y(i,k))$ at t_k and at $(x(i,k-J), y(i,k-J))$ at t_{k-J}. Then the average velocity $\vec{v}_{i,k,J}$ of P_i at t_k is defined by

$$(2.1) \qquad \vec{v}_{i,k,J} = \left(\frac{x(i,k)-x(i,k-J)}{J\Delta t}, \frac{y(i,k)-y(i,k-J)}{J\Delta t}\right).$$

In the examples to be described, we will discuss results for various values of J.

Observe also that the connection between the wall speed V and the Reynolds number Re is given by [17]:

$$(2.2) \qquad Re = |V|B/\nu,$$

in which B is the span, that is, the length CD in Figure 2, and ν is the average kinematic viscocity of water . Thus, for example, for $V = -24\,\mathring{A}\,/ps$, $B = 183.6\ \mathring{A}$ and , at 100° C [20], $\nu = 30\ \mathring{A}^2\,/ps$, then (1.9) yields $Re = 146$.

2.3 Primary Vortex Generation.

Consider first the parameter choices $V = -24\overset{\circ}{A}/ps,$, $J = 20000$, $\Delta t = 0.0001\,ps$. Figures 3-6 show the development of a primary vortex at the respective times $t = 4, 8, 14, 20$. An area of compression which precedes and drives the vortical motion is evident in Figure 3. Other values of J which were studied were 16000, 12000, 9000, and 6000, each of which yielded results similar to those of Figures 3-6. Note that these results agree qualitatively with experimental results for cavity flow in the large [5].

Results similar to those in Figures 3-6, but with larger primary vortices were obtained with $V = -40$, -130, and -600.

2.4 Turbulent Flow Generation.

Turbulence is the most common yet least understood type of fluid flow. Turbulent flows have two well defined characteristics: (1) Many small vortices appear and disappear quickly [14], and (2) A strong current develops across the usual primary direction [18]. Though mathematical fluid dynamicists are aware that the Navier-Stokes equations are not the equations of turbulent flow, engineers continue to generate "turbulent" flows using the Navier Stokes equations with high Reynolds number [16]. It should also be observed that formulas of statistical mechanics are not applicable to turbulent flow [1,2].

The discussion in Section 2.3 for primary vortex generation now leads to the following approach to generating turbulent flows. For a sufficiently large magnitude of the wallspeed V, let us show that turbulence results when, for given Δt, a stable calculation results but no J exists which yields a primary vortex.

Let us then set $V = -3000$, $\Delta t = 2.5(10)^{-6}$. The motion was simulated to $t = 0.8$. Typical results are shown in Figures 7-10 for $J = 80000, 60000, 40000, 20000$. None of these figures shows a primary vortex because a strong current has developed across the primary vortex direction. Figure 11 shows an enlargement of the section of Figure 8 in the range $45 \leq x \leq 91.8$, $10 \leq y \leq 100$, and reveals this crosscurrent clearly.

To support the contention that Figures 7-10 represent fully turbulent flow, we now define the concept of a small vortex. For $3 \leq M \leq 6$, we define a *small vortex* as a flow in which M molecules nearest to an $(M+1)$st molecule rotate either clockwise or counterclockwise about the $(M+1)$st molecule and , in addition, the $(M+1)$st molecule lies *interior* to a simple polygon determined by the given M molecules. With this definition , Figure 12 shows that the flow in Fig. 8 has 185 small vortices at $t = 0.8$, while Figure13 shows that only 0.2 picoseconds before, that is at $t = 0.6$, the resulting flow has 182 small vortice which are completely different than those in Fig. 12.

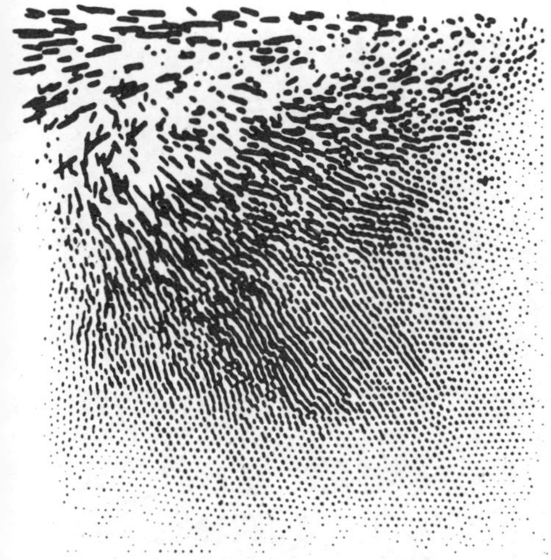

Figure 3 Wallspeed = - 24, J=20000, t=4.0.

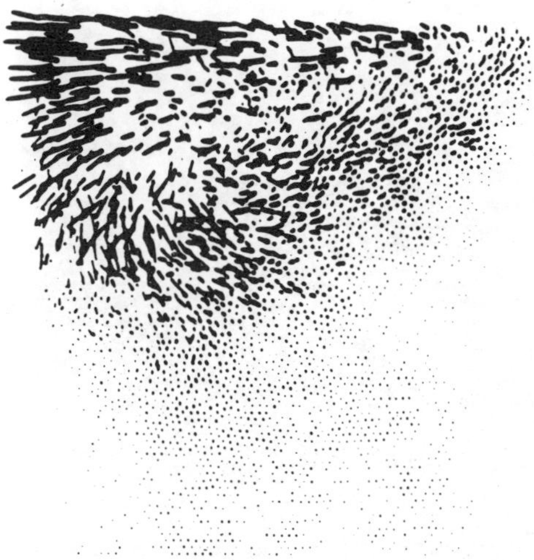

Figure 4 Wallspeed = - 24, J=20000, t=8.0.

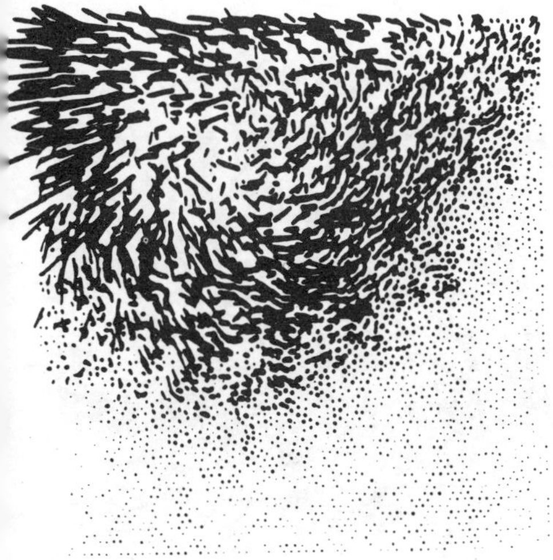

Figure 5 Wallspeed = - 24, J=20000, t=14.0.

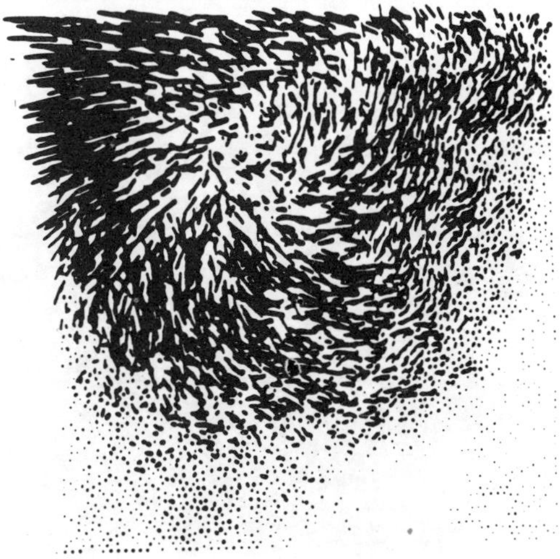

Figure 6 Wallspeed = - 24, J=20000, t=20.0.

Figure 7 Wallspeed = - 3000, J = 80000, t = 0.8.

Figure 8 Wallspeed = - 3000, J = 60000, t = 0.8.

Figure 9 Wallspeed = - 3000, J = 40000, t = 0.8.

Figure 10 Wallspeed = - 3000, J = 20000, t = 0.8.

Figure 11 Blowup of a right Section of Fig. 9.

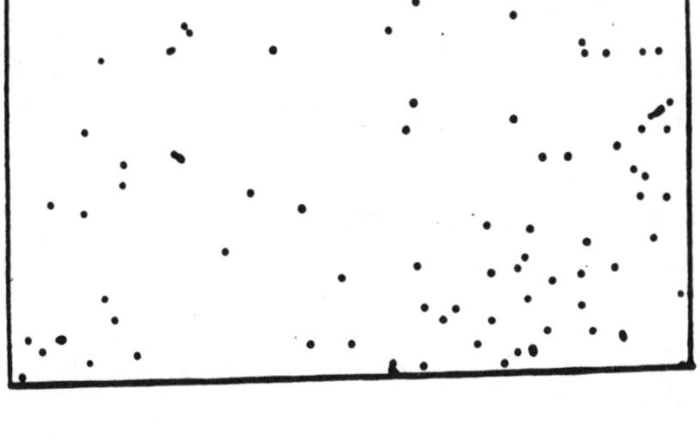

Figure 12 185 small vortices at t = 0.8.

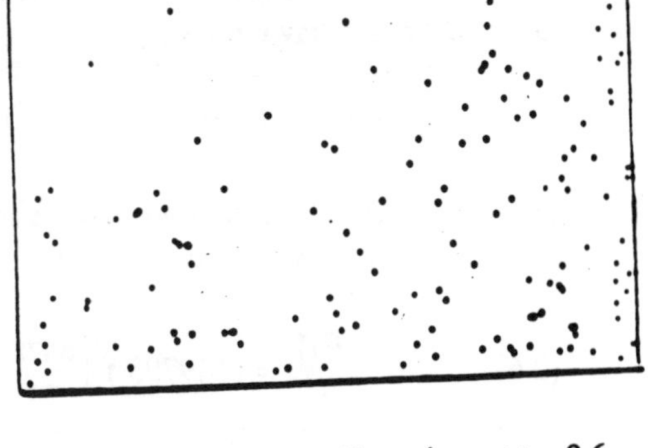

Figure 13 182 small vortices at t = 0.6.

2.5 Primary Vortices and Turbulence for Air

It is rather interesting that even though air is heterogeneous and consists of a variety of atoms and molecules, experimental Lennard-Jones potentials are readily available only for homogeneous air [13]. One such potential is

$$(2.3)\quad \phi(r_{ij}) = (5.36)10^{-14}\Big[\frac{3.617^{12}}{r_{ij}^{12}} - \frac{3.617^{6}}{r_{ij}^{6}}\Big]\,\mathrm{erg}\ .$$

The force $\vec{F}_{ij}$ exerted on P_i by P_j is then

$$(2.4)\quad \vec{F}_{ij} = (5.36)10^{-6}\Big[\frac{12(3.617^{12})}{r_{ij}^{13}} - \frac{6(3.617^{6})}{r_{ij}^{7}}\Big]\frac{\vec{r}_{ji}}{r_{ij}}\ \mathrm{dynes}$$

and the equilibrium value is $r_{ij} = 4.06\ \mathring{A}$. On the molecular level the effective force on P_i is local so that only those molecules within a distance D determined by

$$\frac{dF_{ij}}{dr_{ij}} = 0$$

are considered. The solution to this equation yields $r_{ij} = 4.50\ \mathring{A}$. Thus, for $r_{ij} \geq D = 4.50\ \mathring{A}$, we choose $\vec{F}_{ij} = \vec{0}$.

Before proceeding to dynamical considerations, it is necessary to characterize carefully the hypothetical air molecule to be used. We assume that the air to be used is *nondilute* and *dry*. Dry air consists primarily of 78% N_2, 21% O_2, and 1% Ar, whose respective masses are

$$m(N_2) = 28(1.660)10^{-24}\,gr$$
$$m(O_2) = 32(1.660)10^{-24}\,gr$$
$$m(Ar) = 40(1.660)10^{-24}\,gr.$$

We now characterize an "air" molecule A as consisting of proportionate amounts of N_2, O_2, and Ar and having mass

$$(2.5)\quad m(A)=[0.78(28) + 0.21(32) + 0.01(40)](1.660)10^{-24} = (4.807)10^{-23}\,gr.$$

From (2.4) and (2.5) it follows that the acceleration, in $\mathring{A}/(ps)^2$, of an air molecule P_i due to interaction with an air molecule P_j satisfies the equation

$$(2.6)\qquad \frac{d^2\vec{r}_i}{dt^2} = (149795.)\Big[\frac{4478}{r_{ij}^{13}} - \frac{1}{r_{ij}^{7}}\Big]\frac{\vec{r}_{ji}}{r_{ij}}\ \frac{\mathring{A}}{ps^2}\ ;\ r_{ij} < D.$$

The equations of motion for a system of air molecules are then

$$(2.7)\quad \frac{d^2\vec{r}_i}{dt^2} = (149795.)\sum_{\substack{j\\ j\neq i}}\left[\frac{4478}{r_{ij}^{13}} - \frac{1}{r_{ij}^{7}}\right]\frac{\vec{r}_{ji}}{r_{ij}}\quad \frac{\overset{\circ}{A}}{ps^2}\ ;\ i = 1,2,3,...,N;\ r_{ij} < D.$$

We consider again the cavity problem. Note that for a regular triangle with edge length 4.06 $\overset{\circ}{A}$, which is the equilibrium distance, the altitude is 3.516 $\overset{\circ}{A}$. Using these values we now construct a regular triangular grid of 4235 points on the basin as follows:

$$\begin{aligned}
&x(1) = -121.8, \quad y(1) = 0\\
&x(i) = x(i-1) + 4.06, \quad y(i) = 0, \quad i = 2, 61\\
&x(62) = -119.77, \quad y(62) = 3.516\\
&x(i) = x(i-1) + 4.06, \quad y(i) = 3.516, \ i = 62, 121\\
&x(i) = x(i-121), \quad y(i) = y(i-121) + 7.032, \quad i = 122, 4235.
\end{aligned}$$

At each point $(x(i), y(i))$ we set an air molecule P_i, $i = 1, 4235$. This array is like that shown in Figure 2.2 , except that the square has edge length 243.6 $\overset{\circ}{A}$ and provides the initial positions of the molecules.

To complete the initial data we need the initial velocities of the molecules. In two dimensions at 35° C the speed of an air molecule is 4.207 $\overset{\circ}{A}/ps$. Each molecule is then assigned a speed of 4.207 $\overset{\circ}{A}/ps$ in either the X or Y direction, determined at random, with its sign ($\pm$) also determined at random. Once the wallspeed V is prescribed, we are ready to solve the resulting cavity problem using the leap frog formulas for equations (2.7) with $N = 4235$.

For the parameter choices $V = -40$, $J = 17000$, $\Delta t = 0.0001$, Figures 14-17 show the development of a primary vortex at the respective times $t = 1.7,\ 5.1,\ 6.8,$ and 10.2. Other values of J which were studied were 7000, 9000, 11000, and 15000, but all yielded results entirely similar to those in Figures 14-17. The vortex in Figure 17 is slightly larger than the corresponding one for water vapor at $t = 10.2$. The average speed for water vapor at $t = 10.2$ is 7.52 $\overset{\circ}{A}/ps$ while the average speed in Figure 17 for air is 9.02 $\overset{\circ}{A}/ps$, even though the initial velocity for water was larger than that for air.

For the parameter choices $V = -3000$, $J = 40000$, $\Delta t = (5)10^{-6}$, Figures 18-20 show the transition to turbulence at the respective times $t = 0.4,\ 0.6,\ 1.0$. The flow is entirely analogous to that shown in Figures 7-11. The average speed of the air molecules is 362 $\overset{\circ}{A}/ps$ at $t = 0.6$ while that for water at the same time was 353 $\overset{\circ}{A}/ps$.

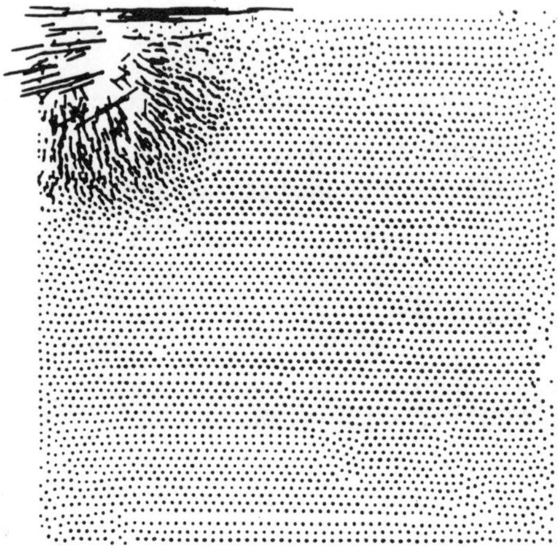

Fig. 14 **Wallspeed = -40, $J = 17000$, $t = 1.7$**

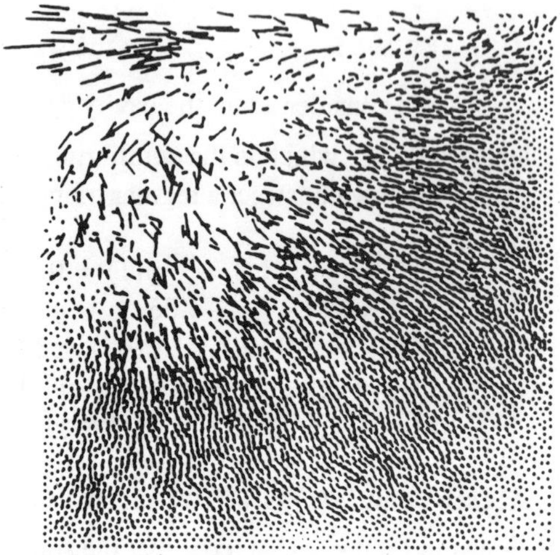

Figure 15 **Wallspeed = -40, $J = 17000$, $t = 5.1$**

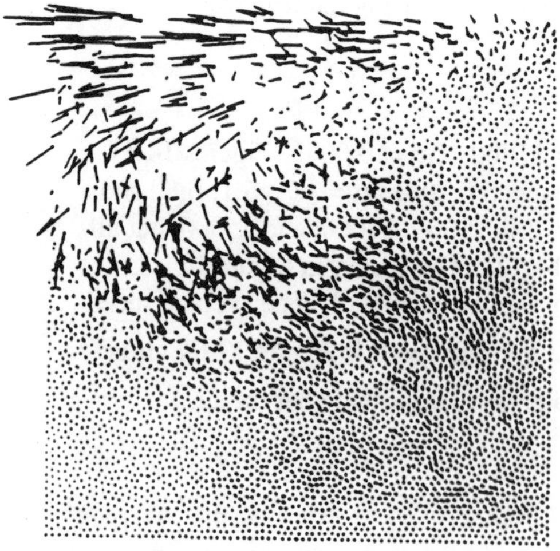

Figure 16 **Wallspeed = -40, $J = 17000$, $t = 6.7$**

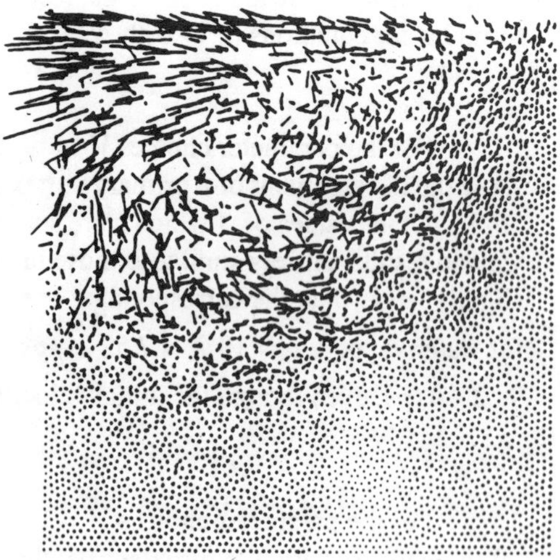

Figure 17 **Wallspeed = -40, $J = 17000$, $t = 10.2$**

igure 18 **Wallspeed =** $-$ **3000,** J **= 40000,** t **= 0.4**

Figure 19 **Wallspeed =** $-$ **3000,** J **= 40000,** t **= 0.6**

Figure 20 **Wallspeed =** $-$ **3000,** J **= 40000,** t **= 1.0.**

3. N Very Large.

3.1 Particle Simulation

Because statistical mechanics formulas are not applicable to turbulent flow , we will use a simple route of reducing this case to the case N large. This will allow us to apply the numerical methodology of the previous section. We will do this by invoking the engineering lumped mass approach, that is, molecules will be aggregated into particles using mass and energy conservation. The particles will then be studied dynamically by means of molecular type formulas, that is, formulas which include both attraction and repulsion. However, it will now be essential to include gravity.

3.2. Particle Arrangement and Equations for Water Vapor.

Consider first a 30 cm by 240 cm rectangle and on it construct a regular triangular grid with 8479 points, shown in Figure 21, by the recursion formulas

$x(1) = -15.0,\ y(1) = 0.0$
$x(i) = x(i-1) + 1.0,\ y(i) = 0.0,\ i = 2, 31$
$x(32) = -14.5,\ y(32) = 0.866$
$x(i) = x(i-1) + 1.0,\ y(i) = 0.866,\ i = 33, 61$
$x(i) = x(i-61),\ y(i) = y(i-61) + 1.732,\ i = 62, 8479.$

The side of each triangle in the grid has length 1 cm. At each grid point $(x(i), y(i))$ set a particle P_i, that is an aggregate of molecules. Thus, the distance between any two immediate neighbors is unity. Since the initial particle positions are now known, we also assign each particle a small initial velocity of 0.00001 in the X or Y direction, determined at random, with its sign ($\pm$) also determined at random

The force between two distinct particles P_i and P_j which are R_{ij} cm apart will be taken to have magnitude F given by

$$(3.1)\quad F(R_{ij}) = -\frac{A}{R_{ij}^3} + \frac{B}{R_{ij}^5} \quad \text{(dynes).}$$

The justification for this choice is as follows. First we wish to choose a formula which is qualitatively like (1.3) so that it too is composed of attraction and repulsion components. The choices of the exponents 3 and 5 guarantees further that the volatile motion between molecules with exponents 7 and 13 will not prevail for the more massive particles.

Thus, from (3.1),

$$(3.2)\quad \phi(R_{ij}) = -\frac{A}{2R_{ij}^2} + \frac{B}{4R_{ij}^4} \quad \text{(ergs).}$$

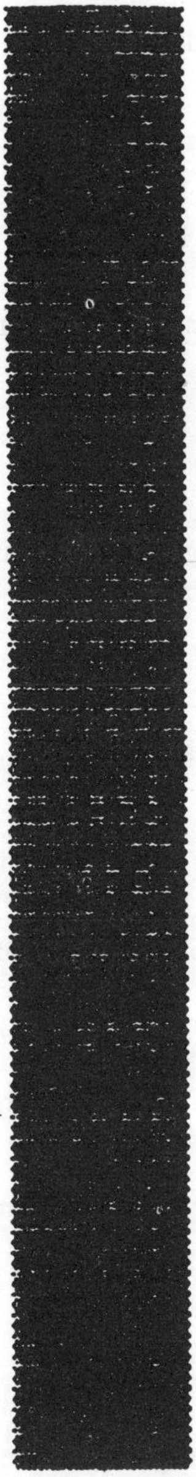

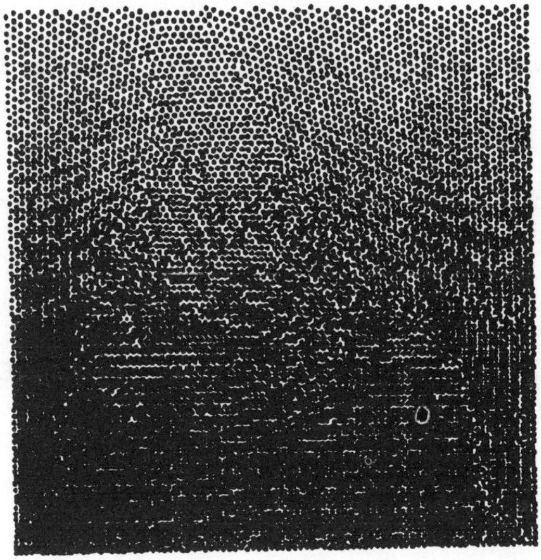

Figure 22 7549 particles

Figure 21 8479 particles

Our first problem is to determine A and B. Assume that $F(1) = 0$, so that from (3.1),

(3.3) $\quad - A + B = 0.$

In order to determine a second equation, some relevant observations must be made first.

Note that the number N of water vapor molecules which can be arranged in the rectangle using a regular triangular grid is

(3.4) $\quad N = \dfrac{30}{(3.06)10^{-8}} \cdot \dfrac{240}{(2.65)10^{-8}} = (8.87)10^{18}.$

Also, note that since the mass of a water molecule is $(30.103)10^{-24}$ gr, the total mass M of the water molecules inside the 30 cm by 240 cm rectangle in Figure 21 is

(3.5) $\quad M = (2.67)10^{-4}$ gr.

Distributing this mass over the 8479 particles for conservation of total mass yields an individual particle mass m of

(3.6) $\quad m = (3.15)10^{-8}$ gr.

From (1.2) and (3.4), the total potential energy E_M of the molecular configuration is, approximately,

(3.7) $\quad E_M = 3\sum_{i=1}^{(8.88)10^{18}} \left\{(1.9646)10^{-13}\left[\left(\frac{2.725}{3.06}\right)^{12} - \left(\frac{2.725}{3.06}\right)^{6}\right]\right\} = -(1.3)10^{6}$ erg.

On the other hand, the total potential energy E_p of the particle configuration is, from (3.2), approximately,

(3.8) $\quad E_p = 3\sum_{i=1}^{8479}\left(-\tfrac{1}{2}A + \tfrac{1}{2}B\right) = 25437\left(-\tfrac{1}{2}A + \tfrac{1}{2}B\right).$

However, the kinetic energies of both the particle and molecular configurations are relatively negligible so that total energy is conserved by setting $E_M = E_p$. Thus the second equation for A and B is

(3.9) $\quad 25437\left(-\tfrac{1}{2}A + \tfrac{1}{2}B\right) = -(1.3)10^{6}.$

The solution of (3.3) and (3.9) is $A = B = 205$. Thus, (3.1) takes the particular form

$$F(R_{ij}) = 205\,(\,-\,\frac{1}{R_{ij}^3} + \frac{1}{R_{ij}^5}\,)$$

We assume next that two particles interact only within the local interaction distance $D = 1.3\,\text{cm}$, which is the solution of the equation $\dfrac{dF}{dR_{ij}} = 0.$

The dynamical equation of motion for each particle P_i of the system is then given by

$$(3.10)\qquad \frac{d^2\vec{R}_i}{dt^2} = -\,980\,\vec{\delta} + \frac{\alpha}{m}\sum_{\substack{j\\ j\neq i}}(205)\Big[\frac{1}{R_{ij}^5} - \frac{1}{R_{ij}^3}\Big]\frac{\vec{R}_{ji}}{R_{ij}} \qquad ;\ R_{ij} < D,$$

in which $\vec{\delta} = (0,1)$, α is a parameter, and $i = 1, 8479$. The reason for the introduction of the parameter α is that particle interaction should be local relative to gravity, that is, gravity must dominate for R_{ij} less than, but close to, unity, which we assume to mean at present for $R_{ij} \geq 0.7$. However, this is the case by choosing $\alpha = m$, since, for $R_{ij} = 0.9,\ 0.8,\ 0.7$, and 0.6, F takes the values 66, 225, 622, and 1687, respectively , the first three of which are less than 980. Thus, the dynamical equations for the particles are

$$(3.11)\qquad \frac{d^2\vec{R}_i}{dt^2} = -\,980\,\vec{\delta} + \sum_{\substack{j\\ j\neq i}}(205)\Big[\frac{1}{R_{ij}^5} - \frac{1}{R_{ij}^3}\Big]\frac{\vec{R}_{ji}}{R_{ij}} \qquad ;\ R_{ij} < D.$$

It should be noted that other choices of α are also under study.

3.3 Particle Equilibrium.

We now allow the 8479 particles in Figure 21 to find their own equilibrium when interacting in accordance with (3.11). We choose $\Delta t = 0.0001$ and use the no slip reflection protocol.

The initial motion of the system is almost one of free fall. So, for the first 20000 time steps, each velocity is damped by the factor 0.2 every 20000 time steps. For the next 20000 time steps, each velocity is damped by the factor 0.4 every 20000 time steps. For the third 20000 time steps, each velocity is damped by the factor 0.7 every 20000 steps. For the final 20000 steps the damping is removed. In this fashion, the particles are eased down into a stable configuration. Finally, to obtain a square set of particles, all particles with $y_i > 30$ are removed to yield the 7549 particle set shown in Figure 22. The positions and velocities of these particles are used as the initial data for the cavity flow examples to be described next.

3.4 Examples.

In the cavity examples which follow, we use the same protocol as was used for molecules in Section 2.2 with respect to wall reflection and averaging velocities.

Example 1. Consider now the cavity problem for the 7459 particles in Figure 22. Let $V = -40$, $\Delta t = 0.00001$ and $J = 20000$. Then Figures 23-26 show the development of a classical primary vortex at the respective times $t = 0.2, 0.6, 1.0, 1.4$. Entirely similar results followed for $J = 14000, 11000, 9000$.

Results analogous to those just described were obtained also with $V = -100$, -250, -600, but the size of the primary vortex increased with the wallspeed.

Example 2. To simulate turbulence, set $V = -3000$, $\Delta t = 5(10)^{-7}$, $J = 80000$. Figures 27-30 show the flow development at the respective times 0.04, 0.08, 0.12.,0.15. Figure 27 shows that the motion begins with a compression wave. Figures 28 and 29 show that the ensuing particle repulsion is up and to the right in the usual primary vortex direction. Figure 30 shows the large crosscurrent over the usual primary vortex direction. Again, to support the contention that Figure 30 represents fully turbulent flow, we use the concept of a small vortex stated in Section 2.4. In searching for small vortices, attention was confined to within a circle of radius 1 cm around each particle. There resulted 355 small vortices which are shown in Figure 31. Moreover, Figure 32 shows the distribution of 349 small vortices at the time $t = 0.18$, so that, after only 0.03 *sec*, this figure shows the rapid change throughout the cavity of the vortex distribution shown in Figure 31.

Entirely similar results for Figures 30-33 followed with $J = 60000, 40000, 20000$.

Other examples of particle models include the stress and crack development in a slotted copper plate [8] and approximation of the contact angle of adhesion of a sessile drop on a solid surface [15]

4. Molecular Parameters Unavailable

4.1 Qualitative Models.

Though quantitative models are always to be preferred, in many interesting cases the molecular parameters may not be known. For such cases, we often develop qualitative models. Such models do provide the researcher with several advantages, including

(a) the development of intuition,

(b) allowance for variation of parameters to determine which parameters are of most significance, and,

(c) development of insights into the mechanisms of material behavior.

With this in mind, we examine next a qualitative model.

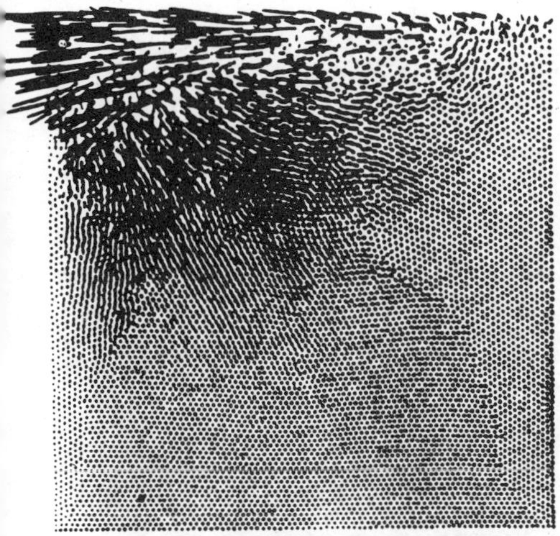

Figure 23 t=0.2

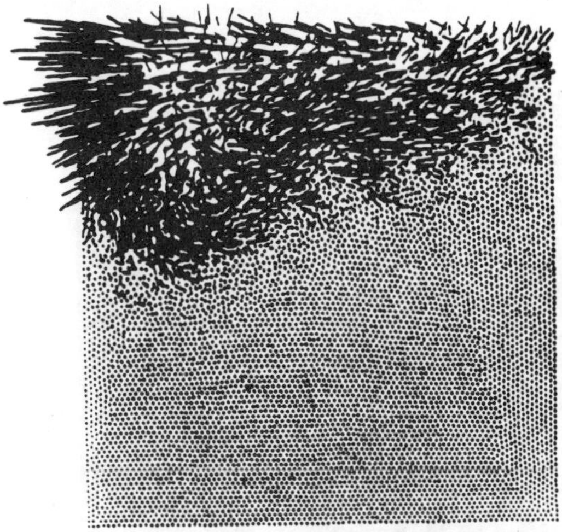

Figure 24 t=0.6

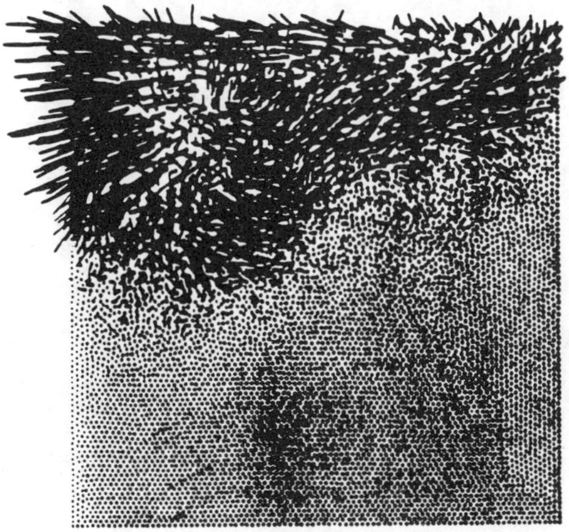

Figure 25 t=1.0

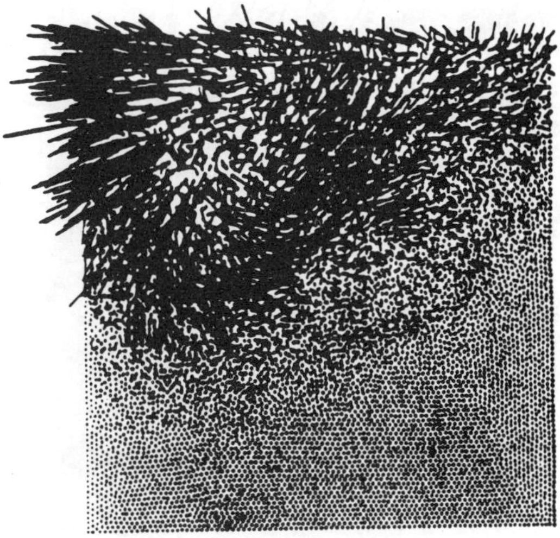

Figure 26 t=1.4

Figure 27 $V = -3000, t = 0.04$

Figure 28 $V = -3000, t = 0.08$

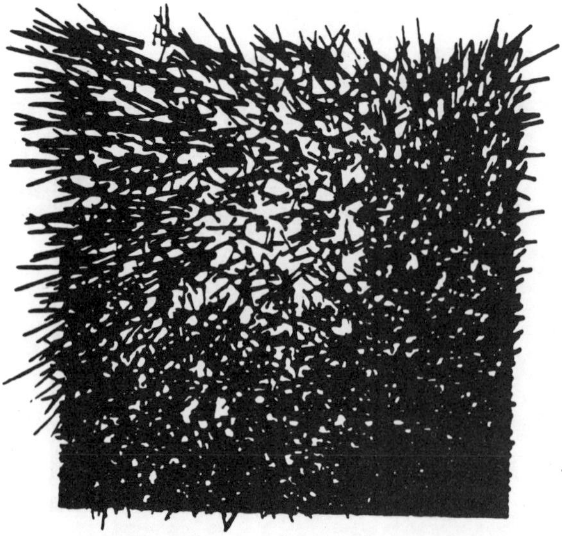

Figure 29 $V = -3000, t = 0.12$

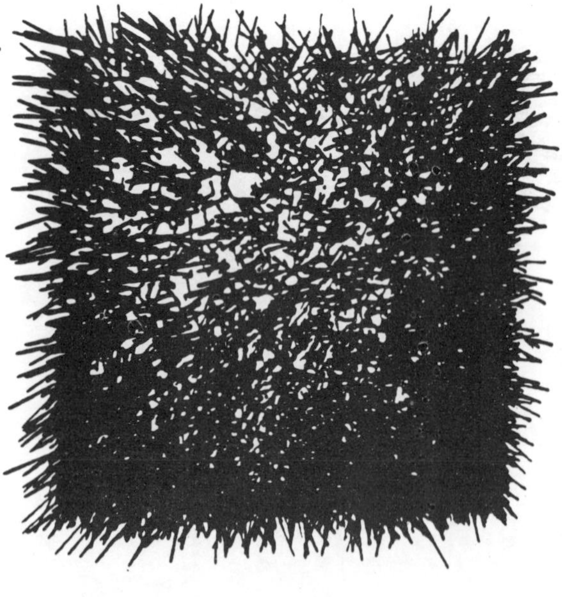

Figure 30 $V = -3000, t = 0.15$

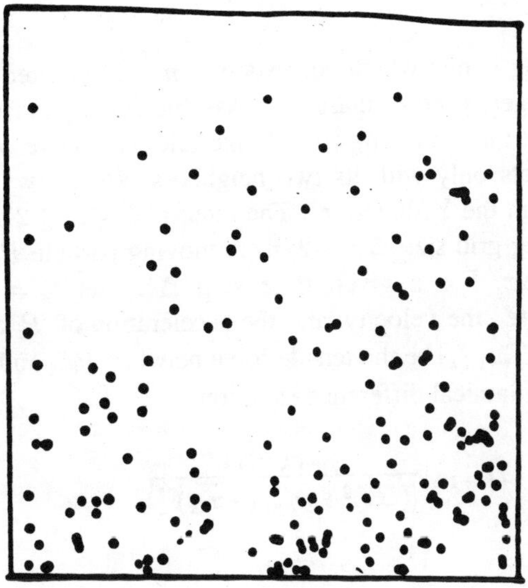

Figure 31 Small vortices (355) at t = 0.15.

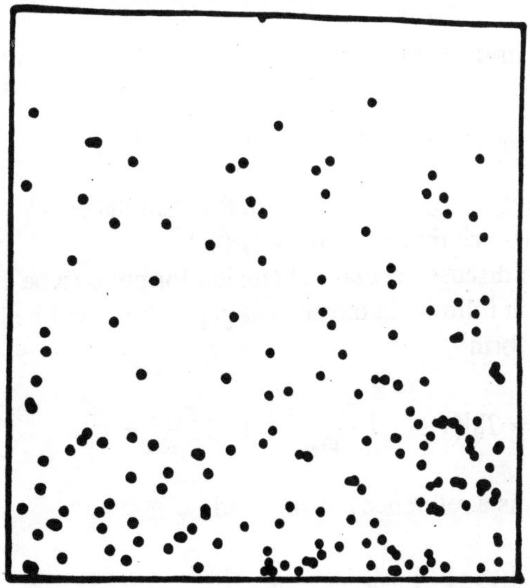

Figure 32 Small vortices (349) at t = 0.18.

Example. Soliton Collision [9]

A discrete string is one which consists of $(n+1)$ ordered particles $P_1, P_2, \ldots, P_{n+1}$, n a positive integer, each of mass m. Assume that P_1 is fixed at $(0,0)$ and P_{n+1} is fixed at $(2,0)$, while the remaining $(n-1)$ particles can move. Assume also that each moving particle interacts only with its two neighbors. We allow the moving particles $P_2 - P_n$ to move only in the Y direction . The interval $0 \le x \le 2$ is now divided into n equal parts, each having grid size $\Delta x = 2/n$. A moving particle P_i is assigned the fixed x coordinate $(i-1)\Delta x$. For a given time step Δt, let $t_k = k\Delta t$, $k = 0, 1, 2, \ldots$. Denote the Y coordinate, the velocity, and the acceleration of P_i at t_k by $y_{i,k}$, $v_{i,k}$, and $a_{i,k}$, respectively. Let $|T_{i-1,i,k}|$ be the tensile force between P_{i-1} and P_i at t_k. Then, each $a_{i,k}$ is defined by the dynamical difference equation

$$(4.1) \qquad m a_{i,k} = |T_{i,i+1,k}| \frac{y_{i+1,k} - y_{i,k}}{[(\Delta x)^2 + (y_{i+1,k} - y_{i,k})^2]^{\frac{1}{2}}}$$

$$- |T_{i-1,i,k}| \frac{y_{i,k} - y_{i-1,k}}{[(\Delta x)^2 + (y_{i,k} - y_{i-1,k})^2]^{\frac{1}{2}}}, \ i = 2, 3, \ldots, n.$$

From (4.1), each $v_{i,k+1}$ is then determined from the formulas

$$(4.2) \qquad \begin{cases} v_{i,1} = v_{i,0} + (\Delta t) a_{i,0} \\ v_{i,k+1} = v_{i,k} + (\Delta t)(1.5 a_{i,k} - 0.5 a_{i,k-1}), \ k = 1, 2, 3, \ldots , \end{cases}$$

while each $y_{i,k+1}$ is determined from

$$(4.3) \qquad y_{i,k+1} = y_{i,k} + \tfrac{1}{2}(\Delta t)(v_{i,k+1} + v_{i,k}), \ k = 0, 1, 2, \ldots .$$

From initial data $y_{i,0}$, $v_{i,0}$, $i = 2, 3, \ldots$, n, the motions of $P_2, P_3, \ldots, P_n$ are then determined recursively at each time step by (4.1)-(4.3).

It remains then to discuss the class of tension formulas to be explored in the computer examples which follow. In the present paper these will be limited to tensile forces of the particular form

$$(4.4) \qquad |T_{i,i+1,k}| = T_0[(1-\epsilon)\left(\frac{r_{i,i+1,k}}{\Delta x}\right) + \epsilon\left(\frac{r_{i,i+1,k}}{\Delta x}\right)^2]$$

in which $0 \le \epsilon \le 1$, T_0 is a reference tension, and

$$(4.5) \qquad r_{i,i+1,k} = [(\Delta x)^2 + (y_{i+1,k} - y_{i,k})^2]^{\frac{1}{2}}.$$

Note that if all the particles are on the X axis, then

$$|T_{i,i+1,k}| = T_0, \ i = 1,2,3, \ldots, 1000.$$

Our parameter choice are $n = 1000$, $m = 0.01, T_0 = 10$, $\Delta x = 0.002$, $\epsilon = 0.0$ and $\Delta t = 0.0001$. Thus the fixed particles are P_0 and P_{1001}, while the moving particles are $P_2 - P_{1000}$. The initial positions of the moving particles are all set on the X axis, so that in each example we need only specify their initial velocities.

A soliton is created now on the left by assigning initial velocities of $P_2 - P_{63}$as follows

$$v_{i,0}(P_2) = 0.5$$
$$v_{i,0}(P_{i+1}) = v_{i,0}(P_i) + 0.5,\ i = 2, 31$$
$$v_{i,0}(P_{33}) = v_{i,0}(P_{32}) = 15.5$$
$$v_{i,0}(P_{i+1}) = v_{i,0}(P_i) - 0.5,\ i = 33, 62.$$

In addition, a smaller soliton is generated on the right as follows:

$$v_{i,0}(P_{1000}) = 0.5$$
$$v_{i,0}(P_{999}) = 1.0$$
$$v_{i,0}(P_{999-i}) = v_{i,0}(P_{1000-i}) + 0.5,\ i = 1, 23$$
$$v_{i,0}(P_{975}) = v_{i,0}(P_{976}) = 12.5$$
$$v_{i,0}(P_{975-i}) = v_{i,0}(P_{976-i}) - 0.5,\ i = 1, 24.$$

All other initial velocities are set to zero. We wish to examine the collision of the two solitons.

Figure 33 shows clearly at the times 0.15, 0.60, 0.75, 0.90, 1.05 that the collision of the two solitons results in their passing through each other. Figure 34shows in greater detail the soliton interactions at the times 0.635, 0.650, 0.650, 0.680, 0.695, 0.710, 0.725, 0.740, 0.755, 0.770, 0.785.

Other qualitative models have simulated the biological self reorganizatuion of separated tissue cells [7,19], the bounce of an elastic ball [10], and the generation of the unstable equilibrium position of an elastic snap through [12]

5. N Small.

Now, if N is small, we would like to do a very good job in solving the N-body problem. By this we mean that we would like not only to solve the problem with accuracy, but we would also like to preserve numerically any basic physical invariants of the system. To do this in detail, we concentrate on the 3-body problem, which is the prototype problem because it contains all the difficulties of the general N-body problem. The entire discussion extends in a natural way to the general problem.

For $i = 1, 2, 3$, let P_i of mass m_i be at $\vec{r_i} = (x_i, y_i, z_i)$ at time t. Let the positive distance between P_i and P_j , $i \neq j$, be $r_{ij} = r_{ji}$. Let $\phi = \phi_{ij} = \phi(r_{ij})$, given in ergs, be the potential for the pair P_i, P_j . Then the force on P_i due to P_j is

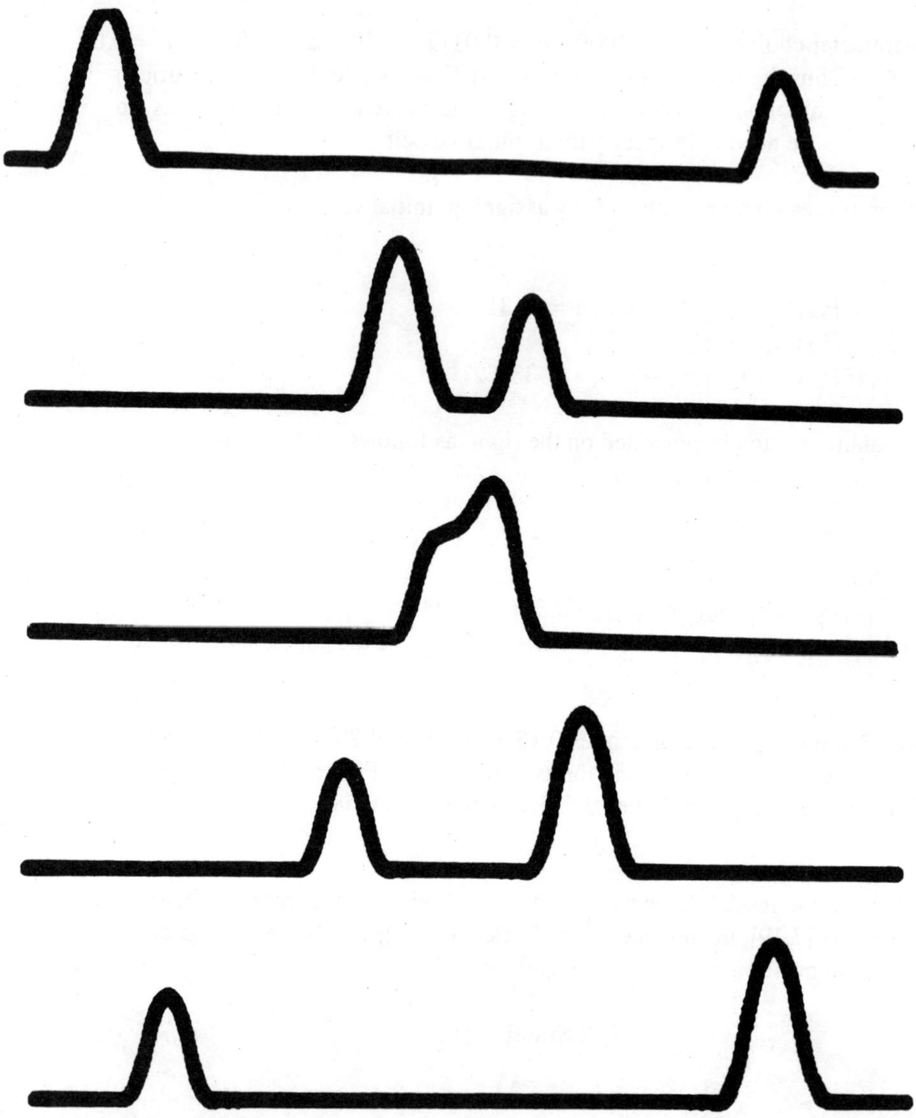

Figure 33

Soliton interaction.

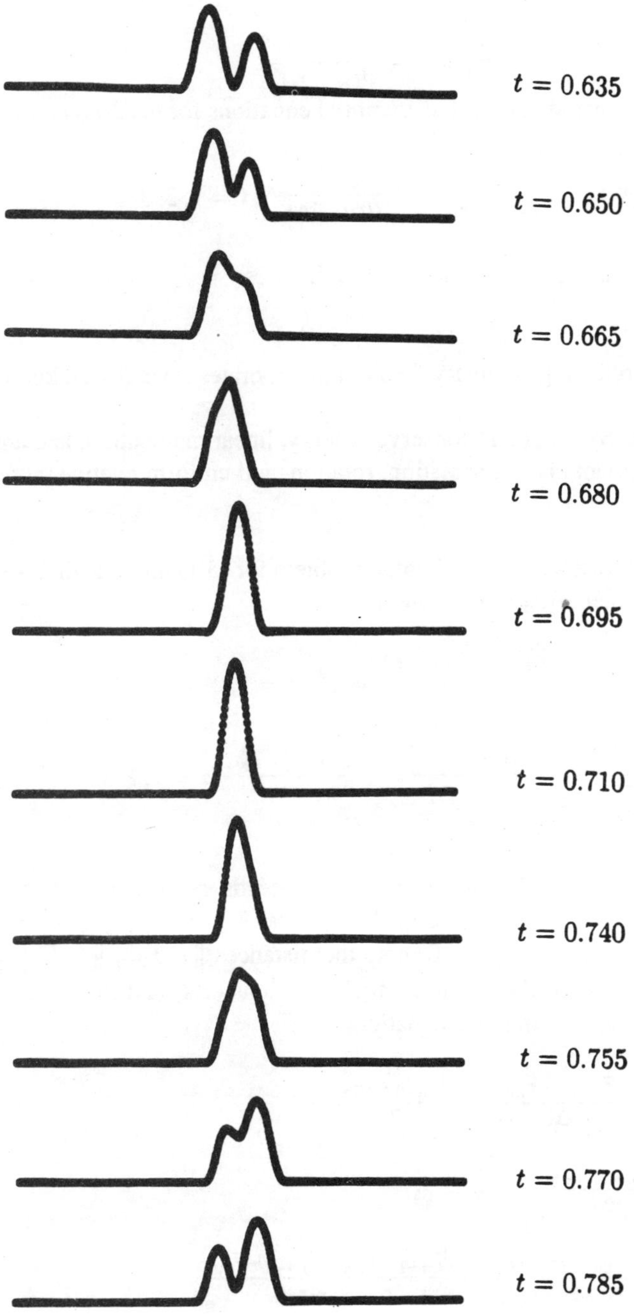

Figure 34

Refined Soliton interaction.

$$\vec{F}_i = -\frac{\partial \phi}{\partial r_{ij}} \frac{\vec{r_i}-\vec{r_j}}{r_{ij}},$$

and the Newtonian dynamical differential equations for the 3-body problem are

$$\text{(5.1)} \quad m_i \frac{d^2\vec{r_i}}{dt^2} = -\frac{\partial \phi}{\partial r_{ij}} \frac{\vec{r_i}-\vec{r_j}}{r_{ij}} - \frac{\partial \phi}{\partial r_{ik}} \frac{\vec{r_i}-\vec{r_k}}{r_{ik}}, i = 1, 2, 3$$

where $i = 1$ implies $j = 2$, $k = 3$; $i = 2$ implies $j = 1, k = 3$; $i = 3$ implies $j = 1$, $k = 2$.

The following summary theorem incorporates several well known results.

Theorem 5.1. System (5.1) conserves energy, linear momentum, and angular momentum. It is also covariant under translation, rotation, and uniform relative motion of coordinate frames [6].

In order to solve an initial value problem for (5.1) numerically, we will first rewrite it as the equivalent first order system

$$\text{(5.2)} \qquad \frac{d\vec{r_i}}{dt} = \vec{v_i}, \; i = 1, 2, 3$$

$$\text{(5.3)} \qquad m_i \frac{d\vec{v_i}}{dt} = -\frac{\partial \phi}{\partial r_{ij}} \frac{\vec{r_i}-\vec{r_j}}{r_{ij}} - \frac{\partial \phi}{\partial r_{ik}} \frac{\vec{r_i}-\vec{r_k}}{r_{ik}}, \; i = 1, 2, 3.$$

Our numerical formulation now proceeds as follows. For $\Delta t > 0$, let $t_n = n(\Delta t)$, $n = 0, 1, 2, \ldots$. At time t_n, let P_i be at $\vec{r}_{i,n} = (x_{i,n}, y_{i,n}, z_{i,n})$ with velocity $\vec{v} = (v_{i,x,n}, v_{i,y,n}, v_{i,z,n})$. Denote the distances $\|P_1P_2\|$, $\|P_1P_3\|$, $\|P_2P_3\|$ by $r_{12,n}$, $r_{13,n}$, $r_{23,n}$, respectively. Differential equations (5.2) and (5.3) are now approximated, respectively, by the difference equations

$$\text{(5.4)} \qquad \frac{\vec{r}_{i,n+1}-\vec{r}_{i,n}}{\Delta t} = \frac{\vec{v}_{i,n+1}+\vec{v}_{i,n}}{2}$$

$$\text{(5.5)} \quad m_i \frac{\vec{v}_{i,n+1}-\vec{v}_{i,n}}{\Delta t} =$$

$$-\frac{\phi(r_{ij,n+1})-\phi(r_{ij,n})}{r_{ij,n+1}-r_{ij,n}} \frac{\vec{r}_{i,n+1}+\vec{r}_{i,n}-\vec{r}_{j,n+1}-\vec{r}_{j,n}}{r_{ij,n+1}+r_{ij,n}}$$

$$-\frac{\phi(r_{ik,n+1})-\phi(r_{ik,n})}{r_{ik,n+1}-r_{ik,n}} \frac{\vec{r}_{i,n+1}+\vec{r}_{i,n}-\vec{r}_{k,n+1}-\vec{r}_{k,n}}{r_{ik,n+1}+r_{ik,n}}$$

Note that the force is approximated, not the potential. We take the very same potential as in continuum mechanics , the significance of which will be seen shortly. Consistency follows immediately as $\Delta t \to 0$.

System (5.4), (5.5) consists of 18 implicit equations for the unknowns $x_{i,n+1}$, $y_{i,n+1}$, $z_{i,n+1}$, $v_{i,x,n+1}$, $v_{i,y,n+1}$, $v_{i,z,n+1}$ in the 18 knowns $x_{i,n}$, $y_{i,n}$, $z_{i,n}$, $v_{i,x,n}$, $v_{i,y,n}$, $v_{i,z,n}$ and is solvable readily by Newton's method [7].

We now state for our numerical method the following theorem which is the difference analogue of Theorem 5.1.

Theorem 5.2. The numerical method conserves exactly the same energy, linear momentum and angular momentum as does differential system (5.1) , and does this independently of the time step Δt. In addition the difference equations are covariant under translation , rotation, and uniform relative motion of coordinate frames [7].

Rather than give the proof of all parts of the theorem, let us show in complete detail the proof of conservation of energy to indicate the methodology used.

Proof. Define

$$W_N = \sum_{n=0}^{N-1} \sum_{i=1}^{3} m_i(\vec{r}_{i,n+1} - \vec{r}_{i,n}) \cdot \frac{(\vec{v}_{i,n+1} - \vec{v}_{i,n})}{\Delta t} \tag{5.6}$$

Then, from (5.4),

$$W_N = \sum_{n=0}^{N-1} \sum_{i=1}^{3} m_i \frac{(\vec{r}_{i,n+1} - \vec{r}_{i,n})}{\Delta t} \cdot (\vec{v}_{i,n+1} - \vec{v}_{i,n})$$

$$= \sum_{n=0}^{N-1} \sum_{i=1}^{3} m_i \frac{(\vec{v}_{i,n+1} + \vec{v}_{i,n})}{2} \cdot (\vec{v}_{i,n+1} - \vec{v}_{i,n})$$

$$= \sum_{n=0}^{N-1} \sum_{i=1}^{3} m_i \left(\frac{v_{i,n+1}^2}{2} - \frac{v_{i,n}^2}{2}\right)$$

$$= \tfrac{1}{2} m_1 v_{1,N}^2 + \tfrac{1}{2} m_2 v_{2,N}^2 + \tfrac{1}{2} m_3 v_{3,N}^2 - \tfrac{1}{2} m_1 v_{1,0}^2 - \tfrac{1}{2} m_2 v_{2,0}^2 - \tfrac{1}{2} m_3 v_{3,0}^2$$

$$= K_N - K_0 .$$

Next, observing first that $\vec{r}_i - \vec{r}_j = \vec{r}_{ji}$, and then substituting (5.5) into (5.6) yields with simple algebraic manipulation that

$$W_N = \sum_{n=0}^{N-1} (- \phi_{12,n+1} - \phi_{13,n+1} - \phi_{23,n+1} + \phi_{12,n} + \phi_{13,n} + \phi_{23,n})$$

$$= - \phi_{12,N} - \phi_{13,N} - \phi_{23,N} + \phi_{12,0} + \phi_{13,0} + \phi_{23,0}$$

$$= - \phi_N + \phi_0 .$$

Hence,

$$K_N - K_0 = - \phi_N + \phi_0$$

or

$$K_N + \phi_N = K_0 + \phi_0 \ , \ N = 0, 1, 2, 3, \ldots$$

Moreover, since K_0 and ϕ_0 depend only on $\vec{r}_{i,0}$ and $\vec{v}_{i,0}$ it follows that $K_0 + \phi_0$ is the same in both the continuous and the discrete cases, so that the energy conserved is exactly that of the continuous system. Note also that the proof was independent of Δt.

QED

Rigid body motion is of fundamental interest in mathematics, science, and engineering. We will consider a discrete dodecahedral body and simulate its motion when it spins like a top on a smooth surface. The approach will not require the use of special coordinates, Cayley-Klein parameters, tensors, dyadics, or related concepts [6]. All that is required is the conservative numerical methodology developed above.

Consider, now, in XYZ space as shown in Figure 35, a dodecahedron $P_1 - P_8$. For the present, let the distances between the vertices be given as follows:

$$\|P_1P_j\| = \|P_jP_8\| = 2\,cm, \quad j = 2 - 7$$

$$\|P_2P_3\| = \|P_3P_4\| = \|P_4P_5\| = \|P_5P_6\| = \|P_6P_7\| = \|P_7P_8\| = 1\,cm$$

$$\|P_2P_5\| = \|P_3P_6\| = \|P_4P_7\| = 2\,cm.$$

The vertices $P_2 - P_7$ are taken to be vertices of a regular, plane hexagon with edge length $1\,cm$. The hexagon is located in the plane $z = \sqrt{3}$ and the vertex coordinates are $P_2(0, 1, \sqrt{3})$, $P_3(\frac{1}{2}\sqrt{3}, \frac{1}{2}, \sqrt{3})$, $P_4(\frac{1}{2}\sqrt{3}, -\frac{1}{2}, \sqrt{3})$, $P_5(0, -1, \sqrt{3})$, $P_6(-\frac{1}{2}\sqrt{3}, -\frac{1}{2}, \sqrt{3})$, $P_7(-\frac{1}{2}\sqrt{3}, \frac{1}{2}, \sqrt{3})$. The geometric center of the hexagon is the point $Q(0, 0, \sqrt{3})$. The point P_8, located on the line through P_1Q, is at $P_8(0, 0, 2\sqrt{3})$

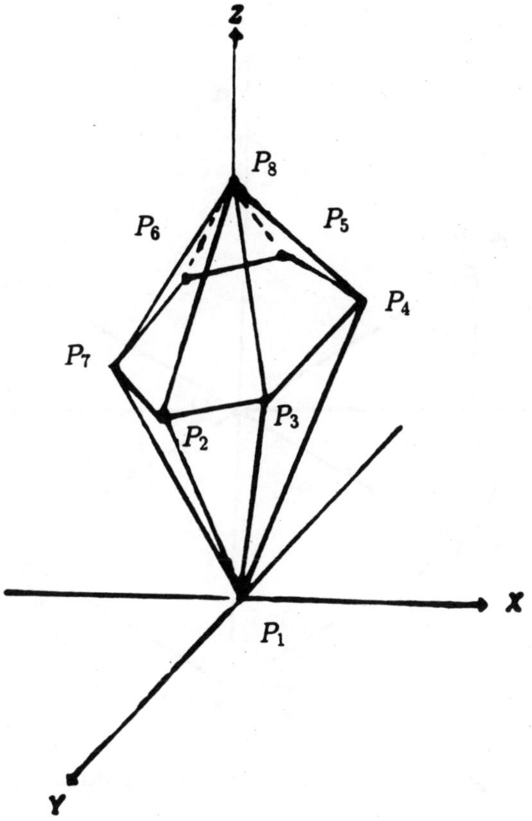

Figure 35 The dodecahedron.

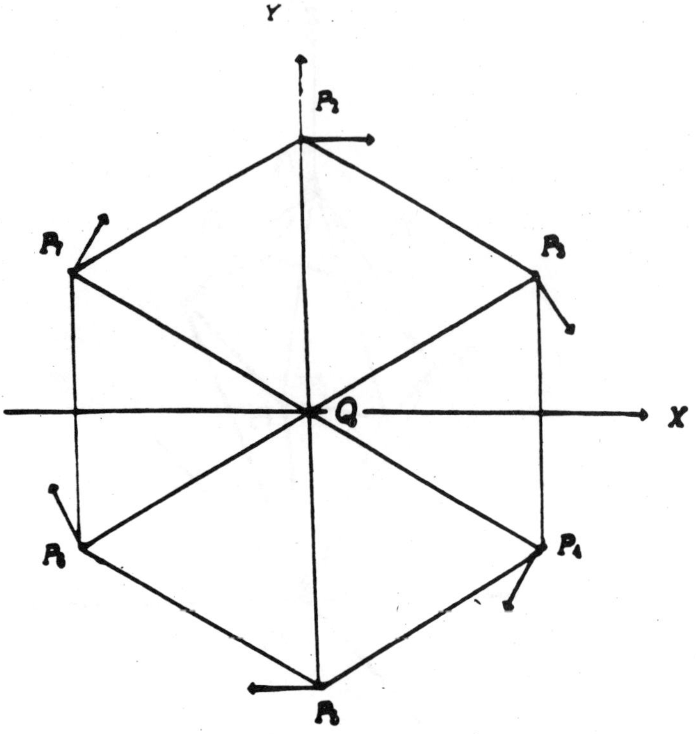

Figure 36 The planar hexagon and initial velocities.

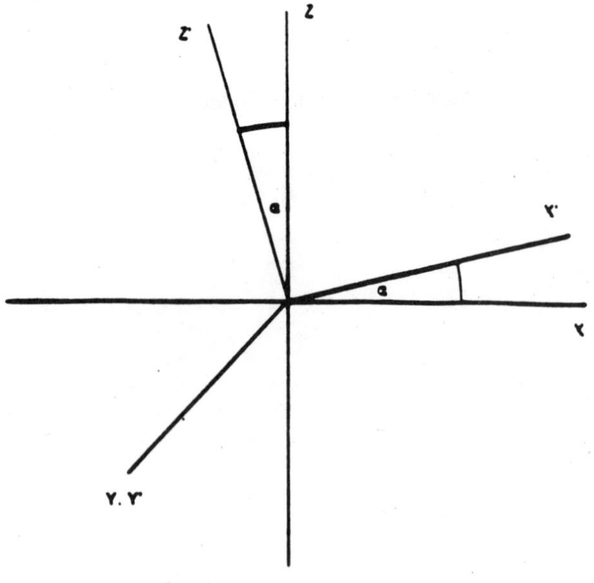

Figure 3.7 The rotation of the XZ plane.

In order to set the top in rotation , the initial velocities of $P_2 - P_7$ are chosen to be, as shown in Figure 36, respectively, $\vec{v}_2 = (v, 0, 0)$, $\vec{v}_3 = (\frac{1}{2}v, -\frac{1}{2}v\sqrt{3}, 0)$, $\vec{v}_4 = (-\frac{1}{2}v, -\frac{1}{2}v\sqrt{3}, 0)$, $\vec{v}_5 = (-v, 0, 0)$, $\vec{v}_6 = (-\frac{1}{2}v, \frac{1}{2}v\sqrt{3}, 0)$, $\vec{v}_7 = (\frac{1}{2}v, \frac{1}{2}v\sqrt{3}, 0)$, so that the velocity of each particle is perpendicular to the line joining it to Q. The points P_1 and P_8 are given zero initial velocities.

Next, we want to tilt the top and this is done by rotating the XZ plane through an angle α, as shown in Figure 37. Thus, the initial positions (x_i', y_i', z_i') and initial velocities $(v_{i,x'}, v_{i,y'}, v_{i,z'})$ of $P_1 - P_8$ are determined finally by

$$x_i' = x_i \cos\alpha + z_i \sin\alpha, \quad y_i' = y_i, \quad z_i' = -x_i \sin\alpha + z_i \cos\alpha,$$

$$v_{ix'} = v_{ix} \cos\alpha + v_{iz} \sin\alpha, \quad v_{iy'} = v_{iy}, \quad v_{iy'} = -v_{ix} \sin\alpha + v_{iz} \cos\alpha.$$

Thus, once the parameters v and α are given, all initial data for a tilted, rotating dodecahedral top are determined, and the resulting 8-body problem will be solved with energy, linear momentum and angular momentum conserved. Note also that because the top is rigid, a description of the motions of P_1 and Q suffice to determine the motion of the entire top.

5. Examples [7].

Example 1. Let $\alpha = 15^o$, $v = 400$. The conservative numerical formulas were then run for 200,000,000 time steps with a printout every 100,000 steps. The resulting motion of P_1 for the first cycle is shown in Figure 38, which contains 1086 points, on the region

$$-0.32 \le x \le 3.1, \ -1.71 \le y \le 1.71, \tag{5.3}$$

and in the enlarged version in Figure 39 on the region

$$-0.196 \le x \le 0.916, \ -0.4678 \le y \le 0.4678.$$

The motion is characterized by a circular, cycloid type trajectory around the central point $(0.4483, 0)$ in the XY plane . The point $Q(x^*, y^*, z^*)$ is always on the line $(0.4483, 0, z)$ with z^* varying in the range $1.668 < z^* < 1.673$.

Example 1 is typical of a variety of others studied in which the parameters v and α were varied. Let us then examine a difficult problem in which certain masses are not all the same.

Example 2. As in Example 1, let $\alpha = 15^o$, $v = 400$, but set $m_2 = 2$, $m_4 = 4$. The conservative numerical formulas were then run for 100,000 time steps with a printout

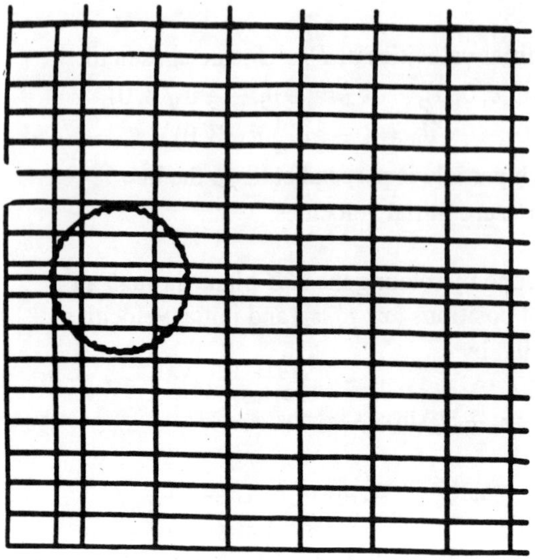

Figure 38

Masses all unity.

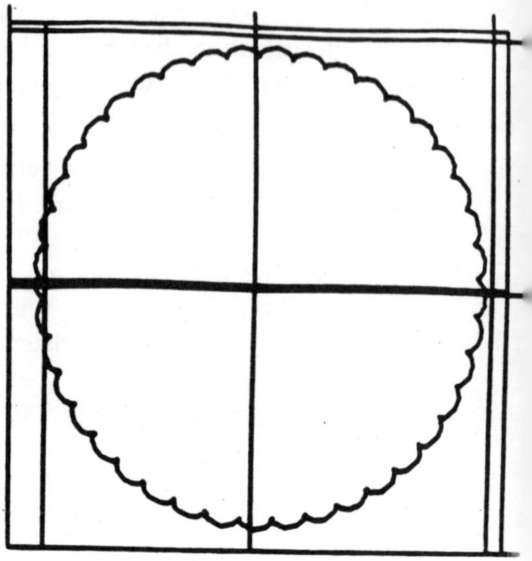

Figure 39

Enlargement of Figure 38.

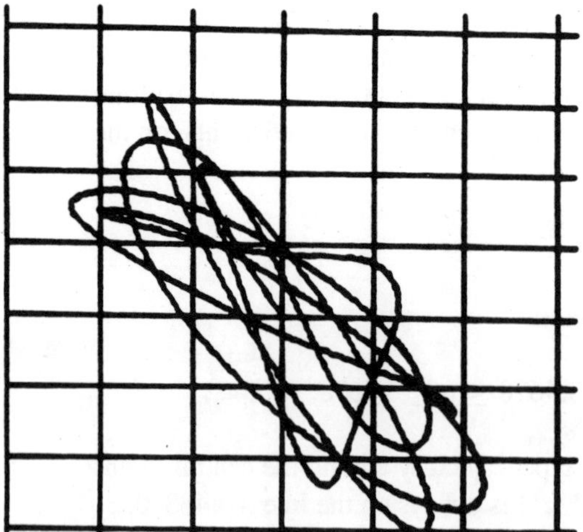

Figure 40

Masses not all unity.

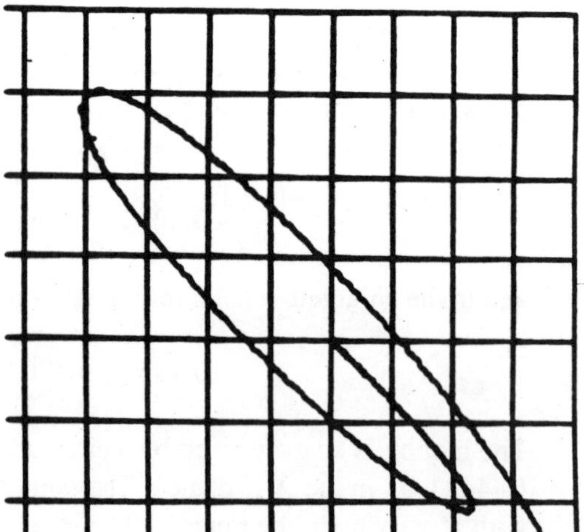

Figure 41

Enlargement of Figure 40

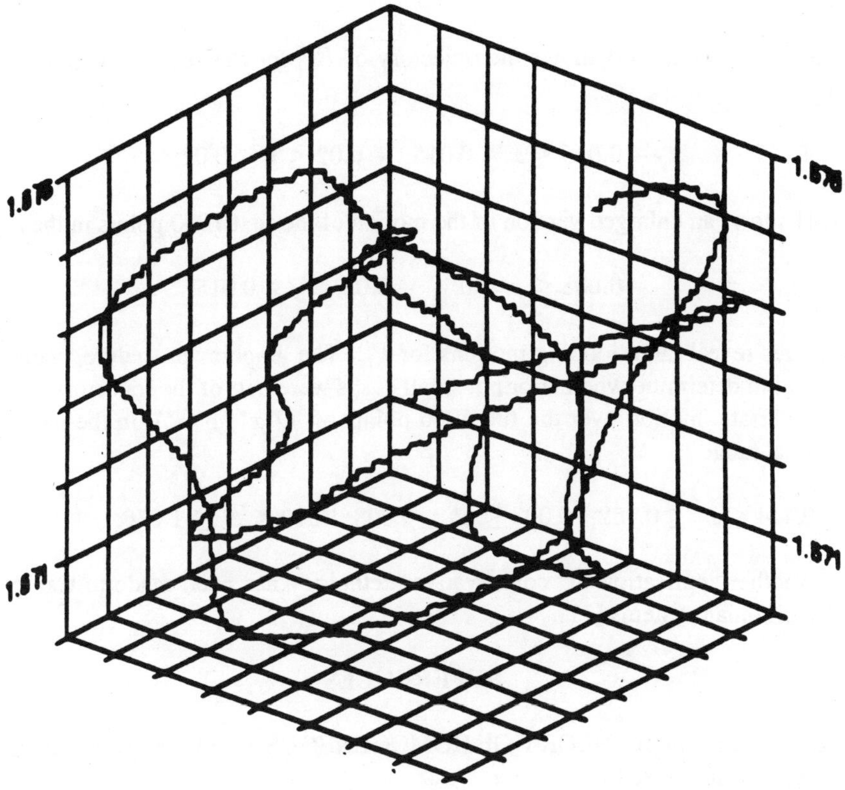

Figure 42

Motion of Q.

every 20 steps. Figure 40 shows the trajectory of P_1, for the first 4000 points, in the region

$$-0.015 \leq x \leq 0.015, \ -0.02 \leq y \leq 0.02.$$

Figure 41 shows an enlarged version of the motion of the first 1000 points in the region

$$-0.008 \leq x \leq 0.1, \ -0.02 \leq y \leq 0.015.$$

Both figures reveal large looping motions for P_1. The graphics procedure available was insufficient to determine whether or not small cusps were part of the trajectory. Figure 42 shows the erratic motion over the full 5000 points of $Q(x^*, y^*, z^*)$ in the narrow three dimensional range

$$0.444 \leq x^* \leq 0.452, \ -0.008 \leq y^* \leq 0.008, 1.670 \leq z^* \leq 1.676.$$

Another application of conservative methology has been made to Calogero and Toda Hamiltonian systems [11].

REFERENCES

1. G. K. Batchelor, THE THEORY OF HOMOGENEOUS TURBULENCE, Cambridge Univ. Press, Cambridge, 1959.
2. P. S. Bernard, "Transition and turbulence. Basic physics", in THE HANDBOOK OF FLUID DYNAMICS (R. W. Johnson, Ed.) CRC Press, Boca Raton, FL, 1998, p. 13-5
3. R. P. Feynman, "There's plenty of room at the bottom", Invited Lecture, Ann. Meet. APS, Cal. Inst. Tech., Pasadena, 1959 (available at www.zyvex.com/nanotech/feynman.html).
4. R. P. Feynman, R. B. Leighton, and M. Sands, THE FEYNMAN LECTURES ON PHYSICS, vol I, Addison-Wesley, Reading, 1963.
5. C. J. Freitas, R. L. Street, A. N. Findikakis and J. R. Koseff, "Numerical simulation of three dimensional flow in a cavity", Int. J. Num. Meth. Fluids, 5, 1985, pp. 561-575.
6. H. Goldstine, CLASSICAL MECHANICS, 2nd Ed., Addison-Wesley, Reading, 1980.
7. D. Greenspan, PARTICLE MODELING, Birkhauser, Boston, 1997.
8. D. Greenspan, "Supercomputer simulation of cracks and fractures by quasimolecular dynamics", Int. J. Phys. Chem. Solids, 50, 1989, pp.1245-1249.
9. D. Greenspan, "Discrete string solitons", Tech. Report 343, Dept. Math., UTA, Arlington, Tx, 76019, May, 2001.
10. D. Greenspan, "Particle simulation in contact mechanics of a bouncing elastic ball", Mathl. Comp. Modelling, 32, 2000, pp.803-812.
11. D. Greenspan, "Conservative difference formulation of Calogero and Toda Hamiltonian systems", Comp. Math. Applic., 19, 1990, pp. 91-95.
12. D. Greenspan and V. Casulli, "Particle modelling of an elastic arch", Appl. Math. Modelling, 9, 1985,pp.215-219.

13. J. O. Hirschfelder, C. F. Curtiss, and R, B. Bird, MOLECULAR THEORY OF GASES AND LIQUIDS, Wiley, N.Y., 1967.
14. A. N. Kolmogorov, "Toward a more precise notion of the structure of the local turbulence in a viscous fluid at elevated Reynolds number", in THE MECHANICS OF TURBULENCE (A. Favre, Ed.), Gordon and Breach, N.Y., 1964, p. 447-458.
15. M. S. Korlie, "Particle modeling of liquid drop formation on a solid surface in 3-D", Comp. Math. Applic., 33, 1997, pp. 97-114
16. O. A. Ladyzhenskaya, THE MATHEMATICAL THEORY OF VISCOUS INCOMPRESSIBLE FLOW, (2nd Ed.), Gordon and Breach, N.Y., 1969.
17. F. Pan and A. Acrivos, "Steady flows in a rectangular cavity", J. Fluid Mech., 28, 1967, pp. 643-655.
18. H. Schlichting, BOUNDARY LAYER THEORY (4th Ed.), McGraw-Hill, N.Y., 1960.
19. M. S. Steinberg, "Reconstruction of tissues by disassociated cells", Science, 141, 1963, pp. 401-403.
20. V. L. Streeter, FLUID MECHANICS, 3rd Ed., McGraw-Hill, N.Y., 1962.

Modern Applied Mathematics
Editor: J.C. Misra

Asymptotic Behaviour of Solutions of Some Nonlinear Partial Differential Equations

P.L. Sachdev

Department of Mathematics, Indian Institute of Science, Bangalore 560 012, India

1 Introduction

Nonlinear partial differential equations (PDEs) with or without reference to initial/boundary conditions do not, in general, admit closed form solutions; these solutions are even more rare when initial/boundary conditions are imposed. There are exceptional circumstances when the PDEs possess certain symmetries: they are invariant to a class of finite or infinitesimal transformations. When this is the case, the PDEs are (exactly) reducible to ordinary differential equations (ODEs) if the original PDEs have two independent variables; the ODEs may occasionally be integrated in a closed form. Alternatively one might study their qualitative properties, and obtain the actual solution numerically with reference to appropriate initial boundary conditions. These solutions, called self-similar and belonging to one of the two classes - first kind and second kind - (Zel'dovich and Raizer (1966); Sachdev (2000)) - have been investigated in much detail. Barenblatt (1996) specifically discussed self-similar solutions of the second kind, referring to them as intermediate asymptotics: "They describe the behaviour of these solutions away from the boundaries of the region of the independent variables. Alternatively, they represent solutions in a region where in a sense the solution is no longer dependent on the details of the initial and/or boundary conditions but is still far from being in a state of equilibrium (Barenblatt and Zel'dovich, 1972)". The solution may depend on the initial conditions in an integral form-representing the memory of the solution, derived from the initial conditions. Thus, a whole class of solutions arising from a certain class of initial conditions may evolve asymptotically in time to the similarity solution of the given system of partial differential equations, Sachdev (1987).

The symmetry properties referred to above may not be possessed by most nonlinear PDEs that occur in applications, thus exact reduction to nonlinear ODEs and hence closed form solution, implicit or explicit, is usually ruled out. One must therefore explore other avenues so that the asymptotic behaviour of solutions may still be determined to a high degree of accuracy, if not exactly. One such approach is called the method of balancing, used extensively by Grundy and his collaborators (1974, 1996, 1997), see also Sachdev (1991), which essentially consists of introducing a special set of variables (the usual choice being one or several similarity variables and time) and picking out from the full PDEs those terms which are of the same order for large time. This often leads to several combinations of terms for different sets of parameters. Most of these reduced equations are simpler than the original ones and admit analysis in a closed form or otherwise. Thus one has access to good approximate solutions for large time which, suitably combined with the numerical solutions, provide ample information about the large time behaviour(s) of the solution for different sets of parameters appearing in the PDEs and the initial/boundary conditions.

The scheme of the present paper is as follows. In section 2 we derive 'exact' large time asymptotic form of the solution from the exact 'formal' solution of initial value problems for the Burgers equation and the Fisher's equation; this becomes possible only for some special PDEs. This asymptotic form is often called 'profile at infinity' which is quite descriptive of the solution. In section 3, we motivate the balancing arguments with reference to a nonlinear ODE. The special first order PDE $u_t + u^\alpha u_x + u^\beta = 0$ is treated exactly and by balancing argument in section 4. An example of nonlinear parabolic equations due to Grundy et al (1994) is discussed in section 5. Finally, the nonplanar Burgers equation with N wave initial conditions is treated in section 6. The conclusions of the present study are contained in section 7.

2 Profile at Infinity

To illustrate the significance of asymptotic solutions, the famous Burgers equation proves handy. We consider three different types of solutions - travelling waves, N-waves and single hump solutions - each arising from different sets of initial conditions and leading to specific special exact solutions in the large time limit.

This situation is peculiar to Burgers equation. These exact solutions are referred to as 'Profiles at Infinity' by Clothier et al (1981), a rather neat description of the asymptotic behaviour.

We begin with the initial value problem for the Burgers equation

$$u_t + uu_x = \frac{\delta}{2}u_{xx}, \tag{2.1}$$

$$u(x,0) = f(x), \quad -\infty < x < \infty. \tag{2.2}$$

The Cole-Hopf transformation

$$u = -\delta\frac{\phi_x}{\phi} \tag{2.3}$$

or

$$\phi(x,t) = \exp\left[-\frac{1}{\delta}\int^x u(\xi,t)d\xi\right] \tag{2.4}$$

changes (2.1) - (2.2) to the initial value problem for the heat equation, namely

$$\phi_t = \frac{\delta}{2}\phi_{xx}, \tag{2.5}$$

$$\begin{aligned}\phi(x,0) &= \exp\left[-\frac{1}{\delta}\int_0^x u(\xi,0)d\xi\right] \qquad (2.6)\\ &= \exp\left[-\frac{1}{\delta}\int_0^x f(\xi)d\xi\right] \equiv \Phi(x), \quad \text{say,}\end{aligned}$$

where we have chosen the lower limit of x to be zero in the definition of $\Phi(x)$. Hopf (1950) proved that, if $f(x)$ is integrable in every finite x interval and

$$\int_0^x f(\xi)d\xi = o(x^2) \tag{2.7}$$

for large x, then

$$u(x,t) = \frac{\int_{-\infty}^{\infty} \frac{x-\xi}{t} \exp\left[-\frac{1}{\delta}F(x,\xi,t)\right] d\xi}{\int_{-\infty}^{\infty} \exp\left[-\frac{1}{\delta}F(x,\xi,t)\right] d\xi} \tag{2.8}$$

with

$$F(x,\xi,t) = \frac{(x-\xi)^2}{2t} + \int_0^{\xi} f(\eta)d\eta, \tag{2.9}$$

is a regular solution of (2.1) in the half plane $t > 0$, that satisfies the initial condition

$$\int_0^x u(\xi,t)d\xi \to \int_0^a f(\xi)d\xi \quad \text{as} \quad x \to a, t \to 0 \tag{2.10}$$

for every a. If in addition, $f(\xi)$ is continuous at $x = a$, then $u(x,t) \to f(a)$ as $x \to a, t \to 0$. A solution of (2.1) which is regular in the strip $0 < t < T$ and which satisfies (2.10) for each value of the number a necessarily coincides with (2.8).

Now, we choose specially

$$u(x,0) = f(x) = \begin{cases} u_1, & x > 0, \\ u_2, & x < 0, \end{cases} \tag{2.11}$$

where $u_2 > u_1$. The transformation (2.4) (and hence (2.6)) take (2.11) to

$$\phi(x,0) = \begin{cases} e^{-u_1 x/\delta}, & x > 0 \\ e^{-u_2 x/\delta}, & x < 0. \end{cases} \tag{2.12}$$

Substituting (2.12) into (2.8) and rearranging the terms we arrive at the solution of (2.1) and (2.11) as

$$u(x,t) = u_1 + \frac{u_2 - u_1}{1 + \left\{\exp\left[\frac{u_2-u_1}{\delta}\left(x - \frac{u_1+u_2}{2}t\right)\right]\right\} \frac{\int_{-(x-u_1 t)}^{\infty} e^{-y^2/2\delta t}dy}{\int_{x-u_2 t}^{\infty} e^{-y^2/2\delta t}dy}}. \tag{2.13}$$

Now, if let t tend to infinity in (2.13) such that $u_1 < x/t < u_2$, we may easily show that it goes to

$$u(x,t) = u_1 + \frac{u_2 - u_1}{1 + \exp \frac{u_2 - u_1}{\delta}(x - Ut)}, \quad U = \frac{u_1 + u_2}{2}. \tag{2.14}$$

This is because in the integrals in (2.13) the lower limit for $u_1 < x/t < u_2$, as t tends to infinity, is minus infinity and, so, the ratio of the integrals tends to one. Now we demonstrate that the asymptotic solution (2.14) is just the travelling wave solution of (2.1) with the end conditions $u \to u_1, u_2$ as $x \to +\infty, -\infty$, respectively, where $u_2 > u_1$.

We seek solution of (2.1) in the travelling wave form

$$u = u(\xi), \quad \xi = x - Ut, \tag{2.15}$$

where U is a constant. Putting (2.15) into (2.1), we get

$$-Uu_\xi + uu_\xi = \frac{\delta}{2}u_{\xi\xi}. \tag{2.16}$$

Since u tends to constant values u_1 and u_2 at $\xi = \mp\infty$, u_ξ tends to zero in these limits; so (2.16) integrates to yield

$$\frac{1}{2}u^2 - Uu + C = \frac{\delta}{2}u_\xi, \tag{2.17}$$

where C is a constant. Using the end conditions at $\xi = \pm\infty$ in (2.17) we get

$$\frac{1}{2}u_1^2 - Uu_1 = \frac{1}{2}u_2^2 - Uu_2 = -C \tag{2.18}$$

leading to the following values of U and C:

$$U = \frac{1}{2}(u_1 + u_2), \quad C = \frac{1}{2}u_1u_2. \tag{2.19}$$

Equation (2.17) can now be written as

$$(u - u_1)(u_2 - u) = -\delta u_\xi. \tag{2.20}$$

On integration, (2.20) gives

$$\frac{\xi}{\delta} = \frac{1}{u_2 - u_1} \ln \frac{u_2 - u}{u - u_1} \tag{2.21}$$

or

$$u = u_1 + \frac{u_2 - u_1}{1 + \exp \frac{u_2 - u_1}{\delta}(x - Ut)}, \quad U = \frac{u_1 + u_2}{2}. \tag{2.22}$$

Equation (2.22) is just (2.14) which was obtained from the solution of IVP (2.1) and (2.11) in the limit $t \to \infty$. Thus the travelling wave solution (2.22) is the asymptotic solution of IVP (2.1) and (2.11) for all finite u_1 and u_2 such that $u_2 > u_1$. Actually there are 'infinity' of initial conditions with the same end states but 'reasonable' behaviour in the middle which also tend, in the limit $t \to \infty$, to the steady 'shock structure' or travelling wave form (2.22). This wave moves with the speed $U = \frac{u_1+u_2}{2}$, the average of end speeds, independent of conditions in the middle. These concepts can be made more precise in the context of solution called intermediate asymptotics (see Sachdev (1987)). The step initial condition (2.11) may be thought of "as arising from a fast piston motion such that the wave form gets away from the piston in a time negligible compared with the time scale of the process of shock wave formation in which we are interested," Lighthill (1956).

We discuss now the N-wave initial profile

$$u(x,0) = \begin{cases} x & \text{if } |x| < \ell_0 \\ 0 & \text{otherwise} \end{cases} \tag{2.23}$$

which, in view of (2.4), leads to the IC

$$\phi(x,0) = \begin{cases} e^{-x^2/2\delta} & \text{if } |x| < \ell_0 \\ e^{-\ell_0^2/2\delta} & \text{if } |x| \geq \ell_0 \end{cases} \tag{2.24}$$

and hence to the solution

$$\begin{aligned} \phi(x,t) &= \frac{1}{(2\pi\delta t)^{1/2}} \left[\int_{-\ell_0}^{\ell_0} e^{-[y^2+(x-y)^2/t]/2\delta} dy \right. \\ &\left. + e^{-\ell_0^2/2\delta} \left\{ \int_{-\infty}^{-\ell_0} e^{-(x-y)^2/2\delta t} dy + \int_{\ell_0}^{\infty} e^{-(x-y)^2/2\delta t} dy \right\} \right] \end{aligned} \tag{2.25}$$

of the heat equation (2.5). The solution (2.25) can, in the limit $t \to \infty$, may be approximated by

$$\phi(x,t) = e^{-\ell_0^2/2\delta} + \frac{e^{-x^2/2\delta t}}{(\pi t)^{1/2}} \left[2 \int_0^{\ell_0/(2\delta)^{1/2}} e^{-z^2} dz - \left(\frac{2}{\delta}\right)^{1/2} \ell_0 e^{-\ell_0^2/2\delta} \right] + O\left(\frac{1}{t}\right) \tag{2.26}$$

if use is made of the asymptotic properties of error function. Therefore, we also have

$$\phi_x(x,t) = -\frac{1}{\delta} \frac{x}{t^{1/2}} \frac{e^{-x^2/2\delta t}}{t} \left[\frac{2}{\pi^{1/2}} \int_0^{\ell_0/(2\delta)^{1/2}} e^{-z^2} dz - \left(\frac{2}{\pi\delta}\right)^{1/2} \ell_0 e^{-\ell_0^2/2\delta} \right] + O\left(\frac{1}{t^2}\right). \tag{2.27}$$

The approximations (2.26)-(2.27) are uniformly valid with respect to $\xi = x/(2\delta t)^{1/2}$ in the limit $t \to \infty$. The solution $u(x,t)$ of (2.1) in view of (2.3) is found to be

$$u(x,t) = \frac{(x/t^{1/2}) \frac{e^{-x^2/2\delta t}}{t} \left[\frac{2}{\pi^{1/2}} \int_0^{\ell_0/(2\delta)^{1/2}} e^{-z^2} dz - (2/\pi\delta)^{1/2} \ell_0 e^{-\ell_0^2/2\delta} \right]}{e^{-l_0^2/2\delta} + \frac{e^{-x^2/2\delta t}}{t^{1/2}} \left[\frac{2}{\pi^{1/2}} \int_0^{\ell_0/(2\delta)^{1/2}} e^{-z^2} dz - (2/\pi\delta)^{1/2} \ell_0 e^{-\ell_0^2/2\delta} \right]} + O(1/t). \tag{2.28}$$

If we write

$$c_0 = \frac{2}{\pi^{1/2}} e^{\ell_0^2/2\delta} \int_0^{\ell_0/(2\delta)^{1/2}} e^{-z^2} dz - (2/\pi\delta)^{1/2} \ell_0 \tag{2.29}$$

then (2.28) can simply written as

$$u(x,t) = \frac{(2\delta)^{1/2}\xi}{t^{1/2}\left[1+\frac{t^{1/2}}{c_0}e^{\xi^2}\right]} + O\left(\frac{1}{t}\right), \quad \xi = x/(2\delta t)^{1/2}$$
$$\equiv u^{\infty}(x,t) + O\left(\frac{1}{t}\right). \tag{2.30}$$

It is easily checked that $u_{\infty}(x,t)$ is itself an exact solution of the Burgers equation (2.1) and describes the asymptotic behaviour of the N-wave initial profile (2.23). This is another curiosity of the Burgers equation. The present author and his collaborators (1993, 1994, 1996, 1999, 2000) make an extensive use of the form $u^{\infty}(x,t)$ in (2.30) to find exact asymptotic solution of several generalised Burgers equations for which Cole-Hopf transformation does not exist. They found analogues of (2.30) for the GBEs. Several transformations of GBEs, inspired by (2.30), lead to nonlinear PDEs for the 'reciprocal' functions whose solutions are found in terms of infinite series in time with coefficient functions depending on the similarity variable. These series were shown to be convergent, and were numerically evaluated. The solution then found were shown to agree very well with the numerical solution of IVP for large time (see Section 6).

We summarise the results for an initial-boundary value problem again with reference to the Burgers equation due to Clothier, Knight, and White (1981). The equation

$$\theta_t = D\theta_{zz} - (A\theta + B)\theta_z \tag{2.31}$$

describes nonhysteretic infiltration in nonswelling soil where θ is the volumetric water content, t is time, and z is depth of the soil. D, A, and B are constants. Equation (2.31) is solved subject to the initial condition

$$t = 0, z > 0, \theta = \theta_n, \tag{2.32}$$

where θ_n is the antecedent water content. The boundary condition for constant flux at $z = 0$ is

$$t \geq 0, z = 0; -D\theta_z + \frac{A}{2}(\theta - \theta_n)^2 = V_0 = \frac{A}{2}(\theta_0 - \theta_n)^2. \tag{2.33}$$

The problem (2.31)-(2.33) was solved via Cole–Hopf transformation. The limit of this exact solution as $t \to \infty$ was found; this limit is referred to as the 'profile at inifinity'. The exact solution of the original problem is rather complicated, but the asymptotic form is simple. However, unlike the previous two examples, the asymptotic solution is *not* exact. It is however shown to be a good approximation to the exact solution after a short time by comparing it with both the exact and numerical solutions; the discrepancy is less than 2 percent.

Joseph and Sachdev (1993) generalised these results by considering the initial and boundary conditions

$$u(x,0) = u_\infty + u_e(x), \tag{2.34}$$

$$-\frac{\epsilon}{2}u_x(0,t) + \frac{u^2}{2}(0,t) = \lambda(t), \tag{2.35}$$

$$\lambda(t) = \lambda_\infty + \lambda_e(t) \tag{2.36}$$

where u_∞ and λ_∞ are constants and

$$\lim_{x\to\infty} u_e(x) = 0, \quad \lim_{t\to\infty} \lambda_e(t) = 0, \tag{2.37}$$

and $u_e(x)$ and $\lambda_e(t)$ are such that the integrals

$$\bar{u} = \int_0^\infty u_e(x)dx, \quad \bar{\lambda} = \int_0^\infty \lambda_e(t)dt \tag{2.38}$$

exist. The 'profiles at infinity' were found for the distinct cases $u_\infty > 0$, $u_\infty = 0$, and $u_\infty < 0$. This work also generalises the analysis of the Calogero and De–Lillo ((1989), (1991)).

We may mention here that Whitham (1974) has shown how the so-called single hump solution of the Burgers equation arises from the delta function initial condition superposed over a constant background. Sachdev and his collaborators (1993, 1994, 1996, 1999, 2000) have found extended forms of this solution for several generalised Burgers equations.

3 Balancing Argument – an Example from Nonlinear ODEs

As we remarked in the introductory section, it is quite rare that nonlinear PDEs (or even nonlinear ODEs) can be solved in closed form. However, to get their asympotic behaviour, one may use what is referred to as balancing argument to sift from the full equation(s) those terms which are of the same order in the limit of the relevant variable tending to infinity, and hence solve the resulting simpler equation. There may be several such balances possible for different sets and parameters. This is best illustrated with the help of an example from nonlinear ODEs.

We consider the equation

$$(x'')^2 = x' + x, \tag{3.1}$$

treated originally by Bellman (1955). He posed the question whether this equation has a solution asymptotic to e^{-t} for large t. It is easy to see that Ce^{-t} is a plausible approximate solution which balancing the two terms on the RHS of (3.1) leaving a small error due to the term on the left; here C is any complex number. To improve upon this solution we let

$$x = Ce^{-t} + u \tag{3.2}$$

so that (3.1) changes to

$$u'' + Ce^{-t} = (u' + u)^{1/2}. \tag{3.3}$$

In the sequel we follow Levinson (1969). To proceed further we must have a term on the right of (3.3) to balance Ce^{-t} on the left. Therefore, we let

$$u = C_1 e^{-2t} \tag{3.4}$$

in (3.3) giving

$$4C_1 e^{-2t} + Ce^{-t} = (-C_1 e^{-2t})^{1/2}.$$

If we choose $C_1 = -C^2$, the term Ce^{-t} is accounted for. Now we may further refine the solution and write

$$x = Ce^{-t} - C^2 e^{-2t} + u \tag{3.5}$$

Substitution of (3.5) into (3.1) leads to

$$\begin{aligned} u'' + Ce^{-t} - 4C^2 e^{-2t} &= (C^2 e^{-2t} + u + u')^{1/2} \\ &= Ce^{-t}\left[1 + \frac{e^{2t}}{C^2}(u + u')\right]^{1/2} \end{aligned} \tag{3.6}$$

or

$$\tilde{L}(u) = u'' - \frac{e^t}{2C}(u' + u) = 4C^2 e^{-2t} + \tilde{R}, \tag{3.7}$$

where

$$\tilde{R} = Ce^{-t}\left\{\left[1 + \frac{e^{2t}}{C^2}(u' + u)\right]^{1/2} - 1 - \frac{e^{2t}}{2C^2}(u' + u)\right\}. \tag{3.8}$$

The exact solution of $\tilde{L}u = 0$ is difficult to obtain. Instead we use the WKB method and find its approximate solutions as

$$\psi_1 = e^{-t}, \quad \psi_2 = \exp\left(\frac{e^t}{2C}\right). \tag{3.9}$$

The operator having ψ_1 and ψ_2 as its exact solution is easily found to be

$$Lu = u'' - \frac{e^t}{2C}u'\left(1 - \frac{4C^2 e^{-2t}}{1 + 2Ce^{-t}}\right) - \frac{e^t}{2C}u\left(1 + \frac{2Ce^{-t}}{1 + 2Ce^{-t}}\right). \tag{3.10}$$

Therefore, (3.7) can be written as

$$Lu = 4C^2 e^{-2t} + R \tag{3.11}$$

where

$$R = \tilde{R} + \frac{2Ce^{-t}u' - u}{1 + 2Ce^{-t}}. \tag{3.12}$$

Now using a general theorem due to Levinson (1969), it is shown that provided $Re\ C > 0$, a solution of (3.11) which is $O(e^{-3t})$ exists and so

$$x(t) = Ce^{-t} - C^2e^{-2t} + O(e^{-3t}). \tag{3.13}$$

The cases $Re\ C = 0$ and $Re\ C < 0$ lead to other types of solutions, which are unstable. The solution (3.13) depends only on one parameter C.

To find a two parameter family of solutions of (3.1), we try another (approximate) plausible solution, namely $x \sim Ct^a$. Substitution of this into (3.1) leads to $a = 4$, $C = \frac{1}{144}$. To improve upon this solution we let

$$x = \frac{t^4}{144} + u \tag{3.14}$$

and get from (3.1)

$$L(u) = u'' - \frac{6u}{t^2} = \frac{t}{6} + R, \tag{3.15}$$

where

$$R = \frac{t^2}{12}\left\{\left[1 + \frac{4}{t} + \frac{144}{t^4}(u+u')\right]^{1/2} - 1 - \frac{2}{t} - \frac{72}{t^4}(u+u')\right\} + \frac{6u'}{t^2}. \tag{3.16}$$

In the present case the operator L in (3.15) is the familiar Euler operator and therefore the homogeneous equation $Lu = 0$ is easily solved. Its two solutions are t^3 and t^{-2}. To find the particular solution one uses variation of parameters formula. The factor multiplying the RHS of (3.15) in the integral in this formula is $\frac{t^3}{5s^2} - \frac{s^3}{5t^2}$. The term $\frac{t}{6}$ on RHS of (3.15) then gives rise to an integral $\frac{t^3}{30}\int \frac{s}{s^2}ds = t^3\left(\frac{\log t}{30}\right)$. Using this fact the solution of (3.15) can be written as

$$u(t) = \frac{t^3 \log t}{30} + C_1t^3 + C_2t^{-2} - \frac{1}{5t^2}\int s^3 R ds - \frac{t^3}{5}\int_t^\infty \frac{R}{s^2}ds. \tag{3.17}$$

Again using the general theorem of Levinson (1969) we can show that

$$u(t) = \frac{t^3 \log t}{30} + C_1t^3 + C_2t^{-2} + O(t^2\log^2 t) \tag{3.18}$$

and so

$$x = \frac{t^4}{144} + t^3\frac{\log t}{30} + C_1t^3 + C_2t^{-2} + O(t^2\log^2 t). \tag{3.19}$$

The solution (3.19) may be refined by iteration. This example shows how the special solutions of truncated forms of the original nonlinear ODE arising from balancing argument, or some simple forms suggested by physical requirements, may be refined, and order of their accuracy determined.

4 A First Order Nonlinear PDE

Murray (1970) carried out an extensive study of the analytic properties of the solutions of the equation

$$u_t + g(u)u_x + h(u) = 0, \tag{4.1}$$

with $g_u(u) > 0$, $h_u(u) > 0$, $h(0) = 0$ and $g(0) = 0$, under the initial conditions

$$u(x,0) = \begin{cases} 0, & x < 0, \\ f(x), & 0 < x < X, \\ 0, & X < x, \end{cases} \tag{4.2}$$

where $f(x)$ is either a C^1 function smoothly joined at the ends $x = 0, X$, or has an initial discontinuity at $x = X$. In the former case a shock may develop at a later time. Murray (1970) considered in detail the asymptotic behaviour of the solution of (4.1). Here we consider a special case of (4.1), namely

$$u_t + (\gamma u^\beta)_x + \lambda u^\alpha = 0, \quad \beta > 1, \tag{4.3}$$

where γ and λ are positive constants. We shall find solutions, both analytic and numerical, of (4.3) and investigate their asymptotic character (see Bukiet, Pelesko, Li, and Sachdev (1996) and Natalini and Tesei (1992)).

First we consider the special case $\lambda = 0$ in (4.3) when it describes a conservation law. We also prescribe the so-called top-hat initial conditions

$$u(x,0) = \begin{cases} 0, & x < 0, \\ h, & 0 < x < X, \\ 0, & X < x. \end{cases} \tag{4.4}$$

The initial conditions (4.4) imply that all the values of u between 0 and h are assumed via a rarefaction wave centered at $x = 0$. The characteristics of (4.3) with $\lambda = 0$ are

$$\frac{dx}{dt} = \beta\gamma u^{\beta-1}, \quad \frac{du}{dt} = 0 \tag{4.5}$$

implying that u is constant along these curves. Since $\beta > 1$, the characteristic carrying the highest value of u, namely h, moves fastest with the speed

$$\frac{dx}{dt} = \beta\gamma h^{(\beta-1)}. \tag{4.6}$$

Its locus is given by

$$x_{FR} = \beta\gamma h^{(\beta-1)}t, \tag{4.7}$$

where FR denotes the front of the rarefaction wave. The initial shock at $x = X$ at $t = 0$ moves according to

$$\begin{aligned}\frac{dx_s}{dt} &= \gamma u_b^{\beta-1} \\ &= \gamma h^{\beta-1},\end{aligned} \tag{4.8}$$

that is,

$$x_s = X + \gamma h^{\beta-1} t \tag{4.9}$$

during the time $0 < t \leq t_{\text{catch up}}$, where $t_{\text{catch up}}$ is the time when the front of the rarefaction wave catches up with the shock; here $u_b = h$ denotes the value of u just behind the shock. Then, for $x_{FR} < x < x_s$, $u = h$. For the solution in the rarefaction wave we integrate the first of equations (4.5) and use the condition $x = 0$ when $t = 0$. We obtain

$$u(x,t) = \left(\frac{x}{\beta\gamma t}\right)^{1/(\beta-1)} \tag{4.10}$$

When the front of the rarefaction wave catches up with the shock, we have $x_s = x_{FR}$ and so from (4.7) and (4.9) we get

$$t_{\text{catch up}} = \frac{X}{\gamma(\beta-1)h^{(\beta-1)}}. \tag{4.11}$$

At this time the location of the shock is found from (4.9) and (4.11):

$$x_{s,\text{catch up}} = \frac{\beta X}{\beta - 1}. \tag{4.12}$$

For the time beyond $t_{\text{catch up}}$ we find the position of the shock from the first of (4.8) and (4.10); the latter gives the value u_b (less than h) just behind the shock. After an integration we have

$$x_s = \beta\left[\gamma t \left(\frac{hX}{\beta-1}\right)^{\beta-1}\right]^{1/\beta}, \tag{4.13}$$

where we have used the initial condition (4.11)–(4.12) at the catch up time. We thus have the solution

$$u(x,t) = \begin{cases} (x/\beta\gamma t)^{1/(\beta-1)}, & 0 \leq x \leq x_{FR}, \\ h & x_{FR} < x < x_s \quad \text{when } t \leq t_{\text{catch up}} \\ 0 & x > x_s, \end{cases} \tag{4.14}$$

$$u(x,t) = \begin{cases} (x/\beta\gamma t)^{1/(\beta-1)}, & \text{for } 0 \leq x \leq x_s, \\ & \qquad\qquad \text{where } t \geq t_{\text{catch up}}, \\ 0 & \text{for } x > x_s. \end{cases} \tag{4.15}$$

It is easy to check that the area under the profiles (4.14) or (4.15) is preserved and is equal to the initial area hX (see figure 4.1 for a sample of the solution at several times for the parametric values $\beta = 2, \gamma = 1/2, h = 1$, and $\lambda = 0$).

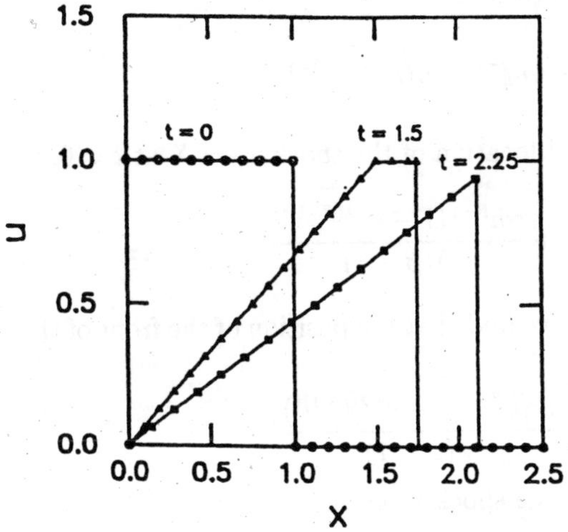

Figure 4.1. Computed and exact solutions to $u_t + (u^2/2)_x = 0$ before and after the rerefaction catches up to the shock, the conservation law case. Open circles, triangles and squares represent the computed solution. ($\Delta x = 0.1$) (Bukiet et al., 1996)

We may observe that the solutions (4.14) and (4.15) may be called either self-similar or separable. We shall presently investigate this class of solutions in some detail.

Now we consider (4.3) with linear damping, namely

$$u_t + (\gamma u^\beta)_x + \lambda u = 0. \tag{4.16}$$

This is a special case of (4.3) with $\alpha = 1$ and admits some explicit analysis. We assume that IC $u(x, 0)$ for (4.16) is such that it vanishes rapidly for large x so that $\int_{-\infty}^{\infty} u dx$ exists. Integrating (4.16) from $x = -\infty$ to $x = +\infty$ and using the vanishing conditions in these limits we have

$$\frac{dA}{dt} = -\lambda A, \quad A(t) = \int_{-\infty}^{\infty} u dx, \tag{4.17}$$

where $A = A(t)$ is the area under the curve at any time. Equation (4.17) yields

$$A(t) = A_0 e^{-\lambda t}, \quad A_0 = \int_{-\infty}^{\infty} u(x, 0) dx. \tag{4.18}$$

To see whether the rarefaction arising from top hat initial conditions (4.4) catches up with the shock in the present case, we proceed as before. The characteristic system of (4.16) is

$$\frac{dx}{dt} = \beta\gamma u^{\beta-1}, \quad \frac{du}{dt} = -\lambda u. \tag{4.19}$$

From (4.18) it follows that the maximum of the top hat IC decays like

$$\max u = he^{-\lambda t}. \tag{4.20}$$

The shock locus in the present case is given by

$$\frac{dx_s}{dt} = \gamma u_b^{\beta-1} = \gamma(he^{-\lambda t})^{\beta-1}. \tag{4.21}$$

Integrating (4.21) and using the initial location of the shock $x_s = X$ at $t = 0$, we obtain

$$x_s = X + \frac{\gamma h^{\beta-1}(1 - e^{-\lambda(\beta-1)t})}{\lambda(\beta - 1)}. \tag{4.22}$$

Using (4.20) in (4.19) and integrating, we find that the position of the front of the rarefaction wave is given by

$$x_{FR} = \frac{\beta\gamma h^{\beta-1}(1 - e^{-\lambda(\beta-1)t})}{\lambda(\beta - 1)}. \tag{4.23}$$

The rarefaction wave catches up with the shock when

$$X + \frac{\gamma h^{\beta-1}(1 - e^{-\lambda(\beta-1)t})}{\lambda(\beta - 1)} = \frac{\beta\gamma h^{\beta-1}(1 - e^{-\lambda(\beta-1)t})}{\lambda(\beta - 1)} \tag{4.24}$$

leading to the catch up time

$$t_{\text{catch up}} = \frac{\ln(1 - \lambda X/\gamma h^{\beta-1})}{-\lambda(\beta - 1)}. \tag{4.25}$$

Therefore, the catch up time exists only if

$$1 - \frac{\lambda X}{\gamma h^{\beta-1}} > 0 \tag{4.26}$$

or

$$h > \left(\frac{\lambda X}{\gamma}\right)^{1/(\beta-1)} \tag{4.27}$$

In this case the location of the shock at $t = t_{\text{catch up}}$ is again $\beta X/(\beta - 1)$.

To get the solution in the rarefaction wave behind its front we put $h = u$ in (4.21) and use the values of u in the range $0 < u < h$. Thus we get the solution in the rarefaction wave as

$$u(x,t) = \left\{\frac{x\lambda(\beta - 1)}{\beta\gamma(e^{\lambda(\beta-1)t} - 1)}\right\}^{1/(\beta-1)} \tag{4.28}$$

After the catch up time, the motion of the shock is given by

$$\frac{dx_s}{dt} = \gamma u_b^{(\beta-1)} = \frac{x_s\lambda(\beta - 1)}{\beta(e^{\lambda(\beta-1)t} - 1)}. \tag{4.29}$$

Integrating (4.29) and using IC (4.22) at the catch up time we have

$$x_s = \left(\frac{\beta}{\beta-1}\right)\left(\frac{\gamma}{\lambda}\right)^{1/\beta}(hX)^{(\beta-1)/\beta}(1-e^{-\lambda(\beta-1)t})^{1/\beta}, t > t_{\text{catch up}}. \tag{4.30}$$

It follows from (4.22) and (4.30) that the shock traverses a finite distance in infinite time:

No catch up:

$$x_{s,\infty} = X + \frac{\gamma h^{(\beta-1)}}{\lambda(\beta-1)}. \tag{4.31}$$

Catch up:

$$x_{s,\infty} = \left(\frac{\beta}{\beta-1}\right)\left(\frac{\gamma}{\lambda}\right)^{1/\beta}(hX)^{(\beta-1)/\beta}. \tag{4.32}$$

In the present case we summarize the result and observe that the solution has a product or self-similar form:

$$u(x,t) = \begin{cases} \left\{\frac{x\lambda(\beta-1)}{\beta\gamma(e^{\lambda(\beta-1)t}-1)}\right\}^{1/(\beta-1)} & 0 \le x \le x_{FR}, \\ he^{-\lambda t} & x_{FR} < x < x_s, \\ 0 & x > x_s, \end{cases} \tag{4.33}$$

where $t \le t_{\text{catch up}}$, and

$$u(x,t) = \begin{cases} \left\{\frac{x\lambda(\beta-1)e^{-\lambda(\beta-1)t}}{\beta\gamma(1-e^{-\lambda(\beta-1)t})}\right\}^{1/(\beta-1)} & 0 \le x \le x_s, \\ 0 & x > x_s, \end{cases} \tag{4.34}$$

where $t \ge t_{\text{catch up}}$.

The area under the profile for all time is given by $hXe^{-\lambda t}$ (see (4.4) and (4.17)).

It is possible to find the catch up time etc. for the more general case provided $\alpha \ne 1$ or $\alpha \ne \beta$ (see Bukiet et al (1996)). In each case it is found that if the front of the rarefaction wave catches up with the shock, the shock position at $t_{\text{catch up}}$ is $\frac{\beta X}{\beta-1}$.

Now we proceed to find exact solutions of (4.3); we may in particular focus on those solutions which are non-negative and bounded C^1 over $\mathcal{D} : \epsilon < x < \infty, \delta < t \le \infty$, where ϵ and δ are positive. Subsequently we shall show that some of these solutions form asymptotics for large time.

a) If we let u be a function of x alone, that is, $u(x,t) = F(x)$, we have from (4.3),

$$\gamma\beta F^{\beta-1}F' + \lambda F^{\alpha} = 0, \tag{4.35}$$

which integrates to give

$$u(x,t) = F(x) = \left[\left(C - \frac{\lambda}{\gamma\beta}x\right)(\beta-\alpha)\right]^{1/(\beta-\alpha)}, \tag{4.36}$$

where C is constant of integration. The solution (4.36) is bounded over $\mathcal{D}$ provided $\beta < \alpha$ and $C < 0$. No bounded solutions over $\mathcal{D}$ exist if $\beta > \alpha$. It is also clear that (4.36) is not valid if $\alpha = \beta$. In this case (4.35) gives

$$u(x,t) = F(x) = Ce^{-(\lambda/\gamma\beta)x}, \tag{4.37}$$

which is clearly bounded.

b) Now if we let $u(x,t) = J(t)$, as function of t alone, then (4.3) becomes

$$J' + \lambda J^{\alpha} = 0. \tag{4.38}$$

Equation (4.38) is solved to yield the following three solutions

(i) $\alpha < 1$.

$$u(x,t) = J(t) = \begin{cases} [(1-\alpha)(C-\lambda t)]^{1/(1-\alpha)}, & t < C/\lambda \\ 0, & t \geq C/\lambda, \end{cases} \tag{4.39}$$

where $C > 0$. This solution decay to zero in a finite time C/λ.

(ii) $\alpha = 1$.

$$u(x,t) = J(t) = Ce^{-\lambda t}, \quad C > 0. \tag{4.40}$$

(iii) $\alpha > 1$.

$$u(x,t) = J(t) = [(\alpha-1)(\lambda t + C)]^{-1/(\alpha-1)}, \quad C > 0. \tag{4.41}$$

The solutions (4.40) and (4.41) decay to zero in infinite time.

(c) It is easy to check that equation (4.3) is invariant under the finite group of transformations

$$x^* = a^n x, \quad t^* = a^m t, \quad u^* = a^r u. \tag{4.42}$$

The exponents n, m and r help to determine the similarity form of the solution

$$u(x,t) = t^{1/(1-\alpha)} H(\eta), \quad \eta = xt^{(\alpha-\beta)/(1-\alpha)} \tag{4.43}$$

where $\alpha \neq 1$ or $\alpha \neq \beta$. Equation (4.3) reduces via (4.43) to the ODE

$$[(\beta-\alpha)\eta + (\alpha-1)\gamma\beta H^{\beta-1}]H' + \lambda(\alpha-1)H^{\alpha} - H = 0. \tag{4.44}$$

First we observe that if $\beta > \alpha$ and $\alpha > 1$, (4.44) has the constant solution

$$H^* = \{\lambda(\alpha-1)\}^{1/(1-\alpha)}. \tag{4.45}$$

The solution, in this case, tends to H^* as $\eta \to \infty$ since $H' < 0$ if $H > H^*$. There are no bounded solutions of (4.44) if $\beta < \alpha$ since the denominator in this case vanishes at some finite η or $H' \sim 1/\eta$ for large η leading to the solution of the form $H \approx O$ ($\log \eta$) as $\eta \to \infty$. The form (4.43) shows that, for $\alpha < 1$, the solutions become unbounded in time.

Actually it is possible to solve (4.44) exactly, though in an implicit form. For simplicity we let $\lambda = \gamma = 1$ in (4.3). First it is easy to check by a balancing argument that (4.44) has the following properties. Let H be a nonnegative solution of (4.44) over $(0, \bar{\eta})$, $\frac{f^{m-1}}{\eta} \to 1$ as $\eta \to 0$, where

$$\bar{\eta} = \begin{cases} \infty & \text{if } \alpha \leq \beta, \\ \frac{(\alpha-1)^{(\alpha-\beta)/(\alpha-1)}}{\alpha-\beta} & \text{if } \alpha > m. \end{cases} \tag{4.46}$$

It is evident that $H = 0$ and $H = \left(\frac{1}{\alpha-1}\right)^{1/(\alpha-1)} \equiv H^*$, say, are stationary solutions of (4.44).
(i) We find from (4.44) with $\lambda = \gamma = 1$ that

$$H' = H^{2-\beta}\left(\frac{1}{\alpha-1} - H^{\alpha-1}\right) \Big/ \left(\beta + \frac{\beta-\alpha}{\alpha-1}\frac{\eta}{H^{\beta-1}}\right) = F(\eta, H). \tag{4.47}$$

It easily follows that

$$\frac{H'}{\eta^{(2-\beta)/(\beta-1)}} \to \frac{1}{\beta-1} \quad \text{as} \quad \eta \to 0. \tag{4.48}$$

In general, we have

$$H^{(k)} \sim \eta^{1/(\beta-1)-k} \quad \text{as} \quad \eta \to 0 (k = 1, 2, \ldots) \tag{4.49}$$

$$\begin{aligned} H(0) &= H'(0) = \ldots = H^{(k-1)}(0) = 0, \\ H^{(k)}(0) &= \begin{cases} \infty, & \text{if } \beta > (k+1)/k, \\ > 0, & \text{if } \beta = (k+1)/k, \\ 0, & \text{if } \beta < (k+1)/k. \end{cases} \end{aligned} \tag{4.50}$$

Now, to find the exact solution, we first consider the special case $\alpha = \beta$. In this case (4.44) is a Bernoulli equation and is immediately integrated:

$$H(\eta) = \left[\frac{1 - \bar{e}^{(\beta-1)\eta/\beta}}{\beta - 1}\right]^{1/(\beta-1)}, \eta \geq 0. \tag{4.51}$$

For $\alpha \neq \beta$, we write (4.44) as

$$\left(H^{\beta-1} + \frac{\beta-\alpha}{(\alpha-1)}\eta\right) dH - H\left(\frac{1}{(\alpha-1)} - H^{\alpha-1}\right) d\eta = 0 \tag{4.52}$$

Equation (4.52) has the integrating factor

$$\mu(H) = -H^{\alpha-\beta-1}\left(\frac{1}{\alpha-1} - H^{\alpha-1}\right)^{(\beta+1-2\alpha)/(\beta-1)} \tag{4.53}$$

and hence can be solved as

$$\left[\frac{1}{\alpha-1}-H^{\alpha-1}\right]^{(\beta-\alpha)/(\alpha-1)}\left[\frac{\eta}{H^{\beta-\alpha}}+\frac{1}{\beta-\alpha}\right]=\frac{1}{\beta-\alpha}\left(\frac{1}{\alpha-1}\right)^{(\beta-\alpha)/(\alpha-1)} \tag{4.54}$$

We may also check that

$$\frac{H^{\beta-1}}{\eta}=1-\frac{\beta-1}{2}\eta^{(\alpha-1)/(\beta-1)}+o(\eta^{(\alpha-1)/(\beta-1)}) \quad \text{as} \quad \eta\to 0. \tag{4.55}$$

(d) If $\alpha\neq 1$ and $\alpha\neq\beta$, we may obtain another form of the similarity solution

$$u(x,t)=x^{1/(\beta-\alpha)}H(\eta), \quad \eta=xt^{(\alpha-\beta)/(1-\alpha)}. \tag{4.56}$$

Actually we may change (4.56) to the form (4.43) by writing it as

$$u(x,t)=[xt^{(\alpha-\beta)/(1-\alpha)}]^{1/(\beta-\alpha)}t^{1/(1-\alpha)}H(\eta). \tag{4.57}$$

However, the form (4.56) may be useful for certain kind of initial/boundary conditions. Equation (4.3) in the present case reduces to thc ODE

$$H'=\frac{(\alpha-1)\{(\beta-\alpha)\lambda H^{\alpha}+\gamma\beta H^{\beta}\}}{\eta(\beta-\alpha)\{(\alpha-\beta)\eta^{(\alpha-1)/(\alpha-\beta)}+\gamma\beta(1-\alpha)H^{\beta-1}\}}. \tag{4.58}$$

One could use the tranformation (4.57) to reduce (4.58) to the form (4.44) and hence solve it in a closed form. It is seen from (4.58) that, for $\beta>\alpha$ and $\alpha>1$, $H(\eta)\to 0$ exponentially as $\eta\to\infty$; so we have bounded solutions in this case. For $\beta<\alpha$ or $\alpha<1$, the denominator on RHS of (4.58) vanishes for some η and so no bounded solutions exist for these cases.
(e) The transformation (4.42) also suggests that separable solutions may exist for $\alpha=1$ or $\alpha=\beta$. For $\alpha=1$, one may verify that no bounded smooth solutions exist.

For $\alpha=\beta$, we have $u(x,t)=F(x)J(t)$ where

$$J(t)=\left(\frac{1}{(C_o+C_1t)(\beta-1)}\right)^{1/(\beta-1)} \tag{4.59}$$

and F is governed by the first order ODE

$$F'=-\frac{(\lambda F^{\beta}-C_1F)}{\gamma\beta F^{\beta-1}}. \tag{4.60}$$

If we put $F^{\beta-1}=G$ in (4.60), it becomes

$$\frac{\gamma\beta}{(\beta-1)}G'+(\lambda G-C_1)=0$$

which is easily solved in the form

$$(\lambda G-C_1)^{\gamma\beta/\lambda(\beta-1)}=\exp(C_2-\eta), \tag{4.61}$$

where C_2 is another constant.

(f) Another set of finite transformations

$$x^* = a^n x, \quad t^* = t + \ln(a), \quad u^* = a^r u \tag{4.62}$$

lead to other similarity forms of solution for (4.3):

$$u(x,t) = e^{rt} H(\eta), \quad \eta = x e^{r(1-\beta)t}, \tag{4.63}$$

where

$$H' = \frac{-(r+\lambda)H}{\gamma\beta H^{\beta-1} - r(\beta-1)\eta} \tag{4.64}$$

or

$$u(x,t) = x^{1/(\beta-1)} H(\eta), \quad \eta = x e^{r(1-\beta)t}, \tag{4.65}$$

where

$$H' = \frac{\lambda H + (\gamma\beta/(\beta-1))H^{\beta}}{\eta\{r(\beta-1) - \gamma\beta H^{\beta-1}\}}. \tag{4.66}$$

The ODEs (4.64) and (4.66) can be transformed to each other (see (4.56)–(4.57)). If we put $\ln\eta = N$ and $H^{\beta-1} = G$ in (4.66), we get

$$\frac{1}{(\beta-1)}\frac{dG}{dN} = \frac{\{\lambda + (\gamma\beta/(\beta-1))\, G\}}{r(\beta-1) - \gamma\beta G}. \tag{4.67}$$

Equation (4.67) is separable and is easily solved.

(g) Finally we have yet another group of finite transformations

$$x^* = x + \ln a, \quad t^* = a^n t, \quad u^* = a^r u,$$

which, on use in (4.3), lead to a similarity form of the solution only if $\alpha = \beta$. In this case

$$u = t^{-1/(\beta-1)} H(\eta), \quad \eta = t e^{r(\beta-1)x} \tag{4.68}$$

where

$$H' = \frac{H - \lambda(\beta-1)H^{\beta}}{(\beta-1)\eta\{1 + \gamma r\beta(\beta-1)H^{\beta-1}\}}. \tag{4.69}$$

Equation (4.69) has the form (4.66) and may be solved exactly in the manner suggested there.

Bukiet et al first checked their numerical scheme by reference to Murray's analysis by choosing both the smooth initial data

$$u(x,0) = \begin{cases} 0 & x < 0 \\ \sin^2(\pi x) & 0 \le x \le 1 \\ 0 & 1 < x \end{cases} \tag{4.70}$$

and the discontinuous data

$$u(x,0) = \begin{cases} 0 & x < 0 \\ h & 0 < x < X \\ 0 & X < x. \end{cases} \tag{4.71}$$

They considered the evolution of initial profiles (4.70)–(4.71) under (4.3) with specific values of the parameters appearing therein. The agreement with Murray's asymptotic results was rather remarkable. They also verified the emergence of the exact solution (4.36), (4.37), (4.39), (4.65) and (4.68) as asymptotics from 'appropriate' classes of initial/boundary conditions. These were the solutions for which the shocks were not expected to form. The asymptotic nature of the other exact solution under suitable initial/boundary conditions may also be checked in the same manner.

5 Asymptotic Solution via Balancing Arguments

As we mentioned in the introduction, Grundy and his collaborators have considered a large class of nonlinear parabolic equations and have obtained their asymptotic behaviour by using the so-called balancing argument; we introduced this idea in the context of nonlinear ODEs in section 3. Here we illustrate this approach by discussing in detail the work of Grundy, Van Duijn and Dawson (1994) which treats the large time behaviour of the equation

$$(u + u^p)_t = Du_{xx} - u_x, \quad -\infty < x < \infty, \quad t \geq 0 \tag{5.1}$$

with pulse type (non-negative) initial conditions. Here D is a constant. In appropriately scaled variables, (5.1) describes the one-dimensional flow of a solute through a porous medium with the solute undergoing adsorption by the solid matrix of the medium. It is assumed that the total mass both adsorbed and in solution is invariant. We shall see how the asymptotic solution of (5.1) changes its character, as p varies. Several distinctive behaviours come out by the use of balancing argument. We seek solution to (5.1) subject to the initial condition

$$u(x,0) = u_0(x) \tag{5.2}$$

where

$$u_0(\pm\infty) = 0, \quad u_0 \geq 0 (\not\equiv 0) \quad \text{on} \quad \mathcal{R} \tag{5.3}$$

and require that $u_0 + u_0^p \in L^1(\mathcal{R})$ so that the total mass of adsorbed and dissoluted concentration is finite. We may eliminate the coefficient D in (5.1) by the simple scaling $\bar{x} = \frac{x}{D}, \bar{t} = \frac{t}{D}$ but it is often convenient to retain it since it helps to consider the limiting hyperbolic case as $D \to 0$.

We begin by transforming (5.1) with $D = 1$ through the obvious choice of variables

$$t, \quad \xi = x - t \tag{5.4}$$

so that it becomes

$$(u + u^p)_t = u_{\xi\xi} + (u^p)_\xi, \tag{5.5}$$

where now $u = u(\xi, t)$. To incorporate the 'spread' of the solution it is usual to introduce the similarity variable

$$\eta = \frac{\xi}{t^\delta} = \frac{(x-t)}{t^\delta} \tag{5.6}$$

where $\delta > 0$, and seek the solution in the form

$$u(\xi, t) = t^\alpha v(\eta, t) \tag{5.7}$$

in an attempt to generalise the similarity solution where it exists; here $\alpha < 0$. Equation (5.1) in terms of v becomes

$$\begin{aligned}(tv_t + \alpha v - \delta\eta v_\eta) + t^{\alpha(p-1)}\{t(v^p)_t + \alpha p v^p - \delta\eta(v^p)_\eta\} \\ = t^{1-2\delta}v_{\eta\eta} + t^{\alpha(p-1)+(1-\delta)}(v^p)_\eta\end{aligned} \tag{5.8}$$

or symbolically

$$A + B = C + D. \tag{5.9}$$

To impose the invariance of the 'mass' or integral

$$M = \int_{-\infty}^{\infty}(u + u^p)dx \tag{5.10}$$

we use (5.7) and require that

$$M = \int_{-\infty}^{\infty}(t^{\alpha+\delta}v + t^{\alpha p+\delta}v^p)d\eta \tag{5.11}$$

is independent of time. Since $p > 1, \alpha < 0, \delta > 0, M \sim t^{\alpha+\delta}$ as $t \to \infty$; therefore we must have

$$\alpha = -\delta \tag{5.12}$$

to the dominant order. We now consider various possibilities in (5.8). We consider the asymptotic limit $t \to \infty$ with the further assumptions that $\eta = O(1)$, and v and its derivatives with respect to η, and tv_t are all bounded in this limit. Since $p > 1$ and $\alpha < 0$, the term A dominates on the LHS of (5.8) while two possibilities arise on the RHS. Either C balances A or D balances A, as $t \to \infty$. This argument is referred to as asymptotic balancing.

We consider below each of these possibilities.

I. C dominates D and balances A on the left. In this case, we have from (5.8)

$$\delta = \frac{1}{2} \tag{5.13}$$

and

$$\alpha(p-1) + 1 - \delta < 0. \tag{5.14}$$

With $\alpha = -\delta = -1/2$ from (5.12)–(5.13), (5.14) requires that $p > 2$. Now (5.8) becomes

$$\begin{aligned} tv_t - \frac{1}{2}(v + \eta v_\eta) + t^{-(p-1)/2}\left\{t(v^p)_t - \frac{p}{2}v^p - \frac{\eta}{2}(v^p)_\eta\right\} \\ = v_{\eta\eta} + t^{-(p-2)/2}(v^p)_\eta. \end{aligned} \tag{5.15}$$

Now we expand the solution as

$$v(\eta, t) = v_0(\eta) + o(1) \tag{5.16}$$

such that $tv_t = o(1)$ in the limit $t \to \infty$ and $\eta = O(1)$. Substituting (5.16) into (5.15), the leading order behaviour is given by

$$v_0'' + \frac{1}{2}(\eta v_0' + v_0) = 0, \tag{5.17}$$

where primes denote differentiation with respect to η. Equation (5.17) is linear and has the general solution

$$v_0 = K_1 e^{-\eta^2/4} - K_2 e^{-\eta^2/4}\int_\eta^\infty e^{s^2/4} ds = K_1 e^{-\eta^2/4} + O(\eta^{-1}) \text{ as } \eta \to \infty, \tag{5.18}$$

where K_1 and K_2 are arbitrary constants. Since we require that (5.11) holds, the intergral of $v_0(\eta)$ from $-\infty$ to $+\infty$ must converge. Thus we must have $K_2 = 0$ in (5.18) and so

$$v_0(\eta) = K_1 e^{-\eta^2/4}. \tag{5.19}$$

The 'mass'

$$M = K_1 \int_{-\infty}^{\infty} e^{-\eta^2/4} d\eta \quad \text{or} \quad K_1 = \frac{M}{2\sqrt{\pi}}. \tag{5.20}$$

Thus, for $p > 2$, we have, to the leading order, the solution

$$u(x,t) = \frac{M}{2\sqrt{\pi}} t^{-1/2} e^{-(x-t)^2/4t}\{1 + o(1)\} \tag{5.21}$$

as $t \to \infty$, and $(x-t)/2\sqrt{t} = O(1)$. The result (5.21) is uniform in x.

To check the veracity of (5.21) as the asymptotic behaviour for large t, Grundy *et al* (1994) solved (5.1) numerically for $p = 3$ with the initial data

$$u_0(x) = H(x+1) - H(x-1), \tag{5.2a}$$

where $H(x)$ is the Heaviside function. Choosing $M = 4$ in (5.10), Grundy *et al* (1994) compared $v(\eta, t) \equiv t^{1/2}u(x,t)$, as obtained from the numerical solution and from (5.21) for various values of t. The agreement was found to be good (See figure 5.1).

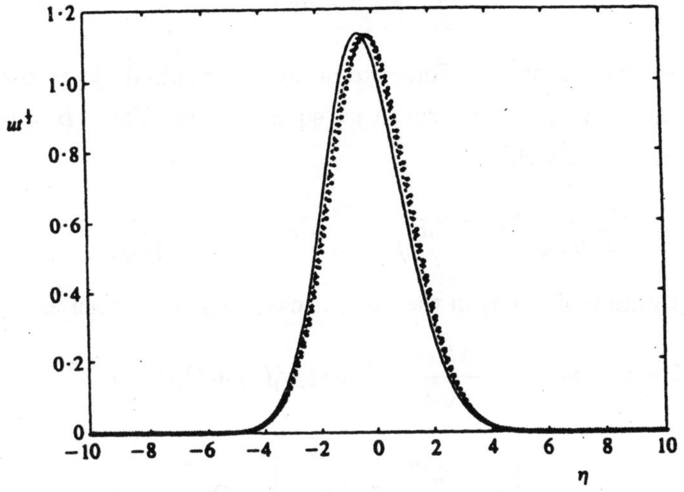

Figure 5.1. Convergence of the numerical solution to the limiting profile for the case $p = 3$. $ut^{1/2}$ plotted as a function for η for —— $t = 80$, - - - $t = 320$, case + + + $t = 1000$, x x x $t = 2000$ (from Grundy (1994).

To improve upon the solution (5.21) (and (5.19)) we write $v = v_0 + \epsilon$ in (5.16) and linearise it. It is then found that the next order term, should be of order $t^{-1/2}$ logt. Indeed one may write the solution of (5.15) as

$$v(\eta, t) = v_0(\eta) + t^{-1/2} \log t \; v_1(\eta) + t^{-1/2} v_2(\eta) + o(t^{-1/2}) \tag{5.22}$$

where v_0 is given by (5.19). Putting (5.22) into (5.15) and equating terms of order $t^{-1/2} \log t$, we have

$$v_1'' + \frac{1}{2}\eta v_1' + v_1 = 0, \quad v_1(\pm\infty) = 0. \tag{5.23}$$

For exponential decay at $\eta \to \pm\infty$, we must choose the solution

$$v_1 = Av_0'(\eta)$$

of (5.23), where A is an arbitrary constant. Again by substituting (5.22) in (5.15) etc., we get the equation for v_2 as

$$v_2'' + \frac{1}{2}\eta v_2' + v_2 = (A - 3v_0^2)v_0', \tag{5.24}$$

which has the solution

$$v_2 = v_0'(\eta) w_2(\eta) \tag{5.25}$$

where

$$w_2' = \frac{Be^{\eta^2/4}}{\eta^2} + \frac{3M^2}{4\pi}\frac{e^{\eta^2/4}}{\eta^2}\int_\eta^\infty s^2 e^{-3s^2/4} ds$$
$$- \frac{Ae^{\eta^2/4}}{\eta^2}\int_\eta^\infty s^2 e^{-s^2/4} ds. \tag{5.26}$$

B in (5.26) is an arbitrary constant. For v_2 to decay exponentially to zero as $\eta \to +\infty$, we must put the exponentially growing term in w_2' equal to zero. Thus, $B = 0$. The same condition in the limit $\eta \to -\infty$ yields

$$A = \frac{3M^2}{4\pi}\int_{-\infty}^\infty s^2 e^{-3s^2/4} ds \Big/ \int_{-\infty}^\infty s^2 e^{-s^2/4} ds = \frac{M^2}{4\pi\sqrt{3}}. \tag{5.27}$$

Thus the asymptotic solution of (5.15) in the present case may be written as

$$v(\eta, t) = v_0(\eta) + \frac{M^2}{4\pi\sqrt{3}} t^{-1/2}(\log t) v_0'(\eta) + O(t^{-1/2}) \tag{5.28}$$

or equivalently

$$v(\eta, t) = v_0\left\{\eta + \frac{M^2}{4\pi\sqrt{3}} t^{-1/2}\log t\right\} + O(t^{-1/2}) \tag{5.29}$$

The representation (5.7) with (5.29) and (5.27) gives a much more accurate asymptotic description to which the solution of IVP (5.1) and (5.2a) approaches quite rapidly.

It is interesting to quote here the more rigorous result due to Escobedo and Zuazua (1991). They proved that, for $p > 2$, the finite mass solutions of (5.5) converge in the L^1 sense to the solutions of the heat equation obtained by neglecting the terms $(u^p)_t$ and $(u^p)_\xi$, therein. The present analysis due to Grundy et al. (1994) takes the nonlinear terms into account and gives an explicit asymptotic answer, which agrees very well with the numerical solution of IVP after a short initial time.

II. Now we turn to the other balance in (5.9) when the term D on the RHS dominates and balances asymptotically as $t \to \infty$ with A on the left. In this case we have

$$\alpha(p-1) + 1 - \delta = 0 \tag{5.30}$$

and

$$1 - 2\delta < 0. \tag{5.31}$$

Equations (5.12) and (5.30) give

$$\alpha = -1/p, \quad \delta = 1/p \tag{5.32}$$

and the inequality (5.31) becomes

$$p < 2. \tag{5.33}$$

Since we have assumed that $p > 1$, we have $1 < p < 2$. With the values of α and δ in (5.32), (5.8) assumes the form

$$tv_t - \frac{1}{p}(v + \eta v_\eta) \;+\; t^{-(p-1)/p}\{t(v^p)_t - v^p - \frac{\eta}{p}(v^p)_\eta\}$$
$$= t^{-(2-p)/p} v_{\eta\eta} + (v^p)_\eta. \tag{5.34}$$

Now we write

$$v = v_0(\eta) + o(1), \tag{5.35}$$

as $t \to \infty$, with $\eta = o(1)$, and the assumption that $tv_t = o(1)$ Grundy et al (1994) refer to this as the outer expansion. To the leading order (5.34) becomes

$$(v_0 + \eta v_0') + p(v_0^p)' = 0, \tag{5.36}$$

where prime again denotes differentiation with respect to η. The general solution of (5.36) is easily found to be

$$v_0^p + \frac{\eta}{p} v_0 = C \tag{5.37}$$

where C is an arbitrary constant.

By differentiating (5.37) etc., we check that, for $C < 0$, $v_0(\eta)$ is double-valued if

$$\eta \le -p^2(-C)^{1/p}/(p-1)^{(p-1)/p}. \tag{5.38}$$

This is not admissible. If $C > 0$, v_0 is single-valued on $-\infty < \eta < \infty$ but behaves like

$$v_0 \sim pC/\eta \tag{5.39}$$

as $\eta \to \infty$. This behaviour makes the mass integral (5.11) diverge so we must choose C equal to zero. Thus we have

$$v_0 = \left(-\frac{\eta}{p}\right)_+^{1/(p-1)} \tag{5.40}$$

The solution (5.40) is defined for all $\eta \in \mathcal{R}$ but again gives infinite mass if it is assumed to hold for all η. We must therefore restrict the validity of (5.40) to the finite interval $\eta_1 \le \eta \le 0$, setting v_0 identically equal to zero outside this interval. Thus we have a discontinuous solution, with the discontinuity located at $\eta = \eta_1$. This solution, in the literature on singular perturbation theory, is referred to as the outer solution. Substituting (5.35) and (5.40) into (5.11) we have, to leading order,

$$M = \int_{\eta_1}^{0} \left(-\frac{\eta}{p}\right)^{1/(p-1)} d\eta$$

or

$$\eta_1 = -p\left(\frac{M}{p-1}\right)^{(p-1)/p} < 0. \tag{5.41}$$

Thus, we have shown that, for $1 < p < 2$, the outer solution in terms of original variables is

$$u(x,t) = t^{-1/p}\left(\frac{t-x}{pt^{1/p}}\right)^{1/(p-1)} \{1 + o(1)\} \tag{5.42}$$

as $t \to \infty$, $(x-t)/t^{1/p} = O(1)$. Since we have $\eta_1 \le \eta \le 0$, we require that

$$\eta_1 t^{1/p} \le x - t < 0. \tag{5.43}$$

Now we consider the borderline cases $p = 2$ and $p = 1$, the latter being just the linear PDE. When $p = 2$, the terms A, C, and D in (5.9) are asymptotically of the same order as $t \to \infty$ and $\eta = O(1)$, and so we have

$$\alpha = -\delta = -\frac{1}{2}. \tag{5.44}$$

Now writing

$$v = v_0(\eta) + o(1) \tag{5.45}$$

and assuming that $tv_t = o(1)$, we are led to the second order ODE for v_0,

$$-\frac{v_0}{2} + \frac{1}{2}\eta(v_0)_\eta = (v_0)_{\eta\eta} + 2v_0(v_0)_\eta, \tag{5.46}$$

with the solution in terms of the error function, namely

$$v_0(\eta) = \frac{e^{\frac{-\eta^2}{4}}}{K + \sqrt{\pi}\,\mathrm{erf}(\eta/2)}, \tag{5.47}$$

where K is an arbitrary constant obtained from the asymptotic mass invariance condition

$$M = \int_{-\infty}^{\infty} v_0(\eta)d\eta. \tag{5.48}$$

This comes out to be

$$K = \sqrt{\pi}(e^M + 1)/(e^M - 1). \tag{5.49}$$

The present asymptotic solution in terms of $v(\eta, t) = t^{1/2}u(x,t)$ was compared by Grundy et al (1994) with the numerical solution. The convergence of the numerical solution to this asymptotic form was found to be slow. However, as for the case $1 < p < 2$, Grundy et al (1994) showed in their appendix A that

$$v(\eta, t) = v_0(\eta) - 0.0639t^{-1/2}(\log t)v_0'(\eta) + O(t^{1/2}) \tag{5.50}$$

may be rewritten as

$$v(\eta, t) = v_0(\eta_1) + O(t^{1/2}) \tag{5.51}$$

where

$$\eta_1 = \eta - 0.0639t^{-1/2}\log t. \tag{5.52}$$

The solution $t^{1/2}u(x,t)$ corresponding to (5.51) was approached much faster by the numerical solution of the same initial value problem, namely (5.1) and (5.2a). We may observe that in the present case ($p = 2$), the self-similar solution (5.45) and (5.47) was also discussed by Escobodo et al (1991) in their analysis of the approach of solution of IVP to the asymptotic linear form.

We may observe in passing that, for $p = 1$, equation (5.5) becomes linear, namely

$$2u_t + u_x - u_{xx} = 0, \tag{5.53}$$

and has the asymptotic profile

$$u(x,t) = \frac{M}{(2\pi t)^{1/2}} \exp\left\{-\frac{1}{2t}\left(x - \frac{t}{2}\right)^2\right\} \quad \text{as } t \to \infty. \tag{5.54}$$

Observe that the travelling wave speed in (5.54) is $\frac{1}{2}$ and not 1 as for $1 < p \leq 2$.

The case $0 < p < 1$ turns out to be rather distinctive. Here Grundy et al (1994), guided by numerical results and some intuition, use a non-translated form of the similarity variable. This choice is compatible with the form for the limiting hyperbolic case $D = 0$ of (5.1).

So, introducing the variable

$$\eta = \frac{x}{t^\nu}, \quad \nu > 0 \tag{5.55}$$

and writing

$$u(x,t) = t^\beta v(\eta, t), \quad \beta < 0, \tag{5.56}$$

equation (5.1) with $D = 1$ transforms to

$$t^{\beta(1-p)}(tv_t + \beta v - \nu\eta v_\eta) + (t(v^p)_t + \beta p v^p - \nu\eta(v^p)_\eta) = t^{\beta(1-p)+1-2\nu}v_{\eta\eta} - t^{\beta(1-p)+1-\nu}v_\eta. \tag{5.57}$$

Since $\beta < 0$ and $p < 1$, the second term on the LHS of (5.57) dominates the first as $t \to \infty$. Considering the possibilities on the RHS it is found that the second term dominates the first and balances the second term on the LHS. Thus, we require that

$$\beta(1-p) + 1 - \nu = 0 \tag{5.58}$$

and

$$\beta(1-p) + 1 - 2\nu < 0. \tag{5.59}$$

The second condition for β and ν is obtained by considering again the mass invariance condition (5.11) which, in the present case, becomes

$$M = \int_{-\infty}^{\infty} \{t^{\beta+\nu} v + t^{\beta p+\nu} v^p\} d\eta. \tag{5.60}$$

For $p < 1$ and $\beta < 0$, M is constant only if

$$\nu = -\beta p. \tag{5.61}$$

This, together with (5.58), determines β and ν:

$$\beta = -1, \quad \nu = p. \tag{5.62}$$

With these values of the parameters, equation (5.57) becomes

$$t(v^p)_t - p(v^p + \eta(v^p)_\eta) + t^{-(1-p)}(tv_t - v - p\eta v_\eta) = t^{-p} v_{\eta\eta} - v_\eta. \tag{5.63}$$

Now assuming that $(tv_t) = o(1)$ as $t \to \infty$ and $\eta = O(1)$, we may write

$$v(\eta, t) = v_0(\eta) + o(1). \tag{5.64}$$

Then, to the leading order, (5.63) gives

$$p\{v_0^p + \eta(v_0^p)'\} = v_0', \tag{5.65}$$

where prime denotes differentiation with respect to η. The general solution of (5.65) is

$$p\eta v_0^p - v_0 = K, \tag{5.66}$$

where K is a constant. An argument similar to that for the case $1 < p < 2$ shows that $K = 0$ and so (5.66) gives

$$v_0 = (p\eta)_+^{1/(1-p)}. \tag{5.67}$$

Here again we must define the solution (5.67) over a finite interval $0 < \eta < \eta_2$ so that the mass invariance condition is not violated. Thus the mass in the present case becomes

$$M = \int_0^{\eta_2} (p\eta)^{p/(1-p)} d\eta \tag{5.68}$$

or

$$\eta_2 = \left\{\frac{M}{1-p}\right\}^{1-p} p^{-p}. \tag{5.69}$$

The outer solution in the present case, $0 < p < 1$, in terms of the original variables x and t may be written as

$$u(x, t) = t^{-1} \left(\frac{px}{t^p}\right)^{1/(1-p)} \{1 + o(1)\}, \tag{5.70}$$

as $t \to \infty$, $x/t^p = O(1)$. The condition $0 < \eta < \eta_2$ defines the range for x:

$$0 < x < \eta_2 t^p. \tag{5.71}$$

As we have pointed out earlier, the outer solutions suffer from two defects – they have a finite support and have discontinuities at the ends – weak or strong. Therefore, these must be supplemented by layers at front and the trailing ends. To that end, we first consider the case $1 < p < 2$. We introduce the variable

$$\eta = \eta_1 + t^{-\mu}\zeta, \ \mu > 0 \tag{5.72}$$

near $\eta = \eta_1$, where $\zeta = O(1)$ and $t \to \infty$. This defines a thin layer on the η scale, which has thickness $t^{-\mu}$ where μ must be found from the analysis. In this layer we seek a solution which varies on the ζ scale. We substitute

$$v(\eta, t) = w(\zeta, t) \tag{5.73}$$

into (5.34) and obtain

$$t^{-\mu}\left\{tw_t - \frac{w}{p} + \left(\mu - \frac{1}{p}\right)\zeta w_\zeta\right\} + t^{-1/p}\{t(w^p)_t + w^p + \mu\zeta(w^p)_\zeta\}$$
$$-\frac{\eta_1}{p}w_\zeta = t^{\mu-\{(2-p)/p\}}w_{\zeta\zeta} + (w^p)_\zeta. \tag{5.74}$$

Typically, in the boundary or trailing layer, the diffusion term is important since it spreads the profile. Assuming w, w_ζ, and $w_{\zeta\zeta}$ are all bounded, we have from (5.74)

$$\mu = \frac{2-p}{p} > 0. \tag{5.75}$$

Writing

$$w(\zeta, t) = w_0(\zeta) + o(1) \tag{5.76}$$

with $tw_t = o(1)$ as $t \to \infty$ and $\zeta = O(1)$ in (5.74), we have, to the leading order,

$$w_0'' + (w_0^p)' + \frac{\eta_1}{p}w_0' = 0 \tag{5.77}$$

where primes denote differentiation with respect to ζ. The matching condition requires that

$$w_0 = \left(-\frac{\eta_1}{p}\right)^{1/(p-1)} \quad \text{as} \quad \zeta \to \infty \tag{5.78}$$

(see (5.40)) while the boundary conditions yield

$$w_0 = w_0' = 0 \quad \text{as} \quad \zeta \to -\infty. \tag{5.79}$$

The solution of (5.77) subject to (5.78) and (5.79) may be found to be

$$w_0 = \left\{ \frac{\eta_1 e^{-(p-1)\eta_1(\zeta-\zeta_0)/p}}{p[1 - e^{-(p-1)\eta_1(\zeta-\zeta_0)/p}]} \right\}^{1/(p-1)} \tag{5.80}$$

This solution is unique within the translational shift ζ_0. This constant seems difficult to obtain analytically and must be obtained by matching the analytic form with the numerical solution.

To determine the leading edge layer near $\eta = 0$, we introduce the variable

$$\chi = \eta t^{\epsilon}, \tag{5.81}$$

where $\epsilon > 0$ will be chosen such that the diffusion term in (5.34) becomes important. To understand how to scale the independent variable v in the trailing edge layer, we write the outer expansion in terms of χ:

$$v(\eta, t) = v_0(\eta) + o(1) = \left(-\frac{\eta}{p}\right)^{1/(p-1)} + o(1) \sim t^{-\epsilon/(p-1)} \left(-\frac{\chi}{p}\right)^{1/(p-1)} \tag{5.82}$$

This suggests the form

$$v(\eta, t) = t^{-\epsilon/(p-1)} W(\chi, t). \tag{5.83}$$

Equation (5.34) now becomes

$$\begin{aligned} tW_t \quad &- \left(\frac{\epsilon}{p-1} + \frac{1}{p}\right) W + \left(\epsilon - \frac{1}{p}\right) \chi W_\chi \\ &+ t^{-1/2} \left\{ t(W^p)_t + \left(1 - \frac{\epsilon p}{p-1}\right) W^p + \left(\epsilon - \frac{1}{p}\right) \chi (W^p)_\chi \right\} \\ &= t^{2\epsilon - (2-p)/p} W_{\chi\chi} + (W^p)_\chi. \end{aligned} \tag{5.84}$$

We assume that $tW_t = o(1)$ as $t \to \infty$, $\chi = O(1)$, and write

$$W(\chi, t) = W_0(\chi) + o(1) \tag{5.85}$$

where we also require that

$$W_0(\chi) \sim (-\chi/p)^{1/(p-1)} \quad \text{as} \quad \chi \to -\infty \tag{5.86}$$

(see (5.82)). Since we insist that the second derivative in (5.84) becomes important as $t \to \infty$ and $\chi = O(1)$, we impose the condition $\epsilon = \frac{2-p}{2p}$. Now (5.84) becomes

$$tW_t - \frac{W}{2(p-1)} - \frac{\chi}{2} W_\chi + t^{-1/2} \left\{ t(W^p)_t + 2W^p - \frac{\chi}{2}(W^p)_\chi \right\} = W_{\chi\chi} + (W^p)_\chi. \tag{5.87}$$

Substituting (5.85) into (5.87) and equating leading order terms we are led to the ODE,

$$W_0'' + (W_0^p)' + \frac{1}{2}\chi W_0' + \frac{W_0}{2(p-1)} = 0, \tag{5.88}$$

where prime denotes derivatives with respect to χ. We must solve (5.88) subject to (5.86) and the condition $W_0 = 0$ as $\chi \to \infty$. It does not seem possible to solve this problem explicitly. However the work of Cazenave and Escobedo (1994) shows that this problem has a family of monotone solutions of which only one has exponential decay of the form $\exp(-\chi^2/4)$ as $\chi \to \infty$. Other solutions have a power law decay in the same limit. Grundy et al (1994) choose the special solution with unique exponential decay as appropriate to the boundary value problem defined above.

The case $0 < p < 1$ also requires finding solutions in the leading-edge and trailing-edge layers. We refer the reader to the original work of Grundy et al (1994) for detials of this case.

It is clear from the present discussion regarding the non-negative mass-invariant solutions of equation (5.1) that considerable analysis in a constructive form may be carried out by using balancing arguments. However, two points may be noted. There is a strong element of intuition in the approach. In order to get a complete picture numerical solutions must aid the analysis. Moreover, the leading and trailing layer solutions must be found to supplement the results when the leading order behaviour single out the so-called outer solution which is valid over a finite interval only. Further, these solutions must be matched appropriately. It may be desirable to seek out composite solution, aided and guided by the approximate solution obtained by the balancing argument. We shall illustrate this kind of approach in the next section.

Grundy et al (1994) carried out detailed numerical study for the three cases that arise, namely, $0 < p < 1$, $p = 1$, $1 < p < 2$, to elucidate and justify the choice of different kinds of variables. They also confirmed the solutions in the leading and trailing layers.

6 Composite Solutions for Nonplanar Burgers Equation

We now consider another equation where the balancing argument distinguishes various limiting behaviours. For each of these behaviours matched asymptotics solutions are found. In one case the outer solution is matched by using singular perturbation approach to the shock layer. In the other balance for which the linear equation gives the dominant behaviour an exact composite solution is found which combines in itself both inviscid and viscous behaviours, appropriately. This approach goes beyond the method of Grundy and his collaborators. In the sequel we follow the work of Sachdev, Srinivasa Rao and Joseph (1999).

We look for the asymptotic (large time) behaviour of the N wave solutions of nonplanar Burgers equation

$$u_t + uu_x + \frac{ju}{2t} = \frac{\delta}{2}u_{xx} \tag{6.1}$$

where $j > -1$, and δ is small; the condition $j > -1$ is imposed by the analysis to follow. The initial conditions relevant to the N wave are

$$\begin{aligned} u(x, t_i) &= (1 - j/2)x/t_i, \quad j < 2, \quad |x| < d_0 \\ &= 0, \qquad \text{otherwise} \end{aligned} \tag{6.2}$$

where d_0 is the initial width of one lobe of the N wave. As in section 5, we introduce the variables

$$t, \eta = xt^a, \quad u = t^c v(\eta, t) \tag{6.3}$$

so that (6.1) now becomes,

$$\left(C + \frac{j}{2}\right) v + a\eta v_\eta + tv_t = \frac{\delta}{2}t^{(2a+1)}v_{\eta\eta} - t^{(a+c+1)}vv_\eta. \tag{6.4}$$

We compare the relative importance of various terms in (6.4) for large t, with the assumption that $tv_t = o(1)$ and $\eta = O(1)$ as $t \to \infty$. We have two possibilities: the first term on the RHS is smaller than the second and vice versa. In the former case we have

$$a + c + 1 = 0, \quad 2a + 1 < 0. \tag{6.5}$$

In this approximation the diffusive term is less important. Motivated by the inviscid N wave solution of (6.1) namely $u = (1 - j/2)x/t$, we choose

$$a = -1, \quad c = 0. \tag{6.6}$$

Equation (6.4) now becomes

$$\frac{j}{2}v - \eta v_\eta + tv_t = \frac{\delta}{2}t^{-1}v_{\eta\eta} - vv_\eta. \tag{6.7}$$

Introducing the large time expansion for (6.7), namely

$$v = v_0(\eta) + o(1) \tag{6.8}$$

we get

$$\frac{j}{2}v_0 - \eta v_0' = -v_0 v_0' \tag{6.9}$$

with the general solution

$$v_0(\eta) = (1 - j/2)\eta + c_1, \tag{6.10}$$

where c_1 is a constant. The anti-symmetry of N wave solution requires that $c_1 = 0$, and so we have

$$v(\eta, t) = (1 - j/2)\eta + o(1), \tag{6.11}$$

as $t \to \infty$, or the corresponding asymptotic behaviour

$$u(x, t) = (1 - j/2)\frac{x}{t} + o(1), \quad j \neq 2 \tag{6.12}$$

We shall pursue this case later, following the work of Crighton and Scott (1979).

For the second case, we again assume that $tv_t = o(1)$, $\eta = O(1)$ as $t \to \infty$, and the first term on the RHS of (6.4) dominates the second, and so

$$2a + 1 = 0, \quad a + c + 1 < 0, \tag{6.13}$$

leading to

$$a = -1/2, c < -1/2. \tag{6.14}$$

In the present case, the convective term in (6.4) is subdominant. Indeed the remaining linear PDE from (6.1) has the N-wave solution

$$u(x, t) = c_1 \frac{x}{t^{(3+j)/2}} \exp(-x^2/2\delta t). \tag{6.15}$$

This solution, when compared with the form (6.3), gives

$$c = -(1 + j/2). \tag{6.16}$$

In the present case the full equation (6.4) becomes

$$-v - \frac{\eta}{2}v_\eta + tv_t = \frac{\delta}{2}v_{\eta\eta} - t^{-(j+1)/2}vv_\eta. \tag{6.17}$$

So, writing the asymptotic solution as

$$v(\eta, t) = v_0(\eta) + o(1) \tag{6.18}$$

and assuming $tv_t = o(1)$, $\eta = O(1)$ as $t \to \infty$, we have from (6.17) the equation governing $v_0(\eta)$:

$$\frac{\delta}{2}v_0'' + \frac{\eta}{2}v_0' + v_0 = 0. \tag{6.19}$$

The general solution of (6.19) is

$$v_0(\eta) = A\eta \exp(-\eta^2/2\delta) + B\eta \exp(-\eta^2/2\delta) \int_0^\eta \frac{\exp(\eta^2/2\delta)}{\eta^2} d\eta, \tag{6.20}$$

where A and B are arbitrary constants. Since the second integral in (6.20) does not converge as $\eta \to \infty$, we choose $B = 0$. We thus have

$$v_0(\eta) = A\eta \exp(-\eta^2/2\delta) \tag{6.21}$$

or, correspondingly, just the exact linear solution (6.15) for $u(x,t)$ in the limit $t \to \infty$.

In the following we improvise upon each of these leading behaviours. To show how the approach due to Grundy and his collaborators may be made more effective, we discuss the work of Crighton and Scott (1979) in detail. This will demonstrate clearly how the leading 'outer' solution (6.12) may be matched to the shock layer. We re-write (6.1) as

$$U_t + UU_X + \frac{jU}{2t} = \frac{\delta}{2} U_{XX} \tag{6.22}$$

and impose the corresponding initial conditions

$$U(X, t_0) = \begin{cases} U_0 X/l_0, & |X| < l_0, \\ 0, & |X| > l_0, \end{cases} \tag{6.23}$$

where U_0 is the amplitude of the initial sawtooth structure and l_0 its wavelength. Now we introduce the variables

$$V = \left(\frac{t}{t_0}\right)^{j/2} \frac{U}{U_0}, \quad x = \frac{X}{l_0}, \tag{6.24}$$

$$T = \begin{cases} 1 + U_0(t - t_0)/l_0, & j = 0, \\ 1 + 2U_0(t_0^{1/2} t^{1/2} - t_0)/l_0, & j = 1, \\ 1 + (U_0 t_0/l_0)\ln(t/t_0), & j = 2. \end{cases} \tag{6.25}$$

Equations (6.22) - (6.23) then change to

$$V_T + VV_x = \epsilon g(T) V_{xx}, \tag{6.26}$$

$$V(x, 1) = \begin{cases} x, & |x| < 1, \\ 0, & |x| > 1, \end{cases} \tag{6.27}$$

where ϵ is an inverse Reynolds number given by

$$\epsilon = \begin{cases} \delta/(2U_0 l_0), & j = 0, \\ \delta/(U_0 l_0 T_0), & j = 1, \\ \delta e^{-1/T_0}/(2U_0 l_0), & j = 2, \end{cases} \tag{6.28}$$

and

$$T_0 = \begin{cases} 2U_0 t_0/l_0, & j = 0, \\ 2U_0 t_0/j l_0, & j = 1, 2. \end{cases} \tag{6.29}$$

The coefficient function $g(T)$ in (6.26) is given by

$$g(T) = 1, \quad \frac{1}{2}(T + T_0 - 1), \quad e^{T/T_0} \tag{6.30}$$

for $j = 0, 1, 2$, respectively. The inviscid solution of (6.26) for all $j = 0, 1, 2$ now simply is

$$V(x,T) = \begin{cases} \frac{x}{T}, & x < T^{1/2} \\ 0, & x > T^{1/2}. \end{cases} \tag{6.31}$$

A formal outer expansion for (6.26) subject to (6.27) quickly shows that all higher order terms $O(\epsilon^n)$, subject to zero initial conditions (see (6.27)), are zero. The shock locus is easily obtained by writing the inviscid form of (6.26) as the conservation law

$$V_T + \left(\frac{1}{2}V^2\right)_x = 0, \tag{6.32}$$

so that

$$\frac{dx}{dT} = \frac{\left[\frac{1}{2}V^2\right]}{[V]} = \frac{1}{2}\frac{x}{T}. \tag{6.33}$$

The symbol [ ] denotes jump in the enclosed quantity. Integration of (6.33) subject to IC $x = 1, T = 1$ gives the shock locus as $x = T^{1/2}$. Now we find the solution in the shock layer and attempt to match it to the outer solution (6.31). For this purpose we introduce the stretched variable centered at the shock location $x = T^{1/2}$,

$$x^* = \frac{x - T^{1/2}}{\epsilon}. \tag{6.34}$$

Equation (6.26) now becomes,

$$\epsilon V_T - \frac{1}{2}\frac{1}{T^{1/2}}V_{x^*} + VV_{x^*} = g(T)V_{x^*x^*}. \tag{6.35}$$

Equation (6.35) has a singular character, it is solved in the form

$$V = V_0^*(x^*, T) + \epsilon V_1^*(x^*, T) + \cdots. \tag{6.36}$$

Substituting (6.36) into (6.35) and equating terms of order ϵ^0 and ϵ^1 on both sides, we have

$$\left(V_0^* - \frac{1}{2T^{1/2}}\right)(V_0^*)_{x^*} = g(T)(V_0^*)_{x^*x^*}, \tag{6.37}$$

$$g(T)(V_1^*)_{x^*x^*} + \left(\frac{1}{2}T^{-1/2} - V_0^*\right)_{x^*}(V_1^*) - V_1^*(V_0^*)_{x^*} = (V_0^*)_T. \tag{6.38}$$

It is straightforward to check that (6.37) has the solution

$$V_0^* = \frac{1}{2}T^{-1/2}\left[1 - \tanh\frac{x^* - A(T)}{4T^{1/2}g(T)}\right], \tag{6.39}$$

which, to zeroth order, matches the outer solution (6.31) written in terms of inner variable, $V_0 = \frac{1}{T^{1/2}} + \epsilon x^*/T$, as $\epsilon \to 0$ and $x^* \to -\infty$, and which tends to zero as $x^* \to +\infty$ according to intermediate limits. The function $A(T)$, which arises as 'constant' of integration when we solve (6.37), must be found from higher order approximations; it gives the shock wave displacement due to viscous diffusion. To solve (6.38), we introduce the variable

$$y = \frac{x^* - A(T)}{4T^{1/2}g(T)} \tag{6.40}$$

instead of x^* (see (6.39)) so that it becomes,

$$\frac{1}{2}(V_1^*)_{yy} + \text{tanh}y(V_1^*)_y + \text{sech}^2 y V_1^* = 8Tg(T)(V_0^*)_T. \tag{6.41}$$

Two independent solutions of the homogeneous part of (6.41) are $\text{sech}^2\, y$ and $y\,\text{sech}^2 y + \text{tanh}y$. A particular integral of (6.41) is found to be

$$\begin{aligned} V_1^* \;=\; & \frac{dA}{dt} - T^{-1/2}g(T)(y^2\text{sech}^2 y + 2y\tanh y - 1 - 2y) \\ & + G(T)(y\text{sech}^2 y + \tanh y) + K(T)\text{sech}^2 y \\ & + 4T^{1/2}\frac{dg}{dT}\Bigg\{ y - (\ln\cosh y)\tanh y + \tanh y \\ & + \text{sech}^2 y\left[y\left(\ln 2 + \frac{1}{2}y\right) + \frac{1}{2}\text{diln}(1+e^{2y})\right]\Bigg\} \end{aligned} \tag{6.42}$$

where

$$\text{diln}(x) = -\int_1^x \frac{\ln t}{t-1}dt,$$

(see Abramowitz and Stegun (1964), p.1004). The functions $G(T)$ and $K(T)$ are found by matching. The solution V_1^* must be matched to $V_1 = 0$ in the intermediate limit, that is, $V_1 = 0$ as $\epsilon \to 0$, $y \to -\infty$ when $x < T^{1/2}$, and again to $V_1 = 0$ as $\epsilon \to 0$ and $y \to \infty$ when $x > T^{1/2}$. Following this procedure the following two equations are obtained for $A(t)$ and $G(T)$:

$$\frac{dA}{dt} - AT^{-1} + T^{-1/2}g(T) - G(T) - 4T^{1/2}(1+\ln 2)\frac{dg}{dT} = 0, \tag{6.43}$$

$$\frac{dA}{dt} + T^{-1/2}g(T) + G(T) + 4(1+\ln 2)T^{1/2}\frac{dg}{dT} = 0. \tag{6.44}$$

The equation for A alone is easily obtained from (6.43) - (6.44), namely

$$\frac{dA}{dT} + T^{-1/2}g(T) - \frac{1}{2}AT^{-1} = 0, \tag{6.45}$$

which, on integration, give

$$A(T) = A(1) - T^{1/2}\int_1^T \frac{g(t)}{t}dt. \tag{6.46}$$

The constant $A(1)$ is found to be zero by Crighton and Scott (1979) by matching (6.42) and (6.46) to an embryonic shock which evolves from the discontinuous initial conditions before it assumes a steady Taylor shock structure. Thus, the function $A(T)$ from (6.46) is found, for different geometries, to be

$$A(T) = -T^{1/2}\ln T \text{ for } j=0, \tag{6.47}$$

$$A(T) = -\frac{1}{2}T^{1/2}\{T-1+(T_0-1)\ln T\} \text{ for } j=1, \tag{6.48}$$

$$A(T) = -T^{1/2}\left\{\mathrm{Ei}\left(\frac{T}{T_0}\right)-\mathrm{Ei}(T_0^{-1})\right\} \text{ for } j=2 \tag{6.49}$$

where Ei is the exponential integral

$$\mathrm{Ei}(x) = \int_{-\infty}^{x} t^{-1}e^{t}dt$$

(Abramowitz and Stegun (1964), p. 228).

Crighton and Scott (1979) state that for planar and cylindrical cases, nothing short of the knowledge of the full solution would be required for further progress, while they are in a position to find the constant in the so-called old-age solution (see (6.15)) for $j=2$ by a more complicated perturbation argument.

Three assumptions underly the singular perturbation analysis of Crighton and Scott (1979): (i) the shock displacement due to diffusion remains small, (ii) the shock remains thin in comparison with the characteristic scale of the inviscid flow, (iii) the Taylor-type shock solution remains valid in the shock layer. For different values of $j = 0,1,2$, one or more of these assumptions are violated at orders of time, defined by $\ln T = O(\epsilon^{-1})$ or $T = O(\epsilon^{-1})$, and then recourse must be made to numerical solution. We refer the reader to the original sophisticated singular perturbation analysis of Crighton and Scott (1979), where they make some further progress for the case of spherical geometry, $j=2$.

Now we turn to another apporach due to Sachdev and his collaborators which works very well except during the initial time when, in effect, the shock is very thin-embryonic or Taylor type. This solution, exact in the sense that the planar solution is exact (see Section 2 for the derivation) holds for all time except some finite initial time. In this sense, this work complements the work of Crighton and Scott (1979). The two approaches together give analytic solutions which span the entire evolution of the N-wave from its initial discontinuous form (6.27).

We begin with the second possibility pointed out by the balancing argument (see (6.14)-(6.15)), namely, when the nonlinear convection term is sub-dominant and the linear solution gives the leading behaviour. This solution is given explicitly by (6.15). First we show how this solution may be improved upon in the manner of Grundy et al. (1994). We recall that, to zeroth order, the solution in terms of the function $v(x/t^{1/2}, t) = t^{(1+j/2)}u(x,t)$, is given simply by $v_0(\eta)$, defined by (6.21). We write

$$v(\eta, t) = v_0(\eta, t) + \epsilon(\eta, t), \tag{6.50}$$

where $\epsilon \to 0$ as $t \to \infty$. Putting (6.50) into (6.17) and ignoring higher order terms we have

$$-\epsilon - \frac{\eta}{2}\epsilon_\eta + t\epsilon_t = \frac{\delta}{2}\epsilon_{\eta\eta} - A^2 t^{-(j+1)/2}\eta(1-\eta^2/\delta)\exp(-\eta^2/\delta). \tag{6.51}$$

If we write

$$\epsilon(\eta, t) = t^{-(j+1)/2} f(\eta), \tag{6.52}$$

equation (6.51) changes to

$$f''(\eta) + \frac{\eta}{\delta}f'(\eta) + \frac{(3+j)}{\delta}f(\eta) = \frac{2A^2}{\delta}\eta(1-\eta^2/\delta)\exp(-\eta^2/2\delta) \equiv R, \text{ say.} \tag{6.53}$$

The homogeneous part of (6.53) has the (linearly independent) solutions

$$f_1(\eta) = \exp(-\eta^2/2\delta)\ {}_1F_1\left(-\frac{j+2}{2}, \frac{1}{2}; \frac{\eta^2}{2\delta}\right), \tag{6.54}$$

$$f_2(\eta) = \frac{\eta}{(2\delta)^{1/2}}\exp(-\eta^2/2\delta)\ {}_1F_1\left(\frac{1-j}{2}, \frac{3}{2}; \frac{\eta^2}{2\delta}\right), \tag{6.55}$$

where

$${}_1F_1(a, c; z) = \sum_{k=0}^{\infty} A_k z^k, \tag{6.56}$$

$$A_k = [a(a+1)\ldots(a+k-1)]/[c(c+1)\ldots(c+k-1)k!], \tag{6.57}$$

is the solution of the confluent hypergeometric equation. Variation of parameters gives a particular solution of (6.53) as

$$f_p(\eta) = f_1(\eta)\int_0^\eta \frac{-f_2(s)R(s)}{w(f_1, f_2)}ds + f_2(\eta)\int_0^\eta \frac{f_1(s)R(s)}{w(f_1, f_2)}ds. \tag{6.58}$$

The general solution of (6.53), therefore, is

$$f(\eta) = c_1 f_1(\eta) + c_2 f_2(\eta) + f_p(\eta). \tag{6.59}$$

However, the N wave solution we seek must be antisymmetric with respect to the origin. Therefore, $c_1 = c_2 = 0$. Equation (6.59) then reduces to

$$f(\eta) = f_p(\eta). \tag{6.60}$$

The solution $v(\eta)$ of (6.17), in the limit $t \to \infty$, has the form

$$v(\eta) = A\eta\exp(-\eta^2/2\delta) + t^{-(j+1)/2} f(\eta), \tag{6.61}$$

where $f(\eta) = f_p(\eta)$ is given by (6.58), we have

$$u(x,t) = t^{-(1+j/2)}\left[A\frac{x}{t^{1/2}}\exp(-x^2/2\delta t) + t^{-(1+j)/2}f(x/t^{1/2}) + O(t^{-(j+1)})\right], \tag{6.62}$$

as $t \to \infty$. It is not difficult to check that the large time behaviour (6.62) of the N wave solution of plane Burgers equation comes out as a special case of (6.62) (see Srinivasa Rao (1999)). We showed in section 2 that the first term in (2.30) referred to as $u^\infty(x,t)$ in the solution of the N-wave initial value problem for the Burgers equation (2.1) is in fact an exact solution of the latter, even though the asymptotic form (2.30) shows an error $O(1/t)$. We first write this solution as

$$\begin{aligned} u^\infty(x,t) &= \frac{x/t}{1+(t/t_0)^{1/2}\exp(x^2/2\delta t)} \\ &= A\frac{x}{t^{3/2}}\exp(-x^2/2\delta t)\left\{\frac{1}{1+\frac{A}{t^{1/2}}\exp(-x^2/2\delta t)}\right\}, \end{aligned} \tag{6.63}$$

where $A = t_0^{1/2}$. If we expand (6.63) for large t, we get

$$u^\infty(x,t) = A\frac{x}{t^{3/2}}\exp(-x^2/2\delta t)\left\{1 - \frac{A}{t^{1/2}}\exp(-x^2/2\delta t) + \cdots\right\} \tag{6.64}$$

which is a special case of (6.62) with $j = 0$.

In a series of papers, Sachdev and his collaborators (1993, 1994, 1996, 1999, 2000) attempted to nonlinearise the linear asymptotic limiting solution and accommodate the exact inviscid solution as well, whenever possible. This was inspired and motivated by the exact solution $u^\infty(x,t)$ (see (2.30)) for the plane Burgers equation. We follow here the work of Sachdev, Srinivasa Rao and Joseph (1999). The inviscid form of (6.1) with $j = m/n$ (where m and n are integers with no common divisors), namely

$$u_t + uu_x + \frac{m}{2nt}u = 0, \tag{6.65}$$

has an exact N wave solution

$$u(x,t) = \frac{1}{(2n/(2n-m))}\frac{x}{t^{1/2}}\frac{1}{t^{1/2}}. \tag{6.66}$$

The linear (old age) form of (6.1)

$$u_t + \frac{m}{2nt}u = \frac{\delta}{2}u_{xx} \tag{6.67}$$

has an anti-symmetric solution

$$u(x,t) = C\frac{x}{t^{1/2}}\frac{1}{t^{(2n+m)/2n}}e^{-x^2/2\delta t}, \tag{6.68}$$

where C is an arbitrary constant. Motivated by the form (2.30) of the N wave solution for the plane Burgers equation with $j = m = 0$, we write

$$u(x,t) = \frac{(2\delta)^{1/2}\xi}{V(\eta,T)}, \tag{6.69}$$

where

$$\xi = x/(2\delta t)^{1/2}, \quad \eta = \xi^2, \quad T = t^{1/2}, \tag{6.70}$$

and

$$V(\eta,T) = \sum_{i=0}^{\infty} f_i(T)\frac{\eta^i}{i!}. \tag{6.71}$$

Substituting (6.69) into (6.1) we have the following PDE for $V(\eta,T)$:

$$V(jV - TV_T) + (2T - V)(V - 2\eta V_\eta) + 3VV_\eta + 2\eta VV_{\eta\eta} - 4\eta V_\eta^2 = 0. \tag{6.72}$$

Now we put (6.71) into (6.72) and equate coefficients of $T^i (i = 0, 1, \ldots k, \ldots)$ to zero. We obtain

$$3f_1 + 2T - Tf_0' + (j-1)f_0 = 0, \tag{6.73}$$

$$5f_0f_2 + 2jf_0f_1 - f_1^2 - T(f_0f_1)' - 2Tf_1 = 0, \tag{6.74}$$

$$7f_0f_3 + [2(j+1)f_0 - Tf_0' - 3f_1 - 6T]f_2 - Tf_0f_2' + 2f_1((j+1)f_1 - Tf_1') = 0, \tag{6.75}$$

$$\begin{aligned}(j-1)\sum_{k=0}^{i} \frac{f_k}{k!}\frac{f_{i-k}}{(i-k)!} &- T\sum_{k=0}^{i}\frac{f_k}{k!}\frac{f_{i-k}'}{(i-k)!} + 2T\frac{f_i}{i!} + 3\sum_{k=0}^{i}\frac{f_k}{k!}\frac{f_{i-k+1}}{(i-k)!} \\ &-4T\frac{f_i}{(i-1)!} + 2\sum_{k=0}^{i-1}\frac{f_k}{k!}\frac{f_{i-k}}{(i-k-1)!} + 2\sum_{k=0}^{i-1}\frac{f_k}{k!}\frac{f_{i-k+1}}{(i-k-1)!} \\ &-4\sum_{k=0}^{i-1}\frac{f_{k+1}}{k!}\frac{f_{i-k}}{(i-k-1)!} = 0, k = 3, 4, 5, \ldots .\end{aligned} \tag{6.76}$$

Equations (6.73) - (6.76) form an infinite system of coupled nonlinear ODEs. This system has a special structure. If f_0, f_1 and f_2 can be found from (6.73)-(6.74) with the help of information gleaned from both the inviscid and the linear limiting behaviours, all other f_i's can be determined successively from (6.76). However, in general, it does not seem possible to accommodate the inviscid behaviour; the scheme does admit nonlinearisation of the (linear) old age solution though. If we introduce the variable

$$\tau = T^{1/n} = t^{1/2n} \tag{6.77}$$

into (6.73)-(6.74), we get

$$3f_1 + 2\tau^n - \frac{\tau}{n}f_0' + \left(\frac{m-n}{n}\right)f_0 = 0, \tag{6.78}$$

$$5f_0f_2 + \frac{2m}{n}f_0f_1 - f_1^2 - \frac{\tau}{n}(f_0f_1)' - 2\tau^n f_1 = 0. \tag{6.79}$$

The inviscid and old age solutions (6.66) and (6.68) suggest that f_0 should start with the term $t^{1/2} = \tau^n$ and end with $t^{(2n+m)/2n} = \tau^{2n+m}$. Thus we write the following form for $f_i(i = 0, 1, 2)$:

$$f_0(\tau) = \tau^n \sum_{i=0}^{m+n} a_i \tau^i, \tag{6.80}$$

$$f_1(\tau) = \tau^n \sum_{i=0}^{m+n} b_i \tau^i, \tag{6.81}$$

$$f_2(\tau) = \tau^n \sum_{i=0}^{m+n} c_i \tau^i. \tag{6.82}$$

Substituting (6.80)-(6.82) into (6.73)-(6.74) and equating coefficients of different powers of τ to zero we obtain the following relations among the coefficients a_i, b_i and c_i:

$$b_0 = c_0 = 0, \tag{6.83}$$

$$a_{n+m} = b_{n+m} = c_{n+m}, \tag{6.84}$$

$$b_i = [(2n + i - m)/3n]a_i, \quad i = 1, 2, \ldots, (m + n), \tag{6.85}$$

and

$$5\alpha_i + [(2m - 2n - i)/n]\beta_i - 2b_i - \gamma_i = 0, \quad i = 1, 2, \ldots (n + m), \tag{6.86}$$

$$5\alpha_i + [(2m - 2n - i)/n]\beta_i - \gamma_i = 0, \quad i = n + m + 1, \ldots 2(n + m) - 1, \tag{6.87}$$

where α_i, β_i and γ_i are defined by

$$\begin{aligned} \alpha_i &= \sum_{q=0}^{i} a_q c_{i-q}, i = 0, 1, \ldots, (n + m) \\ &= \sum_{q=i-(n+m)}^{n+m} a_q c_{i-q}, i = (n + m + 1), \ldots, 2(n + m), \end{aligned} \tag{6.88}$$

$$\begin{aligned} \beta_i &= \sum_{q=0}^{i} a_q b_{i-q}, i = 0, 1, \ldots, (m + n) \\ &= \sum_{q=i-(m+n)}^{n+m} a_q b_{i-q}, i = (n + m + 1), \ldots 2(n + m), \end{aligned} \tag{6.89}$$

$$\gamma_i = \begin{cases} \sum_{q=0}^{i} b_q b_{i-q}, i = 0, 1, \ldots, (n + m), \\ \sum_{q=i-(n+m)}^{n+m} b_q b_{i-q}, i = (n + m + 1), \ldots, 2(n + m). \end{cases} \tag{6.90}$$

Substituting in (6.87) for b_i in terms of c_i via (6.84)-(6.85), we get the following algebraic equations for the coefficients $c_1, c_2, \ldots, c_{n+m-1}$.

$$5 \sum_{q=i-(n+m)}^{n+m} a_q c_{i-q} + \frac{(2m - 2n - i)}{n} \sum_{q=i-(m+n)}^{m+n} \frac{(2n + i - q - m)}{3n} a_q a_{i-q}$$

$$- \sum_{q=i-(n+m)}^{n+m} \frac{(2n+i-q-m)}{3n}\frac{(2n+q-m)}{3n} a_q a_{i-q} = 0,$$
$$i = m+n+1, \ldots, 2(n+m)-1. \quad (6.91)_i$$

Since (6.84) gives $c_{n+m} = a_{n+m}$, we can find c_{n+m-1} from $(6.91)_i$, $i = 2(n+m)-2$, in terms of a_i. In the same manner $(6.91)_i$ gives us $c_{n+m-1}, c_{n+m-2}, \ldots, c_1$ in terms of $a_i, i = 1, 2, \ldots, (m+n)$. The expressions for c_i are used in (6.86) to get $(n+m)$ algebraic equations for $n+m-1$ unknows $a_1, a_2, \ldots, a_{n+m-1}$; the constant $C = a_{n+m}^{-1}$ is the old age constant appearing in the linear solution (6.68) and must be found numerically. By assumption (see (6.80)-(6.82)), the limit $f_i/\tau^{2n+m} \to a_{n+m}$ as $\tau \to \infty$ is true for $i = 0, 1, 2$. It can be shown by an induction argument that $f_i/\tau^{2n+m} \to a_{n+m}$ as $\tau \to \infty$ for all $i \geq 3$ (see Sachdev, Joseph and Nair (1994)). We conclude that the solution so constructed has the appropriate old age limit as $t \to \infty$, provided we identify a_{n+m} with C^{-1} in (6.68). We give details for some special cases.

(i) $j = 1$.

In this case, $\tau = T = t^{1/2}$ and so equations (6.73)-(6.74) become

$$3f_1 - \tau f_0' + 2\tau = 0, \quad (6.92)$$
$$5f_0 f_2 - f_1^2 + 2f_0 f_1 - 2\tau f_1 - \tau(f_0 f_1)' = 0. \quad (6.93)$$

We write the solution of (6.92)-(6.93) as

$$f_0 = a_0\tau + a_1\tau^2 + a_2\tau^3, \quad (6.94)$$
$$f_1 = b_0\tau + b_1\tau^2 + b_2\tau^3, \quad (6.95)$$
$$f_2 = c_0\tau + c_1\tau^2 + c_2\tau^3, \quad (6.96)$$

(see (6.80)-(6.82)). Substituting (6.94)-(6.96) into (6.92) and equating coefficients of different powers of τ to zero, we infer that

$$b_0 = \frac{1}{3}(a_0 - 2), b_1 = 2a_1/3, b_2 = a_2. \quad (6.97)$$

Similarly, from (6.93)-(6.96), we have

$$5a_0c_0 - b_0^2 - 2b_0 = 0, \quad (6.98)$$
$$5a_0c_1 + 5a_1c_0 - 2b_0b_1 - a_0b_1 - a_1b_0 - 2b_1 = 0, \quad (6.99)$$
$$5a_0c_2 + 5a_1c_1 + 5a_2c_0 - b_1^2 - 2b_0b_2 - 2a_0b_2 - 2a_1b_1 - 2b_0a_2 - 2b_2 = 0, \quad (6.100)$$
$$5a_1c_2 + 5a_2c_1 - 2b_2b_1 - 3a_2b_1 - 3a_1b_2 = 0, \quad (6.101)$$
$$5a_2c_2 - b_2^2 - 4a_2b_2 = 0. \quad (6.102)$$

From (6.97) and (6.102) we get

$$c_2 = a_2. \tag{6.103}$$

On using (6.97) and (6.103), (6.101) gives

$$c_1 = \frac{4}{15}a_1 \tag{6.104}$$

From (6.97) and (6.98) we get

$$c_0 = \frac{1}{45a_0}(a_0^2 + 2a_0 - 8). \tag{6.105}$$

Substituting $c_i (i = 0, 1, 2)$ from (6.103)-(6.105) and b_i from (6.97) into (6.99), we get a quadratic in a_1 with the roots $a_1 = 2, 0$. For the choice $a_1 = 2$ we can find from (6.97) and (6.103)-(6.105) the values

$$\begin{aligned} a_0 &= 2, a_1 = \pm 3a_2^{1/2}, \\ b_0 &= 0, b_1 = \pm 2a_2^{1/2}, b_2 = a_2, \\ c_0 &= 0, c_1 = \pm\frac{4}{5}a_2^{1/2}, c_2 = a_2. \end{aligned} \tag{6.106}$$

The solution with negative sign in (6.106) gives a singularity in the Reynolds number (see (6.111) below); we therefore arrive at the solution

$$\begin{aligned} f_0(t) &= 2t^{1/2} + 3a_2^{1/2}t + a_2t^{3/2}, \\ f_1(t) &= 2a_2^{1/2}t + a_2t^{3/2}, \\ f_2(t) &= \frac{4}{5}a_2^{1/2}t + a_2t^{3/2}. \end{aligned} \tag{6.107}$$

The solution functions $f_i (i = 0, 1, 2)$ in (6.107) along with (6.106) involve the old age constant a_2^{-1} which must be found numerically.

Now we turn to the other possibility, namely $a_1 = 0$. Following the procedure for the case $a_1 = 2$, we arrive at the quadratic for a_0, namely

$$2a_0^2 + a_0 - 1 = 0, \tag{6.108}$$

with the roots $a_0 = -1, \frac{1}{2}$. The case $a_0 = -1$ is again ruled out as before. $a_0 = 0$ now gives the other coefficients as

$$\begin{aligned} a_0 &= 1/2, \quad a_1 = 0, \\ b_0 &= -1/2, \quad b_1 = 0, \quad b_2 = a_2, \\ c_0 &= -3/10, \quad c_1 = 0, \quad c_2 = a_2. \end{aligned} \tag{6.109}$$

and hence the coefficient functions

$$\begin{aligned} f_0(t) &= \frac{1}{2}t^{1/2} + a_2 t^{3/2}, \\ f_1(t) &= -\frac{1}{2}t^{1/2} + a_2 t^{3/2}, \\ f_2(t) &= -\frac{3}{10}t^{1/2} + a_2 t^{3/2}. \end{aligned} \tag{6.110}$$

We note that the coefficients in $f_0(t)$ must all be positive since the Reynolds number (see (6.111) below) involves this coefficient and requires $f_0(\tau)$ to have the same sign in $(0, \infty)$ to avoid singularity at a finite value of τ. The quantity that describes the decay of the wave is referred to as the Reynolds number and is defined as

$$Re = \frac{1}{\delta} \int_0^\infty u dx. \tag{6.111}$$

It is essentially (nondimensional) area under one of the lobes of the N wave. Integrating (6.1) from 0 to ∞ and using the condition $u(\infty, t) = 0$ we get

$$\frac{dR}{dt} + \frac{j}{2t}R = -\frac{1}{2}u_x(0,t). \tag{6.112}$$

We use the solution (6.69) to get $u_x(0,t)$ in (6.112) and integrate the latter from t_0 to t.

We get two possible forms of the Reynolds number from the two forms of $f_0(\tau)$, given by (6.107) and (6.110). However, a comparison with the numerical solution shows that the latter expression gives a closer agreement. The explicit expression for $R(t)$ in this case is

$$R(t) = t^{-1/2}(R_0 t_0^{1/2} - h(t) + h(t_0)), \tag{6.113}$$

where

$$h(s) = \left(\sqrt{2}\arctan\left(\sqrt{2a_2 s}\right)/\sqrt{a_2}\right). \tag{6.114}$$

We recall that a_2^{-1} is the old age constant in the linear solution.

Similarly, explicit results for other values of $j = \frac{1}{2}, \frac{1}{3}, \frac{1}{4}$ may be found in Sachdev, Srinivasa Rao and Joseph (1999). The analysis for negative values of j does not seem to work. Instead, it was found that a simple sum of contributions from the inviscid and old age solutions (which is what holds exactly for the planar case $j = 0$) gives adequate results for the Reynolds number. We may also remark that a comparison of singular perturbation solution of Crighton and Scott (1987) with the numerical solution shows that it gives excellent results in the early part of the evolution of the N-wave when none of the assumptions underlying their analysis, referred to earlier, is violated.

We also refer here to the work of Sachdev and Srinivasa Rao (2000) who studied the N-wave problem for the modified Burgers equation

$$u_t + u^n u_x = \frac{\delta}{2}u_{xx}, \quad n \geq 2. \tag{6.115}$$

Here, for n even, the problem is more complicated since the N wave does not evolve symmetrically. However, the series approach similar to one adopted for the nonplanar GBE in this section gives good results after the node of the N-wave (which moves in the early part of the N wave evolution) has 'essentially' come to rest. The early thin shock regime of the N wave for (6.115) was studied by Harris (1996), using the singular perturbation approach of Crighton and Scott (1979) for the nonplanar Burgers equation and discussed in detail earlier in this section. The analysis of Harris (1996) and Sachdev and Srinivasa Rao (2000) fully delineates the evolution of the N wave governed by the modified Burgers equation (6.115).

7 Conclusion

In this review we have shown how 'exact' asymptotic solutions represent limiting behaviour of classes of initial value problems. These solutions may be travelling waves, self-similar solutions, or more 'generalised' form of solutions which arise from the reduction of original nonlinear PDEs to an infinite system of ODEs. Their asymptotic character, as $t \to \infty$, is shown either analytically or numerically. Other approaches include balancing argument and singular perturbation methods which, however, lead to approximate asymptotic forms. These solutions serve a very useful purpose in providing analytic underpinnings to numerical studies. The analytic methods reported here may be extended to systems of nonlinear PDEs for which, however, analysis will become much more complicated.

This work was carried out under Nonlinear Studies Group Programme at the Indian Institute of Science, supported by ISRO and DRDO (India).

References

Abramowitz, M. and Stegun, I.A. (1965). Handbook of Mathematical Functions, Dover, New York.

Barenblatt, G.I. and Zel'dovich, Y.B. (1972). Self-similar solutions as intermediate asymptotics, Ann. Rev. Fluid Mech., 4, 285.

Barenblatt, G.I. (1996). Scaling, Self-similarity and Intermediate Asymptotics. Cambridge University Press, Cambridge.

Bellman, R. (1955). Research problem, Bull. Amer. Math. Soc., 61, 192.

Bukiet, B., Pelesko, J., Li, X.L. and Sachdev, P.L. (1996). A characteristic based numerical method with tracking for nonlinear wave equations, Comp. Math. Applic., 31, 75.

Calogero, F. and De Lillo, S. (1989). The Burgers equation on the semi infinite and finite intervals, Nonlinearity, 2, 37.

Calogero, F. and De Lillo, S. (1991). The Burgers equation on the semiline with general boundary condition at the origin, J. Math. Phys., 32, 99.

Cazenave, and Escadedo, M. (1994). A two-parameter shooting problem for a second order differential equation, J. Diff Eq. 113, 418.

Clothier, B.E., Knight, J.H. and White, I. (1981). Burgers' equation: Application to field constant flux infiltration, Soil Science, 132, 4, 255.

Crighton, D.G. and Scott, J.F. (1979). Asymptotic solutions of model equations in nonlinear acoustics, Phil. Trans. Roy. Soc. A. 292, 101.

Crighton, D.G. (1979). Model equations of nonlinear acoustics A. Rev. Fluid Mech. 11, 11.

Dawson, C.N., Van Duijn and Grundy, R.E. (1996). Large time asymptotics in contaminant transport in porous media, SIAM J. App. Math., 56, 4, 965.

Escobedo, M. and Zuazua, E. (1991). Large time behaviour of a convection diffusion equation in R^N, J. Funct. Anal., 100, 119.

Escobedo, M. and Grundy, R.E. (1996). Asymptotic profiles of a decaying contaminant in transport through a porous medium, Eur. J. Appl. Math. 7, 395.

Grundy, R.E., Sachdev, P.L. and Dawson, C.N. (1994). Large Time Solution of an Initial Value Problem for a Generalized Burgers Equation in Nonlinear Diffusion Phenomenon, (P.L. Sachdev and R.E. Grundy (Eds.)), 68, Narosa Publishing House, New Delhi.

Grundy, R.E., Van Duijn, C.J. and Dawson, C.N. (1994). Asymptotic profiles with finite mass in one-dimensional contaminant transport through porous media: the fast reaction case, Q. J. Mech. Appl. Math., 47, 1, 69.

Harris, S.E. (1996). Sonic shocks governed by the modifed Burgers equation, Eur. J. Appl. Math., 7, 201.

Hopf, E. (1950). The partial differential equation $u_t + uu_x = \mu u_{xx}$, Comm. Pure. Appl. Math., 3, 201.

Joseph, K.T. and Sachdev, P.L. (1993). Exact analysis of Burgers equation on semiline with flux condition at the origin, Int. J. Non-linear Mech., 28, 6, 627.

Levinson, N. (1969). Asymptotic behaviour of solutions of nonlinear differential equations, Stud. Appl. Math., 49, 285.

Lighthill, M.J. (1956). Viscosity Effects in Sound Waves of Finite Amplitude. In Surveys in Mechanics, ed. G.K. Batchelor and R.M. Davies, 250-351, Cambridge University Press.

Murray, J.D. (1970). Perturbation effects on the decay of discontinuous solutions of non-linear first order wave equations, SIAM J. Appl. Math., 19, 135.

Natalini, R. and Tesei, A. (1992). On a class of perturbed conservation laws, Advances in Appl. Math., 13, 429.

Sachdev, P.L. (1987). Nonlinear Diffusive Waves, Cambridge University Press, Cambridge.

Sachdev, P.L. (1991). Nonlinear Ordinary Differential Equations and Their Applications, Marcel Dekker, Inc., New York.

Sachdev, P.L., Joseph, K.T. and Nair, K.R.C. (1994). Exact N-wave solutions for the non-planar Burgers equation, Proc. R. Soc. Lond. A, 445, 501.

Sachdev, P.L., Joseph, K.T. and Mayil Vaganan, B. (1996). Exact N-wave solutions of generalized Burgers equations, Stud. Appl. Math. 97, 349.

Sachdev, P.L., Ch. Srinivasa Rao and Joseph, K.T. (1999). Analytic and numerical study of N-waves governed by the nonplanar Burgers equation, Stud. Appl. Math., 103, 89.

Sachdev, P.L. (2000). Self-similarity and Beyond-exact Solutions of Nonlinear Problems, Chapman & Hall (CRC Press), London.

Sachdev, P.L. and Chi. Srinivasa Rao (2000). N-wave solution of modified Burgers equations, Appl. Math. Letters, 13, 1.

Van Duijn, C.J., Grundy, R.E. and Dawson, C.N. (1997). Large time profiles in reactive solute transport, Transport in Porous Media, 27, 57.

Whitham, G.B. (1974). Linear and Nonlinear Waves, Wiley, New York.

Zel'dovich, Y.B. and Raizer, Y.P. (1966). Physics of Shock Waves and High Temperature Hydro-dynamic Phenomena, Vol. II, Academic Press, New York.

Modern Applied Mathematics
Editor: J.C. Misra

Mixed Boundary Value Problems: Singular Integral Equations and Special Functions

R.P. Srivastav

Applied Mathematics and Statistics, State University of New York at Stony Brook
Stony Brook, New York-11794-3600

Let me begin on a personal note. I wish to thank Professor J.C. Misra for his invitation to contribute an article for the Golden Jubilee Commemoration volume of Indian Institute of Technology, Kharagpur. In a certain sense, the work reported here traces its origin to I.I.T. Kharagpur. It is appropriate to tell that story. More than forty years ago, I was visiting Kharagpur. Dr. Abinash Chandra Srivastava, who was a lecturer in Mathematics at that time, asked me to accompany him to meet a distinguished visitor, Kampe de Feriet. Since I had not heard the name before, I asked Abinash what made Kampe de Feriet "distinguished". Abinash pulled out a book Fourier Transforms by Ian N. Sneddon[16] and showed me two references to Kampe de Feriet. I had not seen that book either, though I was familiar with Titchmarsh's Theory of Fourier Integrals[27]. I bought a copy of Sneddon's book, started reading it and also started dreaming. Several things happened rather rapidly and a few months later I found myself working with Professor Sneddon at University of Glasgow, Scotland. While my stay in Glasgow was relatively short (October 1961-June 1963), it formed the basis for a life-long relationship. Professor Sneddon passed away on November 4, 2000. My own interests have been evolving in other directions; but, I could not resist the temptation to seize this opportunity to pay a tribute to Professor Sneddon for his contribution to mathematics and to the education of many of my generation.

1 Introduction

Mathematical models of many a physical phenomenon lead to Laplace's equation for an auxiliary function. The quantities of interest can be calculated by solving for functions, which satisfy Laplace's equation in the interior of a region and specified boundary conditions. These problems are more amenable to analytical methods when the boundary of the domain can be identified with the coordinate surfaces. The boundary conditions are often referred to as the Dirichlet (boundary values of the function), Neumann(boundary values of normal derivatives) and Robin (a linear relation between the values of the function and its normal derivative) and mixed (the function on a part and the normal derivative on the other part but on the same coordinate surface).[1]

Integral equations have been an important tool for the formulation and solution of a variety of problems in physics and engineering. Of concern here is a special subset called dual integral equations and Cauchy singular integral equations. The former owes its name to the nature of the domain. The latter name derives from the integral being a Cauchy principal value. These equations have been studied intensively for a long time; for, they provide an effective analytical approach to mixed boundary value problems with suitable geometry and qualitatively form the basis for approximate numerical solutions of problems, which are not amenable to mathematical analysis. I shall present here some recent work at Stony Brook and sketch a few topics, which in my opinion, need to be further studied. By design, I talk of ideas, which are half-baked. Our bias is function-theoretic and the style is informal. In particular, we interchange the orders of integration and summation freely without appealing to uniform convergence or Fubini's theorem.

In the next section we begin with a simple-minded approach to Laplace's Equation. A variable-separable solution is described from which it is easy to derive dual integral equations from the boundary conditions. A number of results have been included, which in the author's opinion, are helpful in the study of mixed boundary value problems, even though not everything is used here. In section 3 we present an elementary technique for the solution of dual integral equations given by Sneddon[18], who reduced the problems to an Abel-type integral equation. We use this solution to derive a relation analogous to Poisson's integral formula, which expresses a potential function in terms of its boundary value for the unit circle or a half-plane. Further, we solve the same problem using conformal mapping. As a byproduct, we get an identity involving trigonometric and hyperbolic functions - something I had not seen before. A hybrid technique utilizing conformal mapping and transform calculus developed by Aghili[1] is used to solve mixed boundary value problems for quarter-planes. Section 4 addresses the dual series relations and suggests a new approach to an old problem. Section 5 is devoted to numerical solution of Cauchy Singular Integral Equations.

[1]From Cauchy-Riemann equations $\frac{\partial\phi}{\partial n} = \frac{\partial\psi}{\partial s}$ it is apparent that the normal derivative of a harmonic function is related to the boundary value of conjugate harmonic function.

2 Preliminary Mathematics

2.1 Laplace Equation

The Laplace Equation is a partial differential equation of the second order. In three dimensions and Cartesian coordinates it is usually written as

$$\frac{\partial^2 \phi}{\partial x^2} + \frac{\partial^2 \phi}{\partial y^2} + \frac{\partial^2 \phi}{\partial z^2} = 0. \tag{2.1.1}$$

Often the mathematical model assumes that the function is independent of one space variable, say z, and then the equation simplifies to

$$\frac{\partial^2 \phi}{\partial x^2} + \frac{\partial^2 \phi}{\partial y^2} = 0. \tag{2.1.2}$$

A reduction in dimension also takes place if the problem is assumed to be axisymmetric. Taking the axis of symmetry as the $z-$axis in cylindrical polar coordinates, the Laplace equation

$$\frac{\partial^2 \phi}{\partial r^2} + \frac{1}{r}\frac{\partial f}{\partial r} + \frac{1}{r^2}\frac{\partial^2 \phi}{\partial \theta^2} + \frac{\partial^2 \phi}{\partial z^2} = 0, \tag{2.1.3}$$

since ϕ does not depend on θ due to axisymmetry, reduces to

$$\frac{\partial^2 \phi}{\partial r^2} + \frac{1}{r}\frac{\partial f}{\partial r} + \frac{\partial^2 \phi}{\partial z^2} = 0 \tag{2.1.4}$$

In a boundary value problem, a function ϕ is to be determined so that it satisfies the differential equation in the interior of a region and it takes prescribed values on the boundary of the region. For example, $\phi(x, 0) = f(x), -\infty < x < \infty$ and

$$\frac{\partial^2 \phi}{\partial x^2} + \frac{\partial^2 \phi}{\partial y^2} = 0, y > 0$$

Since the domain is infinite, a regularity condition is required for uniqueness, such as , $\phi(x, y) \to 0$ as $\sqrt{x^2 + y^2} \to 0$. The solution of this classical problem is Poisson's Half-plane integral formula

$$\phi(x, y) = \frac{1}{\pi}\int_{-\infty}^{\infty} \frac{f(t)\, dt}{(x - t)^2 + y^2}. \tag{2.1.5}$$

The term Half-plane is used to distinguish it from the formula for the circle, where the boundary condition is given on the circle and the differential equation is satisfied inside the circle. The Laplace Equation is usually solved by invoking the superposition principle for linear equations and adding solutions obtained on the assumption that the variables are separable,i.e., the solution may be expressed as a product of a function of x only, a function of y only and so on. This gives rise to ordinary differential equations, whose solutions provide the building blocks. For the half-plane problem for example, the equation reduces to

$$\frac{X''}{X} + \frac{Y''}{Y} = 0$$

Since the variables are separated, the only way to satisfy this equation is to choose

$$\frac{X''}{X} = -\xi^2, \frac{Y''}{Y} = \xi^2$$

This leads to the following form for the solution:

$$\phi(x, y) = \int_{-\infty}^{\infty} \psi(\xi) e^{-y|\xi|} e^{ix\xi} d\xi. \tag{2.1.6}$$

2.2 Fourier Transforms

The boundary condition then leads to the problem of finding the inverse Fourier Transform. Of course, we could apply the Fourier transform defined by the relation

$$\Psi(x) = \frac{1}{\sqrt{2\pi}} \int_{-\infty}^{\infty} \psi(\xi) e^{ix\xi} d\xi. \tag{2.2.1}$$

directly to (2.1.2). Formally the inversion formula is

$$\psi(x) = \frac{1}{\sqrt{2\pi}} \int_{-\infty}^{\infty} \Psi(\xi) e^{-ix\xi} d\xi. \tag{2.2.2}$$

For even and odd functions, when the complex exponential is replaced by cosine and sines, the Fourier transform reduces respectively to Fourier cosine and sine transforms defined by the following relations:[2]

$$\Psi(x) = \sqrt{\frac{2}{\pi}} \int_{0}^{\infty} \psi(\xi) \cos x\xi \, d\xi. \tag{2.2.3}$$

[2]Think of a function and its transform or the inverse transform. The variable of integration is unimportant.

$$\Psi(x) = \sqrt{\frac{2}{\pi}} \int_0^\infty \Psi(\xi) \sin x\xi \, d\xi. \tag{2.2.4}$$

The inverse transforms are of the same type.

$$\psi(x) = \sqrt{\frac{2}{\pi}} \int_0^\infty \Psi(\xi) \cos x\xi \, d\xi. \tag{2.2.5}$$

$$\psi(x) = \sqrt{\frac{2}{\pi}} \int_0^\infty \Psi(\xi) \sin x\xi \, d\xi. \tag{2.2.6}$$

2.3 Bessel Functions

Solution of Laplace equation in cylindrical polar coordinates involve Bessel Functions. The separated - variable solution is now given by a combination of Bessel Functions, trigonometric functions and exponential functions. Of special interest is the Bessel Function of order zero of first kind which may be defined by the power series

$$J_0(z) = \sum_0^\infty \frac{(-1)^k z^{2k}}{4^k \, k! \, k!} \tag{2.3.1}$$

Differentiating the above relation, we get the Bessel Function of order one.

$$J_1(z) = \sum_0^\infty \frac{(-1)^k z^{2k+1}}{4^k \, k! \, (k+1)!} \tag{2.3.2}$$

If we replace z by iz, the resulting functions are called modified Bessel functions of the first kind $I_0(z)$ and $I_1(z)$. We have

$$I_0(z) = \sum_0^\infty \frac{z^{2k}}{4^k \, k! \, k!} \tag{2.3.3}$$

$$I_1(z) = \sum_0^\infty \frac{z^{2k+1}}{4^k \, k! \, (k+1)!} \tag{2.3.4}$$

$J_0(z)$ is an entire function of z and the power series converges for all values of z in the complex plane. There is a relationship [3] between trigonometric function

[3] There is a similar relationship between modified Bessel Functions and hyperbolic cosines and sines.All Bessel functions mentioned here are entire functions.

cos x (also sin x) and $J_0(x)$, which proves immensely useful in the solution of mixed boundary value problems. We have

$$J_0(ty) = \frac{2}{\pi}\int_0^t \frac{\cos\, xy\, dx}{\sqrt{(t^2 - x^2)}} \tag{2.3.5}$$

The above relation is easy to derive by term-wise integration of the power series for $\cos(xy)$. Inverting the Fourier cosine transform, we get a special case of Weber-Schafheitlin integral (see Watson[29], p.405) which plays a crucial role in the so-called elementary solution of dual integral equations: For $x \neq t$,

$$\int_0^\infty J_0(ty) \cos\, xy\, dy = \frac{H(t-x)}{\sqrt{t^2 - x^2}} \tag{2.3.6}$$

where H is the Heaviside-function:$H(x) = 1$ for $x \geq 0$ and $H(x) = 0$ elsewhere. Incidentally, Watson's book [29] continues to be the standard source for almost all information related to Bessel functions.

2.4 Integral Equations

Different types of integral equations are encountered in the study of mixed boundary value problems. Of special interest are equations of the Abel-type, Cauchy singular integral equations and Fredholm integral equations of second kind. All the integral transforms are also integral equations. Fredholm integral equations are generally regarded as a suitable point for departure for numerical computation. While the theory for singular integral equations has been shaped by the theory for Fredholm integral equations, direct numerical methods for singular equations is of interest and has been pursued vigorously in the previous century, much of it in the seventies and later. In an integral equation there are one or more functions to be determined. We say that an integral equation is solvable in a closed form, when the solution can be expressed in terms of easily recognized functions. The practical problem is then reduced to one of function evaluation, which may require numerical integration or summation of an infinite series. We include here only those results directly of relevance to the work reported in this article.

2.4.1 Abel's Equation

The classical Abel Equation associated with the problem of Brachistochrone is

$$\int_0^x \frac{g(t)\, dt}{\sqrt{(x-t)}} = f(x), 0 \leq x \leq a. \tag{2.4.1}$$

The function $f(x)$ is known. $g(t)$ is to be determined. We call any equation, which can be reduced to the above form by a change of variable, an equation of Abel-type or just Abel's equation. To minimize the change of variable in applications, a general form is developed in [19]. For an increasing ϕ,

$$\int_a^x \frac{g(t)\,dt}{\mid \phi(x) - \phi(y) \mid^{\mu}} = f(x), a \leq x \leq b, 0 < \mu < 1. \tag{2.4.2}$$

There is a similar form for a decreasing ϕ as well.

The most useful case is $\phi(x) = x^2$.

$$g(t) = \int_0^t \frac{f(s)\,ds}{\sqrt{t^2 - s^2}}. \tag{2.4.3}$$

The solution is expressed by the relation

$$f(x) = \frac{2}{\pi}\frac{d}{dx}\int_0^x \frac{t\,g(t)\,dt}{\sqrt{x^2 - t^2}}. \tag{2.4.4}$$

The pair of equations (2.4.3) and (2.4.4) may be viewed as a pair of an integral transform and its inverse, and hence occasionally called the Abel-transforms.

If we think of the Bessel function in equation (2.3.2) as the Abel-transform of the trigonometric function, the inversion-formula given by (2.4.4)yields the relation

$$\sin xy = \frac{2}{\pi}\int_0^x \frac{t\,J_0(yt)\,dt}{\sqrt{x^2 - t^2}}. \tag{2.4.5}$$

2.4.2 Hilbert Transforms:

The classical Hilbert Transform is defined by the relation

$$F(y) = \frac{1}{\pi}\int_{-\infty}^{\infty} \frac{f(x)\,dx}{x - y}, -\infty < y < \infty.$$

The integral in the above equation is the principal value. A formal inversion formula is a similar integral

$$f(x) = -\frac{1}{\pi}\int_{-\infty}^{\infty} \frac{F(y)\,dy}{y - x}, -\infty < x < \infty.$$

The corresponding finite transform is defined by

$$F(y) = \frac{1}{\pi}\int_{-1}^{1} \frac{f(x)\,dx}{x - y}, -1 < y < 1$$

When $F(y)$ is known and $f(x)$ is to be determined, it is a special case of Cauchy Singular Integral Equation of the form[4]

$$a(s)\,g(s) + \frac{b(s)}{\pi}\int_{-1}^{1}\frac{g(t)dt}{t-s} + \int_{-1}^{1} g(t)K(s,t)\,dt = f(s), \quad -1 < s < 1. \tag{2.4.6}$$

Among others, the finite Hilbert transform has been studied by Tricomi [28]. This integral is the dominant part of many singular integral equations encountered in mixed boundary value problems. In the class of Hölder continuous functions, there is a one-parameter family of solutions given by

$$f(x) = \frac{C}{\sqrt{1-x^2}} + \frac{1}{\pi\sqrt{1-x^2}}\int_{-1}^{1}\frac{\sqrt{1-y^2}F(y)\,dy}{y-x}, -1 < x < .1 \tag{2.4.7}$$

The constant C is usually determined by an auxiliary physical condition ,e.g.,

$$\int_{-1}^{1} f(x)\,dx = N \tag{2.4.8}$$

Exercise: Show that

$$\int_{-1}^{1}\frac{dt}{\sqrt{1-t^2}(t-s)} = 0, \quad -1 < s < 1. \tag{2.4.9}$$

Some times it is useful to split the problem in even and odd parts. We express $f(x)$ as the sum of

$$f_1 = \frac{f(x)+f(-x)}{2} \quad \text{and} \quad f_2 = \frac{f(x)-f(-x)}{2}.$$

Obviously f_1 is even and f_2 is odd. Splitting the range of integration to use the symmetry and changing the variable from x to $-x$ for (-1,0) and rearranging the terms we get the expression

$$\frac{1}{\pi}\int_{-1}^{1}\frac{f(x)\,dx}{x^2-y^2} = \frac{y}{\pi}\int_{0}^{1}\frac{f_1(x)\,dx}{x^2-y^2} + \frac{1}{\pi}\int_{0}^{1}\frac{xf_2(x)\,dx}{x^2-y^2}$$

[4]For the classical treatment see Muskhelishvili [14] or Gakhov[9]. For a different approach see Prössdorf and Silberman [15].

We now equate the odd and the even components of $F(y)$, namely,

$$F_1 = \frac{F(y)+F(-y)}{2} \quad \text{and} \quad F_2 = \frac{F(y)-F(-y)}{2}.$$

This gives us the integral equations

$$\frac{y}{\pi}\int_0^1 \frac{f_1(x)\,dx}{x^2-y^2} = F_1(y). \tag{2.4.10}$$

$$\frac{1}{\pi}\int_0^1 \frac{xf_2(x)\,dx}{x^2-y^2} = F_2(y). \tag{2.4.11}$$

It is a simple exercise in calculus to show that

$$\frac{1}{\pi}\int_0^1 \frac{xf(x)\,dx}{x^2-y^2} = \frac{d}{dy}\int_0^y \frac{t\,dt}{\sqrt{y^2-t^2}}\int_t^1 \frac{f(s)\,ds}{\sqrt{s^2-t^2}} \tag{2,4.12}$$

Moreover,

$$\int_0^1 \frac{dt}{t^2-x^2}\int_t^1 \frac{\phi(s)\,ds}{\sqrt{s^2-t^2}} = \frac{\pi}{2}\int_0^x \frac{\phi(s)\,ds}{\sqrt{x^2-s^2}} \tag{2.4.13}$$

Thus, Cauchy singular integral equations and equations of Abel-type can be interchanged, a connection that has not been explored fully.

2.4.3 Hilbert Kernel:

Another integral equation which is related to the Cauchy Euation is

$$a\phi(x) + \frac{b}{2\pi}\int_0^{2\pi} \phi(t)\cot\frac{t-x}{2}\,dt = f(x), \qquad 0 < x < 2\pi. \tag{2.4.14}$$

This equation has a solution (see Mikhlin[13]) if and only if

$$\int_0^{2\pi} f(x)dx = 0 \tag{2.4.15}$$

We have given here the basic kernels which relate the boundary values of conjugate harmonic functions.

2.5 Orthogonal Polynomials:

In the numerical solution of Cauchy singular integral equations, Chebyshev polynomials and Jacobi polynomials play a special role. We shall consider here only the former because of their intrinsic simplicity.

2.5.1 Definition,Recurrence and Orthogonality

We follow the notation of Erdèlyi et al [6] and define respectively the Chebyshev polynomials of the first and second kind of degree n by the relations

$$T_n(x) = \cos(n \cos^{-1} x), -1 \le x \le 1 \tag{2.5.1}$$

$$U_n(x) = \frac{\sin((n+1)\cos^{-1} x)}{\sin(\cos^{-1} x)}, -1 \le x \le 1. \tag{2.5.2}$$

For $x = \cos\theta$, $T_n(\cos\theta) = \cos n\theta$ and $U_{n-1}(\cos n\theta)\sin\theta = \sin n\theta$. Obviously, $T_0(x) = 1$ and $U_0(x) = 1$. A number of properties of Chebyshev polynomials can be derived from trigonometric identities. The reader is now invited to verify the following relations:

$$T_{n+1}(x) + T_{n-1}(x) = 2x\, T_n(x) \tag{2.5.3}$$

$$T_{n+1}(x) - T_{n-1}(x) = 2(x^2 - 1)\, U_{n-1}(x) \tag{2.5.4}$$

$$U_{n+1}(x) + U_{n-1}(x) = 2x\, U_n(x) \tag{2.5.5}$$

$$U_{n+1}(x) - U_{n-1}(x) = 2\, T_n(x) \tag{2.5.6}$$

There are other useful relations between the Chebyshev polynomials of the first and second kind. For example,

$$\frac{d}{dx} T_n(x) = U_{n-1}(x) \tag{2.5.7}$$

$$\frac{1}{\pi}\int_{-1}^{1} \frac{T_k(x)\,dx}{\sqrt{(1-x^2)}(x-y)} = -U_{k-1}(y), k = 1, 2, \ldots \tag{2.5.8}$$

For $k = 0$, $T_0(x) \equiv 1$ and the integral vanishes.

For nonnegative integers m, n at least one of these different from zero

$$\int_{-1}^{1} \frac{T_m(x) T_n(x) dx}{\sqrt{1-x^2}} = \frac{\pi}{2}\delta_{mn} \tag{2.5.9}$$

where δ_{mn} is the Kronecker symbol with the value 1 for $m = n$ and 0 otherwise. Similarly,

$$\int_{-1}^{1} \sqrt{1-x^2}\, U_m(x) U_n(x) dx = \frac{\pi}{2}\delta_{mn} \tag{2.5.10}$$

We shall denote the zeros of $T_n(x)$ by $\{t_{j,n}\}$ with the zero $t_{1,n}$ nearest to the point $x = 1$ and successive zeros getting farther away with $t_{n,n}$ nearest to $x = -1$. Similarly, $\{s_{j,n}\}$ denote the zeros of $U_n(x)$. Replacing x by cos θ, it is easy to see that

$$t_{j,n} = \cos\left(\frac{j - \frac{1}{2}\pi}{n}\right), s_{j,n} = \cos\left(\frac{j\pi}{n+1}\right), j = 1, 2, \ldots, n. \tag{2.5.11}$$

We shall omit the second subscript, when the degree of the polynomial is clear from the context. Since the polynomials are either odd or even functions, the zeros are distributed symmetrically about the origin.The zeros of Chebyshev polynomials are very useful for interpolation by polynomials and approximate integration. We refer the reader to Rivlin's monograph[25] for a detailed exposition.

2.5.2 Identities involving zeros of Chebyshev polynomials:

In the course of our investigation we came across a number of identities involving zeros of these polynomials. [5]We list some of them here as an exercise for the readers!

$$\sum_1^n t_j = 0, \sum_1^n {t_j}^2 = \frac{n}{2}.$$

$$\sum_1^n \frac{1}{1-t_j} = n^2,$$

$$\sum_1^n \frac{1}{1-{t_j}^2} = n^2, \sum_1^n \frac{1}{(1-t_j)^2} = \frac{n^2(2n^2+1)}{3}.$$

$$\sum_1^{n-1} s_j = 0, \sum_1^{n-1} {s_j}^2 = \frac{n-1}{2}.$$

$$\sum_{j=1}^n \frac{1}{s_k - t_j} = 0, k = 1, 2, \ldots, n-1.$$

$$\sum_{j=1}^n \frac{1}{(s_k - t_j)^2} = \frac{n^2}{1 - s_k^2}, k = 1, 2, \ldots, n-1.$$

$$\prod_{1 \le j < k < \le n} (t_j - t_k) = 2^{-\frac{1}{2}(n-1)^2} . n^{\frac{n}{2}}$$

[5]For more identities see [23].The identities are useful in calculating the variance due to Gaussian error in uncertain data in §5.2.2

$$\prod_{1\le j<k<\le n} (s_j - s_k) = 2^{-\frac{1}{2}(n^2-2n)}.(n+1)^{\frac{n}{2}-1}$$

2.5.3 Gauss-Chebyshev Quadrature

The application of most interest to us is the formula

$$\frac{1}{\pi}\int_{-1}^{1}\frac{f(x)dx}{\sqrt{1-x^2}} \approx \frac{1}{n}\sum_{j=1}^{n} f(t_j). \tag{2.5.12}$$

The formula is exact for polynomials of degree $2n-1$ or lower and particularly suitable for certain singular integral equations. The formula follows from the interpolatory polynomial approximation

$$f(x) \approx \frac{1}{n}\sum_{j=1}^{n}\frac{f(t_j)T_n(x)}{(x-t_j)U_{n-1}(t_j)} \tag{2.5.13}$$

and the integral (2.5.5).

2.5.4 Lobatto-Chebyshev Quadrature

An analogous formula for approximate evaluation is

$$\frac{1}{\pi}\int_{-1}^{1}\frac{f(x)dx}{\sqrt{1-x^2}} \approx \frac{1}{n}\left[\frac{1}{2}f(1)+\frac{1}{2}f(-1)+\sum_{j=1}^{n-1} f(s_j).\right] \tag{2.5.14}$$

Here $\{s_j\}$ are the zeros of the Chebyshev polynomial $U_{n-1}(x)$.

2.6 Meaningful Divergent Integrals

One approach to deal with divergent integrals is to interpret them as "Abel-sums". For example,

$$\int_0^\infty f(x)dx$$

whenever divergent is reinterpreted as

$$\lim_{c\to 0+}\int_0^\infty e^{-cx}f(x)dx$$

In formal analysis using this interpretation, I have not yet encountered any inconsistency. The space of functions in which the solutions are determined changes from the usual functions to *tempered distributions.* (see[22])

Example 1:

$$\int_0^\infty \sin\, ax\,,dx = \frac{1}{a}, a > 0.$$

For,

$$\int_0^\infty e^{-cx} \sin\, ax\,,dx = \frac{a}{c^2 + a^2}, a > 0, c > 0.$$

Hence

$$\lim_{c \to 0+} \int_0^\infty e^{-cx} \sin\, ax\, dx = \lim_{c \to 0+} \frac{a}{c^2 + a^2} = \frac{1}{a}, a > 0.$$

Example 2:

$$\int_0^\infty \cos\, yx\, dx = \frac{\pi}{2} \delta(y).$$

Here δ denotes the famous Dirac's delta function. Formally, $\delta(x)$ is the "derivative" of Heaviside's unit-step function, which has the substitution property:

$$\int_{-\infty}^\infty f(x)\, \delta(x - a) dx = f(a)\, \delta(y).$$

Similar ideas are useful in dealing with divergent series. For example,

$$1 - 1 + 1 - 1 + \ldots$$

i.e. the infinite series $\sum (-1)^n$ could be assigned the sum $\frac{1}{2}$ as $\lim_{x \to 1} \sum (-x)^n$. An alternative is the Cesàro-sum (C,1)-sum derived as the

$$\lim_{n \to \infty} \left(\sigma_n = \frac{S_1 + S_2 + \ldots + S_n}{n} \right)$$

where S_n is the $n-$th partial sum. For convergent series both the sums are the same as $\lim_{n \to \infty} S_n$. The corresponding analogue for the integral is

$$\lim_{\lambda \to \infty} \int_0^\lambda (1 - \frac{x}{\lambda}) f(x)\, dx$$

Example 3:

$$\sum_{n=1}^\infty \sin n\theta = \frac{1}{2} \cot \frac{\theta}{2}, \theta \neq 0 \qquad (2.6.1)$$

3 Mixed boundary value problem for a quarter-plane

3.1 Dual Integral Equations

Let us consider the problem of finding $\phi(x, y)$, which satisfies the Laplace Equation

$$\frac{\partial^2 \phi}{\partial x^2} + \frac{\partial^2 \phi}{\partial y^2} = 0, x > 0, y > 0, \tag{3.1.1}$$

and the following boundary conditions:

$$\phi_x(0, y) = 0. \tag{3.1.2}$$

$$\phi_y(0, x) = f(x), 0 < x < 1; \phi(x, 0) = 0, x > 1. \tag{3.1.3}$$

An additional requirement is that $\phi(x, y) \to 0$ as $\sqrt{x^2 + y^2} \to 0$. We shall resolve the mixed boundary condition by means of conformal mapping later. We present first a formulation based upon dual integral equations and describe an elegant elementary solution of dual equations given by Sneddon [18] . Using separation of variables, we may express the function $\phi(x, y)$ in terms of an auxiliary function ψ :

$$\phi(x, y) = \int_0^\infty \psi(\xi) e^{-\xi y} \cos x\xi \, d\xi. \tag{3.1.4}$$

The function ψ is determined by the following pair of equations:

$$\int_0^\infty \xi \psi(\xi) \cos x\xi \, d\xi = - f(x), 0 < x < 1, \tag{3.1.5}$$

$$\int_0^\infty \psi(\xi) \cos x\xi \, d\xi = 0, x > 1. \tag{3.1.6}$$

Such equations are called *dual integral equations.*[6] We note that the condition (2) as well as the regularity condition are satisfied by $\phi(x, y)$ as given by a large class of functions.

[6]While Weber,in 1873, had solved a pair of equations corresponding to an electrified disk[30], the name seems to have been given by Busbridge(1937);see [3], [27].

3.2 Sneddon's Elementary Solution

Following Sneddon[18], let

$$\psi(\xi) = \frac{2}{\pi}\int_0^1 t\,g(t)\,J_0(\xi t)\,dt, \xi > 0. \tag{3.2.1}$$

Here $g(t)$ is an auxiliary function to be determined so that $\psi(\xi)$ satisfies the pair of dual integral equations. Fortuitously, the second of the two equations is satisfied for all reasonable functions $g(t)$, as can be verified using the Weber-Schafheitlin integral mentioned earlier. The first of the dual equations will be satisfied by $g(t)$ determined by the equation

$$\frac{2}{\pi}\frac{d}{dx}\int_0^x \frac{t\,g(t)\,dt}{\sqrt{x^2-t^2}} = -f(x), \qquad 0 < x < 1. \tag{3.2.2}$$

From the properties of Abel-transforms, it is easily seen that

$$g(t) = -\int_0^t \frac{f(s)\,ds}{\sqrt{t^2-s^2}}. \tag{3.2.3}$$

Hence

$$\phi(x,y) = \frac{2}{\pi}\int_0^1 t\,g(t)\,H(t,x,y)\,dt, \tag{3.2.4}$$

where

$$H(t,x,y) = \int_0^\infty e^{-y\xi} J_0(t\xi)\,\cos(x\xi)\,d\xi. \tag{3.2.5}$$

Since ([16], p.514(20))

$$\int_0^\infty e^{-pt}\,J_0(at)\,dt = \frac{1}{\sqrt{a^2+p^2}}, \; Re(p) > Im(a), \tag{3.2.6}$$

we have

$$H(t,x,y) = \frac{1}{2}\left(\frac{1}{\sqrt{t^2+(y+ix)^2}} + \frac{1}{\sqrt{t^2+(y-ix)^2}}\right). \tag{3.2.7}$$

If we combine equations (3.2.4) and (3.2.7) and interchange the order of integration we get the representation

$$\phi(x,y) = -\frac{2}{\pi}\int_{s=0}^1 f(s)\int_s^1 \frac{tH(t,x,y)\,dt}{\sqrt{(t^2-s^2)}}ds. \tag{3.2.8}$$

Further integration yields an analogue of Poisson's integral formula. We mention this because while, in principle, a variety of mixed boundary value problems have been solved, such relations have not been calculated.

3.3 A Poisson Relation for the Quarter-plane

It is of some interest to express the solution in terms of the functions specifying the nonhomogeneous boundary conditions. Let

$$\chi(z,s) = \int_s^1 \frac{t\,dt}{\sqrt{(t^2-s^2)(t^2+z^2)}}. \tag{3.3.1}$$

χ is an analytic function of the complex variable z in a cut-plane with branch-points at ± 1. In fact,

$$\chi(z,s) = \log\left(\frac{\sqrt{1-s^2}+\sqrt{1-z^2}}{\sqrt{z^2+s^2}}\right). \tag{3.3.2}$$

Evaluating the integral in (3.2.8) we find that

$$\phi(x,y) = -\frac{1}{\pi}\int_0^1 f(s)\log\left(\frac{\sqrt{1-\overline{z}^2}+\sqrt{1-s^2})(\sqrt{1-z^2}+\sqrt{1-s^2})}{\sqrt{(s^2-\overline{z}^2)(s^2-z^2)}}\right)ds \tag{3.3.3}$$

where $z = x + i\,y$ and $\overline{z} = x - i\,y$.

In the next section, we solve this problem using a different technique, and to show that the two solutions are identical, we write $\phi(x,y)$ in an alternative form by changing the variables as follows:

$$s = \sin\frac{1}{2}t, \qquad z = \sin\frac{1}{2}w$$

$$\phi(x,y) = -\frac{1}{2\pi}\int_0^1 f(\sin\frac{1}{2}t)\cos\frac{1}{2}t\log\left(\frac{(\cos\frac{1}{2}\overline{w}+\cos\frac{1}{2}t)(\cos\frac{1}{2}w+\cos\frac{1}{2}t)}{\sqrt{(\sin^2\frac{1}{2}t-\sin^2\frac{1}{2}w)(\sin^2\frac{1}{2}t-\sin^2\frac{1}{2}\overline{w})}}\right)dt \tag{3.3.4}$$

3.4 Conformal mapping into a strip

Conformal mapping has been employed by Aghili[1] to study a variety of boundary conditions for a quarter-plane. The basic idea is to resolve the mixing of boundary conditions on a boundary by mapping it into a geometry where the segments of boundary correspond to different co-ordinate surfaces. The mapping

$$w = 2 \sin^{-1} z \tag{3.4.1}$$

maps the quarter-plane into a strip. Following the conventional notation, let $w = u + i\,v$ and $z = x + i\,y$. Equating real and imaginary parts, we get the relations

$$x = \sin \frac{1}{2}u \cosh \frac{1}{2}v, \quad y = \cos \frac{1}{2}u \sinh \frac{1}{2}v \tag{3.4.2}$$

u, v are determined by the relations

$$\frac{x^2}{\sin^2 \frac{1}{2}u} - \frac{y^2}{\cos^2 \frac{1}{2}u} = 1, \quad \frac{x^2}{\cosh^2 \frac{1}{2}v} + \frac{y^2}{\sinh^2 \frac{1}{2}v} = 1. \tag{3.4.3}$$

Let us recall (see for example Dettman([5],p 253)) that the conformal mapping $w = f(z)$ subject to the condition $f'(z_0) \neq 0$ leads to the relation

$$\left(\frac{\partial^2 \phi}{\partial x^2} + \frac{\partial^2 \phi}{\partial y^2}\right)_{z_0} = \left(\frac{\partial^2 \Phi}{\partial u^2} + \frac{\partial^2 \Phi}{\partial v^2}\right)_{w_0} \mid f'(z_0) \mid^2 . \tag{3.4.4}$$

The problem is now reduced to one of finding $\Phi\,(u, v)$. More importantly, the Laplace equation is now satisfied for the interior of the strip bounded by the edges

1. $u = 0, v > 0$ corresponding to $x = 0$.
2. $v = 0, 0 \leq u \leq \pi$ corresponding to $y = 0, 0 \leq x \leq 1$.
3. $u = \pi, v > 0$ corresponding to $y = 0, x > 1$.

In the $(u, v)-$plane the harmonic function is given by the relation

$$\Phi(u, v) = \phi(\sin \frac{1}{2}u \cosh \frac{1}{2}v, \cos \frac{1}{2}u \sinh \frac{1}{2}v). \tag{3.4.5}$$

Moreover , the transformed boundary conditions are

$$\Phi(\pi, v) = 0. \tag{3.4.6}$$

$$\frac{\partial \Phi}{\partial u} = 0, \text{for} \quad u = 0, v > 0. \tag{3.4.7}$$

$$\frac{\partial \Phi}{\partial v} = \frac{1}{2} f(\sin \frac{1}{2} u) \cos \frac{1}{2} u, \text{for} \quad v = 0, u > 0. \tag{3.4.8}$$

The appropriate separated variable solution is now given by

$$\Phi\,(u, v) = \sum_{n=1}^{\infty} a_n\, e^{-(n-\frac{1}{2}v)} \cos\,(n - \frac{1}{2})u, \tag{3.4.9}$$

where the coefficients $\{a_n\}$ are determined by the relation

$$\sum_{n=1}^{\infty} a_n \cos\,(n - \frac{1}{2})u = \frac{1}{2} f(\sin \frac{1}{2} u)\, \cos(\frac{1}{2} u), \quad 0 \leq u \leq \pi. \tag{3.4.10}$$

From orthogonality of $\{\cos(n - \frac{1}{2})u\}$, it follows that

$$a_n = -\frac{2}{(2n-1)\pi} \int_0^{\pi} f(\sin \frac{1}{2} u) \cos \frac{1}{2} u \cos(n - \frac{1}{2}) u\, du. \tag{3.4.11}$$

We may now calculate $\Phi(u, v)$ by pluging in the above expression for the coefficients in (3.4.9) .

To reconcile the two solutions, it is desirable to derive an amenable integral representation for $\Phi(u, v)$ of the form

$$\phi\,(x, y) = -\frac{1}{2\pi} \int_0^1 f(\sin \frac{1}{2} t) \cos \frac{1}{2} t\, H_1(t, u, v)\, dt \tag{3.4.12}$$

and match the two kernels. Serendepitously, the infinite series in (3.4.9) corresponding to $H_1(t, u, v)$ can be summed using a classical identity ([10], p. 41 (4)), namely,

$$\sum_{k=1}^{\infty} \frac{p^{2k-1} \cos(2k-1)x}{2k-1} = \frac{1}{4} \log\left(\frac{1 + 2p\cos x + p^2}{1 - 2p\cos x + p^2}\right). \tag{3.4.13}$$

Substituting $p = e^{-\frac{1}{2}v}$ we find that

$$H_1(u, t, v) = -\frac{1}{2} \log\left(\frac{\cosh \frac{1}{2}v + \cos \frac{1}{2}(u+t)}{\cosh \frac{1}{2}v - \cos \frac{1}{2}(u+t)} \frac{\cosh \frac{1}{2}v + \cos \frac{1}{2}(u-t)}{\cosh \frac{1}{2}v - \cos \frac{1}{2}(u-t)}\right) \tag{3.4.14}$$

3.4.1 An identity

Our objective is now to show that

$$A = \frac{1}{2}\left(\frac{\cosh\frac{1}{2}v + \cos\frac{1}{2}(u+t)}{\cosh\frac{1}{2}v - \cos\frac{1}{2}(u+t)}\frac{\cosh\frac{1}{2}v + \cos\frac{1}{2}(u-t)}{\cosh\frac{1}{2}v - \cos\frac{1}{2}(u-t)}\right) \tag{3.4.15}$$

and

$$B = \frac{(\cos\frac{1}{2}\overline{w} + \cos\frac{1}{2}t)(\cos\frac{1}{2}w + \cos\frac{1}{2}t)}{\sqrt{(\sin^2\frac{1}{2}t - \sin^2\frac{1}{2}w)(\sin^2\frac{1}{2}t - \sin^2\frac{1}{2}\overline{w})}} \tag{3.4.16}$$

We push the calculation using elementary facts $\cos(it) = \cosh t$ and $z\overline{z} = |\, z\,|^2$ and trigonometric identities involving sines and cosines. Both the expressions A and B in the above equations reduce to

$$\cot\frac{w+t}{2}\cot\frac{\overline{w}+t}{2}\cot\frac{w-t}{2}\cot\frac{\overline{w}-t}{2}.$$

We omit the details as an exercise for the readers!

3.5 Superposition Techniques

To illustrate the application of superposition techniques, we consider next an other boundary value problem for the quarter-plane. In principle, it is similar to Sneddon's solution of the axisymmetric problem of a charged disk inside an earthed cylinder, subsequently applied to a variety of problems. The objective now is to construct a function which is harmonic in the first quadrant as before with the boundary conditions (3.1.3) unchanged, but (3.1.2) replaced by

$$\phi_x(0, y) = \theta\,(y), y > 0. \tag{3.5.1}$$

The variable-separable solution now consists of two elementary solutions added together. Let

$$\phi\,(x, y) = \int_0^\infty A(\xi)e^{-\xi x}\sin y\xi\, d\xi + \int_0^\infty B(\xi)e^{-\xi y}\cos x\xi\, d\xi. \tag{3.5.2}$$

The boundary condition on $x = 0$ gives us

$$-\lim_{x\to 0+}\int_0^\infty A(\xi)e^{-\xi x}\sin y\xi\, d\xi = \theta\,(y), y > 0. \tag{3.5.3}$$

Inverting the Fourier Sine Transform we get the relation

$$\xi\, A(\xi) = -\frac{2}{\pi}\int_0^\infty \theta\,(s)\sin(\xi\, s)\, ds \tag{3.5.4}$$

The conditions on $y = 0$ gives the dual integral equations

$$\int_0^\infty \xi\, B(\xi)\, \cos\,(x\xi)\, d\xi = F(x), 0 < x < 1 \tag{3.5.5}$$

$$\int_0^\infty B(\xi)\, \cos\,(x\xi)\, d\xi = 0, x > 1. \tag{3.5.6}$$

Here $F(x)$ is the following function:

$$\int_0^\infty \xi\, A(\xi) e^{-x\xi}\, d\xi - f(x).$$

Following the same procedure as outlined earlier, we can express

$$B(\xi) = \frac{2}{\pi}\int_0^1 tg(t)\, J_0(\xi\, t)\, dt \tag{3.5.7}$$

where

$$g(t) = -\int_0^t \frac{f(s)\, ds}{\sqrt{t^2 - s^2}} - \int_0^\infty \frac{\theta(s)\, ds}{\sqrt{(s^2 + t^2)}}. \tag{3.5.8}$$

The boundary value $\phi\,(0, y)$ is useful in the solution of problems with a radiation-type condition

$$\phi_x(0, y) + h\phi\,(0, y) = 0. \tag{3.5.9}$$

It is easy to see that

$$\phi\,(0, y) = \int_0^\infty A(\xi) \sin\, y\xi\, d\xi + \int_0^\infty B(\xi) \cos\, x\xi\, d\xi.$$

Substituting for $A(\xi)$ and $B(\xi)$ from (3.5.4) and (3.5.7), changing the order of integration and evaluating the integrals, we find that

$$\phi\,(0, y) = \frac{2}{\pi}\int_0^1 \frac{t\, g(t)\, dt}{\sqrt{(t^2 + y^2)}} - \frac{1}{\pi}\int_0^\infty \theta\,(s)\, \ln\left|\left(\frac{y+s}{y-s}\right)\right|\, ds \tag{3.5.10}$$

Next, we assume that $\phi_x(0, y) = \theta(y)$. $\theta(y)$ would have to satisfy the equation

$$\theta(y) + h\left[\frac{-1}{\pi}\int_0^\infty \theta\,(s)\, \ln\left|\left(\frac{y+s}{y-s}\right)\right|\, ds + \frac{2}{\pi}\int_0^1 \frac{t\, g(t)\, dt}{\sqrt{(t^2 + y^2)}}\right] = 0 \tag{3.5.11}$$

where once again

$$g(t) = -\int_0^t \frac{f(s)\,ds}{\sqrt{t^2 - s^2}} - \int_0^\infty \frac{\theta(s)\,ds}{\sqrt{(s^2 + t^2)}}. \tag{3.5.12}$$

To sum up, the strategy is to generate a sequence of tractable boundary value problems leading to integral equations which may be solved numerically.The integral equations obtained are only formally of the second kind of Fredholm type. It shall lend more confidence if it could be proved that the associated integral operators are indeed compact.

4 Dual Series Relations

4.1 Introduction

A strip of finite width is frequently used to model the physical problems in continuum mechanics. Many of these problems lead to mixed boundary value problems for elliptic equations, most notably the Laplace Equation. Steady-state temperature distribution in conducting solids with uniform conductivity is perhaps the easiest way to connect the mathematical problem with a problem of physical interest. For this purpose, we consider a strip of finite width π occupying $0 \le x \le \pi$, $y \ge 0$ in R^2. The steady-state temperature $U(x, y)$ satisfies the Laplace's Equation,

$$\frac{\partial^2 U}{\partial x^2} + \frac{\partial^2 U}{\partial y^2} = 0, \tag{4.1.1}$$

vanishes as $\sqrt{x^2 + y^2} \to \infty$, and the boundary conditions. For simplicity, we assume that the edges $x = 0$ and $x = \pi$ are either maintained at zero temperature or are insulated perfectly. There are four possible boundary conditions enumerated below.

1. <u>Both edges at uniform temperature:</u>[7]

$$U(0, y) = U(\pi, y) = 0 \tag{4.1.2}$$

2. <u>Both edges insulated</u>

$$U_x(0, y) = U_x(\pi, y) = 0 \tag{4.1.3}$$

3. <u>Left edge insulated</u>

$$U_x(0, y) = U(\pi, y) = 0 \tag{4.1.4}$$

[7]If the temperatures are different, say πu_1 and πu_2, just add $u_1(\pi - x) + u_2 x$

4. Right edge insulated

$$U(0, y) = U_x(\pi, y) = 0 \tag{4.1.5}$$

For the boundary $y = 0$, we assume the mixed boundary conditions

$$U_y(x, 0) = f_1(x), \quad 0 \le x < c, \quad U(x, 0) = f_2(x), \quad c \le x \le \pi \tag{4.1.6}$$

or its complement, namely,

$$U(x, 0) = f_1(x), \quad 0 \le x < c, \quad U_y(x, 0) = f_2(x), \quad c \le x \le \pi \tag{4.1.7}$$

Thus, we have eight different boundary-value problems. If the solution of these problems is attempted using Sturm-Liouville theory [4], we get dual series relations resembling dual integral equations. Separation of variables yields the following representations

$$\sum_{n=1}^{\infty} a_n e^{-ny} \sin nx, \tag{4.1.8}$$

$$\frac{1}{2}a_0 + \sum_{n=1}^{\infty} a_n e^{-ny} \cos nx, \tag{4.1.9}$$

$$\sum_{n=1}^{\infty} a_n e^{-(n-\frac{1}{2})y} \cos(n - \frac{1}{2})x, \tag{4.1.10}$$

$$\sum_{n=1}^{\infty} a_n e^{-(n-\frac{1}{2})y} \sin(n - \frac{1}{2}x). \tag{4.1.11}$$

The mixed boundary conditions (4.1.6) and (4.1.7) lead to dual series equations, in which coefficients $\{a_n\}$ are to be found to satisfy the appropriate conditions.

These problems received considerable attention in the sixties and the classical techniques are extensively described by Sneddon[17]. Special function techniques work like a charm on special problems. We sketch here some of the approaches [8]

To keep the focus we consider the dual series equations

$$\sum_{n=1}^{\infty} n\, a_n \cos nx = f(x), \quad 0 \le x < c \tag{4.1.12}$$

$$\frac{1}{2}a_0 + \sum_{n=1}^{\infty} a_n \cos nx = 0, \quad c \le x \le \pi. \tag{4.1.13}$$

[8] Of interest will be alternative approaches which can be adapted for numerical computation.

A common feature of the techniques generally used is that the sequence$\{a_n\}$ is determined by an auxiliary function by expressing the sum in (4.1.13) for $0 \leq x < c$. The first series is usually interpreted as

$$\frac{d}{dx}\sum_{n=1}^{\infty} a_n \sin nx = f(x), \quad 0 \leq x < c \tag{4.1.14}$$

In the following sections we describe some known analytical approaches; first we derive a closed form expression and then we reduce it to a Fredholm integral equation of the second kind. Finally, we outline a procedure, which may be effective in other situations as well. While the method works for a large class of functions, to illustrate the application of the technique, we have chosen $f(x) = \cos kx$ in the derivation of the closed-form.

4.2 Closed Form Expression

Following [20], let

$$\frac{1}{2}a_0 + \sum_{n=1}^{\infty} a_n \cos x = \cos(\frac{1}{2}x)\int_x^c \frac{g(t)dt}{\sqrt{\cos x - \cos t}}, \quad 0 \leq x < c \tag{4.2.1}$$

using Mehler's integral([31],p.315),

$$P_n(\cos\theta) = \frac{\sqrt{2}}{\pi}\int_0^{\theta} \frac{\cos(n+\frac{1}{2})xdx}{\sqrt{\cos x - \cos\theta}} = \frac{\sqrt{2}}{\pi}\int_{\theta}^{\pi} \frac{\sin(n+\frac{1}{2})xdx}{\sqrt{\cos\theta - \cos x}}, \tag{4.2.2}$$

we get

$$a_0 = \sqrt{2}\int_0^c g(t)dt \tag{4.2.3}$$

$$a_n = \frac{1}{\sqrt{2}}\int_0^c g(t)(P_n(\cos t) + P_{n-1}(\cos t))dt. \tag{4.2.4}$$

Substituting for a_n in (4.1.14) we get

$$\frac{d}{dx}\left(\int_0^c g(t)\left(\frac{1}{\sqrt{2}}\sum_{n=1}^{\infty}(P_n(\cos t) + P_{n-1}(\cos t))\sin nx\right)dt\right) = \cos kx, \quad 0 \leq x < c. \tag{4.2.5}$$

Using ([17], p.59, 2.6.31) we find that

$$\int_0^c \frac{g(t)dt}{\sqrt{\cos t - \cos x}} = \frac{\sin kx}{k}\sec\frac{1}{2}x, \quad 0 \leq x < c \tag{4.2.6}$$

leading to

$$g(t) = \frac{2}{\pi k}\frac{d}{dt}\left(\int_0^t \frac{\sin kx \sin \frac{1}{2}x dx}{\sqrt{\cos x - \cos t}}\right). \tag{4.2.7}$$

Once again using Mehler's integral, we find that

$$a_0 = \frac{1}{k}\left[P_{k-1}(\cos c) - P_k(\cos c)\right] \tag{4.2.8}$$

and that

$$\begin{aligned} a_n &= \frac{\sqrt{2}}{\pi k}\int_0^c (P_n(\cos t) + P_{n-1}(\cos t))\frac{d}{dt}\int_0^t \frac{\sin kx \sin \frac{1}{2}x dx}{\sqrt{\cos x - \cos t}} dt \\ &= \frac{1}{k}\int_0^c \frac{d}{dt}(P_n(\cos t) + P_{n-1}(\cos t))\left(P_{k-1}(\cos t) - P_k(\cos t)\right) dt \end{aligned}$$

Substituting $\cos t = x$,

$$a_n = -\frac{1}{k}\int_{\cos c}^1 \left(P_n'(x) + P_{n-1}'(x)\right)\left(P_{k-1}(x) - P_k(x)\right) dx.$$

Since $P_n(x)$ are Legendre polynomials, for comparison for small values of k and x, this will give exact numerical values for the coefficients. For $k = 1$,

$$a_0 = 1 - \cos c$$

$$a_n = -\int_{\cos c}^1 \left(P_n'(x) + P_{n-1}'(x)\right)(1 - x)dx.$$

For $n = 1$, $c = \pi/2$,

$$a_1 = -\int_0^1 (1 - x)dx = -\frac{1}{2}.$$

$$a_2 = -\frac{1}{2}\int_0^1 \frac{d}{dx}(3x^2 + 2x - 1)(1 - x)dx = -1$$

These numbers are of interest because they may be used to assess the results obtained through purely numerical procedures. If instead of the special form for the sum, one had assumed that

$$\frac{1}{2}a_0 + \sum_{n=1}^{\infty} a_n \cos x = \phi(x), 0 \le x < c \tag{4.2.9}$$

summing the infinite series as (C,1) or Abel-sums, one would get a singular integral equation with the Hilbert kernel.

4.3 Fredholm Integral Equation of the Second Kind

Following [20], suppose that

$$\frac{1}{2}a_0 + \sum_{n=1}^{\infty} a_n \cos nx = \int_x^c \frac{g(t)dt}{\sqrt{t^2 - x^2}}. \quad 0 \leq x < c \tag{4.3.1}$$

Then,

$$\begin{aligned} a_n &= \frac{2}{\pi}\int_0^c \cos nx \int_x^c \frac{g(t)dt}{\sqrt{t^2-x^2}} \\ &= \frac{2}{\pi}\int_0^c tg(t)\left(\int_0^t \frac{\cos nx dx}{\sqrt{t^2-x^2}}\right) dt, \end{aligned}$$

or

$$a_n = \int_0^c tg(t)J_0(nt)dt. \tag{4.3.2}$$

Substitute this expression for $\{a_n\}$ in (4.1.14). This yields the equation

$$\frac{d}{dx}\int_0^c tg(t)\left(\sum_{n=1}^{\infty} \sin nx J_0(nt)\right) dt = f(x), \quad 0 \leq x < c. \tag{4.3.3}$$

From ([17], p.37)

$$\sum_{n=1}^{\infty} \sin nx J_0(nt) = \frac{H(x-t)}{\sqrt{t^2-x^2}} - \int_0^{\infty} \frac{\sinh xy}{\sinh \pi y} I_0(ty)e^{-\pi y}dy. \tag{4.3.4}$$

We may rewrite (4.3.3) as follows:

$$\frac{d}{dx}\int_0^x \frac{tg(t)}{\sqrt{x^2-t^2}} = \int_0^c tg(t)\left(\int_0^{\infty} \frac{y\cosh xy}{\sinh \pi y} I_0(ty)e^{-\pi y}dy\right) dt + f(x), \quad 0 \leq x < c. \tag{4.3.5}$$

Inverting Abel- transform, we get

$$g(t) = F(t) + \frac{2}{\pi}\int_0^c ug(u)\left(\int_0^{\infty} \frac{ye^{-\pi y}}{\sinh \pi y} I_0(ty)I_0(uy)\right) du, \quad 0 \leq t < c. \tag{4.3.6}$$

Here

$$F(t) = \frac{2}{\pi}\int_0^t \frac{f(x)\,dx}{\sqrt{t^2-x^2}}$$

The computational task involves numerical evaluation of the kernel

$$\int_0^{\infty} \frac{ye^{-\pi y}}{\sinh \pi y} I_0(ty) I_0(uy) dy$$

and the numerical solution of Fredholm integral equation. Then, coefficients a_n are computed using Bessel routines and numerical integration. For techniques to solve Fredholm integral equations of the second kind, namely,

$$g(t) - \lambda \int_0^c g(x)k(x,t)dx = f(t) \tag{4.3.7}$$

see Atkinson[2]. Among the possibilities are:

1. Fixed Point iteration:

$$g_0(t) = f(t), \quad g^{(n)}(t) = f(t) + \lambda \int_0^t xg^{(n-1)}(x)k(x,t)dx. \tag{4.3.8}$$

2. Direct Method: Suppose that we use n abscissae points for numerical integration. Then, we have

$$g(x_k) - \lambda \sum_{j=1}^{n} x_j g(x_j) k(x_j, x_k) w_j = f(x_k). \tag{4.3.9}$$

   where w_j is a weight function. If we let M an $n \times n$ matrix whose (k,j)-th element is $-\lambda x_j k(x_j, x_k) w_j$ if $k \neq j$ and $1 - \lambda x_j k(x_j, x_k) w_j$ if $k = j$, $G = (g(x_k)), F = (f(x - k)), k = 1, \cdots, N$, then we have a linear system $MG = F$. We solve the linear system for G.

Numerical integration for (4.3.2) using the values of $g(t)$ obtained from the fixed point iteration or direct method results the coefficients a_n.

4.4 Least Square Error Approximation

While the closed-form expressions are very pleasing, in my opinion, methods which are not based upon things turning out just right are also needed. One candidate is a technique used by Arnold Lent[12] to study Wiener- Hopf problems using intersection of linear manifolds. (See also Srivastav[21]) Lent's ideas can be translated in classical terms as an optimization problem. To simplify the expressions, we consider here the special case $f(x) = 1$. The solution of the dual series equations is equivalent to zero error in a minimization problem, i.e., the objective is to minimize

$$\Phi(a_n) = \int_0^c \left(\sum_{n=1}^{\infty} a_n \sin nx - x\right)^2 dx + \int_c^{\pi} \left(\frac{1}{2}a_0 + \sum_{n=1}^{\infty} a_n \cos nx\right)^2 dx. \qquad (4.4.1)$$

Suppose we decide to retain N terms in the infinite series.

$$\frac{\partial \Phi}{\partial a_0} = \int_c^{\pi} (\frac{1}{2}a_0 + \sum_{n=1}^{N} a_n \cos nx) dx = 0, \qquad (4.4.2)$$

Then,

$$\frac{(\pi - c)a_0}{2} - \sum_{n=1}^{N} a_n \frac{\sin nc}{n} = 0. \qquad (4.4.3)$$

Moreover,

$$\frac{1}{\pi}\frac{\partial \Phi}{\partial a_p} = \frac{2}{\pi}\int_0^c (\sum_{n=1}^{N} a_n \sin nx - x) \sin px dx + \frac{2}{\pi}\int_c^{\pi} (\frac{1}{2}a_0 + \sum_{n=1}^{N} a_n \cos nx) \cos px dx = 0$$

(4.4.4)

and

$$-a_0 \frac{\sin pc}{p} - 2\int_0^c (\sum_{n=1}^{N} a_n \cos(n+p)x) dx + \pi a_p = 2\frac{\sin pc - pc \cos pc}{p^2}, \quad p = 1, 2, \ldots$$

(4.4.5)

Equivalently,

$$-\sum_{n=1}^{N} a_n \frac{\sin nc}{n} \frac{\sin pc}{p(\pi - c)} - 2\sum_{n=1}^{N} a_n \frac{\sin(n+p)c}{n+p} + \pi a_p = \frac{2 \sin pc - pc \cos pc}{p^2}.$$

$$\sum_{n=1}^{N} \left[\pi \delta_{np} - \frac{2 \sin(n+p)c}{n+p} - \frac{\sin nc \sin pc}{np(\pi - c)}\right] a_n == \frac{2 \sin pc - pc \cos pc}{p^2}. \qquad (4.4.6)$$

Let $\alpha = (a_1, a_2, \cdots, a_n)^T$, $\beta = (b_1, b_2, \cdots, b_n)^T$, where

$$b_i = \frac{2 \sin ic - ic \cos ic}{i^2}$$

and A be an $n \times n$ matrix whose (i, j)th element is

$$\pi \delta_{ij} - \frac{2 \sin(i+j)c}{i+j} - \frac{\sin ic \sin jc}{ij(\pi - c)}.$$

Then, (4.4.6) can be written as

$$A\alpha = \beta. \qquad (4.4.7)$$

5 Numerical Solution of Cauchy Singular Integral Equations

5.1 Motivation

As we have already seen, integral equations are very useful in the treatment of mixed boundary value problems. Most of these equations need to be solved numerically. However, if for simpler problems of the same type analytical solutions are available, then it is easier to judge the quality of the numerical solution by comparing the performance of numerical algorithms on these problems. For the integral equation,

$$\frac{1}{\pi}\int_{-1}^{1}\frac{g(x)\,dx}{x-y}=f(y)\,,-1<y<1 \tag{5.1.1}$$

subject to the condition

$$\int_{-1}^{1} g(x)\,dx = N \tag{5.1.2}$$

it turns out that for two well known procedures the linear systems of algebraic equations have coefficient matrices for which inverse matrices can be explicitly written ,i.e., the entries have closed form analytical expressions. We describe these procedures and analyze them in the hope that similar questions will be asked and answered in other situations too.

The condition (5.1.2) is needed when weak singularities are permitted at both end-points. The solution is uniquely determined if it is required to be bounded only on one end-point. The problem is overdetermined when the solution is required to be bounded at both end-points. This trichotomy follows from the classical theory developed by Muskhelishvili[14].For the overdetermined systems, see[11].

5.2 Gaussian Quadrature and Collocation

In linear fracture mechanics the integral equation (5.1.1) with a condition of the type (5.1.2) is of special importance, because it is the singular part of a variety of problems. A numerical scheme utilising the Gauss-Chebyshev formula for numerical integration has been developed by Erdogan and Gupta[7] and a variant using the Lobatto-Chebyshev formula has been proposed by Theocaris and Ioakimidis [26]. In both procedures, the unknown funcion $g(x)$ is of the form $(1-x^2)^{-\frac{1}{2}}\phi(x)$. Of interest is the value $\phi(\pm 1)$, because it can be identified with the stress intensity factors. The former determines the values of ϕ at the zeros of Chebyshev polynomials of the first kind, which are the abscissa of integration. The latter uses the the end-points ± 1 and the zeros of the Chebyshev polynomials of the second kind and thus eliminates the need for extrapolation. The equations are discretized in both the cases by "collocation" at the zeros of appropriate Chebyshev polynomials.

5.2.1 Numerical integration of Cauchy Principal-Value Integrals

To adapt the quadrature formulae for the singular integrals, we define an auxiliary function $F(t,s)$ as follows:

$$F(t,s) = \frac{f(t)-f(s)}{t-s}, \quad t \neq s; \qquad F(t,s) = f'(s), \quad t = s \tag{5.2.1}$$

If $f \epsilon C^k[-1,1]$, then $F(t,s) \epsilon C^{k-1}$ for every fixed s in $-1 < s < 1$. Since

$$\int_{-1}^{1} \frac{dt}{\sqrt{1-t^2}(t-s)} = 0, -1 < s < 1,$$

we have

$$\int_{-1}^{1} \frac{f(t)dt}{\sqrt{1-t^2}(t-s)} = \int_{-1}^{1} \frac{F(t,s)dt}{\sqrt{1-t^2}(t-s)}, -1 < s < 1 \tag{5.2.2}$$

In the equation (5.1.1), with $g(x) = (1-x^2)^{-\frac{1}{2}}\phi(x)$, Gauss-Chebyshev quadrature yields

$$\frac{1}{n}\sum_{k=1}^{n} \frac{\phi(t_k)}{t_k - s} + \phi(s)\frac{U_{n-1}(s)}{T_n(s)} = f(s) + \epsilon_n(s) \tag{5.2.3}$$

where $\epsilon_n(s)$ is the quadrature error and when ignored, at the collocation points s_j, the zeros of U_{n-1}, leads to (n-1) equations

$$\frac{1}{n}\sum_{k=1}^{n} \frac{\phi^*(t_k)}{t_k - s_j} = f(s_j), j = 1, 2, \ldots, n-1 \tag{5.2.4}$$

Moreover,(5.1.2) is similarly reduced to

$$\frac{1}{n}\sum_{k=1}^{n} \phi^*(t_k) = N. \tag{5.2.5}$$

The asterisk emphasizes the fact that this is an approximation and not the value of ϕ. Once this is recognized, it is more convenient to drop it in subsequent text.

Srivastav[23] has shown that the above system of n equations has the closed form solution

$$\phi(t_k) = N + \frac{1}{n}\sum_{j=1}^{n-1}\frac{(1-s_j^2)f(s_j)}{t_k - s_j} \tag{5.2.6}$$

The key is the observation that the $n \times n$ matrix A with elements a_{jk} defined by the relations

$$a_{jk} = \frac{1}{n}.\frac{1}{t_k - s_j}, \quad 1 \le j \le n-1; \quad k = 1, 2, \dots, n;$$

$$a_{nk} = \frac{1}{n}, k = 1, 2, \dots, n;$$

is always nonsingular and its inverse A^{-1} is given by

$$\alpha_{kj} = \frac{1}{n}.\frac{1-s_j^2}{t_k - s_j}, \quad k = 1, 2, \dots, n; \quad j = 1, 2, \dots, n-1$$

$$\alpha_{kn} = 1, k = 1, 2, \dots, n.$$

Similar results are derived in a subsequent paper [24] for the Lobatto- Chebyshev quadrature. It is not just another manifestation of nostalgia that we have described these results. They can be used to illustrate a way of assessing the effectiveness of numerical algorithms. This is done in the next section.

5.2.2 Computing with uncertain data

When more than one procedures are available for the numerical treatment of a problem, how do we decide which is better? If a problem is to be solved once or only a few times, it is really not cost-effective to agonize too much on this issue. We would naturally choose the more accurate for the same effort. For the same accuracy, we would opt for the one with less labor. But, there is another consideration. This relates to stability. The physical data is often corrupted by measurement errors. Consequently, the approximate solution will have additional errors beyond the truncation errors (induced by approximation) and round-off errors due to the floating-point arithmetic of computers. Yet, a good method will keep the errors commensurate to the errors in the data and will prevent a bad situation from getting worse! In this spirit, let us suppose that the right-hand side function f in equation (5.1.1) is contaminated with errors $\{\epsilon_j\}, j = 1, 2, \dots, n-1$, which are independent and identically distributed normally $\approx N(0, \sigma)$.Since the equations are linear, the errors ε_k in the solution-vector $\phi(t_k)$ ignoring the inevitable round-off errors will be given by the relation

$$\varepsilon_k = \frac{1}{n} \sum_{j=1}^{n-1} \frac{(1 - s_j^2)\epsilon_j}{t_k - s_j} \tag{5.2.7}$$

ε_k is a random variable, with expected value zero. Its variance is therefore $E(\varepsilon^2)$. The reader is invited to compute the variance not only for the Gauss-Chebyshev quadrature-collocation scheme, but also for various other numerical schemes. My criterion is that the procedure with the least variance is the most desirable. What is yours?

References

[1] Aghili,A.,MIXED BOUNDARY VALUE PROBLEMS OF THE QUARTER-PLANE,Ph.D. thesis, State University of New York at Stony Brook (1999)

[2] Atkinson,K., A SURVEY OF NUMERICAL METHODS FOR THE SOLUTION OF FREDHOLM INTEGRAL EQUATIONS OF THE SECOND KIND,SIAM(1976)

[3] Busbridge,I.W.,*Dual Integral Equations,*Proc. London Math.Soc., **44**,115-129 (1938)

[4] Coddington,E.A. and Levinson,N, THEORY OF ORDINARY DIFFERENTIAL EQUATIONS,McGraw-Hill,(1955)

[5] Dettman,J.W.,ADVANCED COMPLEX VARIABLES, McGraw-Hill(1965)

[6] Erdèlyi,A (Editor:Bateman Manuscript Project),TABLES OF INTEGRAL TRANSFORMS ,Vol.1,McGraw-Hill(1954)

[7] Erdogan,F and Gupta,G.D.,*On the numerical solution of singular integral equations,*Q.Appl.Math.,**30**,525-534(1972)

[8] Erdogan, F., Gupta, G. D., and Cook, T. S., *Numerical solution of singular equations in methods of analysis, and solution of crack problem, In the Series Mechanics of Fracture, 1 (1973), 368–425.*

[9] Gakhov,F.D.,BOUNDARY VALUE PROBLEMSPergammon Press(1966)

[10] Gradshteyn,I.S. and Ryzhik,I.M.,TABLES OF INTEGRALS,SERIES AND PRODUCTS,Academic Press(1965)

[11] Jen, Erica. and Srivastav, R. P., *Solving singular integral equations using Gaussian quadrature. and overdetermined system. Comp. and Maths. with Appls. 9 (1983), No. 4, 625–632.*

[12] Lent,A.H.,WIENER-HOPF OPERATORS AND FACTORIZATION,Ph.D. thesis, Northwestern University(1970)

[13] Mikhlin,S.G.,INTEGRAL EQUATIONS *and their applications to certain problems in mechanics,mathematical physics and technology,*Pergammon Press(1964)

[14] Muskhelishvili,N.I.,SINGULAR INTEGRAL EQUATIONS *Boundary Problems of function theory and their application to mathematical physics,*Noordhoff(1958)

[15] Prössdorf,S and Silberman,B.,NUMERICAL ANALYSIS FOR INTEGRAL AND RELATED OPERATOR EQUATIONS,Akademie Verlag, Berlin(1991)

[16] Sneddon, I. N.,FOURIER TRANSFORMS,McGraw-Hill(1951)

[17] Sneddon, I. N., *Mixed boundary value problems in potential theory, North-Hol. and Publ. Company, 1966*

[18] Sneddon, I. N., THE ELEMENTARY SOLUTION OF DUAL INTEGRAL EQUATIONS, Proc.Glasgow Math Assoc.,**4**, 108-110(1960)

[19] Srivastav, R. P.,*A note on certain integral equations of Abel-type* , Proc. Edinb. Math. Soc.,**13**, 271-272(1963)

[20] Srivastav, R. P.,*Dual Series Relations.III. Dual Relations Involving Trigonometric Series,* Proc. Royal Soc. Edinb.,**66**, 173-184(1964)

[21] Srivastav, R. P.,*An L_2-Theory of Dual Integral Equations* ,Journal of M.A.C.T.,**9**,1-21 (1976)

[22] Srivastav, R. P.,*Dual Integral Equations with Trigonometric Kernels and Tempered Distributions,* SIAM J. Math. Anal,**3**,No.3,413-421(1972)

[23] Srivastav, R. P., *Numerical Solution of Singular Integral Equations Using Gauss-type Formulae,* IMA J., Numer. Anal. **3** , No. 3, 305–318 (1983)

[24] Srivastav, R. P., and Jen, E., ***Numerical solution of singular integral equations** using Gauss-type formulae, II.* ***Lobatto-Chebyshev quadrature and collocation on** Chebyshev nodes,* IMA Journal of Numerical Analysis, **3**, No. 3, 319–325(1983)

[25] Rivlin,T.J.,THE CHEBYSHEV POLYNOMIALS,John Wiley(1974)

[26] Theocaris, P. S., and Ioakimidis, N. I., ***Numerical integral methods for the solu-** tion of singular integral equation,* ***Quarterly of Applied Mathematics** 35 (1977), 173–183.*

[27] Titchmarsh,E.C.,INTRODUCTION TO THE THEORY OF FOURIER INTEGRALS, Oxford University Press(1937)

[28] Tricomi,F.G.,INTEGRAL EQUATIONS, Interscience(John Wiley)(1957)

[29] Watson,G.N.,A TREATISE ON THE THEORY OF BESSEL FUNCTIONS,Cambridge Univeristy Press(1944)

[30] Weber,H., *Über die Besselschen* ***Funktionen und*** *ihre* ***Anwendung an die*** *Theorie der Elektrischen Ströme,*Journal für Math.,**75**,75-105(1873)

[31] Whittaker,E.T. and Watson,G.N.,A COURSE OF MODERN ANALYSIS, Fourth Edition,Cambridge University Presss (1927)